Thermodynamics of chemical systems

Thermodynamics of CHEMICAL SYSTEMS

SCOTT E. WOOD
Professor Emeritus of Chemistry, Illinois Institute of Technology

RUBIN BATTINO
Professor of Chemistry, Wright State University

The right of the University of Cambridge to print and sell all manner of books was granted by Henry VIII in 1534. The University has printed and published continuously since 1584.

CAMBRIDGE UNIVERSITY PRESS
Cambridge
New York Port Chester Melbourne Sydney

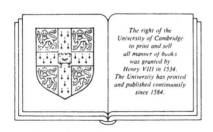

Published by the Press Syndicate of the University of Cambridge
The Pitt Building, Trumpington Street, Cambridge CB2 1RP
40 West 20th Street, New York NY 10011, USA
10 Stamford Road, Oakleigh, Melbourne 3166, Australia

First published 1990

Library of Congress Cataloging in Publication Data
Wood, Scott E. (Scott Emerson), 1910–
Thermodynamics of chemical systems / Scott E. Wood, Rubin Battino.
p. cm.
Includes index.
ISBN 0–521–33041–6
1. Thermodynamics. I. Battino, Rubin. II. Title.
QD504.W66 1989 89-32580
541.3'69–dc20 CIP

British Library Cataloguing in Publication Data
Wood, Scott E.
Thermodynamics of chemical systems.
1. Chemical reactions. Thermodynamics
I. Title II. Battino, Rubin
541.3'69

ISBN 0–521–33041–6 hard covers

Transferred to digital printing 2004

To Our Sons

Edward S. Wood
David Rubin Battino
Benjamin Sadik Battino

Contents

Preface

The systems to which thermodynamics have been applied have become more and more complex. The analysis and understanding of these systems requires a knowledge and understanding of the methods of applying thermodynamics to multiphase, multicomponent systems. This book is an attempt to fill the need for a monograph in this area.

The concept for this book was developed during several years of teaching a one-year advanced graduate course in chemical thermodynamics at the Illinois Institute of Technology. Students who took the course were studying chemistry, chemical engineering, gas technology, or biochemistry. During those years we came to believe that the major difficulty that students have is not with the numerical solution of a problem; the difficulty is with the development of the pertinent relations in terms of experimentally determinable quantities. Moreover, during the initial writing of the book, it became evident that chemical thermodynamics was being applied in many new fields and to systems having more than two or three components. These new fields are so numerous that any attempt to illustrate the application of thermodynamics to each of them would make this book much too long. Therefore, the aim of the book is to develop in a *general* way the concepts and relations that are pertinent to the solution of many thermodynamic problems encountered in multiphase, multicomponent systems. *The emphasis is on obtaining exact expressions in terms of experimentally determinable quantities.* Simplifying assumptions can be made as necessary after the exact expression has been obtained. It is expected that users of this book have some knowledge of physical chemistry and elementary thermodynamics. It is hoped that, once these basic concepts have been developed, the users will be able to apply chemical thermodynamics to any specific problem in their particular field.

The methods of Gibbs are used throughout with emphasis on the chemical potential. The material is presented in a rigorous and mathematical manner. Several topics that are presented briefly or omitted entirely in more-elementary texts are introduced. Among these topics are: the requirements that must be satisfied to define the state of a thermodynamic system, particularly indifferent systems; the use of the Gibbs–Duhem equation in the solution of problems associated not only with simple phase equilibria, but also with phase equilibria in which chemical reactions may occur in one or more phases; the conditions of stability for single-phase, multicomponent systems; and the graphical representation of the thermodynamic functions. Because of the importance of reference states and standard states, special attention is given to their definition and use.

The subject matter is divided into two parts. The first part is devoted to defining the thermodynamic functions and to developing the fundamental relations relevant to chemical systems at equilibrium. The second part is devoted to the application of these relations to real systems and the methods that can be used to obtain additional relations.

The introductory material (Chapters 1–4) is treated briefly. The basic concepts, which are always so difficult to define, are approached from an operational viewpoint and in the classical manner. No attempt is made to use a more general approach.

A bibliography is given after the appendices. The first section lists those books to which reference is made in the text and a few of the more recent and relevant advanced texts in chemical thermodynamics. The other sections give references to data compilations and sources. An abbreviated set of thermodynamic data are given in the appendices for quick reference.

The notation and symbolism used in this text are a combination of those recommended by IUPAC and variations chosen by us for ease and clarity of use. We have chosen the tilde over a symbol, such as $\tilde{V}_1$, to represent the molar volume of component 1, rather than the more cumbersome IUPAC notation of $V_{m,1}$ for the same quantity. Similarly, although the IUPAC notation elegantly represents a partial molar volume, for example, as V_1, we chose the redundant but unambiguous notation of placing a bar over a symbol to indicate this quantity: $\bar{V}_1$. IUPAC recommends the phrase 'amount of substance' for *n* which is commonly called the number of moles. We chose to describe *n* as the 'mole number,' for ease of writing, even though we (and our readers) know that *n* refers to a quantity or amount of substance. In any case, we have been consistent in our usage, and a list of our notation is given after this Preface.

In concluding this Preface, we wish to make some separate and some joint acknowledgments. First, SEW wishes to express his appreciation to those who were influential in stimulating his interest in thermodynamics, and to others who helped in the development of this book: first, to Professors George Scatchard, James A. Beattie, and Louis J. Gillespie, for developing his interest in thermodynamics and for introducing him to the work of J. Willard Gibbs; second, to his many colleagues, but, particularly Dr. Russell K. Edwards, Professor Harry E. Gunning, and Professor Ralph J.

Tykodi, for many long, arduous, and helpful discussions; and third, to the many students who took his course and taught him while they were being taught. RB wishes to express his appreciation to his teachers of thermodynamics: Professors Mark W. Zemansky and John H. Saylor, and to the senior author, Professor Scott E. Wood. We thank Henry E. Sostman for helpful comments and Dr. Thomas W. Listerman for carefully reading Chapter 14. We both express our indebtedness to Dr. Stanley Weissman, who read and critiqued most of the book, and to the secretaries who typed it: Alice Capp, Claudia Hillard, and Mary Alspaugh. Some of the book was written while the senior author was a Senior Fulbright Scholar at the Department of Chemistry at University College, Dublin. Finally, we wish to thank our wives for the patience and encouragement that they showed during the writing of this book.

Scott E. Wood
El Paso, Texas

Rubin Battino
Dayton, Ohio

September 1989

Notation

a area
a activity
A area
A Helmholtz energy
$\mathbf{B}$ magnetic induction
B 1 mole of an unspecified substance
c molarity
c specific heat capacity
C number of components
C heat capacity
C_P heat capacity at constant pressure
C_V heat capacity at constant volume
$đ$ inexact differential
$\mathbf{D}$ electric displacement
E energy
$\mathbf{E}$ electric field strength
$\mathscr{E}$ electromotive force (emf)
f fugacity
$\mathbf{F}$ force, generalized force
$\mathbf{F}_e$ external force
$\mathscr{F}$ Faraday constant
g gaseous state
$\mathbf{g}$ acceleration due to gravity
G Gibbs energy
h height
H enthalpy
$\mathbf{H}$ magnetic field strength
i current
I integration constant

k Henry's law constant
k distribution coefficient
K equilibrium constant
l length
ℓ liquid state
L liter
L length
m mass
m number of units of mass
m molality
$\mathbf{m}$ magnetic moment per unit volume
M molecular mass
$\mathbf{M}$ total magnetic moment
n amount of substance/mole numbers
niA magnetic moment
$\mathbf{p}$ polarization of the medium per unit volume
P pressure
P number of phases
P_k partial pressure of component k
$\mathbf{P}$ total polarization $(=V_{\mathrm{c}}\mathbf{p})$
Q heat
Q electrical charge
r distance
r radius
R gas constant
R number of independent chemical reactions
s solid state
s displacement
S entropy
S number of species
t time
t temperature
t transference number
T Kelvin or thermodynamic temperature
T_{i} inversion temperature (Joule–Thomson)
$\mathbf{v}$ velocity
v specific volume
V volume
V number of variances or number of degrees of freedom
W work
x mole fraction
X_i generalized coordinate
$\tilde{X}_i$ molar property

$\bar{X}_i$ partial molar property
y mole fraction, gas phase
z mole fraction, solid phase
z electrical charge on ion
Z compressibility factor

Greek

α deviations of the real gas or gas mixture from ideal behavior
γ activity coefficient
γ surface tension or interfacial tension
γ fugacity coefficient
Γ surface concentration
ε permittivity
ε_0 permittivity of empty space
$\varepsilon/\varepsilon_0$ dielectric constant
η efficiency of a heat engine
θ arbitrary temperature
μ chemical potential
μ permeability
μ_0 permeability of vacuum
μ/μ_0 relative permeability
μ_{JT} Joule–Thomson coefficient
ν stoichiometric coefficient
π osmotic pressure
ρ density
τ arbitrary temperature
ϕ electrical potential
ϕ function of T, P, μ_1, n_2, . . .
ϕ osmotic coefficient
ϕX apparent molar quantity
Φ gravitational potential, centrifugal potential
Φ electrostatic potential
χ_M magnetic susceptibility
χ_e electric susceptibility
ψ electrical potential
$\boldsymbol{\omega}$ angular velocity
ω acentric factor

Superscripts

E excess change on mixing
M mixture or mixing
P planar
σ defined surface

$^{\ominus}$	standard state
$^{\bullet}$	pure component
$\tilde{}$	molar property
$\bar{}$	partial molar property
$*$	reference state
$^{\infty}$	infinite dilution

Subscripts

1, 2, 3, ...	component 1, 2, 3, ...
b	boiling point
c	condenser
c	critical property
D	dimer
E	excess
f	formation
mp	melting point
mix	for the mixing process
M	monomer
r	reduced property, as in $T_r = T/T_c$
s	mixed solvent (pseudobinary system)
S	space
sat	saturated
tr	transition

1

Introduction

> With thermodynamics, one can calculate almost everything crudely; with kinetic theory, one can calculate fewer things, but more accurately; and with statistical mechanics, one can calculate almost nothing exactly.
>
> Eugene Wigner

The range and scope of thermodynamics is implied in Wigner's epigrammatic quote above, except for the fact that there are a great many phenomena for which thermodynamics can provide quite accurate calculations. Chemical engineering is to a large extent based on thermodynamic calculations applied to practical systems. The phase rule is indispensable to metallurgists, geologists, geochemists, crystallographers, mineralogists, and chemical engineers. Although students sometimes come away with the notion that thermodynamics is remote and abstract, it is actually the most practical of subjects. Part of what we endeavor to do in this book is to illustrate that practicality, the tie to everyday life and utility.

Thermodynamics comprises a field of knowledge that is fundamental and applicable to a vast area of human experience. It is a study of the interactions between two or more bodies, the interactions being described in terms of the basic concepts of heat and work. These concepts are deduced from *experience*, and it is this experience that leads to statements of the first and second laws of thermodynamics. The first law leads to the definition of the energy function, and the second law leads to the definition of the entropy function. With the experimental establishment of these laws, thermodynamics gives an elegant and exact method of studying and determining the properties of natural systems.

The branch of thermodynamics known as the thermodynamics of reversible processes is actually a study of thermodynamic systems at equilibrium, and it is this branch that is so important in the application of thermodynamics to chemical systems. Starting from the fundamental conditions of equilibrium based on the second law, more-practical conditions, expressed in terms of experimentally measurable quantities, have been developed. The application of the thermodynamics of chemical systems thus leads to the determination of the equilibrium properties of macroscopic systems as observed in nature (regardless of the complexity of the systems), to possible limitations, and to the determination of the dependence of the equilibrium properties with changes of the values of various independent variables, such as pressure, temperature, or composition.

Since thermodynamics deals with systems at equilibrium, time is not a thermodynamic coordinate. One can calculate, for example, that if benzene(ℓ) were in equilibrium with hydrogen(g) and carbon(s) at 298.15 K, then there would be very little benzene present since the equilibrium constant for the formation of benzene is 1.67×10^{-22}. The equilibrium constant for the formation of diamond(s) from carbon(s, graphite) at 298.15 K is 0.310; that is, graphite is more stable than diamond. As a final example, the equilibrium constant for the following reaction at 298.15 K is 2.24×10^{-37}:

$$2C(s, gr) + H_2(g) = C_2H_2(g)$$

People do not give away their diamonds or worry about benzene or ethyne spontaneously decomposing, since these substances are *not* in equilibrium with their starting materials. A specific catalyst or infinite time might be required to attain equilibrium conditions.

After the appendices we provide a selected bibliography of general references and data compilations. All cited references appear in the list of general references.

1.1 The language of thermodynamics

In thermodynamics terms mean exactly what we define them to mean. The exact use of language is particularly important in this subject, since calculations and interpretations are directly tied to precisely defined changes of state *and* the way in which those changes of state occur. So, in this introductory chapter the language of thermodynamics is presented.

First, we speak of a *system* and its *surroundings.* The system is any region of matter that we wish to discuss or investigate. The surroundings comprise all other matter in the world or universe that can have an effect on or interact with the system. Thermodynamics, then, is a study of the interaction between a system and its surroundings. These definitions are very broad and allow a great deal of choice concerning the system and its surroundings. The important point is that for any thermodynamic problem we must rigorously define the system with which we are dealing and apply any limitations to the surroundings that appear to be necessary. What we shall consider to be the system and the surroundings is our choice, but it is imperative that we consciously make this choice. As an example, we may be interested in the chemical substances taking part in a chemical reaction. These substances would have to be in a container of some kind. Certainly, the substances taking part in the reaction would comprise at least part of the system, but it is our choice whether we consider the container to be part of the system or part of the surroundings. Although the surroundings were defined generally as all other matter in the world or universe, in all practical cases they are limited in extent. Those parts of the universe that have no or only an insignificant effect on the system are excluded, and only those of the surroundings that actually interact with the system need to be considered. In many experimental studies the surroundings that interact with the system are actually designed for the purpose of controlling the system and making measurements on it.

For any thermodynamic system there is always a *boundary*, sometimes called an *envelope*, which separates the system and the surroundings. The only interactions considered in thermodynamics between a system and its surroundings are those that occur across this boundary. Because of this, it is as important to define the boundary and its properties as it is to define the system and the surroundings themselves. The boundary may be real or hypothetical, but it is considered to have certain properties. It may be rigid, so that the volume of the system remains constant, or it may be nonrigid, so that the volume of the system can change. Similarly, the boundary may be adiabatic or diathermic. The boundary may be semipermeable, under which condition it would be permeable to certain substances but not permeable to others. In actual cases the properties of the boundary are determined by the properties of the system and by our design of the surroundings. In simplified, idealized cases we may endow the boundary with whatever properties we choose. A clear definition of the system and of the boundary that separates the system from its surroundings is extremely important in the solution of many thermodynamic problems.

In addition to the general concept of a system, we define different types of systems. An *isolated system* is one that is surrounded by an envelope of such nature that *no interaction whatsoever* can take place between the system and the surroundings. The system is completely isolated from the surroundings. A *closed system* is one in which *no* matter is allowed to transfer across the boundary; that is, no matter can enter or leave the system. In contrast to a closed system we have an *open system*, in which matter can be transferred across the boundary, so that the mass of a system may be varied. (Flow systems are also open systems, but are excluded in this definition because only equilibrium systems are considered in this book.)

The *state* of the system is defined in terms of certain state variables. The state of the system is then fixed by assigning definite values to sufficient variables, chosen to be independent, so that the values of all other variables are fixed. The number of independent variables depends in general upon the problem at hand and upon the system with which we are dealing. The determination of this number and the type of variables required for the definition of the state of the system is discussed in Chapter 5. When the values of the independent variables that define the state of the system are changed, we speak of a *change of state* of the system. Here we are concerned with the values of the dependent variables, as well as those of the independent variables, at the initial and final states of the system, and not with the way in which the values of the independent variables are changed between the two states. When it is necessary to consider the way in which the values of the independent variables are changed, we speak of the *path*. In a graphical representation the path is any line connecting the two points that depict the two states. The term *process* encompasses both the change of state and the path.

1.2 Thermodynamic properties of systems

We have implied in Section 1.1 that certain properties of a thermodynamic system can be used as mathematical variables. Several independent and different classifications of these variables may be made. In the first place there are many variables that can be evaluated by experimental measurement. Such quantities are the temperature, pressure, volume, the amount of substance of the components (i.e., the mole numbers), and the position of the system in some potential field. There are other properties or variables of a thermodynamic system that can be evaluated only by means of mathematical calculations in terms of the measurable variables. Such quantities may be called *derived* quantities. Of the many variables, those that can be measured experimentally as well as those that must be calculated, some will be considered as *independent* and the others are *dependent*. The choice of which variables are independent for a given thermodynamic problem is rather arbitrary and a matter of convenience, dictated somewhat by the system itself.

Finally, the thermodynamic properties of a system considered as variables may be classified as either *intensive* or *extensive* variables. The distinction between these two types of variables is best understood in terms of an operation. We consider a system in some fixed state and divide this system into two or more parts without changing any other properties of the system. Those variables whose value remains the same in this operation are called intensive variables. Such variables are the temperature, pressure, concentration variables, and specific and molar quantities. Those variables whose values are changed because of the operation are known as extensive variables. Such variables are the volume and the amount of substance (number of moles) of the components forming the system.

1.3 Notation

In this book we use SI units and IUPAC symbolism and terminology as far as possible. The complete set of notation used in this book is given before this chapter. For clarity and consistency we have made some choices of notation that differ from IUPAC recommendations. Notation is defined where first introduced. Amongst other exceptions is the use of the phrase 'amount of substance' for the variable n, which has been traditionally called the 'number of moles' or, as we most frequently call it, the 'mole numbers.'

Some basic reference tables are given in the appendices.

2

Temperature, heat, work, energy, and enthalpy

In this chapter we briefly review the ideas and equations relating to the important concepts of temperature, heat, work, energy, and enthalpy.

2.1 Temperature

The concept of temperature can be defined operationally; that is, in terms of a set of operations or conditions that define the concept. To define a temperature scale operationally we need: (1) one particular pure or defined substance; (2) a specific property of that substance that changes with a naive sense of 'degree of hotness' (i.e., temperature); (3) an equation relating temperature to the specific property; (4) a sufficient number of fixed points (defined as reproducible temperatures) to evaluate the constants in the equation in (3); and (5) the assignment of numerical values to the fixed points. Historically, many different choices have been made with respect to the five conditions listed above, and this, of course, has resulted in many temperature scales.

The ideal gas temperature scale is of especial interest, since it can be directly related to the thermodynamic temperature scale (see Sect. 3.7). The typical constant-volume gas thermometer conforms to the thermodynamic temperature scale within about 0.01 K or less at agreed fixed points such as the triple point of oxygen and the freezing points of metals such as silver and gold. The thermodynamic temperature scale requires only one fixed point and is independent of the nature of the substance used in the defining Carnot cycle. This is the triple point of water, which has an *assigned* value of 273.16 K with the use of a gas thermometer as the instrument of measurement.

The International Practical Temperature Scale of 1968 (IPTS-68) is currently the internationally accepted method of measuring temperature reproducibly. A standard platinum resistance thermometer is the transfer medium that is used over most of the range of practical thermometry.

Interpolation formulas and a defined set of fixed points are used to establish the scale. IPTS-68 is due to be replaced in 1990 or 1991.

The *zeroth law of thermodynamics* is in essence the basis of all thermometric measurements. It states that, if a body A has the same temperature as the bodies B and C, then the temperature of B and C must be the same. One way of doing this is to calibrate a given thermometer against a standard thermometer. The given thermometer may then be used to determine the temperature of some system of interest. The conclusion is made that the temperature of the system of interest is the same as that of any other system with the same reading as the standard thermometer. Since a thermometer in effect measures only its own temperature, great care must be used in assuring thermal equilibrium between the thermometer and the system to be measured.

2.2 Heat and heat capacity

It is observed experimentally that, when two bodies having different temperatures are brought into contact with each other for a sufficient length of time, the temperatures of the two bodies approach each other. Moreover, when we form the contact between the two bodies by means of walls constructed of different materials and otherwise isolate the bodies from the surroundings, the rate at which the two temperatures approach each other depends upon the material used as the wall. Walls that permit a rather rapid rate of temperature change are called *diathermic* walls, and those that permit only a very slow rate are called *adiabatic* walls. The rate would be zero for an ideal adiabatic wall. In thermodynamics we make use of the concept of *ideal adiabatic walls*, although no such walls actually exist.

We describe the phenomenon that the temperatures of the two bodies placed in diathermic contact with each other approach the same value by saying that 'heat' has transferred from one body to another. *This is the only concept of heat that is used in this book.* It is based on the observation of a particular phenomenon, the behavior of two bodies having different temperatures when they are placed in thermal contact with each other.

One way to obtain a quantitative definition of a unit of heat is to choose a particular substance as a standard substance and define a quantity of heat in terms of the temperature change of a definite quantity of the substance. Thus, we can define a quantity of heat, Q, by the equation

$$Q = C(t_2 - t_1) \tag{2.1}$$

where t_2 is the final temperature and t_1 is the initial temperature. The proportionality constant, C, is called the *heat capacity* of the particular quantity of the substance. For the present we will disregard the dependence of C on pressure and volume. Experimental studies have shown that C is an extensive quantity, so that it may be written as mc, where m represents the number of units of mass of the standard substance used and c represents the *specific heat capacity*. The unit of heat is then defined by assigning an arbitrary value to c. A positive value is assigned to c so that the quantity of heat absorbed during an increase of temperature is positive.

The calorie was originally based on 1 g of water. For the purposes of a more exact definition it has been superseded by the joule, so that

$$\text{cal} = \text{cal}_{\text{th}} \equiv 4.184\,\text{joules} \equiv 4.184\,\text{J}$$

by international convention. It is called the 'thermodynamic' calorie. (Many authors omit the subscript 'th.')

Measurements of the heat capacity of all substances have shown that the heat capacity is actually a function of temperature. Therefore, Equation (2.1) must be given as

$$Q = \int_{t_1}^{t_2} C\,\mathrm{d}T = m \int_{t_1}^{t_2} c\,\mathrm{d}T \tag{2.2}$$

where t_2 represents the final temperature and t_1 the initial temperature. If the numerical value of Q is positive, then heat is absorbed by the substance; if it is negative, then heat is emitted by the substance.

The quantity of heat absorbed or emitted by a system for a given change of state depends not only upon the change of temperature, but also upon changes in the values of other independent variables. Here we consider closed systems (those of constant mass) and only either changes in the values of the temperature and pressure or changes in the values of temperature and volume. Differential expressions for the quantity of heat can be written either as

$$đQ = M(T, P)\,\mathrm{d}T + N(T, P)\,\mathrm{d}P \tag{2.3}$$

or

$$đQ = K(T, V)\,\mathrm{d}T + L(T, V)\,\mathrm{d}V \tag{2.4}$$

where $M(T, P)$ and $N(T, P)$ represent functions of the temperature and pressure, and $K(T, V)$ and $L(T, V)$ represent functions of the temperature and volume. Experiment shows that these differential expressions for the heat effect are inexact. The symbol đ is used to distinguish between inexact (đ) and exact (d) differential quantities. *The quantity of heat absorbed by the system for a given change of state thus depends upon the path that is followed from the initial state to the final state.* The symbol Q represents a quantity of heat, taken as *positive* for heat *absorbed* by the system from the surroundings. It may be measured experimentally or determined from the line integral of appropriate inexact differential expressions represented by the symbol $đQ$, such as Equation (2.3) or (2.4).

A general definition of the heat capacity is given by the equation

$$C = \frac{đQ}{dt} = \frac{đQ}{dT} \tag{2.5}$$

where $đQ$ represents the infinitesimal quantity of heat absorbed by a system for a differential increase, dt or dT, in the temperature of the system. However, $đQ$ is an inexact differential quantity and its value is determined only when the path is stated. By relating Equation (2.5) with Equations (2.3) and (2.4) for the systems discussed, we find that $M(T, P)$ is the heat capacity of the system at constant pressure, C_P, and $K(T, V)$ is the heat capacity of the system at constant volume, C_V. Equations (2.3) and (2.4) may then be written as

$$đQ = C_P\,dT + N(T, P)\,dP \tag{2.6}$$

and

$$đQ = C_V\,dT + L(T, V)\,dv \tag{2.7}$$

The heat capacities in Equations (2.1) and (2.2) are also dependent on pressure and volume.

Heat capacities are not limited to those at constant volume and constant pressure. In some cases values must be assigned to other independent variables in addition to the temperature and pressure, or temperature and volume, in order to define the state of the system. For these cases additional terms must be added to Equations (2.3) and 2.4). The heat capacities are then defined at constant values of these variables in addition to the pressure or volume. In all cases the defining equation for the heat capacity is Equation (2.5) with the provision that the path followed for the change in temperature must be given.

When the heat capacity of a system is known as a function of the temperature, the heat absorbed by the system for a given temperature change can be calculated by integration of Equation (2.5). The integral is a line integral and the path for integration must be known.

2.3 Work

The concept of work is developed here from an operational point of view. Mechanical work is discussed first, and then the concept is expanded to more-general interactions. Observation shows that there are actions that, when acting on a body cause a change in the velocity of the body. Such actions are called *forces*. The relation between the force and the change of velocity is expressed by Newton's second law of motion:

$$\mathbf{F} = m\frac{d\mathbf{v}}{dt} \tag{2.8}$$

for nonrelativistic velocities, where **F** represents the force and m the mass of the body. Here t refers to time. The unit of force, the newton (N), is defined in terms of this relation for which m is one kilogram and $d\mathbf{v}/dt$ is one meter per second. The differential quantity of work $đW$, done by a force in operating on a body over a differential displacement, ds, is given by the scalar product of the force and the differential of the displacement, so

$$đW = F \cos \alpha \, ds \tag{2.9}$$

The symbols F and s here refer to scalar rather than vector quantities, and α is the angle between the direction of the force and the direction of the displacement. The unit of work is the joule.

This basic mechanical concept of work must be extended when it is applied in thermodynamics. We are concerned with the interaction of a system and its surroundings across the boundary that separates them. Both the system and the surroundings may exert forces on the boundary, and it is the action of these forces resulting in a displacement of the boundary that constitutes work. The language used to describe the interaction in terms of work must be developed and used with great care.

Because the change of the volume of a system is so important in the application of thermodynamics to chemical systems, the expansion of a gas is used as an example. Consider a known quantity of gas confined in a frictionless piston-and-cylinder, as illustrated in Figure 2.1. The cylinder is fixed in position relative to the Earth. The piston has a mass m and can move in the direction of the gravitational field of the Earth. There is also a known external force, $\mathbf{F}_e$, exerted on the upper surface of the piston. We

Figure 2.1. Piston-and-cylinder arrangement.

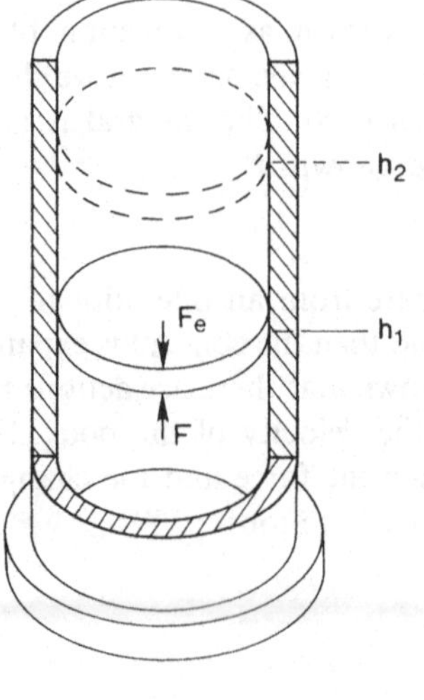

define the system to be the gas contained in the volume within the piston-and-cylinder. The surroundings are the cylinder, piston, and all of the substances and devices that exert the external force on the piston. The boundary is the internal wall of the cylinder and the lower surface of the piston, and it is assumed to be diathermic. The piston is originally clamped in position so that the lower surface of the piston is at the position labeled h_1. The gas exerts a force, $\mathbf{F}$, on this surface. We now assume that, when the clamps holding the piston in position are removed, the gas expands, forcing the piston to move in the upward direction. We also assume that the velocity of the piston can be measured when the lower surface of the piston is at h_2. Whether the process is adiabatic or isothermal is not important for the present discussion.

We can analyze the process by the application of Newton's second law of motion. The net force acting on the boundary, defined as the lower surface of the piston, is $\mathbf{F} - m\mathbf{g} - \mathbf{F}_e$, where $\mathbf{g}$ represents the acceleration due to gravity. Equation (2.9) for this case becomes

$$m\frac{d\mathbf{v}}{dt} = \mathbf{F} - m\mathbf{g} - \mathbf{F}_e \tag{2.10}$$

When we multiply Equation (2.10) by dt and use the relations

$$d\mathbf{v} = \left(\frac{d\mathbf{v}}{dt}\right) dt \tag{2.11}$$

and

$$dt = (1/\mathbf{v})\, dh \tag{2.12}$$

we obtain

$$m\mathbf{v}\, d\mathbf{v} = (\mathbf{F} - m\mathbf{g} - \mathbf{F}_e)\, dh \tag{2.13}$$

On integrating between the limits 0 and v for the velocity and h_1 and h_2 for the distance,

$$\tfrac{1}{2}m\mathbf{v}^2 = \int_{h_1}^{h_2} \mathbf{F}\, dh - m\mathbf{g}(h_2 - h_1) - \int_{h_1}^{h_2} \mathbf{F}_e\, dh \tag{2.14}$$

Finally, we place terms relating to the system on the left-hand side of the equation and terms relating to the surroundings on the right-hand side so that Equation (2.14) becomes

$$\int_{h_1}^{h_2} \mathbf{F}\, dh = \int_{h_1}^{h_2} \mathbf{F}_e\, dh + m\mathbf{g}(h_2 - h_1) + \tfrac{1}{2}m\mathbf{v}^2 \tag{2.15}$$

The term on the left-hand side represents the *quantity of work done by the system and associated with the force* $\mathbf{F}$. The right-hand side represents the effect of this quantity of work on the surroundings. Work has been done against the force $\mathbf{F}_e$ in raising the piston in the gravitational field of the Earth against the force $m\mathbf{g}$, and in imparting kinetic energy to the piston.

The force exerted by the gas on the lower surface of the piston cannot be expressed in terms of the properties of the gas. As soon as the piston has any velocity, pressure and temperature gradients appear in the gas, with the result that the force is something less than the product of the equilibrium pressure of the gas and the area of the piston. The surroundings are therefore devised so that the changes that take place in the surroundings can be measured. The right-hand side of Equation (2.15) represents the *quantity of work done by the system as measured by changes in the surroundings.*

It is informative to consider the same process and the same system and surroundings with the exception that the piston is made a part of the system. Part of the boundary is now defined as the upper surface of the piston rather than the lower surface. Equation (2.15) takes the form

$$\int_{h_1}^{h_2} \mathbf{F}\,dh - m\mathbf{g}(h_2 - h_1) - \tfrac{1}{2}m\mathbf{v}^2 = \int_{h_1}^{h_2} \mathbf{F}_e\,dh \qquad (2.16)$$

under these conditions. The force exerted by the system on the boundary is $(\mathbf{F} - m\mathbf{g})$, but the integral of this force does not represent the work done by the system on the boundary. The only effect in the surroundings is represented by the integral on the right-hand side of Equation (2.16). This integral gives the quantity of work done by the system as determined by changes in the surroundings or against the force $\mathbf{F}_e$. This is the only interaction between the system and the surroundings. The raising of the piston in the gravitational field of the Earth and the imparting of kinetic energy to the piston has no relevance to the interaction of the system and the surroundings.

Further insight into the concept of work, the language used to define work, and the importance of clearly defining the system, surroundings, and boundary is obtained by introducing frictional effects. We consider the same gas and frictionless piston-and-cylinder arrangement as in the previous discussion, but insert a device in the cylinder so that the piston is stopped instantly when the lower surface of the piston reaches the position h_2. We assume that the piston, cylinder, and device are not deformed or stressed by the collision. For the purposes of discussion, we imagine that the piston and cylinder, containing the gas, are separated from the surroundings which exert the external force $\mathbf{F}_e$ by an adiabatic wall. When we compare the result of the expansion in which the piston is stopped and thermal equilibrium is attained, with the result in which the piston is allowed to maintain its velocity at the position h_2, we find that the temperature of the piston, cylinder, and gas is greater for the former case than for the latter. (We assume here that the temperature of the piston, cylinder, and gas can be measured while the piston is moving and is determined when the lower surface of the piston reaches the position h_2.) We can compare the two processes by bringing a body of water into contact with the cylinder by the use of a diathermic wall. The temperature of the piston, cylinder, and gas can now be changed to the temperature when the piston is not stopped by allowing heat to flow into the body of water. The quantity of heat removed is measured by the temperature change of the water. In our idealized experiments we find that the quantity of heat removed is equal to the kinetic energy of the piston at the position h_2, when expressed in the same units. It is only in this operational concept that we can speak of heat resulting from the performance of work.

We can now analyze the quantity of work done by the system associated with the force $\mathbf{F}$ as measured by changes in the surroundings. A quantity of work has been done against the external force $\mathbf{F}_e$ and in raising the piston in the gravitational field when the boundary is taken to be the lower surface of the piston. In addition, a change of temperature has occurred in the piston and cylinder, and the quantity of heat that would be required to cause the temperature change must be considered as an effect of the work done by the system. The temperature change of the gas itself is not included in the consideration of the interaction between the system and the surroundings. If the gas is separated from the piston and cylinder by an adiabatic wall, then all effects of the collision appear in the change of the temperature of the piston and cylinder, and the quantity of heat required to cause the temperature change is considered to be the result of the work done by the system. The analysis is simplified if we choose the gas, piston, and cylinder to be the system. Then all of the effects of the collision are contained within the system and the work done by the system is done only against the external force. In the general case when the effects of friction must be considered, it is convenient to define the system and the envelope in such a way that all frictional effects are included within the system and therefore do not enter into the interaction of the system and its surroundings.

We must recall that the process or processes that we have been discussing have not been completely defined; that is, we have not stated whether the process is adiabatic or isothermal, or whether any specific quantity of heat has been added to or removed from the system during the process. Although we have essentially defined the initial state of the system, we have not defined its final state, neither have we defined the path that we choose to connect the two states. When we do so, *we find by experience that the quantity of work done by the system depends upon the path* and, therefore, the differential quantity of work, $đW$, is an inexact differential quantity.

The basic concept of work is defined by Equation (2.9). However, in thermodynamics this basic concept must be extended, because we are interested in the interaction between a system and its surroundings. We wish to emphasize that *the work done by the system is associated with the force exerted by the system on the boundary separating the system from the surroundings and is measured in general by appropriate changes that take place in the surroundings.* The quantity of work, W, which is the integral of $đW$, also depends upon the path. In this book the symbol W represents a quantity of work, taken as *negative* for work *done by* the system on the boundary. When work is done on the system by use of the surroundings, the *numerical* value of W is *positive*. (Some books use a reverse convention for the sign of work effects—be careful!)

2.4 Quasistatic processes

The force exerted by the system on the lower surface of the piston is not related directly to the equilibrium pressure of the material within the cylinder when the piston is moving. However, there are many times when it is advantageous to approximate this force by the product of the equilibrium pressure and the area of the piston. In so doing we can relate the force to the equilibrium properties of the substance contained within the cylinder. Imagine an idealized process, *approximating* a real process, in which the movement of the piston is controlled so that it never attains an appreciable velocity. The control is obtained by means of friction between the piston and cylinder, or, more simply, by some device that permits a frictionless piston to move only an infinitesimal distance before it is stopped by collision with the device. Figure 2.2 depicts one such possible process. As the steps become smaller, so the incremental increase in volume becomes smaller. The process can be controlled by the size of the steps. Once the piston is stopped, the substance within the cylinder is allowed to attain equilibrium before the piston is permitted to move again. The change of state that takes place in the substance within the cylinder is thus accomplished by a succession of a large number of equilibrium states, each state differing infinitesimally from the previous state. Such processes are called *quasistatic* processes.

The force exerted by the substance within the cylinder on the lower force of the piston under these conditions is the product of the pressure exerted by the substance on the surface of the piston and the area of the piston. Moreover, the product of the area and the differential displacement of the piston is equal to the differential change of volume. The integral $\int_{h_1}^{h_2} \mathbf{F}\, \mathrm{d}h$ is then equal to $\int_{V_1}^{V_2} P\, \mathrm{d}V$. This relation is the only change that is made in Equation (2.15) or a similar equation for quasistatic processes. The frictional effects or the collisions result in a temperature increase either of the surroundings, or of both the system and surroundings as the case may be, or the effects may be interpreted in terms of heat, as discussed above.

The differential of work related to volume changes is inexact for quasistatic processes. First, for quasistatic processes,

$$đW = P(T, V)\, \mathrm{d}V$$

where P is the pressure of the system. However, we know that P is a function of not only the volume, but also the temperature, and therefore we must specify how the temperature of the system is to be changed for a given change of volume. We thus define a path. Second, the volume of a closed system may be taken to be a function of the temperature and pressure, so

$$\mathrm{d}V = \left(\frac{\partial V}{\partial T}\right)_{P,n} \mathrm{d}T + \left(\frac{\partial V}{\partial P}\right)_{T,n} \mathrm{d}P \tag{2.17}$$

When this equation is substituted for $\mathrm{d}V$ in Equation (2.16), the inexact differential expression

$$đW = P\left(\frac{\partial V}{\partial T}\right)_{P,n} \mathrm{d}T + P\left(\frac{\partial V}{\partial P}\right)_{T,n} \mathrm{d}P \tag{2.18}$$

is obtained.

2.5 Other 'work' interactions

We have discussed only two types of interactions between a system and its surroundings. There are many other types, and we must determine

Figure 2.2. The quasistatic process.

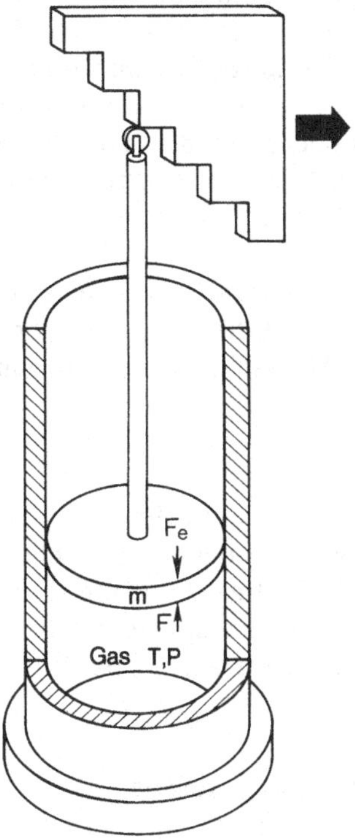

how these interactions may be included in a general concept of work. In order to do so, the concept of *generalized coordinates* and *conjugate generalized forces* must be developed.

There are certain properties of a system whose values change when interactions take place across the boundary between the system and its surroundings. The analysis of the interaction would be extremely difficult if the values of several properties changed for one specific interaction. We seek a set of independent variables or properties so that, for one specific interaction and isolation of the system from all other interactions, the value of only *one* variable is changed while all others remain constant. Such variables or properties are called generalized coordinates, for which we use the symbol X_i in this book. Examples of these coordinates in addition to the volume are: position in a gravitational or centrifugal field, surface area, magnetization in a magnetic field, and polarization in an electrostatic field.

For each generalized coordinate there is a conjugate generalized force. These quantities are the specific properties of the system and the surroundings that cause the interaction and that determine the direction of the change of the value of the conjugate coordinate of the system. The forces of the system and of the surroundings act on the boundary separating the system and its surroundings. The unit of measurement of a force must be defined in terms of the specific phenomenon that is observed. Then the measurement of the force requires a comparison of the force to the unit of measurement. For the measurement, the force being measured *must be in equilibrium* with the force exerted by the measuring instrument. (The requirement for the measurement of a force, that forces can only be measured under conditions of equilibrium, has been suggested as a definition of a force.) Examples of these forces are the gravitational or centrifugal forces, the magnetic force, and the voltage in an electrostatic field.

The differential quantity of work done by a specific force is defined as the product of the force and the differential change of the conjugate coordinate. The differential quantity of work done by the system on the surroundings is expressed as

$$đW_i = -F_i \, \mathrm{d}X_i \tag{2.19}$$

where F_i represents the force exerted by the system at the boundary between the system and the surroundings. (When we consider a force acting on a unit volume of the system as in a magnetic field, the expression for work is complex and an integral over the volume of the system is required (see Ch. 14)). The sign of $đW_i$ is taken to be positive for work done *on* the system.

2.6 The first law of thermodynamics: the energy function

The first law of thermodynamics states that *all experience has shown that, for any cyclic process taking place in a closed system,*

$$\oint (đQ + đW) = 0 \tag{2.20}$$

Because all experience to date has shown this relation to be true, it is assumed that it is true for all possible thermodynamic systems. There is no further proof of such a statement beyond the experiential one.

From mathematics we recognize that the quantity $(đQ + đW)$ is an exact differential, because its cyclic integral is zero for all paths. Then, some function of the variables that describe the state of the system exists. This function is called the *energy function*, or more loosely the *energy*. We therefore have the definition

$$\mathrm{d}E = đQ + đW \tag{2.21}$$

which may be regarded as another mathematical statement of the first law.[1] The energy function is a function of the state of a system. The change in the value of the energy function in going from one state to another state depends *only* upon the two states and not at all upon the path that is used in going from the one state to the other. Its differential is *exact*. The absolute value of the energy function of a system in a given state is not known, however. The absolute value might be given by the integral

$$\Delta E = \int \mathrm{d}E + I \tag{2.22}$$

where I is the integration constant, but the value of I is neither known or determinable (at least at present). However, we can always determine the *difference* in the values of the energy function between two states by using a definite integral, so that

$$\Delta E = E_2 - E_1 = \int_{E_1}^{E_2} \mathrm{d}E \tag{2.23}$$

In order to avoid chaos in reported data, it is convenient to choose some state of a system, which is called the standard state, as the initial state, and then to report the difference between the value of the function in any final state and that in the standard state.

Additional understanding of the first law may be obtained from the integration of Equation (2.21) between two states of a system, to obtain

$$\Delta E = Q + W \tag{2.24}$$

[1] Some texts use the symbol U for the energy, and some texts call it the internal energy. Also, one should be aware that some texts define the energy function as being equal to $(đQ - đW)$. This depends upon the sign convention used for work.

which implies that both Q and W can be determined experimentally. If the system is isolated, both Q and W must be zero, and thus ΔE must be zero. *Therefore, the value of the energy function of an isolated system is constant.* The change in the value of the energy function for a cyclic process taking place in a system that is not isolated must be zero. Therefore, under these conditions Q must be equal to $-W$. This statement is equivalent to the denial of a perpetual-motion machine of the first kind, which states that *no machine can be constructed which , operating in cycles, will perform work without the absorption of an equivalent quantity of heat.*

The symbols $đW$ and W have been used to represent the sum of *all* work terms. It is convenient to express the work term associated with a change of volume separately from other work terms. Equation (2.21) may then be written as

$$dE = đQ - (F/A)\,dV + đW' \tag{2.25}$$

where F is the force exerted by the system on the boundary for a given quasistatic process and A is the area of that portion of the boundary of the system that is moved to cause a change in the volume; W' represents the sum of all other work terms. Equation (2.25) may be written as

$$dE = đQ - P\,dV + đW' \tag{2.26}$$

where P is the pressure of the system.

One use of Equations (2.25) and (2.26) becomes apparent if we consider calorimetric measurements at constant volume. In this case dV and $đW'$ are both zero and, consequently,

$$dE = [dQ]_V \tag{2.27}$$

for an infinitesimal change of state and

$$\Delta E = [Q]_V$$

for a finite change of state. This equation is equivalent to the statement that the change in the value of the energy function for a change of state that takes place in a constant-volume calorimeter is equal to the heat absorbed by the system from the calorimeter.

We can write the differential of the energy function in terms of the differentials of the independent variables that we choose to define the state. We will find in Chapter 5 that only two independent variables need to be used if the system is closed and only the work of expansion is involved. The two most convenient variables to use here are the temperature and volume. The differential of the energy function in terms of T and V is given by

$$dE = \left(\frac{\partial E}{\partial T}\right)_{V,n} dT + \left(\frac{\partial E}{\partial V}\right)_{T,n} dV \tag{2.28}$$

where the subscript n is used to denote that the system is closed. The calculation of the change of the value of the energy function for some change of state then involves the integration of this equation after the coefficients have been evaluated. The discussion of the integration is postponed until Chapter 4, when we will be able to evaluate $(\partial E/\partial V)_{T,n}$. With the use of Equations (2.5) and (2.28), we find that

$$\left(\frac{\partial E}{\partial T}\right)_{V,n} = C_V \tag{2.29}$$

Thus, the derivative of the energy with respect to the temperature of a closed system at constant volume is equal to the heat capacity of the system at constant volume.

2.7 The enthalpy function

The volume appears as an independent variable in the differential expression of the energy function in Equation (2.26). However, the experimental use of the volume as an independent variable is rather inconvenient, whereas the use of the pressure is more convenient for most experimental work. It is thus desirable to change the independent variable from the volume to the pressure. In so doing we define a new function in terms of its differential, dH, based on Equation (2.26) as

$$\mathrm{d}H = \mathrm{d}(E + PV) = đQ + V\,\mathrm{d}P + đW' \tag{2.30}$$

Furthermore, as we are defining a new function, we can assign the value zero to the integration constant involved in the integration of Equation (2.30), and thus define

$$H = E + PV \tag{2.31}$$

This function is called the *enthalpy function*, or more loosely the *enthalpy*.[2] By its definition the enthalpy function is a function of the state of the system. The change in the values of this function in going from one state to another depends only upon the two states, and not at all upon the path. Its differential is exact. Its absolute value for any system in any particular state is not known, because the absolute value of the energy is not known.

We may illustrate one use of Equation (2.30) by considering a change of state taking place under constant pressure and involving no other work. Such a change would take place in an open or constant-pressure calorimeter.

[2] When the pressure of the system is not uniform, as in a gravitational field, the system may be divided into parts in which the pressure is uniform. Then $H = \sum(E + PV) = E + \sum(PV)$, the sum being taken over the individual parts. If the volumes of the regions are infinitesimal, then $\sum(PV)$ may be substituted by $\int P\,\mathrm{d}V$. The integral is taken over the entire volume of the system.

Both dP and $đW'$ would be zero, and consequently

$$dH = [dQ]_P \tag{2.32}$$

for an infinitesimal change and

$$\Delta H = [Q]_P \tag{2.33}$$

for a finite change. Thus, for a change of state taking place under conditions of constant pressure with no other work involved, the change in the value of the enthalpy function is equal to the heat absorbed by the system from the surroundings. *This relation is basic to all calorimetric experiments taking place at constant pressure.*

As with the energy, we are often interested in determining the change in the value of the enthalpy function of a closed system for some change of state without having to measure the heat absorbed and without being confined to constant-pressure processes. We choose the temperature and pressure as the two convenient independent variables to use, and write the differential of the enthalpy as

$$dH = \left(\frac{\partial H}{\partial T}\right)_{P,n} dT + \left(\frac{\partial H}{\partial P}\right)_{T,n} dP \tag{2.34}$$

The calculation of the change in the value of the enthalpy function for some change of state would then involve the integration of this equation after the coefficients have been evaluated in terms of experimentally determined quantities. Again, we cannot evaluate $(\partial H/\partial P)_{T,n}$ until Chapter 4, but we can evaluate $(\partial H/\partial T)_{P,n}$ in the same way that we did for the energy, obtaining

$$\left(\frac{\partial H}{\partial T}\right)_{P,n} = C_P \tag{2.35}$$

The derivative of the enthalpy with respect to the temperature of a closed system at constant pressure is equal to the heat capacity of the system at constant pressure.

2.8 The ideal gas and the first law of thermodynamics

The concept of the ideal gas can be used to illustrate the relations developed in this chapter. The ideal gas is defined by the two relations

$$PV = nRT \tag{2.36}$$

and

$$(\partial E/\partial V)_{T,n} = 0 \tag{2.37}$$

The first of these equations is the familiar ideal gas law, which expresses the relation between the pressure, temperature, volume, and amount of substance of an ideal gas. The second equation expresses the condition that the value of the energy function of a constant mass of an ideal gas is independent of the volume and is a function of the temperature alone. Equation (2.37) must be used in the definition of an ideal gas, since any derivation of it must be based on the Second Law of Thermodynamics (which is developed in Ch. 3) and the identification of the ideal gas temperature scale with the thermodynamic temperature scale. However, the equation itself is used to prove the identity.

The concept behind the defining equation (Eq. (2.37)) comes from an experiment known as the *Joule experiment*, which is illustrated in Figure 2.3. The result of this experiment is known as the *Joule effect*. In this experiment the gas is confined in one part of a closed container and the other part is evacuated. The gas itself is taken to be the substance composing the system. However, the boundary between the system and its surroundings is chosen to be the walls of the container. The volume of the system is the total volume of the container and is not the same as the volume of the gas when it is

Figure 2.3. The Joule experiment.

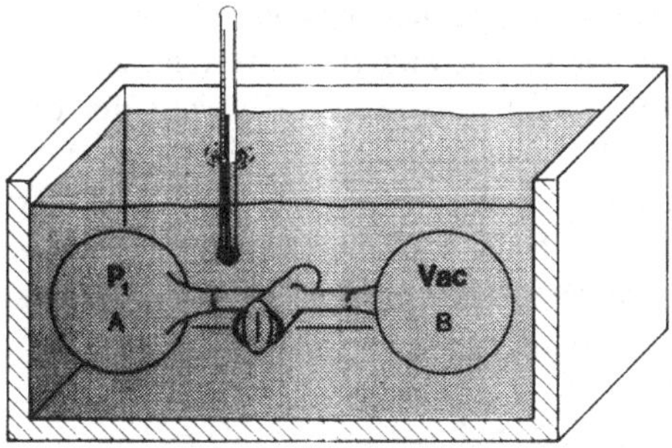

(a)

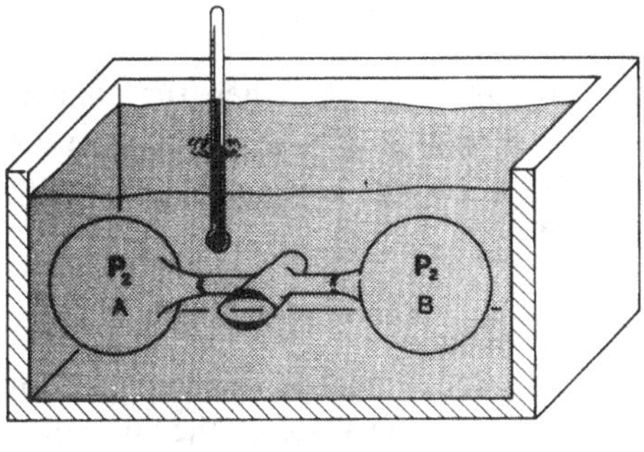

(b)

confined in one part of the container. The gas is allowed to expand adiabatically into the evacuated part of the vessel, and the temperature of the gas is observed before and after the expansion. No work is done by the gas on the surroundings, because the walls are rigid and the volume of the vessel is therefore constant. Also, no heat is absorbed or evolved by the gas. Consequently, the energy of the gas is constant. From Equation (2.28) we obtain that

$$\left(\frac{\partial E}{\partial T}\right)_{V,n} \mathrm{d}T + \left(\frac{\partial E}{\partial V}\right)_{T,n} \mathrm{d}V = 0 \tag{2.38}$$

We are concerned with the three variables E, T, and V. We know that

$$\left(\frac{\partial E}{\partial T}\right)_{V,n} \left(\frac{\partial T}{\partial V}\right)_{E,n} \left(\frac{\partial V}{\partial E}\right)_{T,n} = -1$$

or

$$\left(\frac{\partial T}{\partial V}\right)_{E,n} = -\frac{(\partial E/\partial V)_{T,n}}{(\partial E/\partial T)_{V,n}} = -\frac{(\partial E/\partial V)_{T,n}}{C_V} \tag{2.39}$$

The quantity $(\partial T/\partial V)_{E,n}$ is called the *Joule coefficient*.

The results of early experiments showed that the temperature did not change on the expansion of the gas, and consequently the value of the Joule coefficient was zero. The heat capacity of the gas is finite and nonzero. Therefore, it was concluded that $(\partial E/\partial V)_{T,n}$ was zero. Later and more-precise experiments have shown that the Joule coefficient is not zero for real gases, and therefore $(\partial E/\partial V)_{T,n}$ is not zero. The concept that $(\partial E/\partial V)_{T,n}$ is zero for an ideal gas thus becomes a definition of an ideal gas. This definition is consistent with our present theoretical concepts of the ideal gas.

For an ideal gas Equation (2.28) reduces to

$$\mathrm{d}E = (\partial E/\partial T)_{V,n}\, \mathrm{d}T = C_V\, \mathrm{d}T \tag{2.40}$$

which upon substitution into Equation (2.21) yields

$$đQ = C_V\, \mathrm{d}T - đW \tag{2.41}$$

Thus, the total heat absorbed by an ideal gas when the change of state involves the performance of work as well as a change of temperature is equal to the heat absorbed when the temperature is changed at constant volume plus the work done by the system.

2.9 Representation of changes of state

The state of a chemical system can be defined in terms of the substance or substances that comprise the system, the quantity of each, the states of aggregation, and two of the four state functions: energy, volume, temperature, and pressure. It will usually be convenient to use the temperature and pressure, but it is not necessary to do so. Thus, the symbol $2H_2O[\ell, 300\ K, 1\ bar]$ would indicate a system containing 2 mole of liquid water at 300 K and 1 bar. For solutions we might use $0.1NaCl[m, 100°C, 2\ bar]$ to indicate a solution containing 0.1 mole of sodium chloride at the molality m, 100°C, and 2 bar. Whether the solution is aqueous or not would be understood, but if any question exists, then the solvent must be specified. An alternate symbolism is $0.1NaCl \cdot 200H_2O[\ell, 25°C, 1\ bar]$, where $0.1NaCl \cdot 200H_2O$ represents an aqueous solution containing 0.1 mole of sodium chloride at a concentration of 1 mole of NaCl dissolved in 200 mole of water. Some changes of state would be explicitly characterized by the following:

$$0.1H_2O[\ell, 25°C, 1\ bar] = 0.1H_2O[g, 150°C, 0.5\ bar]$$

$$0.2NaOH \cdot 200H_2O[\ell, 25°C, 1\ bar] + 0.2HCl \cdot 300H_2O[\ell, 25°C, 1\ bar]$$
$$= 0.2NaCl \cdot 501H_2O[\ell, 30°C, 1\ bar]$$

Many choices of independent variables such as the energy, volume, temperature, or pressure (and others still to be defined) may be used. However, only a certain number may be independent. For example, the pressure, volume, temperature, and amount of substance are all variables of a single-phase system. However, there is one equation expressing the value of one of these variables in terms of the other three, and consequently only three of the four variables are independent. Such an equation is called a *condition equation*. The general case involves the Gibbs phase rule, which is discussed in Chapter 5.

3

The second law of thermodynamics: the entropy function

In the development of the second law and the definition of the entropy function, we use the phenomenological approach as we did for the first law. First, the concept of reversible and irreversible processes is developed. The Carnot cycle is used as an example of a reversible heat engine, and the results obtained from the study of the Carnot cycle are generalized and shown to be the same for all reversible heat engines. The relations obtained permit the definition of a thermodynamic temperature scale. Finally, the entropy function is defined and its properties are discussed.

3.1 Heat and work reservoirs

We define a *heat reservoir* as any body, used as part of the surroundings of a particular system, whose *only* interaction with the system is across a diathermic boundary. A heat reservoir is then used to transfer heat to or from a thermodynamic system and to measure these quantities of heat. It may consist of one or more substances in one or more states of aggregation. In most cases a heat reservoir must be of such a nature that the addition of any finite amount of heat to the system or the removal of any finite amount of heat from the system causes only an infinitesimal change in the temperature of the reservoir.

A *work reservoir* is similarly defined as any body or combination of bodies, used as part of the surroundings, whose *only* interaction with the system is one that may be described in terms of work. We may have a different type of reservoir for each mode of interaction other than thermal interaction. A work reservoir then is used to perform work across the boundary separating the reservoir and the thermodynamic system and to measure these quantities of work. In the following we are, in order to simplify the discussion, primarily concerned with mechanical work, but this limitation does not alter or limit the basic concepts. A reservoir for mechanical work may be a set of weights and pulleys in a gravitational field, an idealized spring, or a compressible fluid in a piston-and-cylinder arrangement. In any case the reservoir must be designed so that the work done by it or on it can be measured by changes that take place in the reservoir. It is actually incorrect to consider any storage of heat or work as such, but the concept of a reservoir is introduced for convenience and for lack of a better term.

3.2 Heat engines

A heat engine is any device that is operated in cycles and that performs work on its surroundings. The cyclic operation is required because no system has been designed that can operate continuously in one direction. It is found in Section 3.5 that, in the cyclic operation of a heat engine, a quantity of heat is absorbed by the engine from a heat reservoir at some higher temperature, and a smaller quantity of heat is transferred from the engine to a reservoir at some lower temperature. In chemistry the importance of the study of heat engines lies primarily in the understanding of the relations between heat and work, reversible and irreversible processes, and the definition of the thermodynamic temperature scale. We confine our study to a simple idealized heat engine represented by the Carnot cycle. This cycle is discussed in Section 3.5.

3.3 Reversible and irreversible processes

We know from experience that any isolated system left to itself will change toward some final state that we call a state of equilibrium. We further know that this direction cannot be reversed without the use of some other system external to the original system. From all experience this characteristic of systems progressing toward an equilibrium state seems to be universal, and we call the process of such a change an *irreversible process*. In order to characterize an irreversible process further, we use one specific example and then discuss the general case. In doing so we always use a cyclic process.

We return to the piston-and-cylinder arrangement discussed in Section 2.3. In that discussion we did not completely describe the process because we were interested only in developing the concept of work. Here, to complete the description, we choose an isothermal process and a gas to be the fluid. We then have a gas confined in the piston-and-cylinder arrangement. A work reservoir is used to exert the external force, $\mathbf{F}_e$, on the piston; this reservoir can have work done on it by the expansion of the gas or it can do work by compressing the gas. A heat reservoir is used to make the process isothermal. The piston is considered as part of the surroundings, so the lower surface of the piston constitutes part of the boundary between the system and its surroundings. Thus, the piston, the cylinder, and the two reservoirs constitute the surroundings.

We now consider the isothermal, quasistatic expansion and compression of the gas for several processes. The curve *AE* in Figure 3.1 represents a pressure–volume isotherm of an ideal gas at a given temperature. The expansion of the gas is carried out from the volume V_1 to the volume V_2, and the compression from V_2 to V_1. The initial pressure of the gas is P_1 and the pressure after expansion is P_2. We first expand the gas under the conditions that the work reservoir exerts the constant pressure P_2 (equal to $F_e A$) against the piston. The work done on the gas and associated with the pressure of the gas is given by

$$W = -\int_{V_1}^{V_2} P\,\mathrm{d}V = -P_2(V_2 - V_1) - mg(h_2 - h_1) - |Q'| \tag{3.1}$$

according to Equation (2.15). The work done by the gas is represented by the area $ACDE$. The work done on the work reservoir is $P_2(V_2 - V_1)$, represented by the area $BCDE$. The work done on the piston is $mg(h_2 - h_1)$. Finally $|Q'|$ is the absolute value of the heat that must be transferred to the heat reservoir from the surroundings in order to keep the temperature of the piston and cylinder constant because of the frictional or collisional processes. The area ABE represents the work done on the piston and the quantity of heat transferred to the reservoir. Some heat must also be transferred to the gas in order to keep the temperature constant.

The gas is compressed under the conditions that the work reservoir exerts the pressure P_1 on the piston. The work done on the gas and associated with the pressure of the gas is

$$W = -\int_{V_2}^{V_1} P\,\mathrm{d}V = -P_1(V_1 - V_2) - mg(h_1 - h_2) - |Q''| \tag{3.2}$$

where $|Q''|$ represents the absolute value of the quantity of heat that must

Figure 3.1. Illustration of an irreversible cycle.

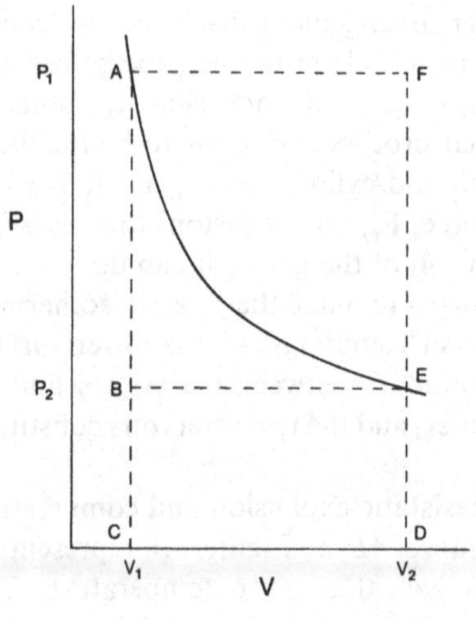

be transferred to the heat reservoir in order to keep the temperature of the piston and cylinder constant. The work done by the work reservoir is $|P_1(V_1 - V_2)|$, represented by the area $ACDF$, and the work done by the gas is again represented by the area $ACDE$. The area AEF represents the difference between $|Q''|$ and $mg(h_2 - h_1)$. Here, in order to keep the temperature of the gas constant, a quantity of heat must be transferred to the heat reservoir.

For the cyclic process, the gas and the piston and cylinder are returned to their original state. The net work done by the gas on the work reservoir is $(P_2 - P_1)(V_2 - V_1)$, which is a negative quantity. Therefore, the work reservoir has actually performed this quantity of work, which is represented by the area $ABEF$. The work done by the gas on expansion is equal to the work done on the gas on compression; moreover, the heat transferred to the gas on expansion is equal to the heat transferred from the gas on compression. Thus, the quantity of heat actually transferred to the heat reservoir in the cyclic process is the sum of $|Q'|$ and $|Q''|$. If we consider the gas, the piston and cylinder, and the two reservoirs as an isolated system, then the cyclic process has resulted in the work reservoir doing some work and the heat reservoir absorbing some heat, and the isolated system has not been returned to its original state. The question is whether the heat may be removed from the heat reservoir and used in some way to cause work to be done on the work reservoir to bring the two reservoirs back to their original state without causing changes in some other system or systems external to the isolated system. *Experience has shown that this is not possible.*

If the expansion is carried out stepwise, first against the constant pressure P_3, then against the constant pressure P_4, and finally against the constant pressure P_2, as illustrated in Figure 3.2, then the amount of work done on

Figure 3.2. The work done by the expansion of a fluid for different conditions.

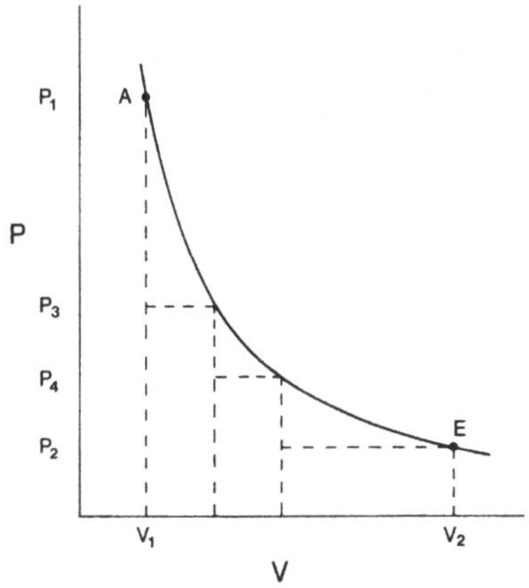

the work reservoir is larger than in the single-step expansion. Also, the total amount of heat removed from the heat reservoir is larger by the same amount. As the number of steps is increased indefinitely, both the work done on the work reservoir and the amount of heat removed from the heat reservoir approach a maximum value equal to the area *ACDE* under the curve. Similarly, if the compression is carried out stepwise (Fig. 3.3), first at the constant pressure P_5, then at the constant pressure P_6, and finally at the constant pressure P_1, then the amount of work done by the work reservoir is less than in the single-step compression. Also, the total amount of heat added to the heat reservoir is less by the same amount. As the number of steps is increased indefinitely, the work done by the work reservoir and the heat added to the heat reservoir approach a minimum value, both equal to the area *ACDE*. We observe than in the limit of expansion and compression along the curve *AE* the quantity of work done on the work reservoir on expansion is equal to the quantity of work done by the work reservoir on compression, and that the heat removed from the heat reservoir on expansion is equal to the heat given to the heat reservoir on compression because both $|Q'|$ and $|Q''|$ become equal to zero. *It is only in this limit that both the work and heat reservoirs are brought back to their original state, as well as the gas.* In this limit the force exerted on the piston by the gas is always equal to the sum of the forces exerted on the lower surface of the piston; that is, the forces that are operating on the boundary between the system and its surroundings are always in balance. Moreover, in the limit no kinetic energy will be imparted to the piston.

In the general case we consider any system that is capable of being used

Figure 3.3. The work done by the compression of a fluid for different conditions.

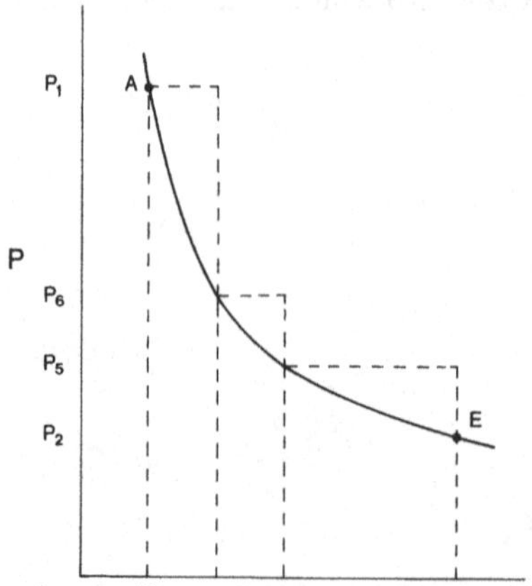

as a heat engine as a part of an isolated system. The other parts are appropriate heat and work reservoirs. These reservoirs interact with the system and are used to measure the heat absorbed by the system from the heat reservoirs and the work done by the system on the work reservoirs for any process that takes place within the isolated system. We start with the system, used as the heat engine, in a given state and carry out a cyclic process. In all actual cases we find that we can return the system to its original state, but always at least one work reservoir has done work on the system and an equivalent amount of heat has been added to at least one heat reservoir. *No method has been found by which the heat can be removed from the heat reservoir and cause work to be done on the work reservoir so that all systems within the isolated system can be returned to their original state without causing changes in systems external to the isolated system.* We could, of course, use a heat reservoir external to the isolated system to return the heat reservoir to its original state and use an external work reservoir to perform work on the work reservoir to return it to its original state. However, such operations do not help, since we would then need to include the external reservoirs within the isolated system and the same situation would prevail. We could also use the heat to operate an external heat engine to do some work on the work reservoir within the system. However, changes occur in such external systems so that they cannot be returned to *their* original state. All processes that occur within an isolated system and produce this result for a cyclic process are called *irreversible*. We find further by experiment that only in the limit in which all forces acting on the envelope between the system and the reservoirs are balanced can we return the entire isolated system to its original state. Processes that occur under these limiting conditions and that produce this result for cyclic processes are called *reversible* processes. Therefore, a reversible process is one in which *all* of the forces involved are balanced by equal and opposite forces; i.e., the process takes place under equilibrium conditions. The direction of the process can be reversed by an *infinitesimal* change in these forces. In an irreversible process none of the conditions hold. The forces are not balanced, the direction of the process cannot be changed by an infinitesimal change in the forces, and the entire isolated systems cannot be brought back to its original state. *All natural processes are irreversible.* If any *one* step in a process is irreversible, then the entire process is irreversible even if all other steps are reversible.

We then come to a general conclusion based on experience. *No isolated system can be returned to its original state when a natural cyclic process takes place in the system.* This statement may be considered as one statement of the Second Law of Thermodynamics. Although there is no rigorous proof of such a statement, all experience in thermodynamics attests to its validity. One concludes, then, that there must be some monotonically varying function that is related to this concept of reversibility. The value of this function for an isolated system can change only in one direction for any cyclic irreversible process and would not change for any cyclic reversible process. This function is the *entropy*, but we consider the Carnot cycle before defining the function.

3.4 The reversible transfer of heat

The transfer of heat between two bodies requires a difference in their temperatures. The reversible transfer of heat between two bodies would require their temperatures to be the same. The question then arises of how we may reversibly add a quantity of heat to or remove a quantity of heat from a system over a temperature range. In order to do so we assume that we have an infinitely large number of heat reservoirs whose temperatures differ infinitesimally. Then, by bringing the system into thermal contact with these reservoirs successively and allowing thermal equilibrium to be obtained in each step, we approach a reversible process.

3.5 The Carnot cycle

The Carnot cycle is only one of many cycles that could be used to illustrate the operation of an ideal, reversible heat engine; however, it is one of the simplest. In a Carnot cycle a heat engine is operated between two heat reservoirs at different temperatures in such a way that all the heat absorbed by the engine is absorbed reversibly from the hotter reservoir and all of the heat that is rejected by the engine is given reversibly to the cooler reservoir. Any type of work (mechanical, electrical, magnetic, etc.) may be involved, but for simplicity and the purposes of illustration we are concerned solely with the work associated with the expansion and compression of a fluid. We therefore consider a given quantity of a fluid contained in a frictionless piston-and-cylinder arrangement. Since the cycle is to be reversible, the force exerted on the lower surface of the piston by this fluid must be opposed by an equal force, which is the sum of the forces exerted by some work reservoir. The cycle consists of four steps and is illustrated in Figure 3.4, which is drawn for an ideal gas, but the results of the Carnot cycle analysis are independent of the substance used or the state of aggregation. The four steps are discussed separately.

1. We start at the state A and carry out an isothermal, reversible expansion at the higher temperature θ_2 to the state B. The symbol θ represents a general temperature without reference to any scale. A quantity of work W_1, equal to the area ABV_2V_1, is done on the system, and to be consistent with our sign conventions this work would have a *negative* value. In order to keep the system isothermal during the expansion, the heat reservoir having the temperature θ_2 has to be in thermal contact with the cylinder and a quantity of heat, Q_2, is transferred from the heat reservoir to the fluid.

2. We next carry out an adiabatic reversible expansion from B to C in which the additional quantity of work, W_2, represented by the area BCV_3V_2 is done on the system (a negative value). During this expansion the cylinder is enclosed by an adiabatic envelope. *No* heat is transferred to the fluid, and its temperature decreases to θ_1.
3. We now compress the fluid isothermally and reversibly from C to D at the temperature θ_1. In this compression the cylinder is in thermal contact with the heat reservoir having this temperature. A quantity of work equal to the area CDV_4V_3 is done and a quantity of heat is transferred from the fluid to the heat reservoir. However, because of our original definition of the symbols W and Q, we state that the work, W_3, is done by the surroundings on the fluid (a positive quantity) and the quantity of heat, Q_1, is transferred from the heat reservoir to the fluid (a negative quantity). The point D or V_4 is not independent. The adiabatic AD is determined by the point A and the isotherm CD is determined by the point C. Point D is the intersection of the two curves.
4. Finally, we compress the fluid adiabatically and reversibly from D to A, thus completing the cycle. In this compression the work, W_4, is done by the surroundings on the fluid (a positive quantity) and no heat is transferred from or to the system.

Figure 3.4. Carnot cycle with an ideal gas.

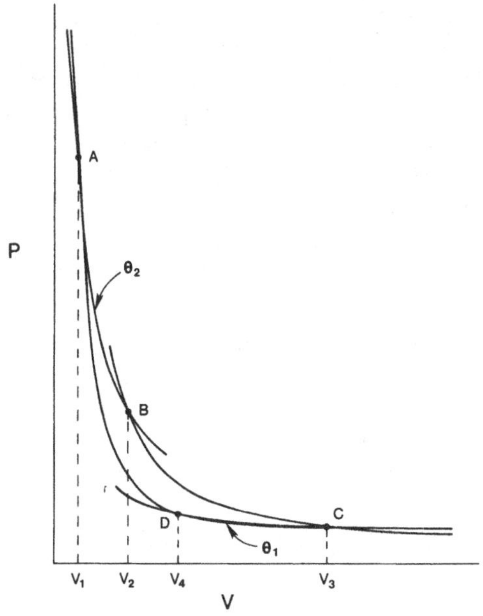

The net work, W, done on the heat engine by the work reservoir is equal to the sum of the work done in each of the steps (a negative quantity); the area $ABCD$ represents this net work. The net heat absorbed by the system is the sum $(Q_2 + Q_1)$. The result of the cycle is illustrated in Figure 3.5, in which the squares represent the heat reservoirs and the circle represents the heat engine. A quantity of heat, Q_2, is absorbed by the fluid from the heat reservoir at the higher temperature, a smaller quantity of heat, $-Q_1$, is rejected by the fluid to the heat reservoir at the lower temperature, and some work, W, is done by the work reservoir. The efficiency of a heat engine, η, is defined as

$$\eta = -W/Q' \tag{3.3}$$

where Q' is the sum of all of the quantities of heat that are actually transferred from the surroundings to the fluid. The change of energy of the heat engine is zero, because the fluid and the piston and cylinder are returned to their original states in the cyclic operation; then $-W = Q_2 + Q_1$. Note that this equation is a consequence of the first law, which has $\Delta E = 0$ for a cycle. Thus, the efficiency of the reversible cycle is expressed as

$$\eta = -W/Q_2 = (Q_2 + Q_1)/Q_2 \tag{3.4}$$

It is important to remember that the numerical value of Q_1 is negative.

3.6 The thermodynamic temperature scale

The temperature scales that have been discussed earlier are quite arbitrary and depend upon the properties of a particular substance. Kelvin was the first to observe that the efficiency of a reversible heat engine operating between two temperatures is dependent only upon the two temperatures and not at all upon the working substance. Therefore, a temperature scale could be defined that is independent of the properties of any substance.

The second law is independent of the first law. Historically, the second law was generally accepted and understood before acceptance of the first law. Therefore, we base the discussion on the efficiency of a reversible heat engine used in a Carnot cycle. If the efficiency of a reversible heat engine

Figure 3.5. Block representation of the result of the operation of a Carnot cycle.

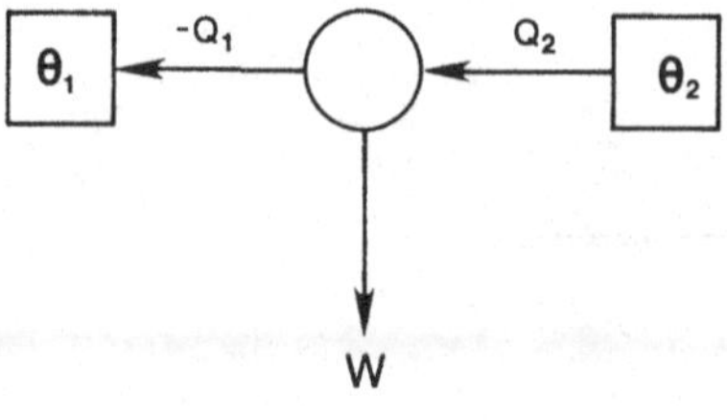

operating between two temperatures in a Carnot cycle is dependent only upon the two temperatures, then it may be expressed as

$$\eta = -W/Q_2 = F(\theta_2, \theta_1) \tag{3.5}$$

where $F(\theta_2, \theta_1)$ is some function of the two temperatures. When the difference between the two temperatures is very small, Equation (3.5) may be written as

$$-\frac{\mathrm{d}W}{Q(\theta)} = F(\theta)\,\mathrm{d}\theta \tag{3.6}$$

or

$$-\mathrm{d}W = Q(\theta)\,F(\theta)\,\mathrm{d}\theta \tag{3.7}$$

because $\mathrm{d}W$ must become zero when $\mathrm{d}\theta$ is zero (refer to Fig. 3.4). The function $F(\theta)$ is called *Carnot's function*. For a reversible cycle operating between the same two adiabatics, $\mathrm{d}W$ is an exact differential and consequently $Q(\theta)\,F(\theta)$ is the derivative of some function $\Theta(\theta)$. Then

$$Q(\theta) = \frac{1}{F(\theta)}\frac{\mathrm{d}\Theta(\theta)}{\mathrm{d}\theta} \tag{3.8}$$

Temperature scales can be defined by assigning specific functions to $F(\theta)$ and the derivatives. Many scales could be defined, but the choice of a suitable scale has depended upon utility and past history. Only the functions used in defining the present thermodynamic or Kelvin scale are discussed here. The value of the derivative is assigned a constant value, k, and $F(\theta)$ is defined as $1/\theta$. Then

$$Q = k\theta \tag{3.9}$$

which states that the heat absorbed along an isothermal path between two adiabatics is proportional to the temperature. The integration of Equation (3.7) between two temperatures yields

$$-W = k(\theta_2 - \theta_1) \tag{3.10}$$

so that the efficiency is given by

$$\eta = -W/Q(\theta_2) = (\theta_2 - \theta_1)/\theta_2 \tag{3.11}$$

The ratio of the two temperatures is related to the efficiency and the ratio of the heats absorbed along the isotherms by

$$\theta_2/\theta_1 = 1/(1-\eta) \tag{3.12}$$

and

$$\theta_2/\theta_1 = |Q_2|/|Q_1| \tag{3.13}$$

We assume that the efficiency of a Carnot cycle operating between the two temperatures θ_1 and θ_2 can be measured, or we assume that the absolute values of the quantities of heat which are absorbed by the working substance can be measured along the isotherms between two adiabatics. Equations (3.12) and (3.13) give one relation between two quantities. One more relation is needed to fix the value of the two quantities. Since 1954 the definition of the Kelvin scale is completed by assigning the value of 273.16 K to the temperature of the triple point of water. This assignment determines the value of the proportionality constant in Equation (3.9). Again, we can determine the value of any other temperature by operating a reversible Carnot cycle between the unknown temperature and the triple point of water. From Equation (3.12)

$$\theta = 273.16(1 - \eta) \tag{3.14}$$

if the unknown temperature is below the triple point of water and

$$\theta = 273.16/(1 - \eta) \tag{3.15}$$

if it is above. From Equation (3.13) we have

$$\theta = 273.16 \frac{|Q_\theta|}{|Q_t|} \tag{3.16}$$

where Q_θ and Q_t are the heat absorbed by the working substance in the isothermal steps at the unknown temperature and the triple point, respectively.

The Kelvin scale is thus defined in terms of an ideal reversible heat engine. At first such a scale does not appear to be practical, because all natural processes are irreversible. In a few cases, particularly at very low temperatures, a reversible process can be approximated and a temperature actually measured. However, in most cases this method of measuring temperatures is extremely inconvenient. Fortunately, as is proved in Section 3.7, the Kelvin scale is identical to the ideal gas temperature scale. In actual practice we use the International Practical Temperature Scale, which is defined to be as identical as possible to the ideal gas scale. Thus, the thermodynamic scale, the ideal gas scale, and the International Practical Temperature Scale are all consistent scales. Henceforth, we use the symbol T for each of these three scales and reserve the symbol θ for any other thermodynamic scale.

3.7 The identity of the Kelvin and ideal gas temperature scales

We prove the identity of the Kelvin scale and the ideal gas scale by using an ideal gas as the fluid in a reversible heat engine operating in a Carnot cycle between the temperatures T_2 and T_1. An ideal gas has been defined by Equations (2.36) and (2.37). Then the energy of an ideal gas depends upon the temperature alone, and is independent of the volume.

Consequently, the energy of the gas is constant for the isothermal reversible expansion or compression and, according to the first law of thermodynamics, the work done on the gas must therefore be equal but opposite in sign to the heat absorbed by the gas from the surroundings. For a reversible process the pressure must be the pressure of the gas itself. Therefore, we have for the isothermal reversible expansion of n moles of an ideal gas between the volumes V and V'

$$-W = \int_V^{V'} P\,dV \tag{3.17}$$

$$-W = nRT \int_V^{V'} dV/V \tag{3.18}$$

where we have substituted the ideal gas law for P, and finally

$$-W = nRT\ln(V'/V) \tag{3.19}$$

Consequently, along the path AB (Fig. 3.4) for the isothermal reversible expansion

$$-W_1 = Q_2 = nRT_2\ln(V_2/V_1) \tag{3.20}$$

and along the path CD for the isothermal reversible compression

$$-W_3 = Q_1 = nRT_1\ln(V_4/V_3) \tag{3.21}$$

(The numerical value of W_3 is positive and that of Q_1 is negative, because V_4 is smaller than V_3.) We note that in Equations (3.20) and (3.21) the heat absorbed by the gas from the surroundings is proportional to the temperature, provided the ratio of the two volumes is always the same. We know that the work done by the gas in the complete cycle is the sum of the heats absorbed, and therefore

$$-W = nRT_2\ln(V_2/V_1) + nRT_1\ln(V_4/V_3) \tag{3.22}$$

We complete the proof by considering the adiabatic reversible expansion of an ideal gas. In this case we know that

$$dE = -P\,dV \tag{3.23}$$

from the first law of thermodynamics, because the process is adiabatic. We also know that

$$dE = C_V\,dT \tag{3.24}$$

because $(\partial E/\partial V)_{T,n}$ has been defined to be zero for an ideal gas. We then have

$$C_V\,dT = -P\,dV \tag{3.25}$$

and for the reversible expansion or compression of an ideal gas

$$C_V\,\mathrm{d}T = -\frac{nRT}{V}\,\mathrm{d}V \tag{3.26}$$

We divide this equation by T and integrate between the limits T and T' for the temperature and V and V' for the volume, because the heat capacity of an ideal gas is a function of the temperature alone. Thus, for an adiabatic reversible expansion of an ideal gas

$$\int_T^{T'} \frac{C_V}{T} = -\int_V^{V'} nR\frac{\mathrm{d}V}{V} \tag{3.27}$$

We then have along the path BC in the Carnot cycle

$$\int_{T_2}^{T_1} \frac{C_V}{T}\,\mathrm{d}T = -nR\ln(V_3/V_2) \tag{3.28}$$

and along the path DA

$$\int_{T_1}^{T_2} \frac{C_V}{T}\,\mathrm{d}T = -nR\ln(V_1/V_4) \tag{3.29}$$

The left-hand side of these two equations are equal but opposite in sign, and therefore

$$V_3/V_2 = V_4/V_1 \tag{3.30}$$

or

$$V_2/V_1 = V_3/V_4 \tag{3.31}$$

When we substitute this equation into Equation (3.22) we obtain

$$-W = (T_2 - T_1)nR\ln(V_2/V_1) \tag{3.32}$$

The efficiency of this cycle using an ideal gas is therefore

$$\eta = \frac{W}{Q_2} = \frac{T_2 - T_1}{T_2} \tag{3.33}$$

This equation is identical to Equation (3.11). Therefore, the Kelvin thermodynamic scale and the ideal gas scale become identical when the temperature of the triple point of water is assigned the value of 273.16 K.

3.8 The efficiency of reversible heat engines: two statements of the second law of thermodynamics

We now discuss the efficiency of a reversible heat engine operating in a Carnot cycle. The efficiency depends upon the difference between the two temperatures. The greater the difference for a fixed T_2 is, the greater the efficiency. However, the efficiency is always less than unity except in the special case where T_1 is zero on the Kelvin scale. Indeed, this relation defines absolute zero macroscopically as the temperature on the Kelvin scale that the cooler heat reservoir used in a Carnot cycle must have for the efficiency of the heat engine to be unity. A more important conclusion is that, if the two temperatures are equal, the efficiency is zero and no work is done by the heat engine on the surroundings. This is exactly the result that we found for the cyclic isothermal expansion and compression of a fluid in Section 3.3.

This conclusion leads to the Kelvin–Planck statement of the second law: *It is impossible to construct a heat engine which operating in cycles will do work on the surroundings at the expense of heat obtained from a single heat reservoir*. This is the denial of the possibility of a perpetual-motion of the second kind. The difference between the first and second laws of thermodynamics is evident. According to the first law a heat engine operating in cycles must absorb from some heat reservoir a quantity of heat that is equal to the work done by the engine on the surroundings, but it does not state how this is done. The second law states that in the cyclic operation the heat engine must absorb from a heat reservoir a quantity of heat greater than the amount of work done on the surroundings, and must give the excess quantity of heat to a heat reservoir at a lower temperature.

A second statement of the second law of thermodynamics can be obtained by considering the operation of the Carnot cycle in the opposite direction; that is, along the path *ADCB*. In this cycle the system in the heat engine absorbs Q_1 units of heat from the heat reservoir at T_1 and absorbs Q_2 units of heat (a negative quantity) from the heat reservoir at the higher temperature T_2. A quantity of work, W, has been done by the work reservoir on the system. This cycle is the *refrigeration cycle* in which heat is removed from a heat reservoir at a lower temperature and given to another heat reservoir at a higher temperature by doing work on the heat engine. According to the first law, the amount of work done on the system must be equal to the sum of the heats absorbed by the system. Hence,

$$-W = Q_2 + Q_1 \quad \text{or} \quad -Q_2 = W + Q_1 \tag{3.34}$$

That is, the amount of heat given to the heat reservoir at the higher temperature is equal to the sum of the heat removed from the heat reservoir at the lower temperature plus the amount of work done by the surroundings on the heat engine. The *coefficient of performance* for a refrigeration cycle is defined as

$$\frac{Q_1}{W} = \frac{Q_1}{-Q_2 - Q_1} = \frac{T_1}{T_2 - T_1} \tag{3.35}$$

Here the numerical value of Q_2 is negative, which yields a positive value for

T_2. This equation may be written as

$$W = Q_1 \frac{T_2 - T_1}{T_1} \tag{3.36}$$

Moreover, if Q_1 is always positive and nonzero, W must be positive and nonzero except in the case when the two temperatures are equal. This observation results in the Clausius statement of the second law of thermodynamics: *Heat of itself will not flow from a heat reservoir at a lower temperature to one at a higher temperature.* It is *in no way* possible for this to occur without the agency of some system operating as a heat engine in which work is done by the surroundings on the system.

3.9 Carnot's theorem: the maximum efficiency of reversible heat engines

The previous discussion and the conclusions that have been obtained are based on the reversible Carnot cycle. Although the results are independent of the fluid used and of the type of work, we must compare the efficiency of a reversible heat engine operating in a Carnot cycle with that of any other engine, reversible or irreversible, operated in such a cycle that its interactions with the surroundings are those of a Carnot cycle. We consider two engines A and B and two thermal reservoirs, one at the temperature T_2 and the other at the temperature T_1, as shown in Figure 3.6. We assume that A, the reversible or irreversible engine, has a greater efficiency than B, the reversible engine operated in a Carnot cycle. When A is operated in a cycle to produce a quantity of work $|W_A|$, $|Q_2|$ joules of heat are absorbed from the reservoir at T_2 and $|Q_1|$ joules of heat are given to the reservoir at T_1. Then,

Figure 3.6. The relative efficiency of two heat engines.

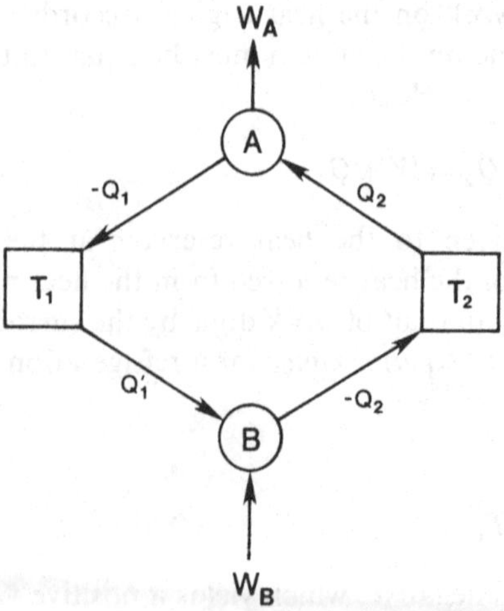

$|W_A| = |Q_2| - |Q_1|$ and the efficiency is $|W_A|/|Q_2|$. We now restore the reservoir at T_2 to its original state by operating the heat engine B in the refrigeration cycle and add $|Q_2|$ joules to the reservoir. In so doing $|Q_1'|$ joules are transferred from the heat reservoir at T_1 to the heat engine B and a quantity of work $|W_B|$ is done on the heat engine. Then $|W_B| = |Q_2| - |Q_1'|$. The efficiency of B operating to produce work is $|W_B|/|Q_2|$. We have assumed that

$$\frac{|W_A|}{|Q_2|} > \frac{|W_B|}{|Q_2|} \tag{3.37}$$

and consequently

$$|W_A| > |W_B| \tag{3.38}$$

The net result is that work is done by the heat engines on the surroundings. When we express this relation in terms of the heats involved, we obtain

$$|Q_2| - |Q_1| > |Q_2| - |Q_1'| \tag{3.39}$$

and, therefore,

$$-|Q_1| > -|Q_1'| \tag{3.40}$$

or

$$|Q_1| < |Q_1'| \tag{3.41}$$

that is, the amount of heat added to the heat reservoir at the temperature T_1 during the work cycle is less than the amount of heat removed from this reservoir during the refrigeration cycle. In this cyclic operation, work is done at the expense of heat obtained from a reservoir at a single temperature. This is contrary to the Kelvin–Planck statement of the second law of thermodynamics. We conclude that no heat engine operating in cycles between two temperatures can have an efficiency greater than that of a reversible engine operating in a Carnot cycle between the same two temperatures. This statement comprises *Carnot's theorem*. Mathematically, we state

$$\eta_A \leqslant \eta_B \tag{3.42}$$

On the basis of Carnot's theorem, we can further prove that two reversible engines operating between the same two temperatures must have the same efficiency. We consider two reversible heat engines B and C operated in a Carnot cycle. When we operate B in a work cycle to drive C in a refrigeration cycle, we have from Equation (3.42) that $\eta_B \leqslant \eta_C$. When C is used in the work cycle to drive B in the refrigeration cycle, we have $\eta_B \geqslant \eta_C$. However, the quantities of work and heat merely change in sign when the direction of a reversible cycle is changed. Therefore, the two efficiencies must be equal. Moreover, the efficiency of a reversible engine operating in a Carnot cycle between two temperatures must be the maximum efficiency obtainable for any engine operating in a cycle between the same two temperatures.

3.10 The entropy function

In Section 3.3 we concluded that an isolated system can be returned to its original state only when all processes that take place within the system are reversible; otherwise, in attempting a cyclic process, at least one work reservoir within the isolated system will have done work and some heat reservoir, also within the isolated system, will have absorbed a quantity of heat. We sought a monotonically varying function that describes these results. The reversible Carnot cycle was introduced to investigate the properties of reversible cycles, and the generality of the results has been shown in the preceding sections. We now introduce the entropy function.

We continue with a reversible heat engine operating in a Carnot cycle, but center our attention on the working substance rather than on the entire system consisting of the heat engine, the work reservoir, and the two heat reservoirs. For such a cycle we can write

$$\frac{Q_2}{T_2} + \frac{Q_1}{T_1} = 0 \tag{3.43}$$

when the absolute values are not used. This equation shows that a certain quantity is zero in the reversible cyclic process; that is, the value of a cyclic integral is zero. From Figure 3.4 we see that, for the steps involving the isothermal expansion and compression of the system,

$$\frac{Q_2}{T_2} = \int_A^B \frac{đQ}{T} \tag{3.44}$$

and

$$\frac{Q_1}{T_1} = \int_E^D \frac{đQ}{T} \tag{3.45}$$

that is, the line integral of $đQ/T$ along the isothermal paths is equal to the heat absorbed by the system from the surroundings during the expansion or compression divided by the temperature at which the heat is absorbed. Similarly, the line integrals along the adiabatic paths are

$$\int_B^C \frac{đQ}{T} = 0 \tag{3.46}$$

and

$$\int_D^A \frac{đQ}{T} = 0 \tag{3.47}$$

The quantity $[(Q_2/T_2)+(Q_1/T_1)]$ is therefore the value of the line integral around the cycle, and we find that this value is zero. We thus have

$$\oint \frac{đQ}{T} = 0 \tag{3.48}$$

However, the cyclic integral of an exact differential is zero and therefore $đQ_r/T$ is an exact differential of some function. The notation $đQ_r$ is used to emphasize that the process is reversible. The new function is called the *entropy function* and is defined in terms of its differential, so

$$dS = đQ_r/T = dQ_r/T \tag{3.49}$$

The differential of the entropy function of a system for an infinitesimal change of state is the infinitesimal quantity of heat absorbed by the system from the surroundings along a reversible path divided by the temperature at which the heat is absorbed. It is defined in terms of a differential as is the energy and, as a result, its absolute value is not known as far as the second law is concerned. (The evaluation of its absolute value is the province of the third law, which is introduced in Chapter 15.) The lack of knowledge of the absolute values is no serious limitation. We are largely concerned with *changes* of state and the accompanying changes in the thermodynamic functions. The differences can be determined experimentally. Mathematically, we are interested in the definite integral of some differential between two limits. Thus, the integration constant cancels and is of no concern.

From its definition the entropy function is a function of the state of the system and its differential is exact. It is important to emphasize that it is defined in terms of a *reversible* process. The change in the value of the entropy function for a change of state may be obtained from its definition only by the integration of Equation (3.49) along a reversible path between the initial and final states. However, because the entropy function is a function of the state of the system, the change of the value of the entropy function of any system between two states is independent of the path. It is the determination of this change, either by experiment or by calculation, that is limited to a reversible path. For any cyclic change of state the change in the value of the entropy function of a system is always zero, whether the path is reversible or irreversible.

3.11 The change in the value of the entropy function of an isolated system along a reversible path

Having defined the entropy function, we must next determine some of its properties, particularly its change in reversible and irreversible processes taking place in isolated systems. (In each case a simple process is considered first, then a generalization.)

For a simple reversible process, consider the isothermal reversible expansion of a fluid along the path AB in Figure 3.4. The fluid confined in the piston-and-cylinder arrangement is at the temperature T_2 and the cylinder is in thermal contact with the heat reservoir at the same temperature. The piston-and-cylinder arrangement, including the fluid, the heat reservoir, and the work reservoir constitute an isolated system. The transfer of heat takes place only between the fluid and the heat reservoir. For any infinitesimal step along the path AB, dQ joules of heat are transferred from the heat reservoir to the fluid. The differential change in the value of the entropy function of the fluid is dQ/T_2 and that of the heat reservoir is $-dQ/T_2$. Thus, the total change in the value of the entropy function of the isolated system is zero. For the entire path from A to B, Q_2 joules of heat are transferred from the heat reservoir to the fluid, and the entropy change of the fluid is Q_2/T_2 and that of the heat reservoir is $-Q_2/T_2$. Thus, the total entropy change of the isolated system along the entire path is also zero. By the same arguments we find that the change in the value of the entropy function of the isolated system for the isothermal reversible compression along the path CD is zero. For the adiabatic reversible expansion or compression along the paths BC and DA the isolated system consists of the fluid in the piston-and-cylinder arrangement and the work reservoir, separated from heat reservoirs by an adiabatic envelope. Then, no heat is transferred from any part of the isolated system to any other part. The total entropy change for these processes is zero. For an isothermal reversible expansion or compression taking place in an isolated system or for an adiabatic reversible expansion or compression taking place in an isolated system, the change in the value of the entropy function is zero.

Consider any number of systems that may do work on each other and also transfer heat from one to another by reversible processes. The changes of state may be of any nature, and any type of work may be involved. This collection of systems is isolated from the surroundings by a rigid, adiabatic envelope. We assume first that the temperatures of all the systems between which heat is transferred are the same, because of the requirements for the reversible transfer of heat. For any infinitesimal change that takes place within the isolated system, the change in the value of the entropy function for the ith system is dQ_i/T, where Q_i is the heat absorbed by the ith system. The total entropy change is the sum of such quantities over all of the subsystems in the isolated system, so

$$dS = \sum_i dQ_i/T = (1/T)\sum_i dQ_i = 0 \tag{3.50}$$

The temperature can be factored from the sum, because it has the same value for all sytems. However, $\sum dQ_i = 0$ because the total system is isolated. Consequently, the change in the value of the entropy function of the isolated system is zero. The change for a finite process is simply the sum of the changes for all of the infinitesimal changes and, therefore, the change in the value of the entropy function for a finite process is also zero.

If several of the systems within the isolated system have different temperatures, we must separate by adiabatic envelopes those systems that have the same temperature from all other systems that have different temperatures. Only in this way can we maintain the reversible transfer of heat. The isolated system is then composed of several regions, each containing those systems that have the same temperature. The separate regions are not isolated from each other, though, because work can still be done by a system in one region on a system in another region. The change in the value of the entropy function of any region is zero for any reversible process taking place by Equation (3.50). The total entropy change for all regions is also zero. We conclude that *the change in the value of the entropy function in any isolated system is zero for any reversible process that takes place within the system.* This is not to say that the value of the entropy function of each system within the isolated system is constant, but only that the increase in the value of the function for some systems is equal to the decrease in the value for other systems.

3.12 The change in the value of the entropy function of an isolated system along an irreversible path

Because the entropy function is defined in terms of a reversible process, the change in the value of this function along an irreversible path cannot be measured directly. We can determine it for any given system that is not necessarily isolated, provided we know the final state as well as the initial state, by devising a reversible path between the two states. However, in many cases the final state of a system undergoing an irreversible process is neither known nor predictable. In order to determine the change of entropy of an isolated system for the general case, we must allow the irreversible process to take place in the system. We then restore the system to its original state by reversible processes. In so doing, the isolation of the system must be removed and at least one extra heat reservoir must be used. The change in the value of the entropy function of the system in the cyclic process must be zero, because the entropy function is a function of the state of the system. Consequently, the entropy change of the system for the irreversible process must be equal and opposite to its entropy change for the reversible process, so

$$\Delta S = -\Delta S_r \tag{3.51}$$

where ΔS is the change of the value of the entropy function of the system for the irreversible process and ΔS_r is that for the reversible process. Since the process restoring the original system to its initial state is reversible, the change of the value of the entropy function for the system must be equal and opposite to that of the heat reservoirs used in the reversible process, so

$$\Delta S_r = -\Delta S_h \tag{3.52}$$

where ΔS_h is the change for the heat reservoirs. Therefore, the change in the value of the entropy function of the isolated system in the original irreversible process must be equal to that of the heat reservoirs used in the reversible process.

The expansion of an ideal gas in the *Joule experiment* will be used as a simple example. Consider a quantity of an ideal gas confined in a flask at a given temperature and pressure. This flask is connected through a valve to another flask, which is evacuated. The two flasks are surrounded by an adiabatic envelope and, because the walls of the flasks are rigid, the system is isolated. We now allow the gas to expand irreversibly into the evacuated flask. For an ideal gas the temperature remains the same. Thus, the expansion is isothermal as well as adiabatic. We can return the system to its original state by carrying out an isothermal reversible compression. Here we use a work reservoir to compress the gas and a heat reservoir to remove heat from the gas. As we have seen before, a quantity of heat equal to the work done on the gas must be transferred from the gas to the heat reservoir. In so doing, the value of the entropy function of the heat reservoir is increased. Consequently, the value of the entropy function of the gas increased during the adiabatic irreversible expansion of gas.

We have already stated that an isolated system cannot be returned to its original state whenever an irreversible process has taken place within it, and that there is always at least one work reservoir that has done work and one heat reservoir that has absorbed heat. To restore the isolated system to its original state, the isolation must be broken. The work reservoir can be returned to its original state by the use of a work reservoir that is external to the originally isolated system. However, the performance of work on the work reservoir does not alter the value of the entropy function of the originally isolated system. The one or more heat reservoirs that have absorbed heat can be returned to their original state by the use of heat reservoirs that are external to the originally isolated system. The reversible transfer of heat from those reservoirs originally within the system to the reservoirs originally external to the system reduces the value of the entropy function of the originally isolated system. Thus, the value of the entropy function of the isolated system increased for the irreversible processes.

The results obtained in the previous two sections are expressed by the differential expression

$$dS \geqslant 0 \quad \text{(isolated system)} \tag{3.53}$$

called the *inequality of Clausius*. This inequality refers to an isolated system for which the differential change of the entropy function is zero for a reversible process and is greater than zero for an irreversible process that may take place in the isolated system. It is important to emphasize that this statement applies to an isolated system. The value of the entropy function of a system that is not isolated may increase or decrease with changes of state. However, in carrying out the changes of state, other systems would have to be used as work or heat reservoirs. Equation (3.53) applies to the isolated systems, which includes the necessary work and heat reservoirs.

3.13 A third statement of the second law of thermodynamics

A third statement of the second law is based on the entropy. In reversible systems all forces must be opposed by equal and opposite forces. Consequently, in an isolated system any change of state by reversible processes must take place under equilibrium conditions. Changes of state that occur in an isolated system by irreversible processes must of necessity be spontaneous or natural processes. For all such processes in an isolated system, the entropy increases. Clausius expressed the second law as: *The entropy of the universe is always increasing to a maximum.* Planck has given a more general statement of the second law: *Every physical and chemical process in nature takes place in such a way as to increase the sum of the entropies of all bodies taking any part in the process. In the limit, i.e., for reversible processes, the sum of the entropies remains unchanged.*

3.14 The dependence of the entropy on temperature

The entropy of a system is a function of the independent variables that are used to define the state of the system. For a closed system in which no work other than that of expansion and compression is involved, we need two variables to define the state. These are usually the temperature and volume or the temperature and pressure. The dependence of the entropy on the volume and the pressure is developed in Chapter 4, but its dependence on the temperature can be obtained with the information that we now have. By definition

$$dS = dQ_r/T \tag{3.54}$$

At constant volume we may write dS as

$$dS = (\partial S/\partial T)_{V,n}\, dT \tag{3.55}$$

Also, at constant volume dQ_r is an exact differential, and consequently

$$dQ_r = (dQ_r/dT)_{V,n}\, dT \tag{3.56}$$

Therefore, on substituting Equations (3.55) and (3.56) into Equation (3.49),

$$\left(\frac{\partial S}{\partial T}\right)_{V,n} = \frac{1}{T}\left(\frac{dQ_r}{dT}\right)_{V,n} \tag{3.57}$$

However, the heat capacity in general is $đQ/dT$, and $(đQ_r/dT)_{V,n}$ is thus the heat capacity of the system at constant volume. We therefore obtain

$$\left(\frac{\partial S}{\partial T}\right)_{V,n} = \frac{C_V}{T} \tag{3.58}$$

where C_V is the heat capacity of the system at constant volume. By a similar argument we obtain, at constant pressure,

$$\left(\frac{\partial S}{\partial T}\right)_{P,n} = \frac{C_P}{T} \tag{3.59}$$

where C_P is the heat capacity of the system at constant pressure.

3.15 The change of entropy for a change of state of aggregation

A reversible change of state of aggregation of a pure substance takes place under the equilibrium conditions of constant temperature and pressure. We can therefore use Equation (3.49) to determine the change of entropy of such a change of state. From this equation,

$$\Delta S = \int \frac{đQ_r}{T} = \frac{1}{T}\int đQ_r \tag{3.60}$$

since the temperature is constant. The integral of $đQ$ is Q, the heat absorbed by the systems during the change of state. Moreover, from Equation (2.33) we know that $Q = \Delta H$, the change of enthalpy for the particular change of state, because the pressure is held constant. Consequently, we find that the change of entropy for a change of state of aggregation of a pure substance is given by

$$\Delta S = \Delta H/T \qquad \text{(change of state of aggregation)} \tag{3.61}$$

4

Gibbs and Helmholtz energy functions and open systems

The only two functions actually required in thermodynamics are the energy function, obtained from the first law of thermodynamics, and the entropy function, obtained from the second law of thermodynamics. However, these functions are not necessarily the most convenient functions. The enthalpy function was defined in order to make the pressure the independent variable, rather than the volume. When the first and second laws are combined, as is done in this chapter, the entropy function appears as an independent variable. It then becomes convenient to define two other functions, the Gibbs and Helmholtz energy functions, for which the temperature is the independent variable, rather than the entropy. These two functions are defined and discussed in the first part of this chapter.

Only closed systems have so far been considered. However, mass can be varied and is an important variable for all thermodynamic functions. The introduction of mass as an independent variable into the basic differential expressions for the thermodynamic functions yields the equations that Gibbs called 'fundamental'. It is on these equations that much of the development of the applications of thermodynamics to chemical systems is based.

Many of the problems that are met in applications require the evaluation of various derivatives and the integration of differential quantities that involve the derivatives. To be of use, the derivatives must be expressed in terms of experimentally determinable quantities. Because the number of derivatives that can occur is extremely large, it is advantageous to be able to derive the necessary relations. Methods to do this are discussed and several examples given in the last part of this chapter.

4.1 The Helmholtz energy

One differential expression for the energy of a system is given by Equation (2.26):

$$dE = đQ - P\,dV + đW'$$

for which the process is quasistatic with respect to the work of expansion or compression. This equation may be used when we are interested in the heat absorbed by the system for some change of state for a specified process. However, if the process is reversible, then $đQ$ becomes dQ_r, which equals $T\,dS$ by the definition of the entropy. The equation

$$dE = T\,dS - P\,dV + dW'_r \tag{4.1}$$

is obtained when this limitation is adopted and the substitution made. In this equation P is the pressure of the system and dW'_r is the sum of all of the work done on the system, excluding the work of expansion or compression, along a reversible path. The integration of Equation (4.1) to determine the difference in energy between two states must be carried out along a reversible path. The energy is still a function of the state of the system, and consequently this difference of energy is independent of the path. It is the determination of this difference that must be made along a reversible path when Equation (4.1) is used. No such limitation is placed on Equation (2.26).

Although Equation (4.1) is useful, particularly for isentropic changes of state for which $dS = 0$, the entropy appears as an independent variable. This circumstance is inconvenient from an experimental point of view, because no meters or devices exist that measure entropy. A new function in which the temperature is an independent variable rather than the entropy is obtained by subtracting

$$d(TS) = T\,dS + S\,dT \tag{4.2}$$

from Equation (4.1). We then obtain

$$d(E - TS) = -S\,dT - P\,dV + dW'_r \tag{4.3}$$

The new function is defined first in terms of its differential, so

$$dA = d(E - TS) \tag{4.4}$$

and then define the integration constant resulting from the integration of Equation (4.4) to be zero. This is justified because we do not know the absolute value of either the energy or the entropy and, consequently, cannot know the absolute value of the new function. Furthermore, as we are defining a new function, we may define it in any way that we please. Therefore, we choose to make the integration constant zero. After the integration, we have the definition[1]

$$A = E - TS \tag{4.5}$$

[1] If the product TS is not uniform throughout the system, the system can be divided into parts in which TS is uniform. Then $\sum TS$, the sum being taken over the individual parts of the system, may be substituted for TS. If the volumes in which TS is uniform are infinitesimal, the $\sum TS$ may be substituted by $\int (TS/V)\,dV$, the integral being taken over the entire volume of the system.

The new function A is called the Helmholtz energy.[2] Since E, T, and S are functions of the state of the system, A is also a function of the state of the system, and its differential is exact. The change in the value of the Helmholtz energy in going from some initial state to some final state is independent of the path. However, the determination of this change must be obtained by the integration of Equation (4.3) along any *reversible* path between the two states.

Any physical interpretation of the Helmholtz energy must be based on interpreting Equation (4.3). Thus, for an isothermal change of state, the equation becomes

$$\mathrm{d}A = -P\,\mathrm{d}V + đW'_\mathrm{r}$$

The algebraic sum of the terms $P\,\mathrm{d}V$ and $đW'_\mathrm{r}$ is the *total* work done reversibly by the system for a differential change of state. Hence, $\mathrm{d}A$ is the negative of the maximum work that the system can do on the surroundings for an *isothermal* differential change of state.

4.2 The Gibbs energy

The differential expression for the enthalpy is

$$\mathrm{d}H = đQ + V\,\mathrm{d}P + đW' \tag{2.30}$$

for quasistatic processes with respect to the work of expansion and compression. If we limit the differential path to reversible processes, we may substitute $T\,\mathrm{d}S$ for $đQ$ and obtain

$$\mathrm{d}H = T\,\mathrm{d}S + V\,\mathrm{d}P + \mathrm{d}W'_\mathrm{r} \tag{4.6}$$

Although the integration of this equation is limited to *reversible* paths, the change of enthalpy in going from one state to another is still independent of the path.

Here, again, the entropy appears as an independent variable. It is therefore convenient to define another function so that the temperature appears as an independent variable. This is done by subtracting Equation (4.2) from Equation (4.6):

$$\mathrm{d}(H - TS) = -S\,\mathrm{d}T + V\,\mathrm{d}P + \mathrm{d}W'_\mathrm{r} \tag{4.7}$$

We define the new function, G, by setting

$$\mathrm{d}G = \mathrm{d}(H - TS) \tag{4.8}$$

[2] Gibbs used ψ for this function; G. N. Lewis called it the work function, with the symbol A; and in many European countries it was called the Helmholtz free energy, with the symbol F. By international agreement the accepted name is the Helmholtz energy and the symbol is A.

and by defining the integration constant of Equation (4.8) to be zero, so

$$G = H - TS \tag{4.9}$$

This function is the Gibbs energy function, or simply the Gibbs energy.[3] The Gibbs energy is a function of the state of the system, and its differential is exact. However, the same restrictions must be put on the use of Equation (4.7) as with other equations containing the entropy. The determination of the change in G in going from some initial state to some final state by the integration of this equation must be made along some *reversible* path. The value of the change so determined is independent of the path.

Any physical interpretation of the Gibbs energy must be based on Equation (4.7). Thus, for *isothermal* and *isopiestic* changes of state the equation becomes

$$\mathrm{d}G = +\mathrm{d}W_r'$$

The term $\mathrm{d}W_r'$ is the *maximum* work done by the surroundings, and thus the negative of the maximum work done by the system, excluding the work of expansion or compression. We emphasize that such an interpretation is valid only for an isothermal and isopiestic reversible process.

4.3 Open systems

All of the equations developed so far are applicable to closed systems, in which the masses of the components of the system are kept constant and no material is added to or removed from the system. Experience has shown that, for many changes of state such as evaporation, fusion, mixing to form a solution, or a chemical reaction, the change of the energy and the entropy depends upon the number of moles of the substance or substances that undergo the change of state. It must be concluded that the energy and entropy, and consequently the enthalpy and the Gibbs and Helmholtz energies, are all functions of the number of moles of the components that form the system, and hence are extensive variables. A variation of the system can be made by changing the mass of any of the components in the system. Such systems, whose mass is variable, are called open systems. (We omit from discussion flow systems, which are also open systems. In such systems the kinetic energy and possibly the potential energy of the mass of material flowing into and out of the volume containing the system must be considered.) The differential expressions for $\mathrm{d}E$, $\mathrm{d}H$, $\mathrm{d}A$, and $\mathrm{d}G$ must be modified to make them applicable to such systems.

The energy of the system may be expressed as a function of the entropy,

[3] Gibbs used the symbol ξ for this function, and G. N. Lewis used F. Sometimes this is referred to as the free energy or the Gibbs free energy. The term 'free enthalpy' has also been used. The IUPAC recommendation, which is followed in this book, is to use the symbol G for the Gibbs energy. For simplicity we omit the word 'function.'

volume, number of moles of each component, and the generalized coordinates associated with the work terms other than the work of expansion or compression, so that

$$E = E(S, V, n_1, n_2, n_3, \ldots, X) \tag{4.10}$$

where X represents the generalized coordinates.[4] The differential of the energy function is then

$$\begin{aligned} dE = {} & \left(\frac{\partial E}{\partial S}\right)_{V,n_1,n_2,n_3,\ldots,X} dS + \left(\frac{\partial E}{\partial V}\right)_{S,n_1,n_2,n_3,\ldots,X} dV \\ & + \left(\frac{\partial E}{\partial n_1}\right)_{S,V,n_2,n_3,\ldots,X} dn_1 + \left(\frac{\partial E}{\partial n_2}\right)_{S,V,n_1,n_3,\ldots,X} dn_2 \\ & + \left(\frac{\partial E}{\partial n_3}\right)_{S,V,n_1,n_2,\ldots,X} dn_3 + \cdots + \sum_j F_j\, dX_j \end{aligned} \tag{4.11}$$

In this equation F_j represents the conjugate generalized forces and the sum is taken over all of the generalized coordinates. The equation must reduce to Equation (4.1) when the number of moles of each component is held constant. Comparison of these two equations shows that $(\partial E/\partial S)_{V,n_1,n_2,n_3,\ldots,X} = T$ and $(\partial E/\partial V)_{S,V,n_1,n_2,n_3,\ldots,X} = -P$. When we use the symbol μ_1 for $(\partial E/\partial n_1)_{S,V,n_2,n_3,\ldots,X}$, etc., Equation (4.11) can be written as

$$dE = T\, dS - P\, dV + \sum_i \mu_i\, dn_i + \sum_j F_j\, dX_j \tag{4.12}$$

where the subscript i represents any component in general and the sum is taken over all components. Differential expressions for dH, dA, and dG for open systems are obtained by the same methods that were used to obtain them for closed systems. We thus have

$$dH = T\, dS + V\, dP + \sum_i \mu_i\, dn_i + \sum_j F_j\, dX_j \tag{4.13}$$

$$dA = -S\, dT - P\, dV + \sum_i \mu_i\, dn_i + \sum_j F_j\, dX_j \tag{4.14}$$

$$dG = -S\, dT + V\, dP + \sum_i \mu_i\, dn_i + \sum_j F_j\, dX_j \tag{4.15}$$

The derivatives of the energy with respect to the number of moles of

[4] Gibbs used mass in his development of thermodynamics, and any unit of mass may be used. The number of moles of a substance is a multiple of a specific unit of mass that is different for each substance: the molecular mass. If there is any question of the molecular mass of the substance, the assumed molecular mass or the assumed chemical formula of the substance must be given; e.g., $AlCl_3$ or Al_2Cl_6.

a component, keeping all other independent variables constant, are so important in thermodynamics that they have been given the separate symbol μ and are called the 'chemical potentials.' They are the rates of change of the energy of the system as the number of moles of a component in the system is increased at constant entropy, volume, and number of moles of all other components, with no work being involved. The problem of keeping the entropy and volume constant as material is added to the system presents no difficulties conceptually, because the temperature and pressure, which are dependent variables according to Equation (4.12), will change as the material is added under the given conditions. Other equivalent conditions of the chemical potential are given in Section 4.5.

In many problems of thermodynamics we are concerned with both open and closed systems simultaneously. As an example we may consider a heterogenous system composed of several phases. The whole system is closed. However, we can consider each phase as a system, and these phases are open because material can be transferred from one phase to another.

4.4 Résumé

We have now defined the five thermodynamic functions: energy, entropy, enthalpy, Helmholtz energy, and Gibbs energy. The energy and entropy are defined in terms of the differentials such that $dE = đQ + đW$ and $dS = dQ_r/T$. The others are defined as $H = E + PV$, $A = E - TS$ and $G = H - TS = E + PV - TS$. Each of these five functions is a function of the state of the system, and consequently their differentials are exact. They are assumed to be continuous functions of the independent variables between certain easily recognized limits; at these limits some of the functions may be still continuous, but others will be discontinuous. These limits are the points where first- or second-order transitions occur. First-order transitions are those in which the Gibbs energy is continuous, but its first and higher derivatives are discontinuous. Such transitions are changes of phase (change of state of aggregation) or of crystal structure. Second-order transitions are those in which the first derivative is also continuous, but the second- and higher-order derivatives are discontinuous. Such transitions occur at the lambda point of helium, at the Curie point of a ferromagnetic material, or at the temperature at which an 'order–disorder' transition takes place in certain alloys and some chemical compounds. Finally, all five functions are homogenous functions of the first degree in the mole numbers (number of moles of each component).

Under the conditions that the system is closed and that we are concerned only with the first law of thermodynamics, we have the two equations

$$dE = đQ + đW \tag{4.16}$$

and

$$dH = đQ + V\,dP + đW' \tag{4.17}$$

Equation (4.17) is applicable to quasistatic processes for the work of expansion and compression. On the introduction of the second law we have four equations for closed systems

$$dE = T\,dS - P\,dV + \sum_j F_j\,dX_j \tag{4.18}$$

$$dH = T\,dS + V\,dP + \sum_j F_j\,dX_j \tag{4.19}$$

$$dA = -S\,dT - P\,dV + \sum_j F_j\,dX_j \tag{4.20}$$

and

$$dG = -S\,dT + V\,dP + \sum_j F_j\,dX_j \tag{4.21}$$

The more general equations for open systems are

$$dE = T\,dS - P\,dV + \sum_i \mu_i\,dn_i + \sum_j F_j\,dX_j \tag{4.22}$$

$$dH = T\,dS + V\,dP + \sum_i \mu_i\,dn_i + \sum_j F_j\,dX_j \tag{4.23}$$

$$dA = -S\,dT - P\,dV + \sum_i \mu_i\,dn_i + \sum_j F_j\,dX_j \tag{4.24}$$

and

$$dG = -S\,dT + V\,dP + \sum_i \mu_i\,dn_i + \sum_j F_j\,dX_j \tag{4.25}$$

The integration of Equations (4.18)–(4.25) is limited to a reversible path between some initial and some final state, but the change in the values of the functions so determined is independent of the path.

We see from these equations that we deal with a large number of variables in thermodynamics. These variables are E, S, H, A, G, P, V, T, each n_i, and two variables for each additional work term. In any given thermodynamic problem only some of these variables are independent, and the rest are dependent. The question of how many independent variables are necessary to define completely the state of a system is discussed in Chapter 5. For the present we take the number of moles of each component, the generalized coordinate associated with each additional work term, and two of the four variables S, T, P, and V as the independent variables. Moreover, not all of the four functions E, H, A, and G are applicable to all problems. The choice of which function to use may be dictated somewhat by the problem, or it may be a matter of individual preference. From Equations (4.22)–(4.25) we have

$$E = E(S, V, n_1, n_2, n_3, \ldots, X_1, X_2, \ldots) \tag{4.26}$$

$$H = H(S, P, n_1, n_2, n_3, \ldots, X_1, X_2, \ldots) \tag{4.27}$$

$$A = A(T, V, n_1, n_2, n_3, \ldots, X_1, X_2, \ldots) \tag{4.28}$$

and

$$G = G(T, P, n_1, n_2, n_3, \ldots, X_1, X_2, \ldots) \tag{4.29}$$

Thus, of the four variables, S, V, T, and P, if S and V are the independent ones we would use the energy, if S and P are the independent variables we would use the enthalpy, if T and V are the independent variables we would use the Helmholtz energy, and if T and P are the independent variables we would use the Gibbs energy. This is not to say that we are limited to these combinations. We could set E as a function of T and V, of T and S, of T and P, of S and P, or of P and V. However, if we do so, the differential expression for $\mathrm{d}E$ is much more complicated. The other three functions (H, A, and G) could also be expressed as functions of the same combinations of the variables, but again more-complicated expressions are obtained for the differentials. The two most important combinations other than those given in Equations (4.22)–(4.25) are to make the energy a function of T and V, and the enthalpy a function of T and P, in addition to the mole numbers and the work variables. These functions are discussed in Section 4.8.

In Equations (4.22)–(4.25) the energy, enthalpy, Gibbs energy, and Helmholtz energy have been taken as dependent variables. However, any of the four equations may be solved for any of the other differentials in that equation. That particular variable would then become dependent and the energy, enthalpy, the Gibbs energy or the Helmholtz energy, as the case may be, would become independent. For example, Equation (4.22) may be written as

$$\mathrm{d}S = \frac{\mathrm{d}E}{T} + \frac{P\,\mathrm{d}V}{T} - \frac{\sum_j F_j\,\mathrm{d}X_j}{T} - \frac{\sum_i \mu_i\,\mathrm{d}n_i}{T} \tag{4.30}$$

in which case the energy is an independent variable and the entropy is dependent.

In this section we have now given all of the basic equations in chemical thermodynamics. Any other relation can be derived from one or more of these. We have also outlined the wide choice of independent variables that may be used. Fortunately, because of the limitation of being able to measure experimentally only the pressure, temperature, volume, mass, and the other work variables, the number of choices is much more limited than is at first apparent.

In the remainder of this book we consider all work terms other than the work of expansion and compression to be zero, except in those specific cases where such terms are definitely considered. This means that the generalized coordinates that appear as differentials in the expression for such work terms are all considered to be constant.

4.5 Dependence of the thermodynamic functions on the independent variables

Each of Equations (4.22)–(4.25) is a differential expression for a certain function written in terms of a number of independent variables. From mathematics we may immediately obtain the partial derivative of the function with respect to each of the independent variables. Thus, from Equation (4.22),[5]

$$\left(\frac{\partial E}{\partial S}\right)_{V,n} = T \tag{4.31}$$

and

$$\left(\frac{\partial E}{\partial V}\right)_{S,n} = -P \tag{4.32}$$

In this case both the temperature and pressure are dependent variables and are functions of the entropy, volume, and mole numbers of the components. From Equation (4.23),

$$\left(\frac{\partial H}{\partial S}\right)_{P,n} = T \tag{4.33}$$

and

$$\left(\frac{\partial H}{\partial P}\right)_{S,n} = V \tag{4.34}$$

Here the temperature and volume are dependent variables and are functions of the entropy, pressure, and mole numbers of the components. From

---

[5] In order to save space, the subscript n in a derivative indicates that the number of moles of *all* components are kept constant except in the case when the variation is one of the components. In that case it indicates that the number of moles of all of the components are kept constant *except* the one that is being varied. Also, it must be remembered that all work terms other than the work of expansion or compression are zero, even though the notation is not made in general practice.

Equation (4.24),

$$\left(\frac{\partial A}{\partial T}\right)_{V,n} = -S \tag{4.35}$$

and

$$\left(\frac{\partial A}{\partial V}\right)_{T,n} = -P \tag{4.36}$$

The entropy and pressure are dependent variables and are functions of the volume, temperature, and the mole numbers of the components. Finally, from Equation (4.25),

$$\left(\frac{\partial G}{\partial T}\right)_{P,n} = -S \tag{4.37}$$

and

$$\left(\frac{\partial G}{\partial P}\right)_{T,n} = V \tag{4.38}$$

In this case the entropy and volume are dependent variables and are functions of the temperature, pressure, and mole numbers of the components. In addition, we have four derivatives that are all equal to the chemical potential, one from each of the four equations, so

$$\mu_i = \left(\frac{\partial E}{\partial n_i}\right)_{S,V,n} = \left(\frac{\partial H}{\partial n_i}\right)_{S,P,n} = \left(\frac{\partial A}{\partial n_i}\right)_{T,V,n} = \left(\frac{\partial G}{\partial n_i}\right)_{T,P,n} \tag{4.39}$$

Throughout this book we use the subscript i to refer to any component in general. When it is necessary to specify a specific component, we use other subscripts such as j, k, l, etc.

Two other important derivatives of the Gibbs and Helmholtz energies with respect to the temperature can be derived from these simpler relations. From the definition of the Gibbs energy and Equation (4.37) we obtain

$$G - T\left(\frac{\partial G}{\partial T}\right)_{P,n} = H \tag{4.40}$$

However, the left-hand side of this equation equals $-T^2(\partial(G/T)/\partial T)_{P,n}$, and consequently

$$\left(\frac{\partial(G/T)}{\partial T}\right)_{P,n} = -\frac{H}{T^2} \tag{4.41}$$

This equation may also be written as

$$\left(\frac{\partial(G/T)}{\partial(1/T)}\right)_{P,n} = H \tag{4.42}$$

because $d(1/T) = -dT/T^2$. Equation (4.41) is sometimes called the *Gibbs–Helmholtz equation*. Similarly, for the Helmholtz energy, we obtain

$$\left(\frac{\partial(A/T)}{\partial T}\right)_{V,n} = -\frac{E}{T^2} \tag{4.43}$$

or

$$\left(\frac{\partial(A/T)}{\partial(1/T)}\right)_{V,n} = E \tag{4.44}$$

4.6 The Maxwell relations and other relations

As stated previously, Equations (4.22)–(4.25) are differential expressions of certain functions. Consequently, the differentials are exact and the conditions of exactness must be satisfied. We can immediately write

$$\left(\frac{\partial T}{\partial V}\right)_{S,n} = -\left(\frac{\partial P}{\partial S}\right)_{V,n} \tag{4.45}$$

$$\left(\frac{\partial T}{\partial P}\right)_{S,n} = \left(\frac{\partial V}{\partial S}\right)_{P,n} \tag{4.46}$$

$$\left(\frac{\partial S}{\partial V}\right)_{T,n} = \left(\frac{\partial P}{\partial T}\right)_{V,n} \tag{4.47}$$

$$\left(\frac{\partial S}{\partial P}\right)_{T,n} = -\left(\frac{\partial V}{\partial T}\right)_{P,n} \tag{4.48}$$

Equations (4.45) and (4.46) may be inverted to read

$$\left(\frac{\partial S}{\partial P}\right)_{V,n} = -\left(\frac{\partial V}{\partial T}\right)_{S,n} \tag{4.49}$$

and

$$\left(\frac{\partial S}{\partial V}\right)_{P,n} = \left(\frac{\partial P}{\partial T}\right)_{S,n} \tag{4.50}$$

These four relations are known as the *Maxwell relations*, and are useful in terms of converting one set of variables to another. They are limited to closed systems, as indicated by the constancy of the mole numbers of the components.

Additional relations can be written when we allow the mole numbers of the components to be varied. Thus, from Equation (4.22),

$$\left(\frac{\partial \mu_i}{\partial S}\right)_{V,n} = \left(\frac{\partial T}{\partial n_i}\right)_{S,V,n} \tag{4.51}$$

and

$$\left(\frac{\partial \mu_i}{\partial V}\right)_{S,n} = -\left(\frac{\partial P}{\partial n_i}\right)_{S,V,n} \tag{4.52}$$

or, from Equation (4.25),

$$\left(\frac{\partial \mu_i}{\partial T}\right)_{P,n} = -\left(\frac{\partial S}{\partial n_i}\right)_{T,P,n} \tag{4.53}$$

and

$$\left(\frac{\partial \mu_i}{\partial P}\right)_{T,n} = \left(\frac{\partial V}{\partial n_i}\right)_{T,P,n} \tag{4.54}$$

4.7 Development of other relations

One of the problems of thermodynamics is to obtain expressions for various derivatives in terms of quantities that can be observed experimentally. The total number of first partial derivatives is exceedingly large and it is useless to attempt to memorize the relations between them. Fortunately, there is only a relatively small number of such relations that are of importance, and some of the simpler ones are easily learned through use. However, it is still important to be able to obtain or derive such relations when they are needed. Some systematic methods have already been developed; of these, probably those most used are by Bridgman [1], and by Shaw [2], who used the methods of Jacobians. The purpose of this section is to illustrate various methods. The discussion is not necessarily complete or systematic, but it does afford training in the use of properties of differentials and derivatives to derive many thermodynamic relations between partial derivatives. It should be remembered that thermodynamics is redundant, in that the enthalpy and the Gibbs and Helmholtz energies have been defined for convenience. Those functions are not absolutely necessary. As a result, there is no single way to obtain a given relation.

(a) Any variable in thermodynamics can always be set as a function of two other variables by keeping all other variables constant. We may be interested in the relation between the three derivatives involving the three variables. As an example, we may be interested in the isentropic change of the volume with respect to the temperature. We may set $S = S(V, T)$ at constant n. Then

$$[dS]_n = \left(\frac{\partial S}{\partial V}\right)_{T,n} [\mathrm{d}V]_n + \left(\frac{\partial S}{\partial T}\right)_{V,n} [\mathrm{d}T]_n \tag{4.55}$$

However, V may be considered as a function of S, T, and n, so

$$[\mathrm{d}V]_n = \left(\frac{\partial V}{\partial S}\right)_{T,n}[\mathrm{d}S]_n + \left(\frac{\partial V}{\partial T}\right)_{S,n}[\mathrm{d}T]_n \tag{4.56}$$

Substituting this equation into Equation (4.55) and setting $[\mathrm{d}S]_n = 0$,

$$0 = \left[\left(\frac{\partial S}{\partial V}\right)_{T,n}\left(\frac{\partial V}{\partial T}\right)_{S,n} + \left(\frac{\partial S}{\partial T}\right)_{V,n}\right][\mathrm{d}T]_n \tag{4.57}$$

The coefficient must be zero because $[\mathrm{d}T]_n$ need not be zero. Therefore

$$\left(\frac{\partial S}{\partial V}\right)_{T,n}\left(\frac{\partial V}{\partial T}\right)_{S,n} + \left(\frac{\partial S}{\partial T}\right)_{V,n} = 0 \tag{4.58}$$

or

$$\left(\frac{\partial V}{\partial T}\right)_{S,n} = -\frac{(\partial S/\partial T)_{V,n}}{(\partial S/\partial V)_{T,n}} = -\frac{C_V}{T(\partial P/\partial T)_{V,n}} \tag{4.59}$$

recalling Equation (4.47). Another form of Equation (4.58) is

$$\left(\frac{\partial S}{\partial V}\right)_{T,n}\left(\frac{\partial V}{\partial T}\right)_{S,n}\left(\frac{\partial T}{\partial S}\right)_{V,n} = -1 \tag{4.60}$$

(b) A number of derivatives may be obtained from Equations (4.22)–(4.25). In these derivatives a new variable, different from the independent variables of the basic equation, is introduced. As an example consider the determination of $(\partial E/\partial V)_{T,n}$. From Equation (4.22),

$$\mathrm{d}E = T\,\mathrm{d}S - P\,\mathrm{d}V + \sum_i \mu_i\,\mathrm{d}n_i \tag{4.22}$$

Because we are interested in the independent variables T and V, we may set E as a function of T and V at constant mole numbers of the components, from which

$$[\mathrm{d}E]_{T,n} = \left(\frac{\partial E}{\partial V}\right)_{T,n}[\mathrm{d}V]_{T,n} \tag{4.61}$$

We also need to express S as a function of T and V at constant mole numbers of the components, to obtain

$$[\mathrm{d}S]_{T,n} = \left(\frac{\partial S}{\partial V}\right)_{T,n}[\mathrm{d}V]_{T,n} \tag{4.62}$$

Equation (4.22) becomes

$$\left(\frac{\partial E}{\partial V}\right)_{T,n}[\mathrm{d}V]_{T,n} = T\left(\frac{\partial S}{\partial V}\right)_{T,n}[\mathrm{d}V]_{T,n} - P[\mathrm{d}V]_{T,n} \tag{4.63}$$

when we substitute Equations (4.61) and (4.62) for dE and dS, respectively. We then divide by dV, so

$$\left(\frac{\partial E}{\partial V}\right)_{T,n} = T\left(\frac{\partial S}{\partial V}\right)_{T,n} - P \qquad (4.64)$$

Finally, from Equation (4.47), we know that $(\partial S/\partial V)_{T,n} = (\partial P/\partial T)_{V,n}$ and, consequently,

$$\left(\frac{\partial E}{\partial V}\right)_{T,n} = T\left(\frac{\partial P}{\partial T}\right)_{V,n} - P \qquad (4.65)$$

(c) Let us determine $(\partial E/\partial n_j)_{T,P,n}$, where the two variables T and P are introduced. We may again start with Equation (4.22):

$$\mathrm{d}E = T\,\mathrm{d}S - P\,\mathrm{d}V + \sum_i \mu_i\,\mathrm{d}n_i \qquad (4.22)$$

The functions E, S, and V can all be expressed as functions of T, P, and all ns. At constant T, P, and all ns except n_j,

$$\mathrm{d}E = \left(\frac{\partial E}{\partial n_j}\right)_{T,P,n} \mathrm{d}n_j \qquad (4.66)$$

$$\mathrm{d}S = \left(\frac{\partial S}{\partial n_j}\right)_{T,P,n} \mathrm{d}n_j \qquad (4.67)$$

and

$$\mathrm{d}V = \left(\frac{\partial V}{\partial n_j}\right)_{T,P,n} \mathrm{d}n_j \qquad (4.68)$$

Substitution of these three equations into Equation (4.22) and division by dn_j yields

$$\left(\frac{\partial E}{\partial n_j}\right)_{T,P,n} = T\left(\frac{\partial S}{\partial n_j}\right)_{T,P,n} - P\left(\frac{\partial V}{\partial n_j}\right)_{T,P,n} + \mu_j \qquad (4.69)$$

(d) So far we have not discussed second derivatives. Of course, the various relations discussed in (a), (b), and (c) above are applicable to the first derivatives, as these are also functions of the same sets of independent variables, as are the original functions. However, in addition, for $z = z(x, y)$ the equality of the two second-order derivatives $[\partial(\partial z/\partial x)_y/\partial y]_x$ and $[\partial(\partial z/\partial y)_x/\partial x]_y$ is of importance. As an example, consider the evaluation of $(\partial C_V/\partial V)_{T,n}$. Since $C_V = (\partial E/\partial T)_{V,n}$, we can write

$$\left(\frac{\partial C_V}{\partial V}\right)_{T,n} = \left[\frac{\partial}{\partial V}\left(\frac{\partial E}{\partial T}\right)_{V,n}\right]_{T,n} \qquad (4.70)$$

and then interchange the order of differentiation, so

$$\left(\frac{\partial C_V}{\partial V}\right)_{T,n} = \left[\frac{\partial}{\partial T}\left(\frac{\partial E}{\partial V}\right)_{T,n}\right]_{V,n} \tag{4.71}$$

We now carry out the indicated differentiation with the use of Equation (4.65). The result is

$$\left(\frac{\partial C_V}{\partial V}\right)_{T,n} = T\left(\frac{\partial^2 P}{\partial T^2}\right)_{V,n} \tag{4.72}$$

4.8 The energy as a function of the temperature and volume, and the enthalpy as a function of the temperature and pressure

Because of the experimental difficulties of using the entropy as an independent variable, it is convenient to use the energy as a function of the temperature, volume, and mole numbers. Thus, for $E = E(T, V, n)$,

$$\mathrm{d}E = \left(\frac{\partial E}{\partial T}\right)_{V,n} \mathrm{d}T + \left(\frac{\partial E}{\partial V}\right)_{T,n} \mathrm{d}V + \sum_i \left(\frac{\partial E}{\partial n_i}\right)_{T,V,n} \mathrm{d}n_i \tag{4.73}$$

With the use of Equations (2.29) and (4.65), we obtain

$$\mathrm{d}E = C_V + \left[T\left(\frac{\partial P}{\partial T}\right)_{V,n} - P\right] \mathrm{d}V + \sum_i \left(\frac{\partial E}{\partial n_i}\right)_{T,V,n} \mathrm{d}n_i \tag{4.74}$$

A similar expression can be derived for the enthalpy as a function of the temperature, pressure, and mole numbers. If we set $H = H(T, P, n)$,

$$\mathrm{d}H = \left(\frac{\partial H}{\partial T}\right)_{P,n} \mathrm{d}T + \left(\frac{\partial H}{\partial P}\right)_{T,n} \mathrm{d}P + \sum_i \left(\frac{\partial H}{\partial n_i}\right)_{T,P,n} \mathrm{d}n_i \tag{4.75}$$

With the use of Equations (4.19) and (4.48), we find that

$$\left(\frac{\partial H}{\partial P}\right)_{T,n} = V - T\left(\frac{\partial V}{\partial T}\right)_{P,n} \tag{4.76}$$

and consequently, with Equation (2.33),

$$\mathrm{d}H = C_P \,\mathrm{d}T + \left[V - T\left(\frac{\partial V}{\partial T}\right)_{P,n}\right] \mathrm{d}P + \sum_i \left(\frac{\partial H}{\partial n_i}\right)_{T,P,n} \mathrm{d}n_i \tag{4.77}$$

Although Equations (4.74) and (4.77) are not as simple as Equations (4.22) and (4.23), they are more important from an experimental viewpoint. This importance arises from the fact that the temperature is the independent variable in these equations, rather than the entropy.

4.9 The integration of Equations (4.74)

Equation (4.74) affords an excellent opportunity to discuss the integration of such equations. When applied to closed systems, it becomes

$$dE = C_V\,dT + \left[T\left(\frac{\partial P}{\partial T}\right)_{V,n} - P\right] dV \tag{4.78}$$

We know that the energy is a function of the state of the system and that dE is an exact differential. Consequently, the function exists and the integration can be performed. In general, C_V and $[T(\partial P/\partial T)_{V,n} - P]$ are functions of both the temperature and the volume. We could integrate C_V with respect to the temperature at constant volume and retain all terms. We would then integrate $[T(\partial P/\partial T)_{V,n} - P]$ with respect to the volume at constant temperature, but would retain only those terms that did not contain the temperature. The sum of the two integrals and the integration constant would then give the energy as a function of the temperature and volume. The integration constant cannot be evaluated, because we do not know the absolute value of the energy. If we wished to determine the difference of the energy of a system in two different states, then we could apply the limits of temperature and volume of the two states to the energy function and obtain the difference on subtraction.

Figure 4.1. Integration along two paths.

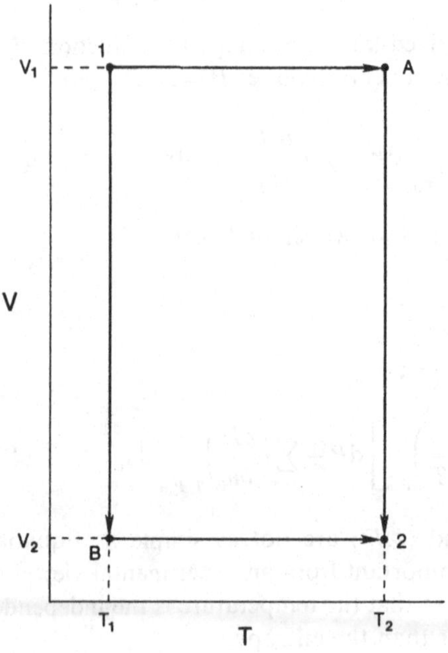

In such an integration $[T(\partial P/\partial T)_{V,n} - P]$ would need to be known as a function of the temperature and volume. Such information can be obtained from an equation of state giving the pressure as a function of the temperature and volume at constant mole numbers of the components. Also, C_V would need to be known as a function of the volume and temperature. With our present knowledge of thermodynamic data, this quantity might be known as a function of the temperature at some fixed volume. If this were the case, then we could still determine C_V as a function of the volume by the integration of Equation (4.72) with respect to the volume at constant temperature when an equation of state is known. The integration constant would be determined by our knowledge of C_V as a function of the temperature at the fixed volume. Again, for this integration an equation of state would be needed.

Even when a limited amount of data is available, we may still be able to perform the integration between two states, (V_1, T_1) and (V_2, T_2). Here we make use of the property of functions; that the change of the energy in going from some initial state to a final state is independent of the path. Consider Figure 4.1. We wish to determine the change in the energy of a closed system in going from state 1 to state 2. We may follow either path *1A2* or path *1B2*. If we choose the first path, we may integrate first along the path *1A*, thus changing the temperature from T_1 to T_2 at the constant volume V_1. For this change of state

$$\Delta E_1 = \int_{T_1}^{T_2} (C_V)_{V_1}\, \mathrm{d}T \tag{4.79}$$

in which C_V is evaluated at V_1. We may then integrate from *A* to *2* at the constant temperature T_2 between the volumes V_1 and V_2. Here the change of energy is

$$\Delta E_2 = \int_{V_1}^{V_2} \left[T\left(\frac{\partial P}{\partial T}\right)_{V,n} - P \right]_{T_2} \mathrm{d}T \tag{4.80}$$

in which $[T(\partial P/\partial T)_{V,n} - P]$ is evaluated at the temperature T_2. The total change of the energy is the sum of these two changes.

If we choose the path *1B2*, we first integrate along the path *1B*, using Equation (4.80), but in this case $[T(\partial P/\partial T)_{V,n} - P]$ is evaluated at T_1. Then we integrate from *B* to *2*, using Equation (4.79); here C_V is evaluated at V_2. The total energy change along the two paths, *1A2* and *1B2*, must of necessity be equal.

5

Conditions of equilibrium and stability: the phase rule

Reversible processes are those processes that take place under conditions of equilibrium; that is, the forces operating within the system are balanced. Therefore, the thermodynamics associated with reversible processes are closely related to equilibrium conditions. In this chapter we investigate those conditions that must be satisfied when a system is in equilibrium. In particular, we are interested in the relations that must exist between the various thermodynamic functions for both phase and chemical equilibrium. We are also interested in the conditions that must be satisfied when a system is stable.

After the conditions of equilibrium have been determined, we can derive the phase rule and determine the number and type of variables that are necessary to define completely the state of a system. The concepts developed in this chapter are illustrated by means of graphical representation of the thermodynamic functions.

5.1 Gibbs' statement concerning equilibrium

Gibbs based his entire treatment of heterogenous equilibrium on the following statement [3].

> The criterion of equilibrium of a material system which is isolated from all external influences may be expressed in either of the following entirely equivalent forms:
>
> 1. For equilibrium of any isolated system it is necessary and sufficient that in all possible variations of the state of the system which do not alter its energy, the variations of its entropy shall either vanish or be negative.
> 2. For equilibrium of any isolated system it is necessary and sufficient that in all possible variations in the state of the system which do not alter its entropy, the variation of its energy shall either vanish or be positive.

The first statement corresponds to the equation

$$(\delta S)_E \leqslant 0 \tag{5.1}$$

and the second statement corresponds to the equation

$$(\delta E)_S \geqslant 0 \tag{5.2}$$

We find in Section 5.3 that the equality sign is applicable when all variations of the system can be carried out in either direction. When any variation of the system can be carried out in only one direction, the inequality sign is applicable.

If the system is isolated and already at equilibrium, then any variation in the state of the system cannot increase the entropy if the energy is kept constant, and cannot decrease the energy if the entropy is kept constant. In other words, for an isolated system at equilibrium, the entropy must have the largest possible value consistent with the energy of the system, and the energy must have the smallest possible value consistent with the entropy of the system. There may be many states of the system that are equilibrium states, but these conditions must be applicable to each such equilibrium state.

The equivalence of the two statements is rather easily proved. Assume that, contrary to Equation (5.1), there is a variation that increases the entropy of the isolated system. We can then decrease the entropy of the varied system to the original value by removing heat from it reversibly. However, in so doing we also decrease the energy. Then, in this new state the energy has been decreased at constant entropy. This is contrary to Equation (5.2). Similarly, assume that, contrary to Equation (5.2), there is a variation that decreases the energy of the system while the entropy is kept constant. We may then increase the energy of the varied system to its original value by adding heat to the system reversibly. However, in so doing we increase its entropy. This is contrary to Equation (5.1). Thus, if the first statement is not satisfied for some possible variation of the system, then there is also a varied state for which the second statement is not satisfied, and if the second statement is not satisfied for some possible variation of the system, then there is a varied state for which the first statement is not satisfied. Therefore, the two statements are equivalent.

The meaning and also the limitation of the term *possible variations* must be considered. For the purposes of discussion, we center our attention on Equation (5.2) and consider a heterogenous, multicomponent system. The independent variables that are used to define the state are the entropy, volume, and mole numbers (i.e., amount of substance or number of moles) of the components. The statements of the condition of equilibrium require these to be constant because of the isolation of the system. Possible variations are then the change of the entropy of two or more of the phases subject to the condition that the entropy of the whole system remains constant, the change of the volume of two or more phases subject to the condition that the volume of the whole system remains constant, or the transfer of matter from one phase to another subject to the condition that the mass of the whole system remains constant. Such variations are virtual or hypothetical, and are not necessarily real variations. If we choose to make the varied system real, then it would be necessary to insert adiabatic walls, rigid walls, or semipermeable walls into the system. There are other variations that are excluded. We do not consider any variation in which the system as a whole is moved through space or in which it acquires kinetic energy. All parts of the system between which matter is transferred must be in contact, and we do not consider the transfer of matter between parts of the system that are separated. We do not consider any change of restrictions placed on the system. As an example, if one part of the system is separated from other parts by means of an adiabatic membrane, then the removal of this membrane is not considered as a possible variable.

Finally, we must consider what Gibbs called *passive resistance*. From experience we recognize many systems that over short time intervals appear to be in equilibrium. However, over long time intervals it is obvious that they are not in equilibrium. As a mechanical example, consider a glass rod supported at its ends with a weight suspended from its center. Over short time intervals such a system appears to be in equilibrium. However, when such a system is observed over a long period, it is apparent that the weight gradually drops in the gravitational field and that the rod becomes curved. For chemical systems we might consider potassium chlorate, ammonium nitrate, or a mixture of hydrogen and oxygen at ordinary temperatures. In each of these cases the system is unstable toward certain products. Potassium chlorate is unstable with respect to potassium chloride and oxygen; ammonium nitrate is unstable with respect to its decomposition products; and the mixture of hydrogen and oxygen is unstable with respect to water. Yet each of these systems is capable of existing under suitable conditions apparently at equilibrium for long periods without any reaction taking place. Operating in each of these systems is some 'resistance' to change, which may be called a *passive resistance*. In our consideration of equilibrium, we do not consider any variation of the system that alters these passive resistances as a possible variation. Thus, if no change takes place in a system in some arbitrary time interval, then we consider the system to be in equilibrium, even though it may be obvious that the system is not in equilibrium when it is studied over long time intervals.

We have pointed out that the possible variations are *virtual*, and not necessarily real. The symbol δ is used in the mathematical relations to emphasize this character of the variations. Moreover, in Equations (5.1) and (5.2) only first-order terms are considered. When second- and higher-order terms are considered, the symbol Δ is used to indicate the increment of the value of the function rather than the differential quantity.[1] Under these

[1] The same symbol is used to indicate a finite, rather than an infinitesimal, change of the value of a function for some change of state. The difference in the meaning of the symbol will be clear from the context.

conditions Equations (5.1) and (5.2) become

$$(\Delta S)_E < 0 \tag{5.3}$$

and

$$(\Delta E)_S > 0 \tag{5.4}$$

respectively; that is, the variation of the value of the entropy function can only be negative for all possible variations of the system at constant energy, and the variation of the value of the energy function can only be positive for all possible variations of the system at constant entropy. The conditions of stability, discussed in Section 5.14, are based on these two equations.

5.2 Proof of Gibbs' criterion for equilibrium [4]

It is a general observation that any system that is not in equilibrium will approach equilibrium when left to itself. Such changes of state that take place in an isolated system do so by irreversible processes. However, when a change of state occurs in an isolated system by an irreversible process, the entropy change is always positive (i.e., the entropy increases). Consequently, as the system approaches equilibrium, the entropy increases and will continue to do so until it obtains the largest value consistent with the energy of the system. Thus, if the system is already at equilibrium, the entropy of the system can only decrease or remain unchanged for any possible variation as discussed in Section 5.1.

We may obtain further insight into the conditions by considering the equation

$$\mathrm{d}E = T\,\mathrm{d}S - P\,\mathrm{d}V + \sum_i \mu\,\mathrm{d}n_i + \mathrm{d}W'_{\mathrm{r}}$$

At constant entropy, volume, and mass,

$$\mathrm{d}E = +\mathrm{d}W'_{\mathrm{r}}$$

If the system is at equilibrium, it is incapable of doing work. If there is a possible variation of the system that would decrease the value of the energy at constant entropy, then the system would be capable of doing work. (The numerical work of $\mathrm{d}W'_{\mathrm{r}}$ would be negative.) However, this is not possible and therefore the energy can only increase or remain constant.

5.3 Conditions of equilibrium for heterogenous systems

Equations (5.1) and (5.2) give the basic criteria of equilibrium, but it is advantageous to obtain more-useful conditions from them. We first consider an isolated heterogenous system composed of any number of phases and containing any number of components at equilibrium. These phases are in thermal contact with each other, and there are no walls separating them.

Each phase may be considered as an open, nonisolated system. The variation of the energy of each phase is given by

$$\delta E' = T'\,\delta S' - P'\,\delta V' + \sum_i \mu_i'\,\delta n_i'$$

$$\delta E'' = T''\,\delta S'' - P''\,\delta V'' + \sum_i \mu_i''\,\delta n_i'' \tag{5.5}$$

$$\delta E''' = T'''\,\delta S''' - P'''\,\delta V''' + \sum_i \mu_i'''\,\delta n_i'''$$

etc. In these equations the primes refer to the phases. For the condition of equilibrium of the entire system we use Equation (5.2), so

$$(\delta E)_S \geqslant 0 \tag{5.6}$$

under the conditions that

$$\delta S = 0 \tag{5.7}$$

$$\delta V = 0 \tag{5.8}$$

$$\delta n_1 = 0 \tag{5.9}$$

$$\delta n_2 = 0 \tag{5.10}$$

$$\delta n_3 = 0 \tag{5.11}$$

$$\vdots$$

$$\delta n_i = 0 \tag{5.12}$$

etc., for all components. However,

$$\delta E = \delta E' + \delta E'' + \delta E''' + \cdots \tag{5.13}$$

$$\delta S = \delta S' + \delta S'' + \delta S''' + \cdots \tag{5.14}$$

$$\delta V = \delta V' + \delta V'' + \delta V''' + \cdots \tag{5.15}$$

$$\delta n_1 = \delta n_1' + \delta n_1'' + \delta n_1''' + \cdots \tag{5.16}$$

$$\delta n_2 = \delta n_2' + \delta n_2'' + \delta n_2''' + \cdots \tag{5.17}$$

etc., for every component. On substituting Equations (5.5) and (5.13) into Equation (5.6),

$$\begin{aligned} T'\,\delta S' + T''\,\delta S'' + T'''\,\delta S''' + \cdots - P'\,\delta V' - P''\,\delta V'' \\ - P'''\,\partial V''' - \cdots + \mu_1'\,\delta n_1' + \mu_1''\,\delta n_1'' + \mu_1'''\,\delta n_1''' + \cdots \\ + \mu_2'\,\delta n_2' + \mu_2''\,\delta n_2'' + \mu_2'''\,\delta n_2''' + \cdots \geqslant 0 \end{aligned} \tag{5.18}$$

The overall variations of the entropy, volume, and number of moles of each component are independent, and we may separate Equation (5.18) into its several parts. Thus,

$$T'\,\delta S' + T''\,\delta S'' + T'''\,\delta S''' + \cdots \geqslant 0 \tag{5.19}$$

$$-P'\,\delta V' - P''\,\delta V'' - P'''\,\delta V''' - \cdots \geqslant 0 \tag{5.20}$$

$$\mu_1'\,\delta n_1' + \mu_1''\,\delta n_1' + \mu_1'''\,\delta n_1''' + \cdots \geqslant 0 \tag{5.21}$$

$$\mu_2'\,\delta n_2' + \mu_2''\,\delta n_2'' + \mu_2'''\,\delta n_2''' + \cdots \geqslant 0 \tag{5.22}$$

etc.

At first consider only Equation (5.19). Of all the variations, $\delta S'$, $\delta S''$, $\delta S'''$, etc., one is dependent and the others are independent because of the condition expressed in Equation (5.7) and the relation given in Equation (5.14).

If we choose to eliminate $\delta S'$, we have

$$\delta S' = -\delta S'' - \cdots \tag{5.23}$$

On substitution of this into Equation (5.19),

$$(T'' - T')\,\delta S'' + (T''' - T')\,\delta S''' + \cdots \geqslant 0 \tag{5.24}$$

in which each variation of entropy is independent. Consequently, we can consider each term separately; so, if we take the first term in Equation (5.24), we require that

$$(T'' - T')\,\delta S'' \geqslant 0$$

The entropy of the double-primed phase may be either increased or decreased, so $\delta S''$ may be either positive or negative. When we consider the equality sign only, we see that $T'' - T'$ must be zero, and consequently T'' must equal T'. When we consider the inequality sign, $(T'' - T')$ must be negative if $\delta S''$ is negative, and consequently T'' would have to be less than T'. However, if $\delta S''$ is positive, $(T'' - T')$ must also be positive and T'' would have to be greater than T'. Also, each phase in the system at equilibrium must have a fixed temperature, and the temperature cannot depend upon the possible variation that we choose. Consequently, a contradiction arises and the inequality sign can have no meaning. The other terms of Equation (5.24) can be studied in exactly the same way. We then conclude that when a system is in equilibrium, the temperatures of all of the phases must be identical. This condition is equivalent to the zeroth law, as indicated in Section 2.1.

Equation (5.20) may be treated in the same way, with the use of Equations (5.8) and (5.15). Thus, we find that the pressure of each phase must be identical at equilibrium.

The argument becomes slightly different when we consider variations in the mole numbers. In some systems not every component is present in every phase. Under such circumstances the variation of the mole number of a component in a phase in which the component does not exist can only be positive. When the variation of the mole number of a component can be both positive and negative, the argument is identical to the previous arguments. Thus, we find that the chemical potential of a component must be the same in each phase of the system in which it exists at equilibrium, by the use of Equations (5.21), (5.22), etc., combined with Equations (5.9), (5.10), etc., and Equations (5.16), (5.17), etc. For the second case we select the term

$$(\mu_1'' - \mu_1')\,\delta n_1'' \geqslant 0$$

with the condition that the first component does not exist in the double-primed phase. If the equality conditions are valid, then the two chemical potentials must have the same value. However, there is no contradiction when we consider the inequality sign. The variation of n_1'' can only be positive, and therefore the value of μ_1'' can be greater than μ_1' under such circumstances.

Gibbs expresses these conclusions as follows [5].

> The potential for each of the component substances must have a constant value in all parts of the given mass of which that substance is an actual component, and have a value not less than this in all parts of which it is a possible component.

As we see below, this inequality does not lead to any difficulty.

In review, the conditions of equilibrium for the system described at the beginning of this section, omitting the inequality sign, are

$$T' = T'' = T''' = T'''' = \cdots \tag{5.25}$$

$$P' = P'' = P''' = P'''' = \cdots \tag{5.26}$$

$$\mu_1' = \mu_1'' = \mu_1''' = \mu_1'''' = \cdots \tag{5.27}$$

$$\mu_2' = \mu_2'' = \mu_2''' = \mu_2'''' = \cdots \tag{5.28}$$

$$\mu_3' = \mu_3'' = \cdots \tag{5.29}$$

$$\mu_4' = \cdots \tag{5.30}$$

The conditions concerning the temperature and pressure are rather obvious. Those concerning the chemical potential are not so obvious, but they are extremely important. It is these conditions that lead to all the thermodynamic relations between different phases at equilibrium.

5.4 Conditions of equilibrium for chemical reactions

When a chemical reaction takes place in a system, new molecular entities, other than the original components, are formed. The components are the substances, a minimum number, that are used to form the system. The term 'species' is used to refer to all of the molecular entities present in the system, including the original components. In all of the equations that have been developed for open systems, we have used the *n*s to indicate the mole numbers (amount of substance) of the components. They can equally be used to represent the mole numbers of species with the condition that not all of the *n*s are independent, because of the conditions of mass balance. The condition of equilibrium for a chemical reaction is obtained when we use the *n*s to represent species.

The argument is again based on the concept that the system is isolated and, consequently, the mass of the system is constant. Rather than considering the general case, we choose a two-phase system containing the components, B_1, B_2, B_4, and B_5. Here each B represents the molecular formula for a substance. We further assume that the reaction

$$\nu_1 B_1 + \nu_2 B_2 = \nu_3 B_3$$

takes place in the system. In this equation ν_1, ν_2, and ν_3 represent the stoichiometric coefficients of the substances that enter into the reaction according to the usual balanced chemical equation. Finally, we assume that all of the species are present in each phase. Then, for equilibrium,

$$\begin{aligned} &T'\,\delta S' + T''\,\delta S'' - P'\,\delta V' - P''\,\delta V'' + \mu_1'\,\delta n_1' + \mu_1''\,\delta n_1'' \\ &\quad + \mu_2'\,\delta n_2' + \mu_2''\,\delta n_2'' + \mu_3'\,\delta n_3' + \mu_3''\,\delta n_3'' + \mu_4'\,\delta n_4' \\ &\quad + \mu_4''\,\delta n_4'' + \mu_5'\,\delta n_5' + \mu_5''\,\delta n_5'' = 0 \end{aligned} \tag{5.31}$$

must be satisfied. This equation is obtained by the same methods that were used to obtain Equation (5.18). The equality sign is used in this equation since all of the variations can take place in either direction. The conditions that must be satisfied are

$$\delta S = 0 \tag{5.32}$$

$$\delta V = 0 \tag{5.33}$$

$$\delta n_1^0 = 0 \tag{5.34}$$

$$\delta n_2^0 = 0 \tag{5.35}$$

$$\delta n_4^0 = 0 \tag{5.36}$$

and

$$\delta n_5^0 = 0 \tag{5.37}$$

Each n^0 represents the original number of moles of each component used to form the system. By use of the condition equation (Eq. (5.32)) we find that the temperature of the two phases must be the same and, according to Equation (5.33), we find that the pressure of each phase must also be the same. According to Equations (5.36) and (5.37), we find that the chemical potential of B_4 must be the same in both phases, and that the chemical potential of B_5 must be the same in both phases. The remaining relation that must be satisfied is

$$\mu_1' \,\delta n_1' + \mu_1'' \,\delta n_1'' + \mu_2' \,\delta n_2' + \mu_2'' \,\delta n_2'' + \mu_3' \,\delta n_3' + \mu_3'' \,\delta n_3'' = 0 \tag{5.38}$$

subject to the conditions of Equations (5.34) and (5.35). According to the chemical equation and an atomic balance,

$$n_1^0 = n_1' + n_1'' + \frac{\nu_1}{\nu_3} n_3' + \frac{\nu_1}{\nu_3} n_3''$$

and

$$n_2^0 = n_2' + n_2'' + \frac{\nu_2}{\nu_3} n_3' + \frac{\nu_2}{\nu_3} n_3''$$

and therefore

$$\delta n_1' + \delta n_1'' + \frac{\nu_1}{\nu_3} \delta n_3' + \frac{\nu_1}{\nu_3} \delta n_3'' = 0 \tag{5.39}$$

and

$$\delta n_2' + \delta n_2'' + \frac{\nu_2}{\nu_3} \delta n_3' + \frac{\nu_2}{\nu_3} \delta n_3'' = 0 \tag{5.40}$$

with the use of Equations (5.34) and (5.35). We may now solve for $\delta n_1'$ and $\delta n_2'$ and substitute into Equation (5.38) to obtain

$$(\mu_1'' - \mu_1') \,\delta n_1'' + (\mu_2'' - \mu_2') \,\delta n_2'' + \left(\mu_3' - \frac{\nu_1}{\nu_3} \mu_1' - \frac{\nu_2}{\nu_3} \mu_1' \right) \delta n_3'$$
$$+ \left(\mu_3'' - \frac{\nu_1}{\nu_3} \mu_1' - \frac{\nu_2}{\nu_3} \mu_2' \right) \delta n_3'' = 0 \tag{5.41}$$

where each δn is now independent. Therefore, the conditions that must be satisfied are

$$\mu_1'' = \mu_1' \tag{5.42}$$

$$\mu_2'' = \mu_2' \tag{5.43}$$

$$\mu_3'' = \mu_3' \tag{5.44}$$

and

$$\nu_3\mu_3 - \nu_1\mu_1 - \nu_2\mu_2 = 0 \tag{5.45}$$

We see that the chemical potential of a substance must be the same in all phases in which it exists, whether this substance is considered as a component or a species. Furthermore, the new condition that must be satisfied in each phase for chemical equilibrium is

$$\nu_1\mu_1 + \nu_2\mu_2 = \nu_3\mu_3$$

for the chemical equation

$$\nu_1 B_1 + \nu_2 B_2 = \nu_3 B_3$$

This relation may be generalized, so

$$\sum_i \nu_i\mu_i = 0 \tag{5.46}$$

for the chemical equation

$$\sum_i \nu_i B_i = 0 \tag{5.47}$$

where the ν_is are taken as positive for products and negative for reagents. We have used a homogenous reaction in this proof, but the same relation must also hold for a heterogenous reaction.

5.5 Conditions of equilibrium for heterogenous systems with various restrictions

The conditions of equilibrium for heterogenous systems in which there were no restrictions involving the transfer of heat between the phases, the changes of volume of the phases, or the transfer of mass between the phases are developed in Section 5.3. Here we consider the conditions of equilibrium in a heterogenous system in which restrictions are introduced. We consider the various phases in the heterogenous system to be separated into a number of groups. Each phase within a group is in contact with the other phases in the group without restrictions. However, each group is separated from the other groups by an envelope or a wall that has certain properties. The envelopes may be adiabatic, so that no heat can transfer from one group to another group; rigid, so that the volumes of the groups are constant; or semipermeable, so that at least one of the components cannot transfer from one group to another.

Consider a heterogenous system composed on P phases. The phases are designated by the number of primes following a symbol, but we indicate this number by the superscript P for simplicity. Thus, P may have any value from 1 to P. We separate the phases into two groups, one group being designated by the superscripts $1, \ldots, k$ and the other by the superscripts $k+1, \ldots, P$. The two groups are separated by some envelope. The condition of equilibrium expressed by Equation (5.18) must still be satisfied subject to the conditions given by Equations (5.7)–(5.12).

If the envelope is adiabatic, then Equation (5.7) must be written as

$$\delta S = \sum_{P=1}^{k} \delta S^P + \sum_{P=k+1}^{P} \delta S^P = 0 \tag{5.48}$$

where the one sum is independent of the other. Consequently, each sum must be zero and the two condition equations

$$\sum_{P=1}^{k} \delta S^P = 0 \tag{5.49}$$

and

$$\sum_{P=k+1}^{P} \delta S^P = 0 \tag{5.50}$$

must be satisfied. By using the same methods used in Section 5.3, we find that the temperature of every phase within a group must be the same, but the temperatures of the two groups of phases are independent of each other and need not be equal.

If the envelope is rigid, then we find the pressure of each phase within a group must be the same, but the pressures of the two groups are independent and need not be equal.

When the envelope is permeable to a component, then the chemical potential of the component must be the same in each phase in which it exists. If the envelope is not permeable to a component, then for each group, the chemical potential of that component must be the same in each phase in which it exists. However, the chemical potential of that component need not be the same in the groups. These conditions apply to species as well as to components.

5.6 Other conditions of equilibrium

The most important conditions of equilibrium are those concerning the temperature, pressure, and chemical potential given in the preceding sections. However, other equivalent conditions are also useful, and it is well to develop them here.

We may be interested in the sign of the change of enthalpy for variations of a system that is already at equilibrium. We do have the knowledge that the pressure is uniform throughout the system. We may use Equation (4.23), from which

$$\delta H = T\,\delta S + V\,\delta P + \sum_i \mu_i\,\delta n_i + \delta W'_r$$

Using the same arguments as in Section 5.3, we find that

$$(\delta H)_{S,P,n} \geqslant 0 \tag{5.51}$$

Similarly, by the use of Equation (4.24) and the knowledge that the temperature of the system is uniform,

$$(\delta A)_{T,V,n} \geqslant 0 \tag{5.52}$$

and finally, by Equation (4.25),

$$(\delta G)_{T,P,n} \geqslant 0 \tag{5.53}$$

Equation (5.53) is of more importance for practical applications than Equations (5.51) and (5.52). One interpretation of this equation is that the Gibbs energy of a system in equilibrium has the smallest possible value consistent with the given temperature and pressure of the system and with the number of moles of each component contained in the system. Moreover, if a change of state can take place within a system under equilibrium or reversible conditions at constant temperature, pressure, and mole numbers of the components, then ΔG must be zero for the change of state. If ΔG is negative for some possible change of state within a system at constant temperature, pressure, and mole numbers, then the change of state will take place unless prevented by passive resistances, because the value of the Gibbs energy of the system in the given state is not the smallest possible value for the given temperature, pressure, and mole numbers. Finally, by the same argument, if ΔG is positive for a possible change of state within a system at constant temperature, pressure, and mole numbers, the change of state will not take place. The inequality sign in these equations has the same limited use as that observed in Section 5.3. In general, the inequality sign will be omitted.

5.7 The chemical potential

We have seen in the preceding sections that the chemical potentials are extremely important functions for the determination of equilibrium relations. Indeed, all of the relations pertaining to the colligative properties of solutions are readily obtained from the conditions of equilibrium involving the chemical potentials. In many of the relations developed in the remainder of this chapter the chemical potentials appear as independent variables. It would therefore be extremely convenient if their values could be determined by direct experimental means. Unfortunately, this is not the case and we must consider them as functions of other variables.

From their definitions (Eq. (4.39)) based on the energy, the enthalpy, and the Gibbs and Helmholtz energies, we may set the chemical potentials to be functions of other variables, as follows:

$$\begin{aligned}\mu_i &= \mu_i(S, V, n_1, n_2, \ldots) = \mu_i(S, P, n_1, n_2, \ldots)\\ &= \mu_i(T, V, n_1, n_2, \ldots) = \mu_i(T, P, n_1, n_2, \ldots)\end{aligned} \tag{5.54}$$

In the majority of applications we use the last relation, because the temperature and pressure are usually the most convenient experimental variables. Therefore, we write

$$d\mu_k = \left(\frac{\partial \mu_k}{\partial T}\right)_{P,n} dT + \left(\frac{\partial \mu_k}{\partial P}\right)_{T,n} dP + \sum_i \left(\frac{\partial \mu_k}{\partial n_i}\right)_{T,P,n} dn_i \tag{5.55}$$

for the differential of the chemical potential of the kth component. Here the sum is taken over all of the components or species, as the case may be. Moreover, we know that the energy, enthalpy, and the Gibbs and Helmholtz energies are all homogenous functions of first degree in the mole numbers. Therefore, the chemical potentials are homogenous functions of zeroth degree in the mole numbers; that is, they are functions of the composition and are intensive variables. We may use any composition variable, but for the present we primarily use the mole fraction. We can then express the differential of the chemical potential as

$$d\mu_k = \left(\frac{\partial \mu_k}{\partial T}\right)_{P,x} dT + \left(\frac{\partial \mu_k}{\partial P}\right)_{T,x} dP + \sum_{i=1}^{C-1} \left(\frac{\partial \mu_k}{\partial x_i}\right)_{T,P,x} dx_i \tag{5.56}$$

as an alternative to Equation (5.55). In Equation (5.56) the sum is taken over all but one of the components or species, because the sum of the mole fractions must be unity. The use of the chemical potentials as independent variables thus introduce the composition variables as independent variables.

5.8 The Gibbs–Duhem equation

The energy of a single-phase system is a homogenous function of the first degree in the entropy, volume, and the number of moles of each component. Thus, by Euler's theorem[2]

$$E = \left(\frac{\partial E}{\partial S}\right)_{V,n} S + \left(\frac{\partial E}{\partial V}\right)_{S,n} V + \sum_i \left(\frac{\partial E}{\partial n_i}\right)_{S,V,n} n_i \tag{5.57}$$

On evaluating the partial differentials, we obtain

$$E = TS - PV + \sum_i \mu_i n_i \tag{5.58}$$

If we differentiate this equation,

$$dE = T\,dS + S\,dT - P\,dV - V\,dP + \sum_i \mu_i\,dn_i + \sum_i n_i\,d\mu_i$$

[2] Euler's theorem is stated here without proof. The statement is that if w is a homogenous function of degree n in the variables x, y, and z, so that $w = w(x, y, z)$, then $nw = x(\partial w/\partial x)_{y,z} + y(\partial w/\partial y)_{x,z} + z(\partial w/\partial z)_{x,y}$.

and by comparison with Equation (4.22), we find that

$$S\,dT - V\,dP + \sum_i n_i\,d\mu_i = 0 \tag{5.59}$$

This equation is extremely important (see Section 5.12 for some applications). It is known as the *Gibbs–Duhem equation*, and such equations as the Duhem–Margules equation may be derived from it. Since no limitation has been put on the type of system considered in the derivation, this equation must be satisfied for every phase in a heterogenous system. We recognize that the convenient independent variables for this equation are the intensive variables: the temperature, the pressure, and the chemical potentials.

5.9 Application of Euler's theorem to other functions

We have seen that for open systems

$$E = E(S, V, n_1, n_2, \ldots)$$

$$H = H(S, P, n_1, n_2, \ldots)$$

$$A = A(T, V, n_1, n_2, \ldots)$$

and

$$G = G(T, P, n_1, n_2, \ldots)$$

The entropy and volume are extensive properties, as are the number of moles of each component, but the temperature and pressure are not. Consequently, we may set H as a homogenous function of the first degree in the entropy and mole numbers if the pressure is kept constant. Then, by Euler's theorem,

$$H = TS + \sum_i n_i\mu_i \tag{5.60}$$

Similarly, we find at constant temperature that

$$A = -PV + \sum_i n_i\mu_i \tag{5.61}$$

and at constant temperature and pressure that

$$G = \sum_i n_i\mu_i \tag{5.62}$$

From mathematics we know that if w is a homogenous function of the nth degree in x, y, and z, then $(\partial w/\partial x)_{y,z}$, $(\partial w/\partial y)_{x,z}$, and $(\partial w/\partial z)_{x,y}$ are homogenous functions of degree $(n-1)$ in x, y, and z. Therefore, many other relations can be obtained, some of which are useful, by the application of Euler's theorem to the various partial derivatives of E, H, A, and G.

5.10 The Gibbs phase rule

The derivation of the phase rule is based upon an elementary theorem of algebra. This theorem states that the number of variables to which arbitrary values can be assigned for any set of variables related by a set of simultaneous, independent equations is equal to the difference between the number of variables and the number of equations. Consider a heterogenous system having P phases and composed of C components. We have one Gibbs–Duhem equation of each phase, so we have the set of equations

$$S'\,\mathrm{d}T' - V'\,\mathrm{d}P' + \sum_i n_i'\,\mathrm{d}\mu_i' = 0$$

$$S''\,\mathrm{d}T'' - V''\,\mathrm{d}P'' + \sum_i n_i''\,\mathrm{d}\mu_i'' = 0 \tag{5.63}$$

$$S'''\,\mathrm{d}T''' - V'''\,\mathrm{d}P''' + \sum_i n_i'''\,\mathrm{d}\mu_i''' = 0$$

etc., for P phases. For the present we let the mole numbers refer to the components. In each equation there are $(C+2)$ intensive variables, and therefore there are a total of $(C+2)P$ intensive variables. However, there are $(C+2)(P-1)$ relations between these variables according to Equations (5.25)–(5.29). In addition, there are the P equations given by Equation (5.63). The difference, V, between the number of variables and the number of condition equations relating them is

$$\begin{aligned} V &= (C+2)P - (C+2)(P-1) - P \\ &= C + 2 - P \end{aligned} \tag{5.64}$$

This is the usual relation given for the phase rule. The difference V is called the number of variances or the number of degrees of freedom. (Note that some texts use $F = C - P + 2$, where F is the same as V.) These are the number of intensive variables to which values may be assigned arbitrarily but within the limits of the condition that the original number of P phases exist. It is evident from Equations (5.63) and (5.56) that the intensive variables are the temperature, pressure, and mole fractions or other composition variables in one or more phases. Note that the Gibbs phase rule applies to *intensive* variables and is not concerned with the amount of each component in each phase.

In this development we have assumed that all of the components are present in each phase. When this is not so, the number of variables present in Equation (5.63) are reduced by one for each instance in which a component is not present in a phase. However, the number of condition equations obtained from Equations (5.25)–(5.29) is also reduced by one, and consequently the phase rule remains the same.

The mole numbers in Equation (5.63) could represent the species present in the system, rather than the components. However, in such a case they are not all independent. For each independent chemical reaction taking place within the system, a relation given by Equation (5.46) must be satisfied. There would thus be $(S - R)$ independent mole numbers for the system if S is the number of species and R is the number of independent chemical reactions. In fact, this relation may be used to define the number of components in the system [6]. If the species are ions, an equation expressing the electroneutrality of the system is another condition equation relating the mole numbers of the species. The total number of components in this case is $C = S - R - 1$.

Modifications of the phase rule are required when restrictions are placed on the system. The changes are obtained from the conditions of equilibrium expressed in Equations (5.25)–(5.29). When there are several groups of phases separated by adiabatic membranes, Equation (5.25) is valid for the phases in each group; however, an additional set of equations must be written for each group. We then find that the number of condition equations is decreased by one and the number of degrees of freedom is increased by one for each adiabatic membrane introduced into the system. The same arguments are applicable to Equations (5.26)–(5.29). We thus observe that the number of degrees of freedom is increased by one for each restriction that is introduced into the system.

The phase rule, as expressed by Equation (5.64) or modifications of it, has proved to be of great practical usefulness. However, more information can often be obtained by the use of the set of equations given in Equation (5.63) together with the condition equations as given in Equations (5.25)–(5.29) or appropriate modifications of them. The method is the solution of the set of simultaneous, independent equations by elimination of as many differential quantities as possible to obtain one equation. The final equation gives one relation between a number of differential quantities; one differential quantity is then dependent and all of the others are independent. The number of independent variables equals the number of degrees of freedom. If the system is univariant, then the final equation yields a derivative of one dependent variable with respect to the one independent variable. Simple examples of this calculation are given in Section 5.12, and a more complete discussion is given in Chapters 10 and 11.

5.11 The variables required to define the state of an isolated system

We are now in a position to determine the type and number of variables that are required to define the state of the system. The definition of a state of a system demands that values be assigned to a sufficient number of independent variables so that the values of all of the other variables may be determined. The energy of each phase is given in differential form by

$$dE' = T'\,dS' - P'\,dV' + \sum_i \mu_i'\,dn_i' \tag{5.65}$$

where the primes are used to emphasize the application to a phase. The differential of the energy of the complete heterogenous system is then given by the sum of such equations. When no restrictions are placed on the system and the conditions of equilibrium are applied, the differential of the energy of the entire system is

$$dE = T\,dS - P\,dV + \sum \mu_i\,dn_i \tag{5.66}$$

because the differential quantities of the energy, entropy, volume, and mole numbers are all additive. The mole numbers, as used here, represent the number of moles of components, because we are interested in the number of independent variables, and the mole numbers in terms of species are not all independent. It is also well to recall that the temperature, pressure, and chemical potentials are derived functions of the energy, and are functions of the entropy, volume, and mole numbers of the components. Therefore, these variables are not independent. We thus find that, according to Equation (5.66), the entropy, volume, and the mole numbers of the components are independent. These quantities are all extensive variables. If there are C components in the system, then the number of independent variables is $(C + 2)$. Thus, in order to define the state of the system we must assign values to $(C + 2)$ extensive variables; C of these variables are the number of moles of each component taken to form the system, and the other two are any two of the three variables energy, entropy, and volume.[3]

When restrictions are placed on a system, values must be assigned to an additional number of extensive variables in order to define the state of a system. If an isolated system is divided into two parts by an adiabatic wall, then the values of the entropy of the two parts are independent of each other. The term $T\,dS$ in Equation (5.66) would have to be replaced by two terms, $T'\,dS'$ and $T''\,dS''$, where the primes now refer to the separate parts. We see that values must be assigned to the entropy of the two parts or to the entropy of the whole system and one of the parts. Similar arguments pertain to rigid walls and semipermeable walls. The value of one additional extensive variable must be assigned for each restriction that is placed on the system.

We must emphasize that this derivation requires that the variables be extensive. We illustrate the necessity of this by considering a one-component system at its triple point. In this case there are no degrees of freedom, and

[3] This statement is similar to Duhem's theorem, which states that values must be assigned to only two independent variables in order to define the state of a closed system for which the original number of moles of each component is known.

the temperature and pressure of the system are fixed. However, the state of the system is not defined even when we know the mass of the system, because we have no information concerning its volume, its entropy, or its energy relative to some fixed state. Only when we assign values to two of the variables—volume, energy, or entropy—do we define the state of the system. However, in some cases we know that intensive variables may be used to define the state of a system. As an example, if we have a one-phase system and know the number of moles of each component in the system, then we may define its state by assigning values to the temperature and pressure.

The Gibbs phase rule provides the necessary information to determine when intensive variables may be used in place of extensive variables. We consider the extensive variables to be the entropy, the volume, and the mole numbers, and the intensive variables to be the temperature, the pressure, and the chemical potentials. Each of the intensive variables is a function of the extensive variables based on Equation (5.66). We may then write (on these equations and all similar ones we use n to denote all of the mole numbers)

$$T = T(S, V, n_1, n_2, \ldots) = T(S, V, n) \tag{5.67}$$

$$P = P(S, V, n) \tag{5.68}$$

and

$$\mu_i = \mu_i(S, V, n) \tag{5.69}$$

If we have one degree of freedom, then we may assign values to one of the intensive variables within limits. When we use the temperature as an example, we can use Equation (5.67) to solve for one of the extensive variables in terms of the temperature and the other extensive variables. Thus, the value of the entropy is determined in principle from the known values of the temperature, the volume, and the mole numbers. If the system has two degrees of freedom, then two of Equations (5.67)–(5.69) may be used to solve for two of the extensive variables in terms of two intensive variables and the remaining extensive variables. We thus find that *for each degree of freedom that the system has, we may substitute one intensive variable for one extensive variable.* This means that, in addition to substituting the temperature and pressure for the entropy and volume, we may substitute the chemical potential of a component for its mole number. This is seldom necessary experimentally, because the determination of the number of moles of each component usually presents no problem. The arguments and results are identical for systems on which restrictions have been placed.

States of closed systems for which it is necessary to assign values to at least one extensive variable in addition to the mole numbers are called *indifferent states*. Thus, all nonvariant and univariant states are indifferent states. These states are discussed in greater detail in Section 5.13.

5.12 Applications of the Gibbs–Duhem equation and the Gibbs phase rule

The Gibbs–Duhem equation is applicable to each phase in any heterogenous system. Thus, if the system has P phases, the P equations of Gibbs–Duhem form a set of simultaneous, independent equations in terms of the temperature, the pressure, and the chemical potentials. The number of degrees of freedom available for the particular systems, no matter how complicated, can be determined by the same methods used to derive the phase rule. However, in addition, a large amount of information can be obtained by the solution of the set of simultaneous equations.

A one-component system may have a maximum of three phases in equilibrium. Therefore, the three possible Gibbs–Duhem equations are

$$S'\,dT - V'\,dP + n'\,d\mu = 0 \tag{5.70}$$

$$S''\,dT - V''\,dP + n''\,d\mu = 0 \tag{5.71}$$

and

$$S'''\,dT - V'''\,dP + n'''\,d\mu = 0 \tag{5.72}$$

when the conditions of equilibrium have been applied. When only one phase is present, the one equation gives the chemical potential as a function of the temperature and pressure, the temperature as a function of the pressure and chemical potential, or the pressure as a function of the temperature and chemical potential. When two phases are present, two of the three equations are applicable. We can then eliminate the chemical potential from Equations (5.70) and (5.71), and obtain

$$\frac{dP}{dT} = \frac{\tilde{S}'' - \tilde{S}'}{\tilde{V}'' - \tilde{V}'} \tag{5.73}$$

This equation is the well-known *Clapeyron equation*, and expresses the pressure of the two-phase equilibrium system as a function of the temperature. Alternatively, we could obtain $d\mu/dT$ or $d\mu/dP$. When three phases are present, solution of the three equations gives the result that dT, dP, and $d\mu$ are all zero. Thus, we find that the temperature, pressure, and chemical potential are all fixed at a triple point of a one-component system.

Similarly, a two-component system may have a maximum of four phases in equilibrium. The four Gibbs–Duhem equations are

$$S'\,dT - V'\,dP + n_1'\,d\mu_1 + n_2'\,d\mu_2 = 0 \tag{5.74}$$

$$S''\,dT - V''\,dP + n_1''\,d\mu_1 + n_2''\,d\mu_2 = 0 \tag{5.75}$$

$$S'''\,dT - V'''\,dP + n_1'''\,d\mu_1 + n_2'''\,d\mu_2 = 0 \tag{5.76}$$

and

$$S'''' \, dT - V'''' \, dP + n_1'''' \, d\mu_1 + n_2'''' \, d\mu_2 = 0 \tag{5.77}$$

If one phase is present, the one equation will give either dT, dP, $d\mu_1$, or $d\mu_2$ in terms of the other three. We thus have three degrees of freedom. When there are two phases present, the solution of two equations after elimination of $d\mu_2$ is

$$\left(\frac{\tilde{S}''}{x_2''} - \frac{\tilde{S}'}{x_2'}\right) dT - \left(\frac{\tilde{V}''}{x_2''} - \frac{\tilde{V}'}{x_2'}\right) dP + \left(\frac{x_1''}{x_2''} - \frac{x_1'}{x_2'}\right) d\mu_1 = 0 \tag{5.78}$$

We thus see that there is one relation between three variables, and hence two of these variables are independent. Of course, we could have eliminated dT, dP, or $d\mu_1$ instead of $d\mu_2$. When three phases are present, two of the variables may be eliminated, leaving one equation expressing a relation between the remaining two variables. If $d\mu_1$ and $d\mu_2$ are eliminated, the solution becomes

$$\left(\frac{(\tilde{S}''/x_2'') - (\tilde{S}'/x_2')}{(x_1''/x_2'') - (x_1'/x_2')} - \frac{(\tilde{S}'''/x_2''') - (\tilde{S}'/x_2')}{(x_1'''/x_2''') - (x_1'/x_2')}\right) dT - \left(\frac{(\tilde{V}''/x_2'') - (\tilde{V}'/x_2')}{(x_1''/x_2'') - (x_1'/x_2')} - \frac{(\tilde{V}'''/x_2''') - (\tilde{V}'/x_2')}{(x_1'''/x_2''') - (x_1'/x_2')}\right) dP = 0 \tag{5.79}$$

which gives the dependence of the pressure of the three-phase system with temperature. Finally, when four phaes are present, dT, dP, $d\mu_1$, and $d\mu_2$ are all zero.

In the previous examples we have assumed that all comonents are present in all of the phases, and we have not introduced any chemical reactions or restrictions. When a component does not exist in a phase, the mole number of that component is zero in that phase and its chemical potential does not appear in the corresponding Gibbs–Duhem equation. As an example, consider a two-component, three-phase system in which two of the phases are pure and the third is a solution. The Gibbs–Duhem equations are then

$$S' \, dT - V' \, dP + n_1' \, d\mu_1 + n_2' \, d\mu_2 = 0 \tag{5.80}$$

$$S'' \, dT - V'' \, dP + n_1'' \, d\mu_1 = 0 \tag{5.81}$$

and

$$S''' \, dT - V''' \, dP + n_2''' \, d\mu_2 = 0 \tag{5.82}$$

From these equations we see that we have four variables and three equations, and hence one degree of freedom. Moreover, two of the variables can be eliminated from the three equations, leaving one equation relating the two remaining variables. Again, if we eliminate $d\mu_1$ and $d\mu_2$, the resultant equation is

$$(\tilde{S}' - x_1'\tilde{S}'' - x_2'\tilde{S}''')\,\mathrm{d}T - (\tilde{V}' - x_1'\tilde{V}'' - x_2'\tilde{V}''')\,\mathrm{d}P = 0 \tag{5.83}$$

which gives the pressure as a function of temperature.

If one or more chemical reactions are at equilibrium within the system, we can still set up the set of Gibbs–Duhem equations in terms of the components. On the other hand, we can write them in terms of the species present in each phase. In this case the mole numbers of the species are not all independent, but are subject to the condition of mass balance and to the condition that $\sum_i \nu_i \mu_i$ must be equal to zero for each independent chemical reaction. When these conditions are substituted into the Gibbs–Duhem equations in terms of species, the resultant equations are the Gibbs–Duhem equations in terms of components. Again, from a study of such sets of equations we can easily determine the number of degrees of freedom and can determine the mathematical relationships between these degrees of freedom.

Finally, consider a two-phase, two-component system in which the two phases are separated by an adiabatic membrane that is permeable only to the first component. In this case we know that the temperatures of the two phases are not necessarily the same, and that the chemical potential of the second component is not the same in the two phases. The two Gibbs–Duhem equations for this system are

$$S'\,\mathrm{d}T' - V'\,\mathrm{d}P + n_1'\,\mathrm{d}\mu_1 + n_2'\,\mathrm{d}\mu_2' = 0 \tag{5.84}$$

and

$$S''\,\mathrm{d}T'' - V''\,\mathrm{d}P + n_1''\,\mathrm{d}\mu_1 + n_2''\,\mathrm{d}\mu_2'' = 0 \tag{5.85}$$

In this set there are six variables and two equations, and consequently four degrees of freedom. Either $\mathrm{d}\mu_1$ or $\mathrm{d}P$ may be eliminated from these two equations. When $\mathrm{d}\mu_1$ is eliminated the resultant equation is

$$\frac{\tilde{S}'}{x_1'}\,\mathrm{d}T' - \frac{\tilde{S}''}{x_1''}\,\mathrm{d}T'' - \left(\frac{\tilde{V}'}{x_1'} - \frac{\tilde{V}''}{x_1''}\right)\mathrm{d}P + \frac{x_2'}{x_1'}\,\mathrm{d}\mu_2' - \frac{x_2''}{x_1''}\,\mathrm{d}\mu_2'' = 0 \tag{5.86}$$

This equation shows the dependence of one of the remaining variables on the other four. If three of the five independent variables were held constant, the system would be univariant. In this particular case it is also apparent that, if the second component did not appear in one of the phases, then the number of degrees of freedom would be reduced by one.

From this discussion it is apparent that the number of degrees of freedom for any particular case can be determined rather easily by examination of the set of Gibbs–Duehm equations that apply to the case. Moreover, the solution of this set of simultaneous independent equations yields much more information than the mere determination of the number of degrees of freedom.

5.13 Indifferent states [7]

The concept of indifferent states of systems is introduced in Section 5.11. There we define indifferent states as those states that required the assignment of a value to at least one extensive variable in addition to the number of moles of the components in order to define the state. Such systems are considered in more detail in this section.

Consider a definite number of moles of a pure component at the triple point for which the solid, liquid, and vapor states are in equilibrium. This is a nonvariant system and, consequently, the temperature and pressure of the system are fixed. One state of such a system may be that in which most of the substance is in the solid phase, with only insignificant amounts of the vapor and liquid phases being present. Now we may add heat to the system and change its volume, taking care always to have the three phases present. The changes of state that take place are the melting of the solid to the liquid or the sublimation of the solid to the vapor under the conditions that the temperature, pressure, and total mass remain constant. We may consider such changes of state to be the transfer of matter from one phase to another. With the addition of heat to the system and the change of volume, the entropy and the energy of the system change. Therefore, we must assign values of two extensive variables in order to define completely the state of the system.

When we consider a one-component, two-phase system, of constant mass, we find similar relations. Such two-phase systems are those in which a solid–solid, solid–liquid, solid–vapor, or liquid–vapor equilibrium exists. These systems are all univariant. Thus, the temperature is a function of the pressure, or the pressure is a function of the temperature. As a specific example, consider a vapor–liquid equilibrium at some fixed temperature and in a state in which most of the material is in the liquid state and only an insignificant amount in the vapor state. The pressure is fixed, and thus the volume is fixed from a knowledge of an equation of state. If we now add heat to the system under the condition that the temperature (and hence the pressure) is kept constant, the liquid will evaporate but the volume must increase as the number of moles in the vapor phase increases. Similarly, if the volume is increased, heat must be added to the system in order to keep the temperature constant. The change of state that takes place is simply a transfer of matter from one phase to another under conditions of constant temperature and pressure. We also see that only one extensive variable—the entropy, the energy, or the volume—is necessary to define completely the state of the system.

Univariant systems containing two components are exemplified by the equilibrium at a eutectic or peritectic. In each case a liquid phase is in equilibrium with two solid phases. Since such systems are univariant, the temperature is a function of the pressure, or the pressure is a function of the temperature. Thus, if the pressure is fixed, so is the temperature. Now consider the cooling of a system of such composition that the eutectic or peritectic equilibrium will be observed. Just as the eutectic or peritectic temperature is reached, the system will consist in general of a liquid phase and a solid phase with only an insignificant quantity of the second solid phase. As we continue to remove heat from the system at the chosen pressure, the temperature remains constant but the relative amounts of the three phases change. The volume will not remain constant, but it will always be known. Again we have a transfer of matter between phases, but in such a manner that the compositions of all three phases remain constant. Moreover, one extensive variable, presumably the entropy or energy, is required to fix the state of the system.

Multivariant systems may also become indifferent under special conditions. In all considerations the systems are to be thought of as closed systems with known mole numbers of each component. We consider here only divariant systems of two components. The system is thus a two-phase system. The two Gibbs–Duhem equations applicable to such a system are

$$\tilde{S}'\,\mathrm{d}T - \tilde{V}'\,\mathrm{d}P + x_1'\,\mathrm{d}\mu_1 + x_2'\,\mathrm{d}\mu_2 = 0 \tag{5.87}$$

and

$$\tilde{S}''\,\mathrm{d}T - \tilde{V}''\,\mathrm{d}P + x_1''\,\mathrm{d}\mu_1 + x_2''\,\mathrm{d}\mu_2 = 0 \tag{5.88}$$

where each equation has been divided by the total number of moles in the corresponding phase, so that the mole fractions rather than moles appear in the equations. When we eliminate $\mathrm{d}\mu_2$ from these equations, we obtain Equation (5.78). This equation shows the divariant nature of this system. However, under the special condition that the mole fractions of the two phases become identical, the equation becomes

$$\left(\frac{\tilde{S}''}{x_2''} - \frac{\tilde{S}'}{x_2'}\right)\mathrm{d}T - \left(\frac{\tilde{V}''}{x_2''} - \frac{\tilde{V}'}{x_2'}\right)\mathrm{d}P = 0 \tag{5.89}$$

which is univariant. Such conditions are found in azeotropic systems and in systems in which the two phases are a liquid and a solid solution and which exhibit a maximum or minimum in the freezing-point curves. At the composition of the azeotrope or the maximum or minimum of the freezing point, matter may be transferred from one phase to another without change of the composition of either phase. In order to do so, heat must be rmoved from or added to the system, and at the same time the volume of the system will change. Thus, again we see that we must assign values so at least one extensive variable in order to define the state of the system at the indifferent composition.

Other divariant systems composed of two phases and containing two components that become univariant and hence indifferent under special conditions are those in which a compound in one phase is in equilibrium with another phase in which the compound is partially or completely dissociated. Such situations are frequently met in the freezing-point diagrams of binary systems. As an example we assume that a compound, designated by the subscript C, is formed in the solid phase from the two components, designated by the subscripts 1 and 2. We assume for simplicity that the mole ratio of the components in this compound is 1:1. We know that such a compound has a definite melting point, and at this point the solid phase is in equilibrium with one liquid phase. We assume further that the compound is partially dissociated into its two components in the liquid phase; the result is the same whether the compound is only partially or completely dissociated in the liquid phase. The two Gibbs–Duhem equations applicable to such a system, when expressed in terms of species, are

$$S'\,\mathrm{d}T - V'\,\mathrm{d}P + n_1'\,\mathrm{d}\mu_1 + n_2'\,\mathrm{d}\mu_2 + n_C'\,\mathrm{d}\mu_C = 0 \tag{5.90}$$

and

$$S''\,\mathrm{d}T - V''\,\mathrm{d}P + n_C''\,\mathrm{d}\mu_C = 0 \tag{5.91}$$

The chemical potentials are subject to the further condition that $\mathrm{d}\mu_C = \mathrm{d}\mu_1 + \mathrm{d}\mu_2$. When we eliminate $\mathrm{d}\mu_C$ from these equations, the resultant equations become

$$S'\,\mathrm{d}T - V'\,\mathrm{d}P + (n_1' + n_C')\,\mathrm{d}\mu_1 + (n_2' + n_C')\,\mathrm{d}\mu_2 = 0 \tag{5.92}$$

and

$$S''\,\mathrm{d}T - V''\,\mathrm{d}P + n_C''\,\mathrm{d}\mu_1 + n_C''\,\mathrm{d}\mu_2 = 0 \tag{5.93}$$

On a molar basis these equations are the same as Equations (5.87) and (5.88). However, for the specific case that we have postulated, $x_1' = x_1''$ and $x_2' = x_2''$, and therefore

$$(\tilde{S}'' - \tilde{S}')\,\mathrm{d}T - (\tilde{V}'' - \tilde{V}')\,\mathrm{d}P = 0 \tag{5.94}$$

The system is thus univariant at this composition, and the temperature is a function of the pressure alone. At a fixed pressure we can transfer matter from one phase to another at constant temperature without changing the composition of the two phases. In doing so, heat must be added to or removed from the system, thus changing the entropy and energy of the system. The volume is determined from equations of state at the fixed temperature and pressure. We find that in such a case we again have to assign values to at least one extensive variable in order to define the state of such a system.

In every example that we have discussed, we have pointed out that it is possible to transfer matter at constant pressure and temperature between the phases present without causing any change in the composition of any phase. This observation permits another definition of an indifferent state of a system: *an indifferent state is a state in which it is possible to transfer matter between the phases present at a constant temperature and pressure without changing the composition of any phase.* For such systems it is obviously necessary, according to the previous definition of indifferent states, to assign values to at least one extensive variable in addition to the mole numbers of the components in order to define the state of the system.

In this discussion of indifferent states we have always used the entropy, energy, and volume as the possible extensive variables that must be used, in addition to the mole numbers of the components, to define the state of the system. The enthalpy or the Helmholtz energy may also be used to define the state of the system, but the Gibbs energy cannot. Each of the systems that we have considered has been a closed system in which it was possible to transfer matter between the phases at constant temperature and pressure. The differentials of the enthalpy and the Helmholtz and Gibbs energies under these conditions are

$$\mathrm{d}H = T\,\mathrm{d}S \qquad \mathrm{d}A = -P\,\mathrm{d}V \qquad \text{and} \qquad \mathrm{d}G = 0$$

Thus, the enthalpy is a function of the entropy and the Helmholtz energy is a function of the volume, and each function may be used in place of the other variable. However, the Gibbs energy is a constant for any closed system at constant temperature and pressure, and therefore its value is invariant with the transfer of matter within the closed system.

5.14 The Gibbs–Konovalow theorems

Equation (5.78) is applicable to the systems that have been described in the immediate neighborhood of the indifferent state. It shows the divariant character of these systems. If the pressure is held constant, this equation becomes

$$\frac{\mathrm{d}T}{\mathrm{d}\mu_1} = \left(\frac{(x_1''/x_2'') - (x_1'/x_2')}{(\tilde{S}''/x_2'') - (\tilde{S}'/x_2')}\right) \tag{5.95}$$

Then, at the composition of the system at which it becomes indifferent, the derivative becomes zero and thus the temperature has either a minimum or maximum value at this composition. Similar arguments show that the value of the pressure of such systems at constant temperature passes through either a maximum or a minimum at the composition of the indifferent state of the system. These observations, which have been obtained for divariant systems, are specific cases of two more-general theorems that are known as the Gibbs–Konovalow theorems [8]. These theorems are applicable to all systems that have two or more degrees of freedom. One theorem states that if the system passes through an indifferent state when its composition is varied at constant temperature, the pressure passes through an extreme value. The second theorem states that if the system passes through an indifferent state when its composition is varied at constant pressure, then the temperature passes through an extreme value. The converse of these theorems is also true.

5.15 Conditions of stability

We have established the conditions that must be satisfied at equilibrium, but we have not discussed the conditions that determine whether a single-phase system is stable, metastable, or unstable. In order to do so, we consider the incremental variation of the energy of a system, ΔE, rather than the differential variation of the energy, δE, for continuous virtual variations of the system. Higher-order terms must then be included. The condition of stability is that

$$(\Delta E)_S > 0 \tag{5.4}$$

for all virtual variations of the system.

We consider an isolated, homogenous system, and imagine that a part of the single phase is separated from the rest of the phase by a diathermal, nonrigid, permeable wall. By this device we can consider variations of the entropy, volume, and mole numbers of the two parts of the system subject to the conditions of constant entropy, volume, and mole numbers of the entire system. The condition of stability then becomes

$$\Delta E' + \Delta E'' > 0 \tag{5.96}$$

subject to the conditions

$$\delta S' + \delta S'' = 0 \tag{5.97}$$

$$\delta V' + \delta V'' = 0 \tag{5.98}$$

$$\delta n_1' + \delta n_1'' = 0 \tag{5.99}$$

$$\delta n_2' + \delta n_2'' = 0 \tag{5.100}$$

$$\delta n_3' + \delta n_3'' = 0 \tag{5.101}$$

$$\vdots$$

$$\delta n_C' + \delta n_C'' = 0 \tag{5.102}$$

when there are C components. The primes are used to designate the two parts. The increments of the energy of the two parts for continuous variation can be expanded in a Taylor's series so that, for the primed part

$$\Delta E' = \delta E' + \delta^2 E' + \delta^3 E' + \cdots \tag{5.103}$$

where

$$\delta E' = \left(\frac{\partial E'}{\partial S'}\right)_{V',n'} \delta S' + \left(\frac{\partial E'}{\partial V'}\right)_{S',n'} \delta V' + \sum_{i=1}^{C} \left(\frac{\partial E'}{\partial n_i'}\right)_{S',V',n'} \delta n_i' \tag{5.104}$$

$$\delta^2 E' = \frac{1}{2}\left\{\left(\frac{\partial^2 E'}{\partial S'^2}\right)_{V',n'}(\delta S')^2 + \left(\frac{\partial^2 E'}{\partial V'^2}\right)_{S',n'}(\delta V')^2 \right.$$
$$+ 2\left[\frac{\partial}{\partial V'}\left(\frac{\partial E'}{\partial S'}\right)_{V',n'}\right]_{S',n'} \delta S'\, \delta V'$$
$$+ 2\sum_{i=1}^{C}\left[\frac{\partial}{\partial n'_i}\left(\frac{\partial E'}{\partial S'}\right)_{V',n'}\right]_{S',V',n'} \delta S'\, \delta n'_i$$
$$+ 2\sum_{i=1}^{C}\left[\frac{\partial}{\partial n'_i}\left(\frac{\partial E'}{\partial V'_i}\right)_{S',n'}\right]_{S',V',n'} \delta V'\, \delta n'_i$$
$$\left. + \sum_{j=1}^{C}\sum_{i=1}^{C}\left[\frac{\partial}{\partial n'_j}\left(\frac{\partial E'}{\partial n'_i}\right)_{S',V',n'}\right] \delta n'_j\, \delta n'_i \right\} \tag{5.105}$$

$$\delta^3 E = \frac{1}{3!}\left\{\left(\frac{\partial^3 E'}{\partial S'^3}\right)_{V',n'}(\delta S')^3 + \left(\frac{\partial^3 E'}{\partial V'^3}\right)_{S',n'}(\delta V')^3 \right.$$
$$+ 3\left[\frac{\partial}{\partial V'}\left(\frac{\partial^2 E'}{\partial S'^2}\right)_{V',n'}\right]_{S',n'} (\delta S')^2\, \delta V'$$
$$\left. + 3\left[\frac{\partial}{\partial S'}\left(\frac{\partial^2 E'}{\partial V'^2}\right)_{S',n'}\right]_{V,n'} (\delta V')^2\, \delta S' + \cdots \right\} \tag{5.106}$$

Identical equations can be written for the double-primed part. The combination of Equation (5.103) and the similar equation for the double-primed part with Equation (5.96) results in the condition that

$$(\delta E' + \delta E'') + (\delta^2 E' + \delta^2 E'') + (\delta^3 E' + \delta^3 E'') + \cdots > 0 \tag{5.107}$$

To satisfy this condition the first term in the series that is not zero must be greater than zero. The conditions of equilibrium require that $(\delta E' + \delta E'')$ must be taken as zero. Therefore, we must examine the conditions based on Equation (5.105) that make

$$(\delta^2 E' + \delta^2 E'') > 0 \tag{5.108}$$

The use of the conditions expressed by Equations (5.97)–(5.102) and the fact that each second partial derivative in Equation (5.105) is identical to the same derivative for the double-prime part shows that $\delta^2 E' = \delta^2 E''$, and therefore Equation (5.108) can be written as

$$2(\delta^2 E') > 0 \tag{5.109}$$

The primes are omitted in the rest of the discussion, because no distinction need be made between the two parts.

The condition expressed in Equation (5.109) requires that the sum of terms in Equation (5.105) must be, in mathematical terms, positive-definite.

The specific requirements are determined more easily when the quadratic form of Equation (5.105) is changed to a sum of squared terms by a suitable change of variables. The general method is to introduce, in turn, a new independent variable in terms of the old independent variables. The coefficients in the resultant equations are simplified in terms of the new variables by a standard mathematical method. First, the entropy is eliminated by taking the temperature as a function of the entropy, volume, and mole numbers, so

$$\delta T = \left(\frac{\partial T}{\partial S}\right)_{V,n} \delta S + \left(\frac{\partial T}{\partial V}\right)_{S,n} \delta V + \sum_{i=1}^{C} \left(\frac{\partial T}{\partial n_i}\right)_{S,V,n} \delta n_i \tag{5.110}$$

or, in terms of the energy

$$\delta T = \left(\frac{\partial^2 E}{\partial S^2}\right)_{V,n} \delta S + \left[\frac{\partial}{\partial V}\left(\frac{\partial E}{\partial S}\right)_{V,n}\right]_{S,n} \delta V + \sum_{i=1}^{C} \left[\frac{\partial}{\partial n_i}\left(\frac{\partial E}{\partial S}\right)_{V,n}\right]_{S,V,n} \delta n_i \tag{5.111}$$

Upon squaring Equation (5.111) and substituting the result for $(\delta S)^2$ in Equation (5.105), we obtain

$$\begin{aligned}
2(\delta^2 E) = {} & \frac{1}{(\partial^2 E/\partial S^2)_{V,n}} (\delta T)^2 + \left(\left(\frac{\partial^2 E}{\partial V^2}\right)_{S,n} - \frac{\left[\frac{\partial}{\partial V}\left(\frac{\partial E}{\partial S}\right)_{V,n}\right]_{S,n}^2}{\left(\frac{\partial^2 E}{\partial S^2}\right)_{V,n}}\right)(\delta V)^2 \\
& + 2\sum_{i=1}^{C}\left(\left[\frac{\partial}{\partial n_i}\left(\frac{\partial E}{\partial V}\right)_{S,n}\right]_{S,V,n} - \frac{\left[\frac{\partial}{\partial V}\left(\frac{\partial E}{\partial S}\right)_{V,n}\right]_{S,n}\left[\frac{\partial}{\partial n_i}\left(\frac{\partial E}{\partial S}\right)_{V,n}\right]_{S,V,n}}{\left(\frac{\partial^2 E}{\partial S^2}\right)_{V,n}}\right)\delta V\,\delta n_i \\
& + \sum_{j=1}^{C}\sum_{i=1}^{C}\left(\left[\frac{\partial}{\partial n_j}\left(\frac{\partial E}{\partial n_i}\right)_{S,V,n}\right]_{S,V,n} - \frac{\left[\frac{\partial}{\partial n_i}\left(\frac{\partial E}{\partial S}\right)_{V,n}\right]_{S,V,n}\left[\frac{\partial}{\partial n_j}\left(\frac{\partial E}{\partial S}\right)_{V,n}\right]_{S,V,n}}{\left(\frac{\partial^2 E}{\partial S^2}\right)_{V,n}}\right)\delta n_i\,\delta n_j
\end{aligned} \tag{5.112}$$

All terms involving the variation of the entropy have been eliminated and one term involving $(\delta T)^2$ has been introduced.

Every coefficient in Equation (5.112) except that of $(\delta T)^2$ can be expressed more simply as a second derivative of the Helmholtz energy. We take only the coefficient of $(\delta V)^2$ as an example. Both $(\partial E/\partial V)_{S,n}$ and $(\partial A/\partial V)_{T,n}$ are equal to $-P$. The differential of $(\partial E/\partial V)_{S,n}$ is expressed in terms of the entropy, volume, and mole numbers and the differential of $(\partial A/\partial V)_{T,n}$ is expressed in terms of the temperature, volume, and mole numbers, so that at constant mole numbers

$$\mathrm{d}\left(\frac{\partial E}{\partial V}\right)_{S,n} = \left[\frac{\partial}{\partial S}\left(\frac{\partial E}{\partial V}\right)_{S,n}\right]_{V,n} \mathrm{d}S + \left[\frac{\partial}{\partial V}\left(\frac{\partial E}{\partial V}\right)_{S,n}\right]_{S,n} \mathrm{d}V \tag{5.113}$$

and, at constant temperature and mole numbers,

$$\mathrm{d}\left(\frac{\partial A}{\partial V}\right)_{T,n} = \left[\frac{\partial}{\partial V}\left(\frac{\partial A}{\partial V}\right)_{T,n}\right]_{T,n} \mathrm{d}V \tag{5.114}$$

However,

$$\mathrm{d}S = \left(\frac{\partial S}{\partial V}\right)_{T,n} \mathrm{d}V \tag{5.115}$$

at constant temperature and mole numbers. When Equation (5.115) is substituted into Equation (5.113) and Equations (5.113) and (5.114) are equated we obtain

$$\left[\frac{\partial}{\partial V}\left(\frac{\partial A}{\partial V}\right)_{T,n}\right]_{T,n} = \left[\frac{\partial}{\partial V}\left(\frac{\partial E}{\partial V}\right)_{S,n}\right]_{S,n} + \left[\frac{\partial}{\partial S}\left(\frac{\partial E}{\partial V}\right)_{S,n}\right]_{V,n}\left(\frac{\partial S}{\partial V}\right)_{T,n} \tag{5.116}$$

We find from Equation (5.111) that

$$\left(\frac{\partial S}{\partial V}\right)_{T,n} = -\frac{\left[\frac{\partial}{\partial V}\left(\frac{\partial E}{\partial S}\right)_{V,n}\right]_{S,n}}{\left[\frac{\partial}{\partial S}\left(\frac{\partial E}{\partial S}\right)_{V,n}\right]_{V,n}} \tag{5.117}$$

Therefore, the coefficient is equal to $(\partial^2 A/\partial V^2)_{T,n}$. All other coefficients can be converted by similar methods and Equation (5.112) written as

$$\begin{aligned}2(\delta^2 E) = {} & \frac{1}{(\partial^2 E/\partial S^2)_{V,n}}(\delta T)^2 + \left(\frac{\partial^2 A}{\partial V^2}\right)_{T,n}(\delta V)^2 \\ & + 2\sum_{i=1}^{C}\left[\frac{\partial}{\partial n_i}\left(\frac{\partial A}{\partial V}\right)_{T,n}\right]_{T,V,n}\delta V\,\delta n_i \\ & + \sum_{j=1}^{C}\sum_{i=1}^{C}\left[\frac{\partial}{\partial n_j}\left(\frac{\partial A}{\partial n_i}\right)_{T,V,n}\right]_{T,V,n}\delta n_i\,\delta n_j\end{aligned} \tag{5.118}$$

In order to eliminate all terms containing δV, the pressure is taken as a function of the volume and mole numbers at constant temperature, so

$$\delta P = \left(\frac{\partial P}{\partial V}\right)_{T,n} \delta V + \sum_{i=1}^{C} \left(\frac{\partial P}{\partial n_i}\right)_{T,V,n} \delta n_i \tag{5.119}$$

or

$$\delta P = -\left(\frac{\partial^2 A}{\partial V^2}\right)_{T,n} \delta V - \sum_{i=1}^{C} \left[\frac{\partial}{\partial n_i}\left(\frac{\partial A}{\partial V}\right)_{T,n}\right]_{T,V,n} \delta n_i \tag{5.120}$$

When Equation (5.120) is squared and substitution is made for $(\delta V)^2$,

$$2(\delta^2 E) = \frac{1}{(\partial^2 E/\partial S^2)_{V,n}} (\delta T)^2 + \frac{1}{(\partial^2 A/\partial V^2)_{T,n}} (\delta P)^2 + \sum_{j=1}^{C} \sum_{i=1}^{C} \left(\left[\frac{\partial}{\partial n_j}\left(\frac{\partial A}{\partial n_i}\right)_{T,V,n}\right]_{T,V,n} - \frac{\left[\frac{\partial}{\partial n_i}\left(\frac{\partial A}{\partial V}\right)_{T,n}\right]_{T,V,n}\left[\frac{\partial}{\partial n_j}\left(\frac{\partial A}{\partial V}\right)_{T,n}\right]_{T,V,n}}{\left(\frac{\partial^2 A}{\partial V^2}\right)_{T,n}} \right) \delta n_i\, \delta n_j \tag{5.121}$$

is obtained. The coefficients contained in the brackets are expressible as second derivatives of the Gibbs energy, so that Equation (5.121) can be written as

$$2(\delta^2 E) = \frac{1}{(\partial^2 E/\partial S^2)_{V,n}} (\delta T)^2 + \frac{1}{(\partial^2 A/\partial V^2)_{T,n}} (\delta P)^2 + \sum_{j=1}^{C} \sum_{i=1}^{C} \left[\frac{\partial}{\partial n_j}\left(\frac{\partial G}{\partial n_i}\right)_{T,P,n}\right]_{T,P,n} \delta n_i\, \delta n_j \tag{5.122}$$

The method used is similar to that discussed in the previous paragraph.

The next step is to eliminate all terms containing δn_1. In order to do so we use the chemical potential of the first component as a function of the mole numbers at constant temperature and pressure. Then

$$\delta \mu_1 = \left(\frac{\partial \mu_1}{\partial n_1}\right)_{T,P,n} \delta n_1 + \sum_{\substack{i=2 \\ i \neq 1}}^{C} \left(\frac{\partial \mu_1}{\partial n_i}\right)_{T,P,n} \delta n_i \tag{5.123}$$

or

$$\delta\mu_1 = \left(\frac{\partial^2 G}{\partial n_1^2}\right)_{T,P,n} \delta n_1 + \sum_{\substack{i=2\\i\neq 1}}^{c} \left[\frac{\partial}{\partial n_i}\left(\frac{\partial G}{\partial n_1}\right)_{T,P,n}\right]_{T,P,n} \delta n_i \tag{5.124}$$

The substitution of $\delta\mu_1$ for δn_i in Equation (5.122) yields

$$2(\delta^2 E) = \frac{1}{(\partial^2 E/\partial S^2)_{V,n}}(\delta T)^2 + \frac{1}{(\partial^2 A/\partial V^2)_{T,n}}(\delta P)^2 + \frac{1}{(\partial^2 G/\partial n_1^2)_{T,P,n}}(\delta\mu_1)^2$$

$$+ \sum_{\substack{j=2\\j\neq 1}}^{c}\sum_{\substack{i=2\\i\neq 1}}^{c}\left(\left[\frac{\partial}{\partial n_j}\left(\frac{\partial G}{\partial n_i}\right)_{T,P,n}\right]_{T,P,n} - \frac{\left[\frac{\partial}{\partial n_j}\left(\frac{\partial G}{\partial n_1}\right)_{T,P,n}\right]_{T,P,n}\left[\frac{\partial}{\partial n_i}\left(\frac{\partial G}{\partial n_1}\right)_{T,P,n}\right]_{T,P,n}}{\left(\frac{\partial^2 G}{\partial n_1^2}\right)_{T,P,n}}\right)\delta n_i\,\delta n_j \tag{5.125}$$

Again, the coefficients of the terms within the brackets can be written as appropriate second derivatives of another function, but in this case the function is a new function for which the independent variables are the temperature, the pressure, the chemical potential of the first component, and the mole numbers of all components except the first. If this new function function is given the symbol ϕ, then

$$\phi = \phi(T, P, \mu_1, n_2, \ldots) \tag{5.126}$$

$$= G - \mu_1 n_1 \tag{5.127}$$

$$= E - TS + PV - \mu_1 n_1 \tag{5.128}$$

$$= \sum_{\substack{i=2\\i\neq 1}}^{c} \mu_i n_i \tag{5.129}$$

and

$$d\phi = -S\,dT + V\,dP - n_1\,d\mu_1 + \sum_{\substack{i=2\\i\neq 1}}^{c} \mu_i\,dn_i \tag{5.130}$$

In order to change the coefficients in the brackets in Equation (5.125), we note that

$$\left(\frac{\partial\phi}{\partial n_i}\right)_{T,P,\mu_1,n} = \left(\frac{\partial G}{\partial n_i}\right)_{T,P,n} = \mu_i \tag{5.131}$$

where the subscript n on the derivative of ϕ refers to all mole numbers except n_i and n_1. By using similar methods to those that have been used before, we can obtain

$$2(\delta^2 E) = \frac{1}{(\partial^2 E/\partial S^2)_{V,n}}(\delta T)^2 + \frac{1}{(\partial^2 A/\partial V^2)_{T,n}}(\delta P)^2 + \frac{1}{(\partial^2 G/\partial n_1^2)_{T,P,n}}(\delta \mu_1)^2 + \sum_{\substack{j=2\\ j\neq 1}}^{C}\sum_{\substack{i=2\\ i\neq 1}}^{C}\left[\frac{\partial}{\partial n_j}\left(\frac{\partial \phi}{\partial n_i}\right)_{T,P,\mu_1,n}\right]\delta n_i\,\delta n_j \tag{5.132}$$

All terms containing δn_2 can be eliminated by use of

$$\delta\mu_2 = \left(\frac{\partial^2 \phi}{\partial n_2^2}\right)_{T,P,\mu_1,n}\delta n_2 + \sum_{\substack{i=3\\ i\neq 2\\ i\neq 1}}^{C}\left[\frac{\partial}{\partial n_i}\left(\frac{\partial \phi}{\partial n_2}\right)_{T,P,\mu_1,n}\right]_{T,P,\mu_1,n}\delta n_i \tag{5.133}$$

which yields

$$\begin{aligned}2(\delta^2 E) = {} & \frac{1}{(\partial^2 E/\partial S^2)_{V,n}}(\delta T)^2 + \frac{1}{(\partial^2 A/\partial V^2)_{T,n}}(\delta P)^2 + \frac{1}{(\partial^2 G/\partial n_1^2)_{T,P,n}}(\delta \mu_1)^2 \\ & + \frac{1}{(\partial^2 \phi/\partial n_2^2)_{T,P,\mu_1,n}}(\delta\mu_2)^2 + \sum_{\substack{j=3\\ j\neq 2\\ j\neq 1}}^{C}\sum_{\substack{i=3\\ i\neq 2\\ i\neq 1}}^{C}\left(\left[\frac{\partial}{\partial n_i}\left(\frac{\partial \phi}{\partial n_j}\right)_{T,P,\mu_1,n}\right]_{T,P,\mu_1,n}\right. \\ & \left. - \frac{\left[\frac{\partial}{\partial n_j}\left(\frac{\partial \phi}{\partial n_2}\right)_{T,P,\mu_1,n}\right]_{T,P,\mu_1,n}\left[\frac{\partial}{\partial n_i}\left(\frac{\partial \phi}{\partial n_2}\right)_{T,P,\mu_1,n}\right]_{T,P,\mu_1,n}}{\left(\frac{\partial^2 \phi}{\partial n_2^2}\right)_{T,P,\mu_1,n}}\right)\delta n_i\,\delta n_j\end{aligned} \tag{5.134}$$

The substitution that has been made here results in another new function that is now a function of the temperature, the pressure, the chemical potentials of the first and second components, and the mole numbers of the other components. The process can be continued for $(C-1)$ components, after which the quadratic expression given in Equation (5.105) consists of a sum of squared terms.

We can now consider the conditions of stability for pure substances, binary systems, and ternary systems based on Equations (5.122), (5.132), and (5.134), respectively. In order to satisfy the conditions, the coefficient of each term (except the last term in each applicable equation, which is zero) must be positive. If any one of the terms is negative for a hypothetical homogenous system, that system is unstable and cannot exist.

First, for any homogenous system containing any number of components, both $(\partial^2 E/\partial S^2)_{V,n}$ and $(\partial^2 A/\partial V^2)_{T,n}$ must be positive. The first derivative requires that

$$C_V > 0 \tag{5.135}$$

because

$$\left(\frac{\partial^2 E}{\partial S^2}\right)_{V,n} = \left(\frac{\partial T}{\partial S}\right)_{V,n} = \frac{T}{C_V} > 0$$

and the temperature is always positive. The second condition requires that

$$-\left(\frac{\partial P}{\partial V}\right)_{T,n} > 0 \tag{5.136}$$

that is, the isothermal coefficient of compressibility must always be positive.

For binary systems the additional condition of stability is

$$\left(\frac{\partial^2 G}{\partial n_1^2}\right)_{T,P,n_2} > 0 \tag{5.137}$$

This condition can be expressed in two equivalent ways. First,

$$\left(\frac{\partial^2 G}{\partial n_1^2}\right)_{T,P,n_2} = \left(\frac{\partial \mu_1}{\partial n_1}\right)_{T,P,n_2} \tag{5.138}$$

and

$$\left(\frac{\partial \mu_1}{\partial n_1}\right)_{T,P,n_2} = \left(\frac{\partial \mu_1}{\partial x_1}\right)_{T,P}\left(\frac{\partial x_1}{\partial n_1}\right)_{T,P,n_2} = \frac{x_2}{n_1 + n_2}\left(\frac{\partial \mu_1}{\partial x_1}\right)_{T,P} \tag{5.139}$$

Therefore,

$$\left(\frac{\partial \mu_1}{\partial x_1}\right)_{T,P} > 0 \tag{5.140}$$

This condition requires that the slope of the curve for μ_1 plotted against x_1 be positive.

The second form is

$$\left(\frac{\partial^2 \tilde{G}}{\partial x_1^2}\right)_{T,P} > 0 \tag{5.141}$$

The condition in this form requires that the curve of $\tilde{G}$ against x_1 be concave upwards. The relation is obtained from

$$\left(\frac{\partial \tilde{G}}{\partial x_1}\right)_{T,P} = \mu_1 - \mu_2 \tag{5.142}$$

and, therefore,

$$\left(\frac{\partial^2 \tilde{G}}{\partial x_1^2}\right)_{T,P} = \left(\frac{\partial \mu_1}{\partial x_1}\right)_{T,P} - \left(\frac{\partial \mu_2}{\partial x_1}\right)_{T,P} \tag{5.143}$$

However, from the Gibbs–Duhem equation

$$\left(\frac{\partial \mu_2}{\partial x_1}\right)_{T,P} = -\frac{x_1}{x_2}\left(\frac{\partial \mu_1}{\partial x_1}\right)_{T,P} \tag{5.144}$$

so

$$\left(\frac{\partial^2 \tilde{G}}{\partial x_1^2}\right)_{T,P} = \frac{1}{x_2}\left(\frac{\partial \mu_1}{\partial x_1}\right)_{T,P} \tag{5.145}$$

For ternary systems the conditions of stability in addition to those given by Equations (5.135) and (5.136) are

$$\left(\frac{\partial^2 \tilde{G}}{\partial n_1^2}\right)_{T,P,n_2,n_3} > 0 \tag{5.146}$$

and

$$\left(\frac{\partial^2 \phi}{\partial n_2^2}\right)_{T,P,\mu_1,n_3} > 0 \tag{5.147}$$

The inequality given in Equation (5.147) can be written in two alternate forms. The first is based on Equation (5.130) and is

$$\left(\frac{\partial \mu_2}{\partial n_2}\right)_{T,P,\mu_1,n_3} > 0 \tag{5.148}$$

The second is obtained from this equation by changing from constant μ_1 to constant n_1. At constant T, P, and n_3, μ_2 is a function of n_1 and n_2, so

$$\left(\frac{\partial \mu_2}{\partial n_2}\right)_{T,P,\mu_1,n_3} = \left(\frac{\partial \mu_2}{\partial n_1}\right)_{T,P,n_2,n_3}\left(\frac{\partial n_1}{\partial n_2}\right)_{T,P,\mu_1,n_3} + \left(\frac{\partial \mu_2}{\partial n_2}\right)_{T,P,n_1,n_3} \tag{5.149}$$

However,

$$\left(\frac{\partial n_1}{\partial n_2}\right)_{T,P,\mu_1,n_3} = -\frac{(\partial \mu_1/\partial n_2)_{T,P,n_1,n_3}}{(\partial \mu_1/\partial n_1)_{T,P,n_2,n_3}} \tag{5.150}$$

and, therefore,

$$\left(\frac{\partial \mu_2}{\partial n_2}\right)_{T,P,\mu_1,n_3} = \left(\frac{\partial \mu_2}{\partial n_2}\right)_{T,P,n_1,n_3} - \frac{(\partial \mu_2/\partial n_1)_{T,P,n_2,n_3}\,(\partial \mu_1/\partial n_2)_{T,P,n_1,n_3}}{(\partial \mu_1/\partial n_1)_{T,P,n_2,n_3}} \tag{5.151}$$

The inequality may then be written as

$$\left(\frac{\partial\mu_2}{\partial n_2}\right)_{T,P,n_1,n_3}\left(\frac{\partial\mu_1}{\partial n_1}\right)_{T,P,n_2,n_3}-\left(\frac{\partial\mu_2}{\partial n_1}\right)^2_{T,P,n_2,n_3}>0 \tag{5.152}$$

which requires that

$$\left(\frac{\partial\mu_2}{\partial n_2}\right)_{T,P,n_1,n_3}>0 \tag{5.153}$$

because, from Equation (5.146), $(\partial\mu_1/\partial n_1)_{T,P,n_2,n_3}$ must be positive. We thus have two equivalent sets of conditions for ternary systems: the inequalities given by Equations (5.146) and (5.148) or those given by Equations (5.146), (5.152), and (5.153).

The determination of the conditions of stability for systems containing more than three components requires only the continuation of the methods that have been discussed. Additional conditions of stability of a homogenous system in an electrical or magnetic field can be obtained by use of the same methods discussed here with the introduction of the appropriate variables.

The Helmholtz and Gibbs energy functions can also be used to establish some of the conditions of stability instead of the energy function. The remaining terms in Equation (5.118), when the variation of the temperature is zero, are just those obtained for the second-order variation of the Helmholtz energy at constant temperature. Similarly, the remaining terms in Equation (5.122), when the variation of both temperature and pressure is zero, are those obtained for the second-order variation of the Gibbs energy at constant temperature and pressure. Indeed, the requirements of stability in terms of the increment of the Helmholtz energy is

$$(\Delta A)_{T,V,n}>0 \tag{5.154}$$

and in terms of the increment of the Gibbs energy is

$$(\Delta G)_{T,P,n}>0 \tag{5.155}$$

The allowed variations are then the volume and mole numbers at constant temperature for the Helmholtz energy, and only the mole numbers at constant temperature and pressure for the Gibbs energy.

5.16 Critical phenomena

The conditions of stability developed in Section 5.15 suggest a boundary between stable and unstable systems. This boundary is determined by the conditions that one of the quantities that determine the stability of a system becomes zero at the boundary; at one side of the boundary the appropriate derivative has a value greater than zero, whereas on the other side its value is less than zero. The derivative is a function of the independent variables used to define the state of the system, and therefore the state of the system at which the derivative becomes zero is dependent on the same variables. The loci of such states in a graphical representation map out the boundary between stable and unstable states. In some systems the boundary terminates, so that beyond a set of values of the independent variables, all states are stable. Critical phases appear at such terminations.

As an example, consider the P–V isotherms of a pure fluid above and below the critical isotherm. Curve I in Figure 5.1 illustrates the isotherm above the critical temperature. Both the curve itself and the slope are continuous, and the slope is negative, consistent with the condition that $(\partial P/\partial V)_T$ must be negative. Curve II along $ABEF$ represents an isotherm below the critical temperature. The system exists in a single phase along AB and EF, and in two phases (gas and liquid) along the horizontal line BE. The portions AB and EF of the curve have negative slope, consistent with the conditions of stability. The slope of the curve at B and at E is discontinuous, resulting from a discontinuous change of phase. The extension of the curves beyond the discontinuities, BC and DE, have the correct slope and represent stable phases although such phases are metastable with respect to the two-phase system. At points C and D, $(\partial P/\partial V)_T = 0$. A mathematical expression for the pressure of a fluid given as a function of the temperature and volume is usually a continuous function with continuous derivatives. We may assume that such is the case here, and connect points C and D with

Figure 5.1. Schematic representation of stable, metastable, and unstable states in a pure fluid.

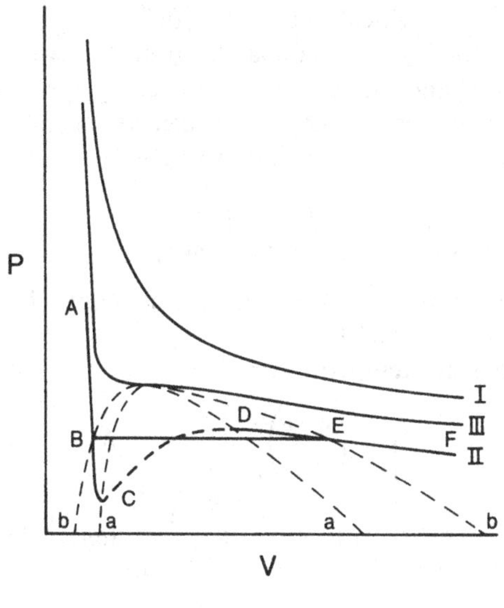

a continuous curve as represented by the broken curve. Along this portion of the curve, $(\partial P/\partial V)_T > 0$ and hence the states of the system along the curve are unstable. The loci of points C and D for other isotherms form the boundary between stable and unstable states. Such curves are indicated by the broken curves labeled a in the figure. The loci of points B and E for various temperatures represent the boundary between metastable and stable states. These curves are represented by the broken curves labeled b. The area between the curves a and b represents a region of metastable states, and the area within the curve marked a represents a region of unstable states. As the temperature is increased toward the critical temperature (curve III), the four points B, C, D, and E approach each other and become identical at the critical point. Moreover, the point of inflection along curves C and D also coincides with the four points at the critical point. The critical isotherm thus has a horizontal point of inflection, for which both $(\partial P/\partial V)_T$ and $(\partial^2 P/\partial V^2)_T$ are equal to zero, at the critical point. Moreover, at this point the boundary between stable and unstable states terminates. The phase at this point is called the *critical phase*.

The stability of the critical phase can be discussed most easily in terms of the Helmholtz energy and the condition expressed in Equation (5.154). By use of the method used in Section 5.15, the second-order variation at constant temperature is expressed as

$$2(\delta^2 A) = \left(\frac{\partial^2 A}{\partial V^2}\right)_{T,n} (\delta V)^2 + 2\left[\frac{\partial}{\partial n}\left(\frac{\partial A}{\partial V}\right)_{T,n}\right]_{T,V} \delta V\, \delta n + \left(\frac{\partial^2 A}{\partial n^2}\right)_{T,V} (\delta n)^2 \tag{5.156}$$

The chemical potential is constant at constant temperature and volume, and therefore $(\partial^2 A/\partial n^2)_{T,V} = 0$. For the critical phase, $(\partial^2 A/\partial V^2)_{T,n} = 0$ because $(\partial P/\partial V)_{T,n} = 0$. That $[(\partial/\partial n)(\partial A/\partial V)_{T,n}]_{T,V} = 0$ may be argued in two ways. First, if the phase is stable, $\partial^2 A$ cannot be negative. However, by appropriate choices of ∂V and ∂n the term may be negative if the coefficient is not zero, and therefore the coefficient must be zero. Second, we have

$$\left[\frac{\partial}{\partial n}\left(\frac{\partial A}{\partial V}\right)_{T,n}\right]_{T,V} = -\left(\frac{\partial P}{\partial n}\right)_{T,V} = -\left(\frac{\partial P}{\partial V}\right)_{T,n}\left(\frac{\partial V}{\partial n}\right)_{T,P}$$

and the term is zero because $(\partial P/\partial V)_{T,n} = 0$ at the critical point. Here we have shown that for the critical phase $\delta^2 A = 0$.

The third-order term at constant temperature is

$$\delta^3 A = \frac{1}{3!}\left\{\left(\frac{\partial^3 A}{\partial V^3}\right)_{T,n} (\delta V)^3 + 3\left[\frac{\partial}{\partial n}\left(\frac{\partial^2 A}{\partial V^2}\right)_{T,n}\right]_{T,V} (\delta V)^2\, \delta n \right.$$
$$\left. + 3\left[\frac{\partial^2}{\partial n^2}\left(\frac{\partial A}{\partial V}\right)_{T,n}\right]_{T,V} \delta V\, (\delta n)^2 + \left(\frac{\partial^3 A}{\partial n^3}\right)_{T,V} (\delta n)^3\right\} \tag{5.157}$$

Because of the odd order of the variations, the term could have negative values for appropriately chosen variations of the volume and mole number. However, according to the condition of stability, negative values of ΔA cannot occur and therefore the third-order term must be zero. The coefficient $(\partial^3 A/\partial n^3)_{T,V}$ is zero because of the chemical potential. The other coefficients become zero when $(\partial^2 P/\partial V^2)_{T,n}$ and $(\partial P/\partial V)_{T,n}$ are zero. This condition is consistent with the horizontal points of inflection at the critical point.

We finally have to consider the fourth-order term, which is

$$(\delta^4 A)=\frac{1}{4!}\left\{\left(\frac{\partial^4 A}{\partial V^4}\right)_{T,n}(\delta V)^4+4\left[\frac{\partial}{\partial n}\left(\frac{\partial^3 A}{\partial V^3}\right)_{T,n}\right]_{T,V}(\delta V)^3\,\delta n\right.$$
$$+6\left[\frac{\partial^2}{\partial n^2}\left(\frac{\partial^2 A}{\partial V^2}\right)_{T,n}\right]_{T,V}(\delta V)^2(\delta n)^2$$
$$\left.+4\left[\frac{\partial^3}{\partial n^3}\left(\frac{\partial A}{\partial V}\right)_{T,n}\right]_{T,V}\delta V\,(\delta n)^3+\left(\frac{\partial^4 A}{\partial n^4}\right)_{T,V}(\delta n)^4\right\} \tag{5.158}$$

For this term to be positive, $(\partial^3 P/\partial V^3)_{T,n}$ must be negative. If $(\partial^3 P/\partial V^3)_{T,n}=0$, the entire term becomes zero. Under such circumstances the fifth-order term must be zero and the sixth-order term would have to be studied for its positive character. Thus, the conditions of stability of the critical phase in a one-component system must be

$$\left(\frac{\partial^2 A}{\partial V^2}\right)_{T,n}=\left(\frac{\partial^3 A}{\partial V^3}\right)_{T,n}=0 \tag{5.159a}$$

and

$$\left(\frac{\partial^4 A}{\partial V^4}\right)_{T,n}\geqslant 0 \tag{5.159b}$$

The conditions

$$\left(\frac{\partial P}{\partial V}\right)_{T,n}=\left(\frac{\partial^2 P}{\partial V^2}\right)_{T,n}=0 \tag{5.160a}$$

and

$$\left(\frac{\partial^3 P}{\partial V^3}\right)_{T,n}\leqslant 0 \tag{5.160b}$$

are, of course, equivalent.

Other conditions at the critical point of a single-component system can be obtained from the condition that $(\partial P/\partial V)_{T,n} = 0$. From

$$\left(\frac{\partial V}{\partial T}\right)_{P,n} = -\left(\frac{\partial P}{\partial T}\right)_{V,n}\left(\frac{\partial V}{\partial P}\right)_{T,n} \tag{5.161}$$

it is apparent that $(\partial V/\partial T)_{P,n}$ becomes infinite. Similarly, the heat capacity at constant pressure becomes infinite, because

$$\left(\frac{\partial S}{\partial T}\right)_{P,n} = \frac{C_P}{T} = -\left(\frac{\partial S}{\partial P}\right)_{T,n}\left(\frac{\partial P}{\partial T}\right)_{S,n} = \left(\frac{\partial V}{\partial T}\right)_{P,n}\left(\frac{\partial P}{\partial T}\right)_{S,n} \tag{5.162}$$

Critical solution phenomena in a binary system are essentially identical to the critical phenomena in a single-component system. Curve I in Figure 5.2 represents the chemical potential of the first component as a function of its mole fraction at a constant temperature and pressure when only one phase exists over the whole range of concentration. The slope is consistent with the condition given in Equation (5.140). Curve II illustrates the behavior of the chemical potential as a function of the mole fraction at the same pressure but a different, though constant, temperature when partial immiscibility occurs. Along the portions of the curve marked AB and EF a stable, single-phase system occurs. Along the horizontal line BE the system exists in two phases. Along the portions of the curve BC and DE a single-phase system would be stable although it is metastable with respect to the two-phase system. At the points C and D, $(\partial \mu_1/\partial x_1)_{T,P} = 0$. If the curve is assumed to be continuous as illustrated by the broken curve between C and D, states of the system along this portion of the curve would be unstable. The loci of points C and D as the temperature is changed at constant pressure represent

Figure 5.2. Critical solution phenomenon in a binary system.

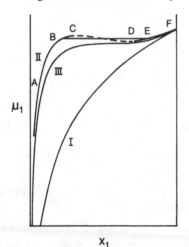

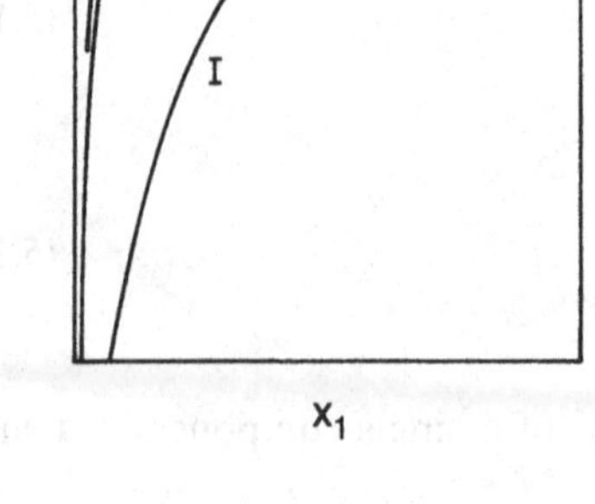

the boundary between stable and metastable systems. Curve III illustrates the behavior of the chemical potential along the critical solution isotherm, and at the critical point both $(\partial\mu_1/\partial x_1)_{T,P}$ and $(\partial^2\mu_1/\partial x_1^2)_{T,P}$ are zero. The same conditions are valid when the pressure is changed at constant temperature.

The conditions of stability for the critical solution phase can be discussed in terms of the Gibbs energy at constant temperature and pressure in terms of Equation (5.155). The second-order term is

$$2(\delta^2 G) = \left(\frac{\partial^2 G}{\partial n_1^2}\right)_{T,P,n_2} (\partial n_1)^2 + 2\left[\frac{\partial}{\partial n_2}\left(\frac{\partial G}{\partial n_1}\right)_{T,P,n_2}\right]_{T,P,n_1} \partial n_1\, \partial n_2 + \left(\frac{\partial^2 G}{\partial n_2^2}\right)_{T,P,n_1} (\partial n_2)^2 \tag{5.163}$$

When $(\partial\mu_1/\partial x_1)_{T,P}=0$, $(\partial^2 G/\partial n_1^2)_{T,P,n_2}=0$ according to Equation (5.138). The entire term then becomes zero for the critical solution phase. The third-order term must be zero when the second-order term is zero, because otherwise some variations would result in negative values of the term. The condition to make the third term zero is that $(\partial^3 G/\partial n_1^3)_{T,P,n_2}$ or the equivalent $(\partial^2\mu_1/\partial x_1^2)$ is zero. If the fourth-order term is positive, then $(\partial^4 G/\partial n_1^4)_{T,P}$ is positive. Thus, the conditions of stability for the critical solution phase are

$$\left(\frac{\partial^2 G}{\partial n_1^2}\right)_{T,P,n_2} = \left(\frac{\partial^3 G}{\partial n_1^3}\right)_{T,P,n_2} = 0 \tag{5.164a}$$

and

$$\left(\frac{\partial^4 G}{\partial n_1^4}\right)_{T,P,n_2} \geqslant 0 \tag{5.164b}$$

or the equivalent

$$\left(\frac{\partial \mu_1}{\partial x_1}\right)_{T,P} = \left(\frac{\partial^2 \mu_1}{\partial x_1^2}\right)_{T,P} = 0 \tag{5.165a}$$

and

$$\left(\frac{\partial^3 \mu_1}{\partial x_1^3}\right)_{T,P} \geqslant 0 \tag{5.165b}$$

An additional problem arises when the stability of the critical phase involving the gas–liquid equilibrium in a binary system is studied. The conditions of stability of a homogenous system at constant pressure are $(\partial^2\tilde{A}/\partial\tilde{V}^2)_{T,x}>0$ and $(\partial^2\tilde{G}/\partial x_1^2)_{T,P}>0$ from Equations (5.136) and (5.141), respectively. The question arises of which of the two conditions becomes zero first as the boundary between stable and unstable phases is approached.

The inequality

$$\left(\frac{\partial^2 A}{\partial x_1^2}\right)_{T,V}\left(\frac{\partial^2 \tilde{A}}{\partial \tilde{V}^2}\right)_{T,x} - \left[\frac{\partial}{\partial x_1}\left(\frac{\partial \tilde{A}}{\partial \tilde{V}}\right)_{T,x}\right]^2_{T,V} > 0 \tag{5.166}$$

is equivalent to $(\partial^2\tilde{G}/\partial x_1^2)_{T,P}>0$. We see that, if $(\partial^2\tilde{A}/\partial\tilde{V}^2)_{T,x}=0$, the expression in Equation (5.166) is already negative. We conclude that this expression becomes zero and hence determines the boundary between stable and unstable phases while $(\partial^2\tilde{A}/\partial\tilde{V}^2)_{T,x}$ is still positive. This latter quantity plays no role in determining the boundary between stable and unstable systems. The conditions of stability of the critical phases can be obtained by the requirement that the third-order term must be zero when the second-order term is zero. The conditions thus obtained are

$$\left(\frac{\partial^2 \tilde{G}}{\partial x_1^2}\right)_{T,P} = 0 \tag{5.167}$$

$$\left(\frac{\partial^3 \tilde{G}}{\partial x_1^3}\right)_{T,P} = 0 \tag{5.168}$$

and

$$\left(\frac{\partial^4 \tilde{G}}{\partial x_1^4}\right)_{T,P} \geqslant 0 \tag{5.169}$$

Equivalent conditions are

$$\left(\frac{\partial \mu_1}{\partial x_1}\right)_{T,P} = 0 \tag{5.170}$$

$$\left(\frac{\partial^2 \mu_1}{\partial x_1^2}\right)_{T,P} = 0 \tag{5.171}$$

and

$$\left(\frac{\partial^3 \mu_1}{\partial x_1^3}\right)_{T,P} \geqslant 0 \tag{5.172}$$

Problems concerning the conditions of stability of homogenous systems for critical phases in ternary systems are very similar to those for the gas–liquid phenomena in binary systems, because of two independent variables at constant temperature and pressure. The conditions for stability are $(\partial^2 G/\partial n_1^2)_{T,P,n_2,n_3}>0$ and $(\partial^2\phi/\partial n_2^2)_{T,P,\mu_1,n_3}>0$, given by Equations (5.146) and (5.147), respectively. Inspection of the condition equivalent to Equation (5.148) given by Equation (5.152) shows that $(\partial^2\phi/\partial n_2^2)_{T,P,\mu_1,n_3}$ must go to zero before $(\partial^2 G/\partial n_1^2)_{T,P,n_2,n_3}$ and, therefore, it is the condition expressed by Equation (5.147) or (5.150) that determines the boundary between stable and unstable homogenous systems. By the same arguments used previously, the conditions of stability of the critical phases are

$$\left(\frac{\partial^2 \phi}{\partial n_2^2}\right)_{T,P,\mu_1,n_3} = 0 \tag{5.173a}$$

$$\left(\frac{\partial^3 \phi}{\partial n_2^3}\right)_{T,P,\mu_1,n_3} = 0 \tag{5.173b}$$

and

$$\left(\frac{\partial^4 \phi}{\partial n_2^4}\right)_{T,P,\mu_1,n_3} \geqslant 0 \tag{5.173c}$$

or the equivalent conditions

$$\left(\frac{\partial \mu_2}{\partial n_2}\right)_{T,P,\mu_1,n_3} = 0 \tag{5.174a}$$

$$\left(\frac{\partial^2 \mu_2}{\partial n_2^2}\right)_{T,P,\mu_1,n_3} = 0 \tag{5.174b}$$

and

$$\left(\frac{\partial^3 \mu_2}{\partial n_2^3}\right)_{T,P,\mu_1,n_3} \geqslant 0 \tag{5.174c}$$

5.17 Graphical representation

Many of the concepts developed in this chapter can be illustrated by graphical representations of the thermodynamic functions, and further insight into these concepts may be obtained from such graphs. We consider here: (a) the surface in three-dimensional space representing the energy function of a one-component system as a function of the entropy and volume; (b) the molar Gibbs energy of a one-component system as a function of the pressure and temperature; and (c) the molar Gibbs energy of a binary system as a function of the temperature and mole fraction at constant pressure.

The energy–entropy–volume surface

For the energy surface we consider a closed system having only one component. The energy is then a function of the entropy and volume alone, and the differential of the energy is given by

$$\mathrm{d}E = T\,\mathrm{d}S - P\,\mathrm{d}V$$

We then consider a set of rectilinear coordinates representing the energy, entropy, and volume; we take the vertical axis to represent the energy. Experience has shown that the thermodynamic functions are single-valued.

Thus, for each set of values assigned to the entropy and volume, there is one value for the energy. Such points, when plotted, map out a surface in the three-dimensional space. Our problem is to investigate the properties of this surface.

First consider a single-phase system, for which there is a single surface. The properties of this surface are obtained by determining the shape of the curve formed by the intersection of the surface and planes drawn perpendicular to the volume axis and to the entropy axis. We consider first the energy–entropy curve at constant volume. We know that both $(\partial E/\partial S)_V$ and $(\partial^2 E/\partial S^2)_V$ must be positive, so the slope of the tangent at any one point on the curve must be positive, and the slopes must become more positive as the entropy is increased. These two conditions require the curve to be concave upward as shown by the heavy curve in Figure 5.3. The lighter line illustrates the tangent. When we consider the energy–volume curve at constant entropy, we are concerned with the sign of $(\partial E/\partial V)_S$ and $(\partial^2 E/\partial V^2)_S$. The first derivative must be negative because the pressure is positive, and the second derivative must be positive from the condition of stability. These two conditions show that the curve must also be concave upward, but that the slope of a tangent to the curve must be negative. The curve is illustrated by the heavier line and the tangent by the lighter line in Figure 5.4.

The entire surface must then be concave upward. A plane, tangent to the surface at any point must lie wholly below the surface. The slope of the line obtained by the intersection of the tangent plane and a plane drawn perpendicular to the volume axis represents the temperature. The slope of

Figure 5.3. The energy as a function of the entropy at constant volume.

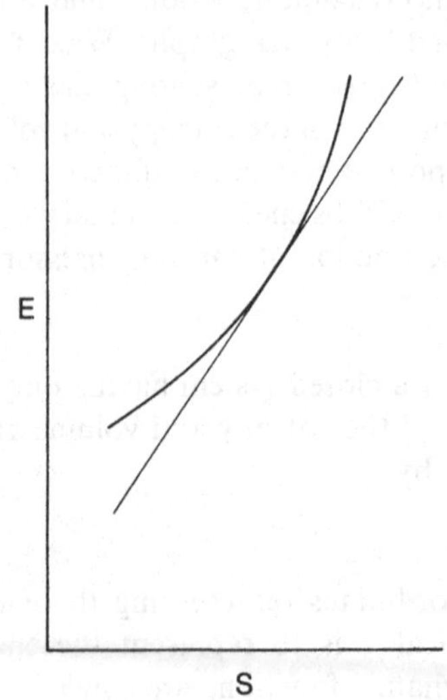

the line obtained by the intersection of the tangent plane and a plane drawn perpendicular to the entropy axis represents the negative of the pressure. Thus, a point on the energy surface is fixed by assigning values to the entropy and volume of the system. The temperature and the negative of the pressure of the system for these values of the entropy and volume are determined from the principal slopes of the plane drawn tangent to the surface at the given point. Such a surface must exist for each phase of the one-component system. Gibbs called the entire surface represented by these surfaces the *primitive surface*.

Next consider the triple point of the single-component system at which the solid, liquid, and vapor phases are at equilibrium. The description of the surfaces and tangent planes at this point are applicable to any triple point of the system. At the triple point we have three surfaces, one for each phase. For each surface there is a plane tangent to the surface at the point where the entire system exists in that phase but at the temperature and pressure of the triple point. There would thus seem to be three tangent planes. The principal slopes of these planes are identical, because the temperatures of the three phases and the pressures of the three phases must be the same at equilibrium. The three planes are then parallel. The last condition of equilibrium requires that the chemical potential of the component must be the same in all three phases. At each point of tangency all of the component must be in that phase. Consequently, the condition

$$n\mu(\mathrm{g}) = n\mu(\ell) = n\mu(\mathrm{s}) \tag{5.175}$$

Figure 5.4. The energy as a function of the volume at constant entropy.

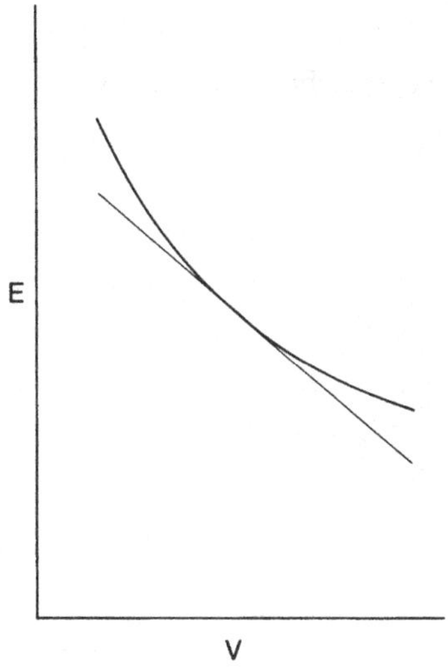

where n represents the number of moles of the component in the system, must be satisfied. According to Equation (5.58),

$$n\mu = E - TS + PV \tag{5.176}$$

and, consequently,

$$E(\mathrm{g}) - TS(\mathrm{g}) + PV(\mathrm{g}) = E(\ell) - TS(\ell) + PV(\ell) = E(\mathrm{s}) - TS(\mathrm{s}) + PV(\mathrm{s}) \tag{5.177}$$

Equation (5.176) is the equation of a plane in three-dimensional space and Equation (5.177) shows that the three equations are identical. Thus, the three possible planes are actually the same plane, which is tangent to all three surfaces at the triple point. The principal slopes of this single plane give the temperature and pressure of the system at the triple point. It is helpful in further discussion to consider the projection of this plane on the entropy–volume plane, as illustrated in Figure 5.5. The three points of tangency are designated s, ℓ, and v for the solid, liquid, and vapor states, respectively. It is assumed that the molar volume of the liquid is greater than the molar volume of the solid. The molar entropy of the liquid is greater than the molar entropy of the solid, and the molar entropy of the vapor is greater than the molar entropies of both the liquid and the solid. Consider a system whose state is determined by designating its entropy and volume so that the point lies within the triangular area *slv*. The energy of this system is given by the value of the energy on the tangent plane at this point. As the entropy and volume of the system are changed within the triangular area, so the energy of the system changes along the tangent plane. Thus, the surface of the tangent plane within the three points of tangency represents the equilibrium

Figure 5.5. The projection of the equilibrium energy surface on the entropy–volume plane.

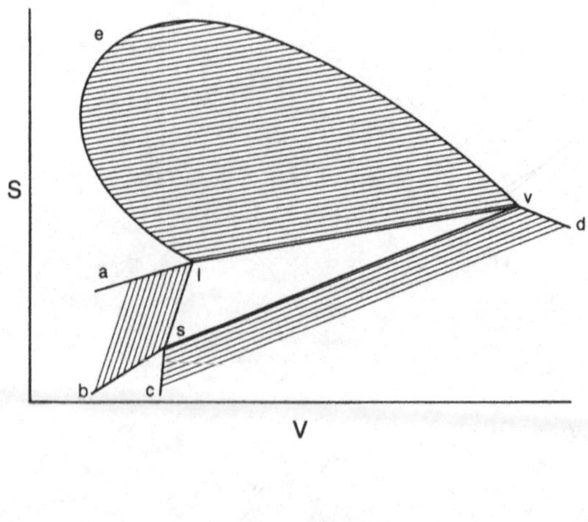

states of the system at its triple point. Gibbs called such a portion of the equilibrium surface a *derived surface*.

This graphical representation of the thermodynamic properties of a single-component system at a triple point gives an excellent illustration of the concept of an indifferent state of a system. The value of the energy of a system at the given triple point must lie on the surface of the tangent plane and within the triangular area determined by the three points of tangency. However, a knowledge of only the temperature and pressure at the triple point and the total number of moles of the component contained in the system is not sufficient to determine its state. The values of the energy, entropy, and volume of the system and of the number of moles of the component in each phase of the system are unknown. It is evident that an infinite number of states of the system lie in the triangular area. Only by assigning values to the entropy and volume of the system do we determine the point on the tangent plane that represents the energy of the system.

The conditions of stability may also be discussed in terms of this graphical representation. Even though the assigned values of the entropy and volume are such that the point lies within the triangular area *slv*, the value of the energy might be such that the point lies on one of the primary surfaces instead of the tangent plane. The system would then exist in a single phase. According to the conditions of stability, such a system would be stable if the primary surface were concave upward. However, the primary surface lies wholly above the tangent plane. Thus, the energy of the system on the primary surface is greater than the energy of the three-phase system that lies on the tangent plane. The condition of equilibrium requires the energy to have its smallest possible value for a given entropy and volume. Thus, the single-phase system, although stable with respect to the conditions of stability, is metastable with respect to the three-phase system.

Of the three possible two-phase equilibria of a single-component system, we discuss only the solid–liquid equilibrium and the liquid–vapor equilibrium; the principles of the solid–vapor equilibrium are identical in all respects to the solid–liquid equilibrium. For the liquid–vapor equilibrium we start with a plane tangent to the three primary surfaces at the triple point. We then move the plane so that it ceases to be tangent to the primary surface representing the solid phase, but remains tangent to the two surfaces representing the liquid and gas phases. That such a surface, tangent to two primary surfaces, is possible may be proved in the same manner that the three possible planes tangent to the three possible planes tangent to the three primary surfaces at the triple point were proved to be identical. At any one state of the vapor–liquid equilibrium there is a plane tangent to each surface, but the conditions that the temperature and pressure of the phases and the chemical potential of the component in each phase must be the same requires that the two planes be identical. As the plane is rolled along the two surfaces, the points of tangency approach each other and become identical at the critical point. The loci of the points of tangency are then a continuous curve, the projection of which on the entropy–volume plane is represented by the curve *lev* in Figure 5.5. A ruled surface can be generated if a line is drawn between the two points of tangency at each position of the plane. The projection of this ruled surface on the entropy–volume plane is represented by the shaded area *lev* in Figure 5.5. This ruled surface must lie wholly below the primary surfaces, as does the tangent plane. Therefore, this surface represents the equilibrium surface for all possible states of the vapor–liquid equilibria in the one-component system. Again, this surface is called a *derived surface*.

The state of the system in the two-phase vapor–liquid equilibrium may be determined in two ways. If values are assigned to the entropy and volume of the system such that the point lies in the area *lev*, the energy of the system is represented by the corresponding point on the derived surface. Only one plane can pass through this point and be tangent to the two primary surfaces. Thus, this plane determines the points of tangency and the point representing the energy of the system must lie on the line connecting the two points of tangency. The principal slopes of this one tangent plane are equal to the temperature and the negative of the pressure of the system. With knowledge of the equation of state for the two phases, the values of all of the thermodynamic functions and the number of moles of the component in each phase can be determined. In the second method we might choose either the temperature or the pressure of the system. This choice would fix one of the principal slopes of the plane, and there can be only one plane with this slope that is tangent to the two surfaces. Consequently, the other principal slope is determined. However, the energy of the system is represented by any point lying on the line connecting the two points of tangency. The assignment of a value to either the entropy or the volume of the system determines a point on this line. Thus, the state of the system is completely defined. Such a two-phase system is again an example of an *indifferent* system. In this case values must be assigned to at least one extensive variable in addition to the mole number in order to define the state of the system completely.

The conditions of stability for the two-phase system can be discussed by reference to the intersection of the energy surface with a plane passing through the line connecting two points of tangency and perpendicular to the entropy–volume plane. Such an intersection is illustrated in Figure 5.6. The intersection of this plane with the plane tangent to the two surfaces is represented by the line tangent to the curve at the points *a* and *b*. Experimental studies do not determine whether the two primary surfaces intersect as shown by the dotted lines, or whether the surface itself as well as its first derivatives are continuous as shown by the solid line. Continuous first derivatives are usually obtained from theoretical studies. If these derivatives are continuous, then a portion of the energy surface must be concave downward. Two points of inflection, represented by *c* and *d* in Figure 5.6, separate that part of the curve which is concave upward from that which is concave downward. Then, according to the conditions of stability, any system whose value of the energy lies on that portion of a primary surface lying between the points *a* and *b* is concave upward is metastable with respect to the two-phase system whose energy lies on the tangent line. However, the conditions of stability are not obeyed along the portion of the primary surface between *c* and *d*, and hence this portion represents states that are inherently unstable. Thus, if the energy surface and its first derivatives are continuous, points *c* and *d* represent the limits of stable states of the one-phase system.

Throughout this discussion we have considered the gas or vapor phase and the liquid phase to be distinct phases, and that for each phase there is a primary energy surface. Because of the existence of the critical point, the entire surface whose projection on the entropy–volume plane is represented by the area above the lines *alevd* in Figure 5.5 is actually one surface. Any plane can be tangent to only one point at a time on this entire surface. The two phases can be distinguished only at the triple point or in the two-phase region. It therefore seems more correct to speak of a *single* fluid phase when one phase is present.

The characteristics of the primary surfaces and the derived surface for the solid–liquid equilibria and the solid–vapor equilibria are identical to those for the liquid–vapor equilibria with the exception that no critical phenomena have ever been observed in these two equilibria. Thus, a plane, tangent to the solid and liquid surfaces, may be rolled along these surfaces. The projections of the loci of the points of tangency are represented by the lines *al* and *bs* in Figure 5.5. If the points of tangency are connected by a straight line lying in the plane at any position of the plane, a ruled surface is derived

Figure 5.6. Energy relations for the vapor–liquid equilibrium.

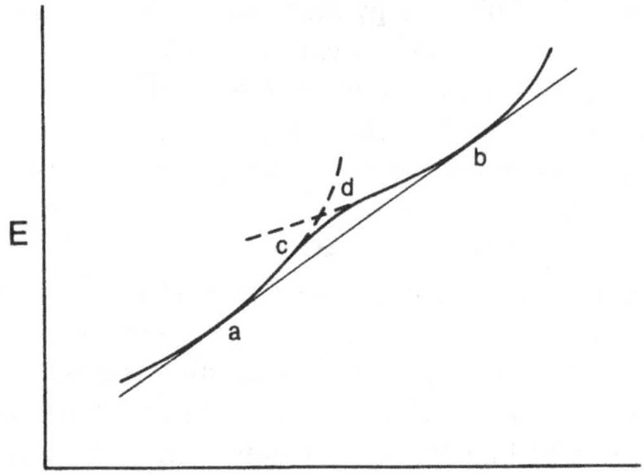

as the plane rolls along the surfaces. This ruled surface then represents all the equilibrium states of the two-phase system. The primary surfaces between the points of tangency lie wholly above this ruled surface. The conditions of stability are identical, although there is essentially no evidence to show whether the primary surface is continuous or discontinuous in its first derivatives. If it is continuous, there must be a region where the primary surface is concave downward, and hence a region where the system is essentially unstable. Similarly, the projections of the points of tangency of the rolling tangent plane for the solid–vapor equilibrium are represented by the lines *vd* and *sc* in Figure 5.5. The ruled surface formed by connecting the points of tangency represents all of the equilibrium states of the solid–vapor system. The primary surfaces lie wholly above the derived, ruled surface, and represent the metastable and unstable states of the system, as the case may be.

In summary, we refer to Figure 5.5, which may be considered as the projection of the entire equilibrium surface on the entropy–volume plane. All of the equilibrium states of the system when it exists in the single-phase fluid state lie in the area above the curves *alevd*. All of the equilibrium states of the system when it exists in the single-phase solid state lie in the area bounded by the lines *bs* and *sc*. These areas are the projections of the primary surfaces. The two-phase systems are represented by the shaded areas *alsb*, *lev*, and *csvd*. These areas are the projections of the derived surfaces for these states. Finally, the triangular area *slv* represents the projection of the tangent plane at the triple point, and represents all possible states of the system at the triple point. This area also is a projection of a derived surface.

The Gibbs energy–temperature–pressure surface

The Gibbs energy for a closed system may be represented by

$$dG = -S\,dT + V\,dP \tag{5.178}$$

In this discussion we consider only a single-component system. It is evident from Equation (5.178) that the Gibbs energy may also be represented as a surface in a three-dimensional coordinate system, in which the rectangular coordinates represent the Gibbs energy, the temperature, and the pressure. The characteristics of the surface for a single-phase system are determined as before from the first and second derivatives of the Gibbs energy with respect to the temperature at constant pressure, and the first and second derivatives with respect to the pressure at constant temperature. Thus, the slope of the curve formed by the intersection of the surface with a plane drawn perpendicular to the pressure axis must be negative, because the entropy is always positive. Furthermore, this slope must decrease with increasing temperature, because both the heat capacity and the temperature are always positive. The curve in Figure 5.7 represents such an intersection, and the straight line represents the tangent to the curve at a given point. The slope of the curve formed by the intersection of the surface with a plane drawn perpendicular to the temperature axis must be positive, because the volume is always positive and the slope must decrease with increasing pressure because $(\partial V/\partial P)_{T,n}$ is always negative. The curve in Figure 5.8 represents such an intersection and the straight line represents the tangent to the curve at a given point. It is evident from these considerations that the surface must

Figure 5.7. The Gibbs energy as a function of the temperature at constant pressure.

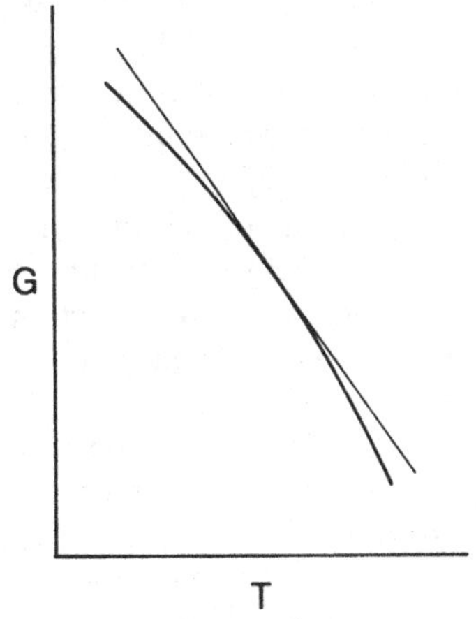

Figure 5.8. The Gibbs energy as a function of the pressure at constant temperature.

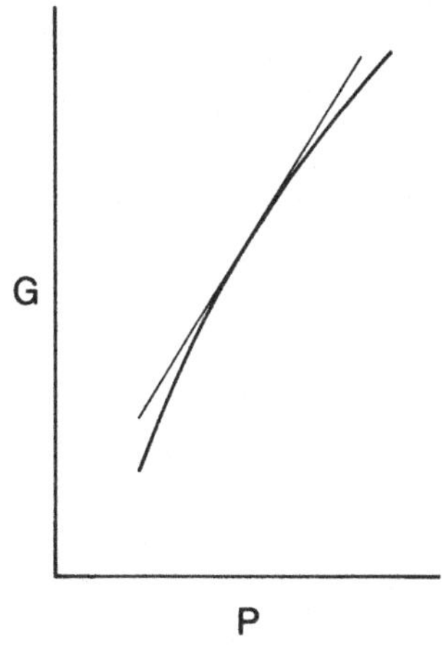

be everywhere concave downward. If the points of tangency in Figures 5.7 and 5.8 are the same point, then the tangent lines represent the principal slopes of a plane tangent to the surface at this point. Thus, the surface in the neighborhood of a given point lies wholly below the plane tangent to the surface at the point. The point representing the Gibbs energy of the single-phase system at a given temperature and pressure must therefore lie on the surface for this phase. The principal slopes of the plane tangent to the surface at the point represents the volume and the negative of the entropy of the system at this point.

Two such surfaces must be considered when two phases are in equilibrium. From the conditions of equilibrium we know that the temperature of the two phases must be the same and that the pressure of the two phases must be the same. Moreover, from the phase rule, a choice of the temperature determines the pressure. We then choose a temperature at which the two-phase system is at equilibrium. Such a choice determines a point on the pressure–temperature plane, and the Gibbs energy of the system must lie on the line drawn perpendicular to this plane at the given point. The third condition of equilibrium requires that the chemical potential of the component in each phase must be the same. We point out in Section 5.13 that this condition requires the Gibbs energy for a one-component, two-phase system to be constant at a given temperature and pressure, and independent of the distribution of the component between the two phases. Consequently, the point representing the Gibbs energy of the system must lie on both surfaces at the given temperature and pressure, and the two surfaces must intersect at the equilibrium point. Figure 5.9 represents the curves formed by the intersection of the two surfaces and a plane perpendicular to the

Figure 5.9. Representation of the Gibbs energy at constant pressure for a two-phase equilibrium.

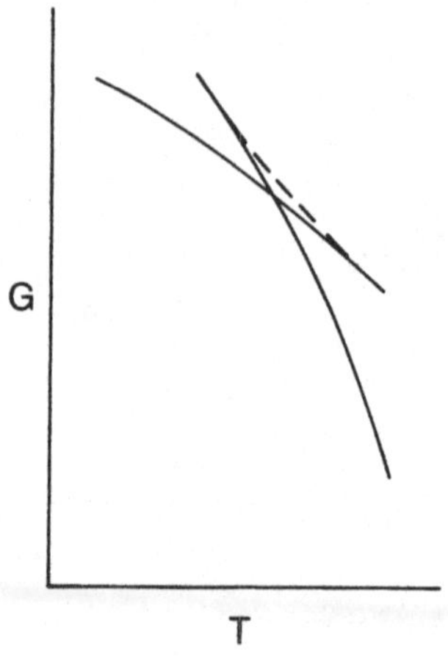

pressure axis through an equilibrium point, and Figure 5.10 represents the curves formed by the intersection of the surfaces with a plane drawn perpendicular to the temperature axis. The point of intersection on each figure represents the point at which the two phases are at equilibrium at the given temperature and pressure.

The same problem of discussing the gas phase and liquid phase as two separate phases or as different manifestations of the same phase arises with the Gibbs energy surface as it did for the energy surface. The intersection of the surfaces for the gas and liquid phases, as we have discussed them here, stops at the critical point. At higher temperatures there is no intersection of two surfaces, and indeed there is only one surface with no discontinuities. We could refer to a single fluid phase except along the line where two fluid phases—the gas and liquid phases—are in equilibrium. The Gibbs energy surface for the fluid phase would then be a continuous single surface with continuous first derivatives except along the vapor pressure curve. Along this curve the surface would still be continuous, but the first derivatives would be discontinuous up to the critical temperature.

In the three-dimensional diagram, the curve formed by the intersection of the two surfaces represents all of the equilibrium points of the two-phase system. Such a curve is obtained for each type of a two-phase equilibrium existing in a single-component system. At any triple point of the system three such curves meet at a point, giving the temperature and pressure of the triple point.

The projection of the three-dimensional surface on the pressure–temperature plane gives the familiar pressure–temperature diagram of a one component system. The projection for only the solid, liquid, and vapor phases

Figure 5.10. Representation of the Gibbs energy at constant temperature for a two-phase equilibrium.

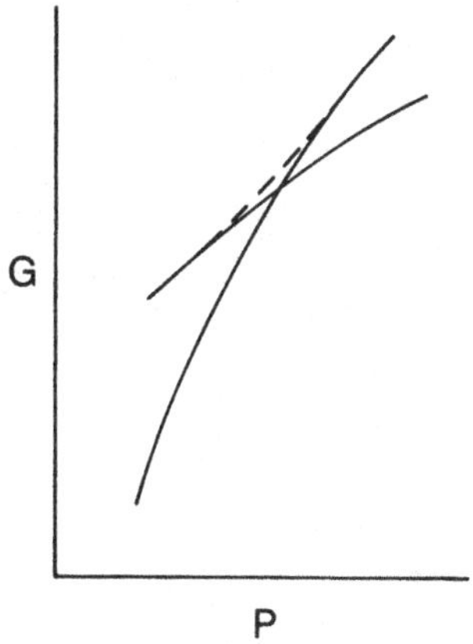

is illustrated in Figure 5.11. In this diagram we have assumed that the slope of the melting-point line is positive. Each line on this diagram is the projection of the intersection of two of the three surfaces. The areas between the lines are the projections of the surfaces for each phase. The triple point is the projection of the intersection of the three lines as well as the intersection of the three surfaces.

The stability of a system may be discussed in terms of Figures 5.9 and 5.10. According to Equation (5.53) the Gibbs energy of a system at equilibrium must have the smallest possible value at a given temperature and pressure. Therefore, the lower curves on either side of the point of intersection represent the stable states of a one-phase system, and the upper curves represent metastable states. For any essentially unstable state the surface would have to be concave upward. The cut of this surface and the plane perpendicular to the appropriate axis is shown by the broken line in Figures 5.9 and 5.10. These curves are obtained when it is assumed that the equation of state and its derivatives are continuous as shown in Figure 5.1.

The Gibbs energy–mole fraction diagrams

When we introduce composition variables, we have more than three variables and consequently must apply certain restrictions in order to illustrate the behavior of the Gibbs energy in two- or three-dimensional graphs. Here we consider only binary systems at constant pressure. The three variables that we use are the molar Gibbs energy, the temperature, and the mole fraction of one of the components. We also limit the discussion to systems having a single phase or two phases in equilibrium.

First consider a system at constant temperature. According to Equation

Figure 5.11. Projection of the Gibbs energy surface on the pressure–temperature plane.

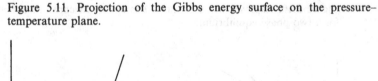

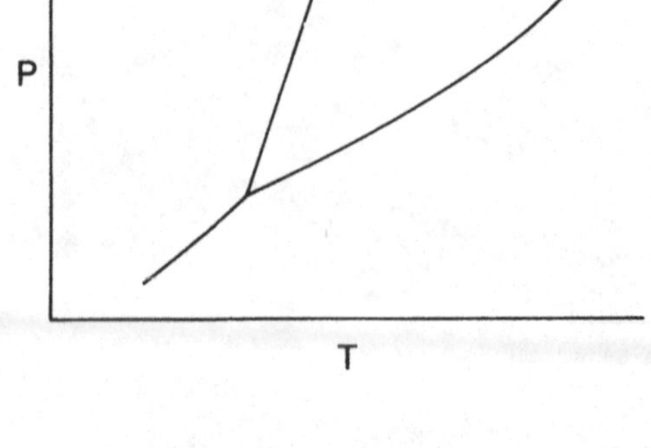

(5.14) $(\partial^2\tilde{G}/x_1^2)_{T,P}$ must be positive, and consequently the curve of the molar Gibbs energy as a function of the mole fraction for a single phase must be concave upward as illustrated by the solid curve in Figure 5.12. There may be another phase, metastable with respect to the stable phase, that would have a higher Gibbs energy at this temperature. The molar Gibbs energy of such a phase is illustrated by the dotted curve in the same figure. At another temperature these two curves may intersect or combine as illustrated in Figure 5.13. In this case a line may be drawn tangent to the two curves at the points

Figure 5.12. Representation of phase equilibria of a binary system in terms of the Gibbs energy at constant temperature and pressure.

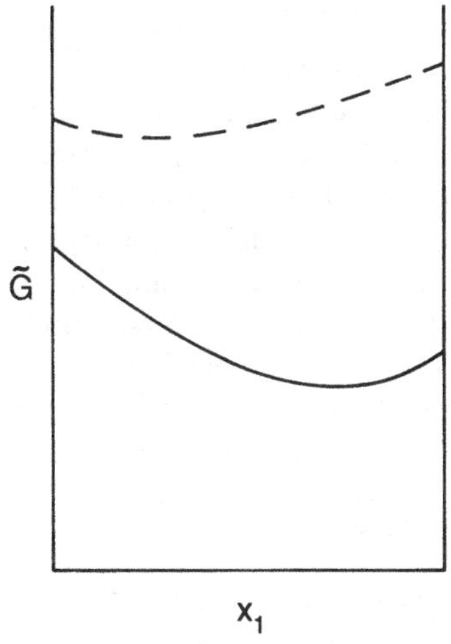

Figure 5.13. Representation of phase equilibria of a binary system in terms of the Gibbs energy at constant temperature and pressure: two-phase equilibrium.

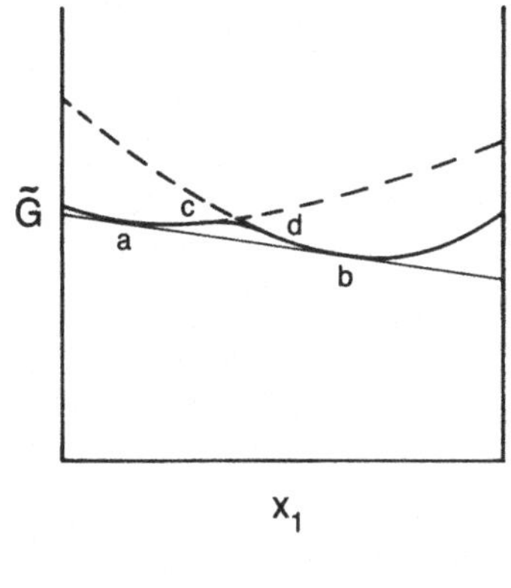

a and *b*. This line lies wholly below the two curves. However, for equilibrium the Gibbs energy must have the smallest possible value consistent with the given temperature, pressure, and mole fraction. It is evident from the figure that between the points *a* and *b* the Gibbs energy of the system on the tangent line is less than that on either of the curves. In this case the system exists in two phases, one phase having the composition represented by the point *a* and the other phase having the composition represented by the point *b*. The shape of the curve between the two points of tangency is uncertain. If the two curves intersect as indicated by the dotted extensions of the two curves, then a point on either of the two curves represents a system that is metastable with respect to the two-phase system. On the other hand, if the first derivatives of the curves are continuous, then a portion of the curve must be concave downward. This portion of the curve lies between the two points of inflection *c* and *d*. A point on this portion of the curve represents a system that is unstable. If we again change the temperature, the two curves may no longer intersect or combine and the phase that was stable at the original temperature may now be metastable with respect to the second phase, as shown in Figure 5.14. We point out that these three figures are actually cuts at different temperatures of the Gibbs energy surface drawn as a function of the temperature and mole fraction. The projection on the temperature–mole fraction plane of the loci of the points of tangency obtained at different temperatures gives the usual temperature–mole fraction phase diagram for the two phases.

Figure 5.14. Representation of phase equilibria of a binary system in terms of the Gibbs energy at constant pressure and temperature.

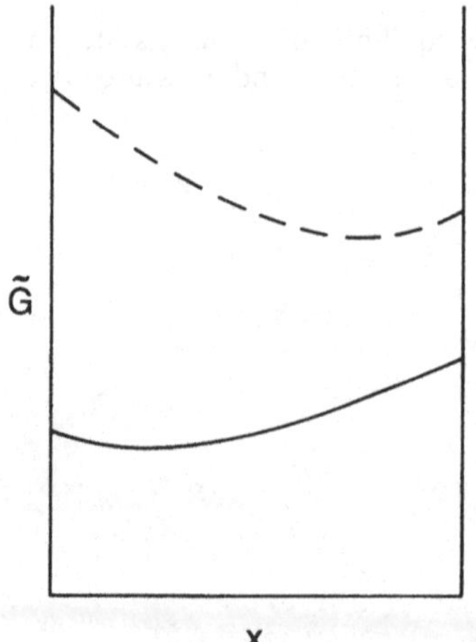

6

Partial molar quantities

The subject of partial molar quantities needs to be developed and understood before considering the application of thermodynamics to actual systems. Partial molar quantities apply to any extensive property of a single-phase system such as the volume or the Gibbs energy. These properties are important in the study of the dependence of the extensive property on the composition of the phase at constant temperature and pressure; e.g., what effect does changing the composition have on the Helmholtz energy? In this chapter partial molar quantities are defined, the mathematical relations that exist between them are derived, and their experimental determination is discussed.

6.1 Definition of partial molar quantities

We let X represent any extensive property of a single phase, and consider it to be a function of the temperature, pressure, and the mole numbers of the components, so that $X = X(T, P, n)$. The differential of X is thus

$$\mathrm{d}X = \left(\frac{\partial X}{\partial T}\right)_{P,n} \mathrm{d}T + \left(\frac{\partial X}{\partial P}\right)_{T,n} \mathrm{d}P + \sum_{i=1}^{C} \left(\frac{\partial X}{\partial n_i}\right)_{T,P,n} \mathrm{d}n_i \tag{6.1}$$

The quantity $(\partial X/\partial n_i)_{T,P,n}$ is defined as the *partial molar quantity* of the ith component. It occurs so frequently in the thermodynamics of solutions that it is given a separate symbol, $\bar{X}_i$. Consequently, *partial molar quantities are defined by the relation*

$$\bar{X}_i = \left(\frac{\partial X}{\partial n_i}\right)_{T,P,n} \tag{6.2}$$

where n refers to all of the *other* components present. Two important points are to be noted in this definition: (1) the differentiation is *always* made at constant temperature and pressure; and (2) it is an extensive quantity that is differentiated, and not a specific quantity. As an example of this second point we differentiate the volume of a solution that contains n_1, n_2, n_3, etc., moles of components rather than the molar volume or the specific volume. Although these quantities have been defined in terms of components, they can also be defined in terms of the species present in the solution. In such a case the mole numbers of all species rather than components must be held constant except the one that is used in the differentiation; the change of the mole numbers of the reacting species caused by an accompanying shift of equilibrium is not considered.

Any physical interpretation of a partial molar quantity must be consistent with its definition. It is simply the change of the property of the *phase* with a change of the number of moles of one component keeping the mole numbers of all the other components constant, in addition to the temperature and pressure. *It is a property of the phase and not of the particular component.* One physical concept of a partial molar quantities may be obtained by considering an infinite quantity of the phase. Then, the finite change of the property on the addition of 1 mole of the particular component of this infinite quantity of solution at constant temperature and pressure is numerically equal to the partial molar value of the property with respect to the component.

6.2 Relations concerning partial molar quantities

The extensive quantities are functions of the temperature, pressure, and mole numbers, and, consequently, so are the partial molar quantities, so that $\bar{X}_k = \bar{X}_k(T, P, n)$. The differential is then written directly as

$$d\bar{X}_k = \left(\frac{\partial \bar{X}_k}{\partial T}\right)_{P,n} dT + \left(\frac{\partial \bar{X}_k}{\partial P}\right)_{T,n} dP + \sum_{i=1}^{C} \left(\frac{\partial \bar{X}_k}{\partial n_i}\right)_{T,P,n} dn_i \qquad (6.3)$$

where the sum is taken over all of the components. The partial molar quantities are derivatives themselves, and therefore the order of differentiation of the derivatives expressed in Equation (6.3) may be interchanged. Thus,

$$\left(\frac{\partial \bar{X}_k}{\partial T}\right)_{P,n} = \left[\frac{\partial}{\partial n_k}\left(\frac{\partial X}{\partial T}\right)_{P,n}\right]_{T,P,n} \qquad (6.4)$$

$$\left(\frac{\partial \bar{X}_k}{\partial P}\right)_{T,n} = \left[\frac{\partial}{\partial n_k}\left(\frac{\partial X}{\partial P}\right)_{T,n}\right]_{T,P,n} \qquad (6.5)$$

and

$$\left(\frac{\partial \bar{X}_k}{\partial n_i}\right)_{T,P,n} = \left[\frac{\partial}{\partial n_k}\left(\frac{\partial X}{\partial n_i}\right)_{T,P,n}\right]_{T,P,n} = \left(\frac{\partial \bar{X}_i}{\partial n_k}\right)_{T,P,n} \qquad (6.6)$$

Various expressions for the differential of the partial molar quantities can be obtained by the use of these expressions.

The extensive properties are homogenous functions of the first degree in the mole numbers at constant temperature and pressure. Then, the partial molar quantities are homogenous functions of zeroth degree in the mole numbers at constant temperature and pressure; that is, they are functions of the composition. We use the mole fractions here, but we could use the molality, mole ratio, or any other composition variable that is zeroth degree in the mole numbers. For mole fractions, $\bar{X}_k = \bar{X}_k(T, P, x)$ and the differential of $\bar{X}_k$ may be written as

$$\mathrm{d}\bar{X}_k = \left(\frac{\partial \bar{X}_k}{\partial T}\right)_{P,x} \mathrm{d}T + \left(\frac{\partial \bar{X}_k}{\partial P}\right)_{T,x} \mathrm{d}P + \sum_{i=1}^{C-1} \left(\frac{\partial \bar{X}_k}{\partial x_i}\right)_{T,P,x} \mathrm{d}x_i \tag{6.7}$$

The sum in the last term of this equation is taken over all of the components of the system except one. The sum of the mole fractions must be unity, and consequently for C components in the system there are $(C-1)$ independent mole fractions; or for S species, there are $(S-1)$ independent mole fractions in terms of the species. The proof of the identity of Equations (6.3) and (6.7) is not difficult.

One of the most important applications of Equation (6.7) is to the chemical potentials. From their definition in terms of the Gibbs energy we know that they are functions of the temperature, pressure, and the mole fractions. The differential of the chemical potential of the kth component is then given by

$$\mathrm{d}\mu_k = \left(\frac{\partial \mu_k}{\partial T}\right)_{P,x} \mathrm{d}T + \left(\frac{\partial \mu_k}{\partial P}\right)_{T,x} \mathrm{d}P + \sum_{i=1}^{C-1} \left(\frac{\partial \mu_k}{\partial x_i}\right)_{T,P,x} \mathrm{d}x \tag{6.8}$$

By the combination of Equations (4.37), (4.38), (6.4), and (6.5) we find that

$$(\partial \mu_k/\partial T)_{P,n} = -\bar{S}_k \quad \text{and} \quad (\partial \mu_k/\partial P)_{T,n} = \bar{V}_k$$

and therefore

$$\mathrm{d}\mu_k = -\bar{S}_k\,\mathrm{d}T + \bar{V}_k\,\mathrm{d}P + \sum_{i=1}^{C-1} \left(\frac{\partial \mu_k}{\partial x_i}\right)_{T,P,x} \mathrm{d}x_i \tag{6.9}$$

This equation then gives the differential of the chemical potential of a component in terms of the experimentally determined variables: the temperature, pressure, and mole fractions. It is this equation that is used to introduce the mole fraction into the Gibbs–Duhem equation as independent variables, rather than the chemical potentials. The problem of expressing the chemical potentials as functions of the composition variables, and consequently the determination of $(\partial \mu_k/\partial x_i)_{T,P,x}$, is discussed in Chapters 7 and 8.

We have already said that the extensive quantities are homogenous functions of the first degree in the mole numbers, and consequently the partial molar quantities are homogenous functions of zeroth degree in the mole numbers, at both constant temperature and constant pressure. We may then make use of Euler's theorem to obtain additional relations. For the extensive quantities Euler's theorem gives

$$X = \sum_{i=1}^{C} n_i \bar{X}_i \tag{6.10}$$

or

$$\tilde{X} = \sum_{i=1}^{C} x_i \bar{X}_i \tag{6.11}$$

at constant temperature and pressure, where $\tilde{X}$ is the *molar* quantity or the value of the quantity per mole of solution. (For ease of interpretation we use the tilde to represent a molar quantity.) We see from these relations that any extensive quantity is additive in the partial molar quantities of the components. For the partial molar quantities, Euler's theorem gives, at constant temperature and pressure,

$$\sum_{i=1}^{C} n_i \left(\frac{\partial \bar{X}_k}{\partial n_i}\right)_{T,P,n} = \sum_{i=1}^{C} n_i \left(\frac{\partial \bar{X}_i}{\partial n_k}\right)_{T,P,n} = 0 \tag{6.12}$$

This equation is very similar to the Gibbs–Duhem equation under the condition that the temperature and pressure are constant. A more general relation can be obtained by differentiating Equation (6.10) and comparing the result with Equation (6.1). The differentiation of Equation (6.10) gives

$$\mathrm{d}X = \sum_{i=1}^{C} n_i \, \mathrm{d}\bar{X}_i + \sum_{i=1}^{C} \bar{X}_i \, \mathrm{d}n_i \tag{6.13}$$

Comparison with Equation (6.1) yields

$$-\left(\frac{\partial X}{\partial T}\right)_{P,n} \mathrm{d}T - \left(\frac{\partial X}{\partial P}\right)_{T,n} \mathrm{d}P + \sum_{i=1}^{C} n_i \, \mathrm{d}\bar{X}_i = 0 \tag{6.14}$$

The similarity to the Gibbs–Duhem equation is quite apparent, and indeed this equation is the Gibbs–Duhem equation if X refers to the Gibbs energy. We should note that the differential $\mathrm{d}\bar{X}_i$, the differential that appears in Equations (6.13) and (6.14), depends upon the differential quantities of the temperature, the pressure, and the mole fractions as expressed in Equation (6.7). At constant temperature and pressure Equation (6.12) becomes a special case of Equation (6.14).

6.3 Algebraic determination of partial molar quantities

As stated in the introduction to this chapter, the methods used and the equations developed in the previous sections are formally applicable to all extensive properties. However, there are two classes of properties: those whose absolute values are known or determinable and those whose absolute values are not known. The volume and heat capacity are examples of the first class, and the Gibbs energy, the Helmholtz energy, the energy, the enthalpy, and the entropy (excluding the third law of thermodynamics) are examples of the second class. For these latter properties it is necessary to make use of unknown constants that are defined in terms of standard states. These relations are discussed in later chapters. In the remainder of this chapter we use either the volume or the heat capacity as examples.

The partial molar properties are not measured directly *per se*, but are readily derivable from experimental measurements. For example, the volumes or heat capacities of definite quantities of solution of known composition are measured. These data are then expressed in terms of an intensive quantity—such as the specific volume or heat capacity, or the molar volume or heat capacity—as a function of some composition variable. The problem then arises of determining the partial molar quantity from these functions. The intensive quantity must first be converted to an extensive quantity, then the differentiation must be performed. Two general methods are possible: (1) the composition variables may be expressed in terms of the mole numbers before the differentiation and reintroduced after the differentiation; or (2) expressions for the partial molar quantities may be obtained in terms of the derivatives of the intensive quantity with respect to the composition variables. In the remainder of this section several examples are given with emphasis on the second method. Multicomponent systems are used throughout the section in order to obtain general relations.

(A) Let the molar volume be given as a function of the mole fractions, so

$$\tilde{V} = \tilde{V}(x_1, x_2, x_3, \ldots) \tag{6.15}$$

at constant temperature and pressure.

We must multiply the molar volume by the total number of moles in order to obtain the volume of the solution containing a total of $\sum_{i=1} n_i$ moles; thus

$$V = \tilde{V}\sum_i n_i \tag{6.16}$$

The partial molar volume of the kth component is then given by

$$\bar{V}_k = \left(\frac{\partial(\tilde{V}\sum_{i=1}^{C} n_i)}{\partial n_k}\right)_{T,P,n} \tag{6.17}$$

According to the first method the indicated multiplication would be performed, the mole fractions expressed in terms of the mole numbers and the differentiation carried out. According to the second method the indicated differentiation is performed first, to obtain

$$\bar{V}_k = \tilde{V} + \left(\sum_{i=1}^{C} n_i\right)\left(\frac{\partial \tilde{V}}{\partial n_k}\right)_{T,P,n} \tag{6.18}$$

The molar volume is a function of $(C-1)$ independent mole fractions and its differential at constant temperature and pressure is

$$d\tilde{V} = \sum_{i=1}^{C-1} \left(\frac{\partial \tilde{V}}{\partial x_i}\right)_{T,P,x} dx_i \tag{6.19}$$

where the sum is taken over the $(C-1)$ components. Then, the derivatives in Equation (6.18) are given by

$$\left(\frac{\partial \tilde{V}}{\partial n_k}\right)_{T,P,n} = \sum_{i=1}^{C-1} \left(\frac{\partial \tilde{V}}{\partial x_i}\right)_{T,P,x} \left(\frac{\partial x_i}{\partial n_k}\right)_n \tag{6.20}$$

Substitution into Equation (6.18) gives

$$\bar{V}_k = \tilde{V} + \left(\sum_{i=1}^{C} n_i\right)\left[\sum_{i=1}^{C-1} \left(\frac{\partial \tilde{V}}{\partial x_i}\right)_{T,P,x} \left(\frac{\partial x_i}{\partial n_k}\right)_n\right] \tag{6.21}$$

When each $(\partial x_i/\partial n_k)_n$ is evaluated, the expression for $\bar{V}_k$ becomes

$$\bar{V}_k = \tilde{V} - \sum_{i=1}^{C-1} x_i \left(\frac{\partial \tilde{V}}{\partial x_i}\right)_{T,P,x} \tag{6.22}$$

if the mole fraction of the kth component is the dependent mole fraction. When the mole fraction of one of the components other than the kth component is dependent, the expression becomes

$$\bar{V}_k = \tilde{V} + \left(\frac{\partial \tilde{V}}{\partial x_k}\right)_{T,P,x} - \sum_{i=1}^{C-2} x_i \left(\frac{\partial \tilde{V}}{\partial x_i}\right)_{T,P,x} \tag{6.23}$$

(B) Let the specific volume be given as a function of the molality (m) of the solutes, so

$$v = v(m_2, m_3, m_4, \ldots) \tag{6.24}$$

at a constant temperature and pressure. (An italic m will be used for molality.) We use the subscript 1 to designate the solvent. If there are C components in the solution, there are $(C-1)$ independent molalities, the molalities of the solutes. In order to convert the specific volume to the extensive variable we must multiply by the mass of a solution containing n_i moles of the components, so

$$V = v \sum_{i=1}^{C} n_i M_i \tag{6.25}$$

where M_i represents the molecular mass of the ith component. Then the partial molar volume of the kth component is given by

$$\bar{V}_k = \left(\frac{\partial(v\sum_{i=1}^{C} n_i M_i)}{\partial n_k}\right)_{T,P,n} \tag{6.26}$$

If an expression for the partial molar volume is desired in terms of the partial derivatives of the specific volume with respect to the molalities, the indicated differentiation must first be performed to give

$$\bar{V}_k = vM_k + \left(\sum_{i=1}^{C} n_i M_i\right)\left(\frac{\partial v}{\partial n_k}\right)_{T,P,n} \tag{6.27}$$

The evaluation of $(\partial v/\partial n_k)_{T,P,n}$ is carried out as in part A of this section. The differential of the specific volume is given by

$$dv = \sum_{i=2}^{C} \left(\frac{\partial v}{\partial m_i}\right)_{T,P,m} dm_i$$

from which

$$\left(\frac{\partial v}{\partial n_k}\right)_{T,P,n} = \sum_{i=2}^{C} \left(\frac{\partial v}{\partial m_i}\right)_{T,P,m} \left(\frac{\partial m_i}{\partial n_k}\right)_n \tag{6.28}$$

Here the sum is taken over all of the solutes. If the subscript k refers to one of the solutes, then each $(\partial m_i/\partial n_k)_n$ is zero except when $i = k$; in this case

$$\left(\frac{\partial m_k}{\partial n_k}\right)_n = \frac{1000}{n_1 M_1} \tag{6.29}$$

When Equations (6.28) and (6.29) are substituted into Equation (6.27) and the mole ratios are expressed in terms of the molalities, the final relation

$$\bar{V}_k = vM_k + \left(\sum_{i=1}^{C} m_i M_i\right)\left(\frac{\partial v}{\partial m_k}\right)_{T,P,m} \tag{6.30}$$

is obtained, where the sum is taken over all components of the system and $m_1 = 1000/M_1$.

If the subscript k refers to the solvent, then

$$\left(\frac{\partial m_i}{\partial n_1}\right)_n = -\frac{m_i}{n_1} \tag{6.31}$$

When Equations (6.28) and (6.31) are substituted into Equation (6.27) and the mole ratios are again expressed in terms of molalities, the partial molar volume of the solvent is given by

$$\bar{V}_i = vM_1 - \frac{M_1}{1000}\left(\sum_{i=1}^{C} m_i M_i\right)\left[\sum_{i=2}^{C} m_i \left(\frac{\partial v}{\partial m_i}\right)_{T,P,m}\right] \tag{6.32}$$

(C) As a third example, assume that the density, ρ, of the solution is given as a function of the molarity (c) of all the solutes, so that

$$\rho = \rho(c_2, c_3, c_4, \ldots) \tag{6.33}$$

at constant temperature and pressure. Here the units of ρ and c_i are mass per volume of solution and moles per liter of solution, respectively. The volume of the solution is given by

$$V = \frac{\sum_{i=1}^{C} n_i M_i}{\rho}$$

where, again, the sum is taken over all of the components. The partial molar volume of the kth component is then given by

$$\bar{V}_k = \frac{\rho M_k - (\sum_{i=1}^{C} n_i M_i)(\partial \rho / \partial n_k)_{T,P,n}}{\rho^2} \tag{6.34}$$

If we center our attention on the second method, the problem becomes one of expressing the partial derivatives of the density with respect to the mole numbers in terms of the partial derivatives of the density with respect to the molarity. The differential of the density is

$$\mathrm{d}\rho = \sum_{i=2}^{C} \left(\frac{\partial \rho}{\partial c_i}\right)_{T,P,c} \mathrm{d}c_i \tag{6.35}$$

at constant temperature and pressure, from which

$$\left(\frac{\partial \rho}{\partial n_k}\right)_{T,P,n} = \sum_{i=2}^{C} \left(\frac{\partial \rho}{\partial c_i}\right)_{T,P,c} \left(\frac{\partial c_i}{\partial n_k}\right)_{T,P,n} \tag{6.36}$$

As in part B, we must make a distinction between the solutes and the solvent. We first consider that the subscript k refers to one of the solutes. Moreover, since

$$c_i = 1000 n_i \rho \Big/ \sum_{i=1}^{C} n_i M_i \tag{6.37}$$

when ρ is expressed in $\mathrm{g\,cm^{-3}}$ and c_i in $\mathrm{mol\,dm^{-3}}$ or $\mathrm{mol\,L^{-1}}$, we must make a distinction between c_i and c_k in order to determine the sum expressed in Equation (6.37). Thus, if $i \neq k$, then

$$\left(\frac{\partial c_i}{\partial n_k}\right)_{T,P,n} = \frac{c_i}{\rho}\left(\frac{\partial \rho}{\partial n_k}\right)_{T,P,n} - \frac{c_i M_i}{\sum_{i=1}^{C} n_i M_i} \tag{6.38}$$

and if $i = k$, then

$$\left(\frac{\partial c_k}{\partial n_k}\right)_{T,P,n} = \frac{c_k}{n_k} + \frac{c_k}{\rho}\left(\frac{\partial \rho}{\partial n_k}\right)_{T,P,n} - \frac{c_k M_k}{\sum_{i=1}^{C} n_i M_i} \tag{6.39}$$

When Equations (6.38) and (6.39) are substituted into Equation (6.36), we obtain

$$\left(\frac{\partial\rho}{\partial n_k}\right)_{T,P,n} = \frac{c_k}{n_k}\left(\frac{\partial\rho}{\partial c_k}\right)_{T,P,c} + \frac{1}{\rho}\left(\frac{\partial\rho}{\partial n_k}\right)_{T,P,n}\sum_{i=2}^{C} c_i\left(\frac{\partial\rho}{\partial c_i}\right)_{T,P,c} - \frac{M_k}{\sum_{i=1}^{C} n_i M_i}\sum_{i=2}^{C} c_i\left(\frac{\partial\rho}{\partial c_i}\right)_{T,P,c} \tag{6.40}$$

from which

$$\left(\frac{\partial\rho}{\partial n_k}\right)_{T,P,n} = \frac{\dfrac{c_k}{n_k}\left(\dfrac{\partial\rho}{\partial c_k}\right)_{T,P,c} - \dfrac{M_k}{\sum_{i=1}^{C} n_i M_i}\sum_{i=2}^{C} c_i\left(\dfrac{\partial\rho}{\partial c_i}\right)_{T,P,c}}{1 - \dfrac{1}{\rho}\sum_{i=2}^{C} c_i\left(\dfrac{\partial\rho}{\partial c_i}\right)_{T,P,c}} \tag{6.41}$$

When this equation is substituted into Equation (6.34), we have

$$\bar{V}_k = \frac{M_k}{\rho} - \frac{1000\rho(\partial\rho/\partial c_k)_{T,P,c} - M_k\sum_{i=2}^{C} c_i(\partial\rho/\partial c_i)_{T,P,c}}{\rho^2 - \rho\sum_{i=2}^{C} c_i(\partial\rho/\partial c_i)_{T,P,c}} \tag{6.42}$$

which, on rearrangement, becomes

$$\bar{V}_k = \frac{M_k - 1000(\partial\rho/\partial c_k)_{T,P,c}}{\rho - \sum_{i=2}^{C} c_i(\partial\rho/\partial c_i)_{T,P,c}} \tag{6.43}$$

If the subscript k refers to the solvent, then

$$\left(\frac{\partial c_i}{\partial n_1}\right)_{T,P,n} = \frac{c_i}{p}\left(\frac{\partial\rho}{\partial n_1}\right)_{T,P,n} - \frac{c_i M_1}{\sum_{i=1}^{C} n_i M_i} \tag{6.44}$$

When Equation (6.44) is substituted into Equation (6.36), we have

$$\left(\frac{\partial\rho}{\partial n_1}\right)_{T,P,n} = \frac{1}{\rho}\left(\frac{\partial\rho}{\partial n_i}\right)_{T,P,n}\sum_{i=2}^{C} c_i\left(\frac{\partial\rho}{\partial c_i}\right)_{T,P,c} - \frac{M_1}{\sum_{i=1}^{C} n_i M_i}\sum_{i=2}^{C} c_i\left(\frac{\partial\rho}{\partial c_i}\right)_{T,P,c} \tag{6.45}$$

which on rearrangement becomes

$$\left(\frac{\partial\rho}{\partial n_1}\right)_{T,P,n} = -\frac{\dfrac{M_1}{\sum_{i=1}^{C} n_i M_i}\sum_{i=2}^{C} c_i\left(\dfrac{\partial\rho}{\partial c_i}\right)_{T,P,c}}{1 - \dfrac{1}{\rho}\sum_{i=2}^{C} c_i\left(\dfrac{\partial\rho}{\partial c_i}\right)_{T,P,c}} \tag{6.46}$$

Finally, when Equation (6.46) is substituted into Equation (6.34), we obtain

$$\bar{V}_i = \frac{M_1}{\rho} + \frac{M_1 \sum_{i=2}^{C} c_i(\partial\rho/\partial c_i)_{T,P,c}}{\rho^2[1 - (1/\rho)\sum_{i=2}^{C} c_i(\partial\rho/\partial c_i)_{T,P,c}]} \tag{6.47}$$

which, on rearrangement, becomes

$$\bar{V}_1 = \frac{M_1}{\rho - \sum_{i=2}^{C} c_i(\partial\rho/\partial c_i)_{T,P,c}} \tag{6.48}$$

(D) Experimental results are frequently reported in terms of deviation functions. The usefulness of these functions arises from the fact many properties of various systems obey approximate laws. Thus, we speak of the deviations from ideal gas behavior or deviations from the ideal solution laws. The advantage of such deviation functions is that their values are usually much smaller than the whole value, and consequently greater accuracy can be obtained with simpler calculations, either graphically or algebraically. As an example, the molar volume of a mixture of liquids is approximately additive in the mole fractions, so that we may write

$$\tilde{V} = \sum_{i=1}^{C} x_i \tilde{V}_i^{\bullet} + \Delta\tilde{V}^{\mathrm{M}} \tag{6.49}$$

where $\tilde{V}_i^{\bullet}$ is the molar volume of the pure ith component and $\Delta\tilde{V}^{\mathrm{M}}$ is the change of the molar volume of the system on mixing the liquid components at constant temperature and pressure. (The superscript dot is used to denote a pure component.) In order to obtain the extensive variable we must multiply by the total number of moles, so

$$V = \sum_{i=1}^{C} n_i \tilde{V}_i^{\bullet} + \Delta\tilde{V}^{\mathrm{M}}\left(\sum_{i=1}^{C} n_i\right) \tag{6.50}$$

From this equation we then obtain

$$\bar{V}_k = \tilde{V}_k^{\bullet} + \left(\frac{\partial[\Delta\tilde{V}^{\mathrm{M}}(\sum_{i=1}^{C} n_i)]}{\partial n_k}\right)_{T,P,n} \tag{6.51}$$

for the partial molar volume of the kth component. Thus, the partial molar volume of this component is given by the sum of its molar volume and the value of the derivative expressed in Equation (6.51). The evaluation of this derivative requires the same methods that have been discussed in this section.

We have limited the discussion here to three different quantities each with a different composition variable. Of course, there are other intensive quantities, such as the volume of a solution containing a fixed quantity of solvent, that could be used to express the same data. Moreover, each intensive quantity can be expressed as a function of any of the composition variables. Thus, there are many different relations capable of expressing the same data, but for any specific case the same general methods given in this section are used to calculate the partial molar quantities

6.4 Apparent molar quantities

Apparent molar quantities (sometimes called apparent molal quantities) are used primarily for binary systems, and are usually expressed as functions of the molality. These quantities are used because in many cases their values are readily obtained by experiment. The general apparent molar quantity, ϕX, is defined as

$$\phi X = \frac{X - n_1 \tilde{X}_1^{\bullet}}{n_2} \tag{6.52}$$

for a binary system. (Note that ϕX is used as a *single* symbol and is *not* a product. When seen together ϕX is the symbol for the apparent molar value of the property X.) It is thus the difference in the value of the quantity for the solution and that for the *pure* solvent per mole of solute. For the volume of a binary system we have

$$\phi V = \frac{V - n_1 \tilde{V}_1^{\bullet}}{n_2} \tag{6.53}$$

and, from this definition,

$$V = n_1 \tilde{V}_1^{\bullet} + n_2\, \phi V \tag{6.54}$$

The partial molar volumes of the two components can then be obtained by differentiation, and are given by

$$\bar{V}_1 = \tilde{V}_1^{\bullet} + n_2 \left(\frac{\partial\, \phi V}{\partial n_1}\right)_{T,P,n} \tag{6.55}$$

and

$$\bar{V}_2 = \phi V + n_2 \left(\frac{\partial\, \phi V}{\partial n_2}\right)_{T,P,n} \tag{6.56}$$

The apparent molar volume is given as a function of one of the composition variables. The methods of evaluating the partial derivatives in Equations (6.55) and (6.56) are identical to those discussed in Section 6.3. The final expressions for the partial molar volumes of the two components in terms of the molality are

$$\bar{V}_1 = \tilde{V}_1^{\bullet} - \frac{M_1 m^2}{1000} \left(\frac{\partial\, \phi V}{\partial m}\right)_{T,P} \tag{6.57}$$

and

$$\bar{V}_2 = \phi V + m\left(\frac{\partial \phi V}{\partial m}\right)_{T,P} \tag{6.58}$$

Figure 6.1 illustrates the concept of the apparent molar volume. *AB* is the portion of the total volume *AC* for n_2 moles of solute that is attributed to the pure solvent. Then, the volume *BC* is *apparently* due to the solute. The slope of the line passing through point *C* and $\tilde{V}_1^\bullet$ is the apparent molar volume. The slope of the curve of the total volume at point *C* is the partial molar volume of component 2. Indeed, the slope of the total volume curve at any point is the partial molar volume of component 2 at that concentration. It is obvious that partial molar properties and apparent molar properties are both functions of concentration.

Apparent molar quantities for multicomponent systems, other than binary systems, have been used to only a limited extent because of the difficulty of dealing with more than two mole numbers. Actually, there is no single definition of an apparent molar quantity for such systems. If we limit the discussion to the volume and refer to the definition of the apparent molar volume of the binary system (Eq. 6.53)), then we see that for multicomponent systems we must consider the difference between the volume of the solution and some other volume, and the meaning of the *molar* quantity. There appear to be two possible cases for these definitions. In the first case we could choose

Figure 6.1. The apparent molar quantity illustrated.

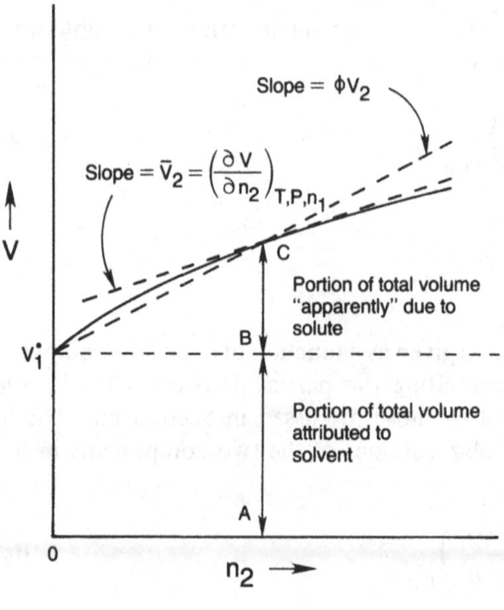

the other volume to be the volume of the pure solvent. We would then have to use the total number of moles of all solutes to determine the molar quantity. The apparent molar volume of all solutes would then be defined by

$$\phi V = \frac{V - n_1 \tilde{V}_1^{\bullet}}{\sum_{i=2}^{C} n_i} \tag{6.59}$$

In the second case we might choose a solution of fixed mole ratios of the components and consider the change in the volume of the solution as another solute is added to it. The apparent molar volume of the added solute would then be defined by

$$\phi V_k = \frac{V - V'}{n_k} \tag{6.60}$$

where V' is the volume of the original solution and n_k is the number of moles of the kth solute in the solution. In either case the apparent molar volumes are functions of all of the composition variables needed to describe the solution. For C components there are $(C - 1)$ such variables. The calculation of the partial molar volumes of the components from these relations follow the same methods discussed in this section and Section 6.3.

6.5 Graphical determination of the partial molar volumes of binary systems

In many instances it is possible to determine the partial molar volumes of binary systems by graphical means. Some cases simply require the application of calculus to the equations developed. As an example we consider Equations (6.57) and (6.58). We plot ϕV against m, determine the slope of the curve at a given m, and then determine $\bar{V}_1$ and $\bar{V}_2$ by means of these equations.

A different method is obtained if the molar volume is given as a function of the mole fraction. Equation (6.22) may be written as

$$\tilde{V} = \bar{V}_2 + x_1 \left(\frac{\partial \tilde{V}}{\partial x_1} \right)_{T,P} \tag{6.61}$$

and as

$$\tilde{V} = \bar{V}_1 + x_2 \left(\frac{\partial \tilde{V}}{\partial x_2} \right)_{T,P} \tag{6.62}$$

However, since $dx_2 = -dx_1$, Equation (6.62) may also be written as

$$\tilde{V} = \bar{V}_1 - (1 - x_1) \left(\frac{\partial \tilde{V}}{\partial x_1} \right)_{T,P} \tag{6.63}$$

If we plot $\tilde{V}$ against x_1 as in Figure 6.2, we see that the equation for the tangent to the curve at a given value of x_1 is given by both Equations (6.62) and (6.63). It is evident then that $\bar{V}_2$ is the intercept of this tangent on the zero axis, and $\bar{V}_1$ is the intercept of the tangent on the unit axis. If we start at $x_1 = 0$ and roll the tangent on the curve, we see that $\bar{V}_2$ starts at $\tilde{V}_2^\bullet$ and steadily increases, whereas $\bar{V}_1$ starts at some quantity greater than $V_1^\bullet$ and decreases to this quantity.

By similar considerations we may plot $\Delta\tilde{V}^{\mathrm{M}}$ defined by Equation (6.49) against the mole fraction and determine the difference between $\bar{V}_2$ and $\tilde{V}_2^\bullet$ and between $\bar{V}_1$ and $\tilde{V}_1^\bullet$ at a given mole fraction from the intercepts of the tangent to the curve at this mole fraction.

6.6 Integration of Equation (6.12) for a binary system

Equation (6.12) may be written as

$$x_1\left(\frac{\partial \bar{V}_1}{\partial x_1}\right)_{T,P} + x_2\left(\frac{\partial \bar{V}_2}{\partial x_1}\right)_{T,P} = 0 \qquad (6.64)$$

for a binary system when it is applied to volumes. If either $\bar{V}_1$ or $\bar{V}_2$ is known as a function of the mole fraction, then the other may be determined by integration. Equation (6.64) may be written in either of the two forms

$$\int \mathrm{d}\bar{V}_1 = \int \left(\frac{\partial \bar{V}_1}{\partial x_1}\right)_{T,P} \mathrm{d}x_1 = -\int \frac{x_2}{x_1}\left(\frac{\partial \bar{V}_2}{\partial x_1}\right)_{T,P} \mathrm{d}x_1 \qquad (6.65)$$

or

$$\int \mathrm{d}\bar{V}_2 = \int \left(\frac{\partial \bar{V}_2}{\partial x_1}\right)_{T,P} \mathrm{d}x_1 = -\int \frac{x_1}{x_2}\left(\frac{\partial \bar{V}_1}{\partial x_1}\right)_{T,P} \mathrm{d}x_1 \qquad (6.66)$$

Figure 6.2. Graphical determination of partial molar volumes in a binary system.

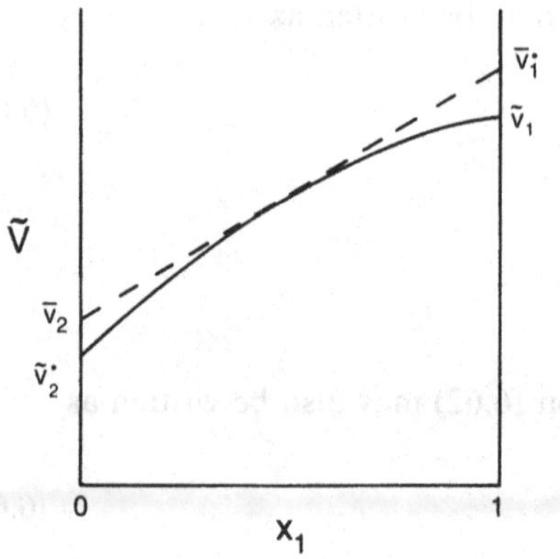

However, in order to obtain the integral, the integration constant or the values of the lower limits must be known.

No difficulty exists for gases and for liquid and solid solutions that have no region of immiscibility or limited solubility. In this case $(\partial \bar{V}_2/\partial x_1)_{T,P}$ (Eq. (6.65)) or $(\partial \bar{V}_1/\partial x_1)_{T,P}$ (Eq. (6.66)) can be determined over the whole range of composition. The lower limit of integration of Equation (6.65) is $x_1 = 1$, for which $\bar{V}_1 = \tilde{V}_1^{\bullet}$, and the lower limit of integration of Equation (6.66) is $x_1 = 0$, for which $\bar{V}_2 = \tilde{V}_2^{\bullet}$.

In the case of limited solubility, for example an aqueous solution of sucrose, no difficulty arises in the use of Equation (6.65), if we use the subscript 1 to refer to the solvent and the subscript 2 to refer to the solute. Here $\bar{V}_2$ could be determined as a function of the mole fraction from the very dilute solution (pure solvent) to the saturation point. The lower limit of integration would then be $\tilde{V}_1^{\bullet}$ when $x_1 = 1$. With Equation (6.66), $\bar{V}_1$ could be determined as a function of the mole fraction from the very dilute solution to the saturation point. However, in order to determine the value of the integration constant, $\bar{V}_2$ would have to be known at some mole fraction. This is usually not the case, and so the integration cannot be completed. This problem is clarified by reference to Figure 6.3, where the molar volume of a system having limited solubility is plotted as a function of the mole fraction of the solvent. The value of the mole fraction at the saturation point is indicated by the vertical broken line; $\bar{V}_1$ is indicated by the solid curve from the saturation value to the pure solvent. The form of the curve of $\bar{V}_2$ can be obtained by integration of Equation (6.66), but its absolute value cannot be determined. The broken curves represent a family of curves that differ by a constant value. All of

Figure 6.3. Integration of Equation (6.12) for a binary system.

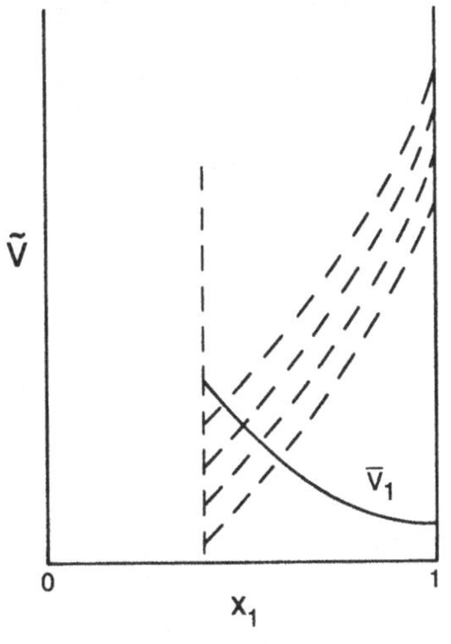

these curves could represent $\bar{V}_2$, but the particular curve that actually gives $\bar{V}_2$ is unknown.

The problem discussed here is common to all partial molar quantities. It is for this and similar reasons, which we have already indicated, that reference and standard states are defined (see Chapter 8).

7

Ideal gases and real gases

The conditions of equilibrium expressed by Equations (5.25)–(5.29) and (5.46) involve the temperature, pressure, and chemical potentials of the components or species. The chemical potentials are functions of the temperature, pressure or volume, and composition, according to Equations (5.54) and (5.56). In order to study the equilibrium properties of systems in terms of these experimentally observable variables, expressions for the chemical potentials in terms of these variables must be obtained. This problem is considered in this chapter and in Chapter 8.

The thermodynamic functions for the gas phase are more easily developed than for the liquid or solid phases, because the temperature–pressure–volume relations can be expressed, at least for low pressures, by an algebraic equation of state. For this reason the thermodynamic functions for the gas phase are developed in this chapter before discussing those for the liquid and solid phases in Chapter 8. First the equation of state for pure ideal gases and for mixtures of ideal gases is discussed. Then various equations of state for real gases, both pure and mixed, are outlined. Finally, the more general thermodynamic functions for the gas phase are developed in terms of the experimentally observable quantities: the pressure, the volume, the temperature, and the mole numbers. Emphasis is placed on the virial equation of state accurate to the second virial coefficient. However, the methods used are applicable to *any* equation of state, and the development of the thermodynamic functions for any given equation of state should present no difficulty.

7.1 The ideal gas

An ideal gas is defined as one whose properties are given by the two equations (Sect. 2.8)

$$PV = nRT \tag{7.1}$$

and

$$\left(\frac{\partial E}{\partial V}\right)_{T,n} = 0 \tag{7.2}$$

These two equations are applicable to mixtures of ideal gases as well as to pure gases, provided n is taken to be the total number of moles of gas. However, we must consider how the properties of the gas mixture depend upon the composition of the gas mixture and upon the properties of the pure gases. In particular, we must define the Dalton's pressures, the partial pressures, and the Amagat volumes. Dalton's law states that each individual gas in a mixture of ideal gases at a given temperature and volume acts as if it were alone in the same volume and at the same temperature. Thus, from Equation (7.1) we have

$$P_l = n_l RT/V \tag{7.3}$$

where P_l is the pressure exerted by n_l moles of the lth gas at the volume V and temperature T. The quantity P_l is called the Dalton's pressure. The addition of such equations for a mixture of gases yields

$$\sum_{l=1}^{C} P_l = \left(\sum_{l=1}^{C} n_l\right)\frac{RT}{V} \tag{7.4}$$

and, by comparison with Equation (7.1), we see that

$$P = \sum_{l=1}^{C} P_l \tag{7.5}$$

Thus, the total pressure of a mixture of ideal gases is the sum of the Dalton pressures.

The partial pressure of a gas in a gas mixture, whether such a mixture is ideal or real, is defined by

$$P_l = y_l P \tag{7.6}$$

where y represents the mole fraction in the gas phase. The sum of the partial pressures must always equal the total pressure, because the sum of the mole fractions must always be equal to unity. By the combination of Equations (7.3)–(7.5), we see that the partial pressure and Dalton's pressure of a component in a mixture of ideal gases are identical. This statement is not true in general for mixtures of real gases.

Amagat's law is very similar to Dalton's law, but deals with the additivity of the volumes of the individual components or species of the gas mixture when mixed at constant temperature and pressure. The Amagat volume may be defined as the volume that n_l moles of the pure lth gas occupies when measured at the temperature T and pressure P, so

$$V_l = n_l RT/P \tag{7.7}$$

and

$$V = \sum_{l=1}^{C} V_l = \left(\sum_{l=1}^{C} n_l \right) \frac{RT}{P} \tag{7.8}$$

Thus, the volume of a mixture of ideal gases is the sum of the volumes of the individual gases when measured at the same pressure and temperature.

7.2 Pure real gases

No actual gas follows the ideal gas equation exactly. Only at low pressures are the differences between the properties of a real gas and those of an ideal gas sufficiently small that they can be neglected. For precision work the differences should never be neglected. Even at pressures near 1 bar these differences may amount to several percent. Probably the best way to illustrate the deviations of real gases from the ideal gas law is to consider how the quantity $P\tilde{V}/RT$, called the *compressibility factor*, Z, for 1 mole of gas depends upon the pressure at various temperatures. This is shown in Figure 7.1, where the abscissa is actually the reduced pressure and the curves are for various reduced temperatures [9]. The behavior of the ideal gas is represented by the line where $P\tilde{V}/RT = 1$. For real gases at sufficiently low temperatures, the $P\tilde{V}$ product is less than ideal at low pressures and, as the pressure increases, passes through a minimum, and finally becomes greater than ideal. At one temperature, called the *Boyle temperature*, this minimum

Figure 7.1. Deviations of real gases from ideal gas behavior.

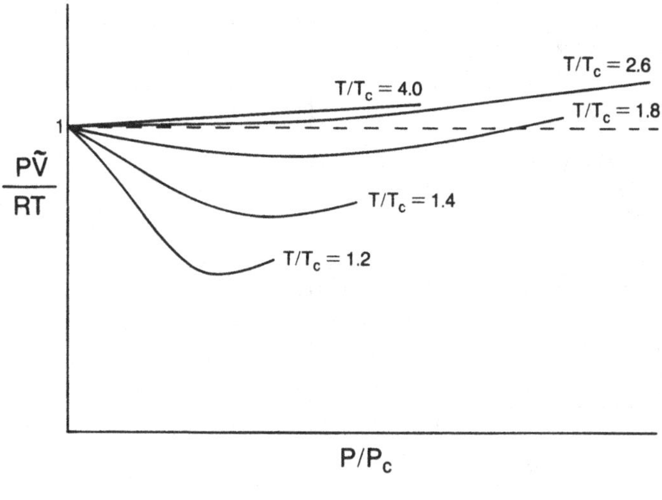

occurs at zero pressure. The *Boyle temperature* is therefore defined as the temperature at which

$$\lim_{P \to 0} \left[\frac{\partial(PV)}{\partial P} \right]_{T,n} = 0 \tag{7.9}$$

Above the Boyle temperature the deviations from ideal behavior are always positive and increase with the pressure. The initial slope of the PV curves at which $P = 0$ is negative at low temperatures, passes through zero at the Boyle temperature, and then becomes positive. A maximum in the initial slope as a function of temperature has been observed for both hydrogen and helium, and it is presumed that all gases would exhibit such a maximum if heated to sufficiently high temperatures. This behavior of the initial slope with temperature is illustrated in Figure 7.2 [10].

Many equations have been suggested to express the behavior of real gases. In general, there are those equations that express the pressure as a function of the volume and temperature, and those that express the volume as a function of the pressure and temperature. These cannot usually be converted from one into the other without obtaining an infinite series. The most convenient thermodynamic function to use for those in which the volume and temperature are the independent variables is the Helmholtz energy. The

Figure 7.2. The second virial coefficient of carbon dioxide.

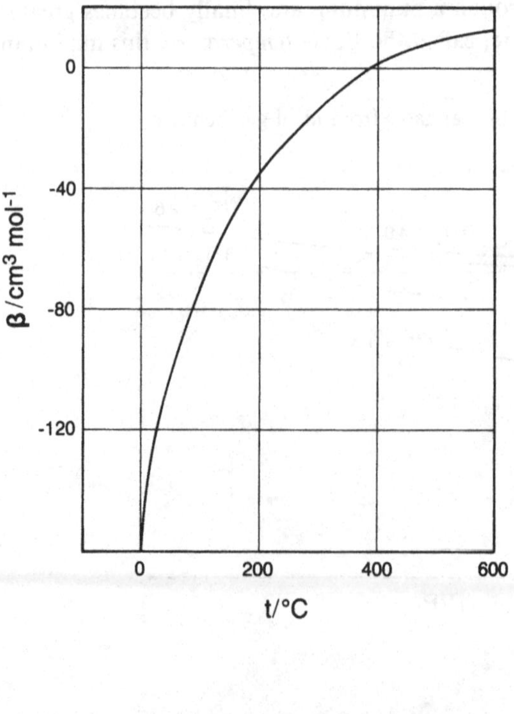

Gibbs energy is the most convenient function to use for those in which the pressure and temperature are the independent variables.

The virial equation of state, first suggested by Kammerlingh-Ohnes, is probably one of the most convenient equations to use, and is used in this chapter to illustrate the development of the thermodynamic equations that are consistent with the given equation of state. The methods used here can be applied to any equation of state.

The virial equation gives the pressure as a power series of the molar density, so that either

$$PV = nRT\left[1 + B\left(\frac{n}{V}\right) + C\left(\frac{n}{V}\right)^2 + D\left(\frac{n}{V}\right)^3 + \cdots\right] \tag{7.10}$$

or

$$P = \frac{nRT}{V} + B'\left(\frac{n}{V}\right)^2 + C'\left(\frac{n}{V}\right)^3 + D'\left(\frac{n}{V}\right)^4 + \cdots \tag{7.11}$$

The volume may also be expressed as a power series of the pressure, so

$$V = \frac{nRT}{P}(1 + \beta' P + \gamma' P^2 + \delta' P^3 + \cdots) \tag{7.12}$$

or

$$V = n\left(\frac{RT}{P} + \beta + \gamma P + \delta P^2 + \cdots\right) \tag{7.13}$$

The quantities B, B', β, and β' are called second virial coefficients; C, C', γ, and γ' third virial coefficients; and D, D', δ, and δ' fourth virial coefficients. These coefficients are functions of the temperature and are characteristic of the individual gas. These equations reproduce the pressure–volume–temperature relations of gases accurately only for low pressures or small densities.

The relationships between B and B', C and C', D and D', β and β', γ and γ', and δ and δ' are quite obvious from the defining equations. The two sets of virial coefficients, however, are not independent, and the relations between these sets are

$$\beta = B \tag{7.14}$$

$$\gamma = (C - B^2)/RT \tag{7.15}$$

and

$$\delta = (D - 3BC + 2B^3)/(RT)^2 \tag{7.16}$$

In general the dependence of the virial coefficients upon the temperature must be determined experimentally. The temperature dependence of β is illustrated in Figure 7.2 because, according to Equation (7.13), β is the limiting slope of the PV isotherms as the pressure goes to zero.

Dymond and Smith [11] give an excellent compilation of virial coefficients of gases and mixtures. Cholinski *et al.* [12] provide second virial coefficient data for individual organic compounds and binary systems. The latter book also discusses various correlational methods for calculating second virial coefficients. Mason and Spurling [13] have written an informative monograph on the virial equation of state.

The search for better equations of state continues apace, since computers make the computational part relatively easy. Stryjek and Vera [14], for example, have developed a complex equation of state based on that of Redlich and Kwong, but which is suitable for representing the P–V–T behavior of liquid and gas phases over a wide range of temperature and pressure for pure nonpolar compounds.

7.3 Mixtures of real gases and combinations of coefficients[1]

The virial equation of state discussed in Section 7.2 is applicable to gas mixtures with the condition that n represents the total moles of the gas mixture; that is, $n = \sum_{l=1}^{C} n_l$. The constants and coefficients then become functions of the mole fractions. These functions can be determined experimentally, and actually the pressure–volume–temperature properties of some binary mixtures and a few ternary mixtures have been studied. However, sometimes it is necessary to estimate the properties of gas mixtures from those of the pure gases. This is accomplished through the combination of constants.

Experimental measurements and statistical mechanics both show that the dependence of the virial coefficients on the mole fractions may be expressed by

$$B = \sum_{i=1}^{C} \sum_{j=1}^{C} y_i y_j B_{ij} \tag{7.17}$$

$$C = \sum_{i=1}^{C} \sum_{j=1}^{C} \sum_{k=1}^{C} y_i y_j y_k C_{ijk} \tag{7.18}$$

$$D = \sum_{i=1}^{C} \sum_{j=1}^{C} \sum_{k=1}^{C} \sum_{l=1}^{C} y_i y_j y_k y_l D_{ijkl} \tag{7.19}$$

etc., where the summations are taken over all of the components or species in the gas mixture for each index. The summations in Equations (7.17) and

---

[1] In this section we give, for completeness, the combination of virial coefficients higher than the second order. However, for the purposes of illustration, we go up to only the second virial coefficient. The references cited at the end of Section 7.2 can be consulted for more-detailed information.

(7.18) for a ternary system, for example, are

$$B = y_1^2 B_{11} + y_2^2 B_{22} + y_3^2 B_{33} + 2y_1 y_2 B_{12} + 2y_1 y_3 B_{13} + 2y_2 y_3 B_{23} \tag{7.20}$$

and

$$\begin{aligned} C = {} & y_1^3 C_{111} + y_2^3 C_{222} + y_3^3 C_{333} + 3y_1^2 y_2 C_{112} + 3y_1 y_2^2 C_{122} \\ & + 3y_1^2 y_3 C_{113} + 3y_1 y_3^2 C_{133} + 3y_2^2 y_3 C_{223} + 3y_2 y_3^2 C_{233} \\ & + 6y_1 y_2 y_3 C_{123} \end{aligned} \tag{7.21}$$

The assumption is made in these equations that the coefficients having different combinations of the same indices are identical, thus $C_{112} = C_{121} = C_{211}$. With the omission of the coefficients, the expansion of these sums are the square, cube, etc., of the sum of the mole fractions of the components or species. With these expressions, the problem of the combination of constants becomes one of expressing the coefficients having unlike indices in terms of those having like indices.

Three methods have been used for the combination of these constants.

1. Linear combination. In this method the coefficients having unlike indices are assumed to be the arithmetical mean of those having like indices. If a represents any of these coefficients, then

$$a_{ij} = \tfrac{1}{2}(a_{ii} + a_{jj}) \tag{7.22}$$

$$a_{ijk} = \tfrac{1}{3}(a_{iii} + a_{jjj} + a_{kkk}) \tag{7.23}$$

$$a_{ijkl} = \tfrac{1}{4}(a_{iiii} + a_{jjjj} + a_{kkkk} + a_{llll}) \tag{7.24}$$

etc. With these relations,

$$\sum_{i=1}^{C} \sum_{j=1}^{C} y_i y_j a_{ij} = \sum_{i=1}^{C} y_i a_{ii} \tag{7.25}$$

$$\sum_{i=1}^{C} \sum_{j=1}^{C} \sum_{k=1}^{C} y_i y_j y_k a_{ijk} = \sum_{i=1}^{C} y_i a_{iii} \tag{7.26}$$

$$\sum_{i=1}^{C} \sum_{j=1}^{C} \sum_{k=1}^{C} \sum_{l=1}^{C} y_i y_j y_k y_l a_{ijkl} = \sum_{i=1}^{C} y_i a_{iiii} \tag{7.27}$$

2. Quadratic combination. This method assumes that the coefficients having unlike indices are the geometric means of those having like indices, thus

$$a_{ij} = (a_{ii} a_{jj})^{1/2} \tag{7.28}$$

$$a_{ijk} = (a_{iii} a_{jjj} a_{kkk})^{1/3} \tag{7.29}$$

$$a_{ijkl} = (a_{iiii}a_{jjjj}a_{kkkk}a_{llll})^{1/4} \tag{7.30}$$

etc. With these relations,

$$\sum_{i=1}^{C}\sum_{j=1}^{C} y_i y_j a_{ij} = \left(\sum_{i=1}^{C} y_i a_{ii}^{1/2}\right)^2 \tag{7.31}$$

$$\sum_{i=1}^{C}\sum_{j=1}^{C}\sum_{k=1}^{C} y_i y_j y_k a_{ijk} = \left(\sum_{i=1}^{C} y_i a_{iii}^{1/3}\right)^3 \tag{7.32}$$

$$\sum_{i=1}^{C}\sum_{j=1}^{C}\sum_{k=1}^{C}\sum_{l=1}^{C} y_i y_j y_k y_l a_{ijkl} = \left(\sum_{i=1}^{C} y_i a_{iiii}^{1/4}\right)^4 \tag{7.33}$$

etc.

3. Lorentz combination. This method is defined by

$$a_{ij} = \tfrac{1}{8}(a_{ii}^{1/3} + a_{jj}^{1/3})^3 \tag{7.34}$$

$$a_{ijk} = \tfrac{1}{27}(a_{iii}^{1/3} + a_{jjj}^{1/3} + a_{kkk}^{1/3})^3 \tag{7.35}$$

and

$$a_{ijkl} = \tfrac{1}{64}(a_{iiii}^{1/3} + a_{jjjj}^{1/3} + a_{kkkk}^{1/3} + a_{llll}^{1/3})^3 \tag{7.36}$$

etc. Then,

$$\sum_{i=1}^{C}\sum_{j=1}^{C} y_i y_j a_{ij} = \frac{1}{8}\sum_{i=1}^{C}\sum_{j=1}^{C} y_i y_j (a_{ii}^{1/3} + a_{jj}^{1/3})^3 \tag{7.37}$$

$$\sum_{i=1}^{C}\sum_{j=1}^{C}\sum_{k=1}^{C} y_i y_j y_k a_{ijk} = \frac{1}{27}\sum_{i=1}^{C}\sum_{j=1}^{C}\sum_{k=1}^{C} y_i y_j y_k \times (a_{iii}^{1/3} + a_{jjj}^{1/3} + a_{kkk}^{1/3})^3 \tag{7.38}$$

and

$$\sum_{i=1}^{C}\sum_{j=1}^{C}\sum_{k=1}^{C}\sum_{l=1}^{C} y_i y_j y_k y_l = \frac{1}{64}\sum_{i=1}^{C}\sum_{j=1}^{C}\sum_{k=1}^{C}\sum_{l=1}^{C} y_i y_j y_k y_l \times (a_{iiii}^{1/3} + a_{jjjj}^{1/3} + a_{kkkk}^{1/3} + a_{llll}^{1/3})^3 \tag{7.39}$$

etc.

Two further crude approximations have been used for the virial equation of state. The first is that the virial coefficients combine linearly. This combination of constants results in an equation of state that is additive in the properties of the pure components. In such a mixture Dalton's and Amagat's laws still hold, and the mixture may be called an *ideal mixture of real gases*. The assumption is probably the crudest that can be used and is not strictly valid even at low pressures. The second approximation assumes that the coefficients combine quadratically. With this approximation Dalton's and Amagat's laws are no longer obeyed. A third approximation may be introduced from the knowledge that the virial coefficients are themselves functions of the temperature. The use of such functions introduces sets of constants that are independent of the temperature and pressure or of the temperature and volume, but which are functions of the composition. The problem of obtaining an equation of state for gas mixtures then becomes one of determining the combination rules for these constants, rather than those for the virial coefficients themselves.

As more experimental data have been obtained, empirical correlations have been suggested for determining the values of the cross coefficients, such as a_{ij}, from the properties of the mixtures rather than from those of the pure components.

7.4 The Joule effect

The Joule effect is discussed in Section 2.8 in conjunction with the definition of an ideal gas. When Equations (2.40) and (4.64) are combined, the expression for the Joule coefficient becomes.

$$\left(\frac{\partial T}{\partial V}\right)_{E,n} = \frac{P - T(\partial P/\partial T)_{V,n}}{C_V} \tag{7.40}$$

Equation (7.11) may be used to evaluate the numerator of the right-hand term, so

$$P - T\left(\frac{\partial P}{\partial T}\right)_{V,n} = \left[B' - T\left(\frac{\mathrm{d}B'}{\mathrm{d}T}\right)\right]\left(\frac{n}{V}\right)^2 + \left[C' - T\left(\frac{\mathrm{d}C'}{\mathrm{d}T}\right)\right]\left(\frac{n}{V}\right)^3 + \cdots \tag{7.41}$$

We conclude from Equation (7.41) that the Joule coefficient is a function of both the temperature and the volume. Moreover, the value of the coefficient goes to zero as the molar volume approaches infinity.

7.5 The Joule–Thomson effect

In an idealized Joule–Thomson experiment (also called the Joule–Kelvin experiment) a gas is confined by pistons in a cylinder that is divided into two parts by a rigid porous membrane (see Fig. 7.3). The gas, starting at pressure P_1 and temperature T_1, is expanded adiabatically and quasi-statically through the membrane to pressure P_2 and temperature T_2. The two pressures are kept constant during the experiment. If V_1 is the initial volume of the number of moles of gas that pass through the membrane and V_2 is the final volume of this quantity of gas, then the work done by the gas is $P_2V_2 - P_1V_1$. The change in energy is $E_2 - E_1$ and is equal to the work. Consequently, we find that $E_2 + P_2V_2 = E_1 + P_1V_1$, and therefore the enthalpy of the gas is constant. We are interested then in the quantity $(\partial T/\partial P)_{H,n}$, the *Joule–Thomson coefficient* (also given the symbol μ_{JT}), which is given by

$$\left(\frac{\partial T}{\partial P}\right)_{H,n} = \mu_{JT} = \frac{T(\partial V/\partial T)_{P,n} - V}{C_P} \tag{7.42}$$

For an ideal gas this quantity is zero. The Joule–Thomson process is *isenthalpic*. When Equation (7.13) is used for the equation of state, the numerator of the right-hand term becomes

$$T\left(\frac{\partial V}{\partial T}\right)_{P,n} - V = \left(T\frac{d\beta}{dT} - \beta\right) + \left(T\frac{d\gamma}{dT} - \gamma\right)P + \cdots \tag{7.43}$$

We conclude that the Joule–Thomson coefficient is a function of both the temperature and the pressure, but, unlike the Joule coefficient, it does not go to zero as the pressure goes to zero. The inversion temperature, the temperature at which $\mu_{JT} = 0$, is also a function of the pressure. The value usually reported in the literature is the limiting value as the pressure goes to zero.

Joule–Thomson experiments can yield quite practical information. The experiments (see Fig. 7.3) involve starting at a fixed temperature and pressure (T_i, P_i) and measuring the final temperature for a series of pressures less than P_i. This results in an isenthalp for each set of starting pressures and temperatures as shown in Figure 7.4. The heavy line in Figure 7.4 is drawn through the points of zero slope of the individual isenthalps. At those points $\mu_{JT} = 0$ and the slope changes from positive (cooling) to negative (heating).

Figure 7.3. The Joule–Thomson experiment.

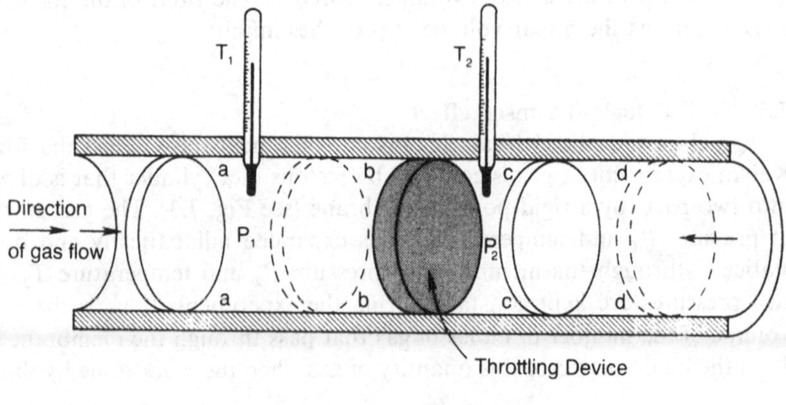

The temperatures on the envelope where $\mu_{JT}=0$ are called *inversion temperatures*, T_i. At any given pressure, up to a maximum pressure, a given gas exhibits *two* inversion temperatures. The Joule–Thomson effect is important in refrigeration and in the liquefaction of gases. Modern refrigeration uses the larger effect of the evaporation of working fluids such as the chlorofluorocarbons.

Some useful relationships may be derived involving the Joule–Thomson coefficient. Let us start with $H = H(T, P)$, to get

$$dH = \left(\frac{\partial H}{\partial P}\right)_T dP + \left(\frac{\partial H}{\partial T}\right)_P dT \tag{7.44}$$

For $dH = 0$ we can rearrange this to

$$\left(\frac{\partial T}{\partial P}\right)_H = -\left(\frac{\partial H}{\partial P}\right)_T \Big/ \left(\frac{\partial H}{\partial T}\right)_P \tag{7.45}$$

and this is just

$$\mu_{JT} = -\frac{1}{C_P}\left(\frac{\partial H}{\partial P}\right)_T \tag{7.46}$$

Equation (7.46) permits us to evaluate the derivative $(\partial H/\partial P)_T$ for real gases (it is zero for an ideal gas) from experimentally determinable properties. Upon substituting the definition of the enthalpy ($H = E + PV$) in Equation (7.46), we get

$$\mu_{JT} = -\frac{1}{C_P}\left[\left(\frac{\partial E}{\partial P}\right)_T + \left(\frac{\partial (PV)}{\partial P}\right)_T\right] \tag{7.47}$$

Figure 7.4. Isenthalps for nitrogen and the Joule–Thomson effect.

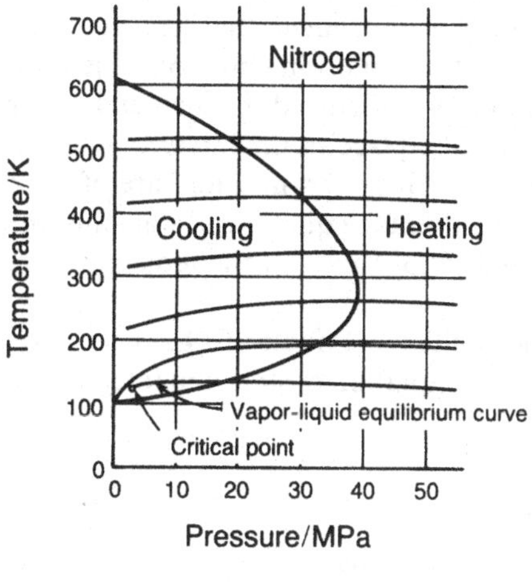

Expanding $(\partial E/\partial P)_T$, we get

$$\mu_{JT} = -\frac{1}{C_P}\left[\left(\frac{\partial E}{\partial V}\right)_T\left(\frac{\partial V}{\partial P}\right)_T + \left(\frac{\partial (PV)}{\partial P}\right)_T\right] \tag{7.48}$$

Equations (7.47) and (7.48) may be used to evaluate the important derivatives, $(\partial E/\partial P)_T$ and $(\partial E/\partial V)_T$, for real gases from experimentally determinable properties and an appropriate equation of state.

There are two variations of the basic set-up of the Joule–Thomson experiment which both yield practical information. In the *isothermal Joule–Thomson experiment* the temperature is held constant with a downstream heater, and the resultant heat input for the pressure decrease permits an experimental evaluation of $(\partial H/\partial P)_T$, the isothermal Joule–Thomson coefficient. In the other variation there is no throttling device used, and the pressure is held constant. For the steady-state flow of gas the temperature change is measured for measurable inputs of heat. This experiment, of course, yields $(\partial H/\partial T)_P$, or C_P. Thus, the variations of this constant-flow experiment can yield all three of the important terms in Equation (7.46).

7.6 Thermodynamic functions for ideal gases

The equation of state for a mixture of ideal gases containing n_1, n_2, $n_3, \ldots, n_i, \ldots$ moles of each component or species is

$$PV = \left(\sum_{l=1}^{C} n_l\right)RT$$

where the sum is taken over all components or species of the gas mixture. When no chemical reaction takes place in the gas mixture, the *n*s represent the mole numbers of the components. When chemical reactions take place between the components in the mixture, the *n*s refer to the individual species. The mole numbers of the species are not all independent, but depend on the mole numbers of the components and on the condition equation, Equation (5.46). The problems of chemical reactions are discussed in Chapter 11. For the present we assume that no reactions take place in the mixture and that the mole numbers refer to the components. However, in a mixture of ideal gases when chemical reactions are considered, the equation of state and the following equations still apply for a fixed set of values of the mole numbers of the species.

At constant temperature and mole numbers, Equation (4.25) becomes $\mathrm{d}G = V\,\mathrm{d}P$. On substitution of the ideal gas equation of state and integration, we obtain

$$G = \left(\sum_{l=1}^{C} n_l\right)RT \ln P + g(T, n) \tag{7.49}$$

where $g(T, n)$ represents a function of the temperature and the mole numbers. The first term of this equation is additive in the mole numbers, and it is necessary to investigate the dependence of the second term on the mole numbers. According to Dalton's law, each gas in a gas mixture confined in a given volume at a temperature T acts as if it were there alone. The Gibbs energy of a pure gas under these conditions is

$$G_k = n_k[RT \ln P_k + g_k(T)] \tag{7.50}$$

where $g_k(T)$ is now a molar quantity, a function of the temperature alone, and characteristic of the individual kth substance. The Gibbs energy of the gas mixture is the sum of the Gibbs energies of the individual gases, since each gas acts independently. Therefore,

$$G = \sum_{l=1}^{C} G_l = \sum_{l=1}^{C} n_l[RT \ln P_l + g_l(T)] \tag{7.51}$$

This equation is consistent with the concept that the Gibbs energy is a homogenous function of the first degree in the mole numbers. Comparison of this equation with Equation (7.49) together with the relation that $P_i = Py_i$ shows that

$$g(T, n) = \sum_{l=1}^{C} n_l[RT \ln y_l + g_l(T)] \tag{7.52}$$

The function $g_l(T)$ is essentially an unknown function and, consequently, only the differences in the Gibbs energy of two different states at the same temperature can be determined, for under such conditions these functions cancel.[2] We discuss the form of these functions in Section 7.12. In order to obtain consistency in reporting the differences in the Gibbs energies of two states, it is advantageous to define one state at each temperature as a standard state to which all measurements are referred. From Equation (7.50) we see that $g_k(T)$ is the molar Gibbs energy of the pure kth gas at 1 bar pressure and temperature T. We use either of the symbols, $\tilde{G}_l^{\ominus}(T)$ or $\mu_l^{\ominus}(T)$, in place of $g_l(T)$ in Equation (7.51) and write[3]

$$G[T, P, n] = \sum_{l=1}^{C} n_l(RT \ln(Py_l) + \mu_l^{\ominus}[T, 1 \text{ bar}, y_l = 1]) \tag{7.53}$$

Other thermodynamic functions can be derived from this equation. With

[2] The third law of thermodynamics, by which absolute values of the entropy can be determined, permits the calculation of the change of values of the Gibbs and Helmholtz energies with temperature. This subject is discussed in Chapter 15.

[3] In order to keep clearly in mind the states to which the symbols refer, it is advantageous to indicate the states specifically. Brackets are used throughout this book to indicate states, and parentheses are used to indicate the independent variables of a function.

the use of the ideal equation of state and the relation between the Helmholtz and the Gibbs energies, the expression for the Helmholtz energy is

$$A[T, V, n] = \sum_{l=1}^{C} n_l(RT\ln(n_l/V) + a_l[T]) \tag{7.54}$$

where $a_l[T] = g_l(T) - RT + RT\ln(RT)$. The quantity, $a_l(T)$, may be interpreted as the molar Helmholtz energy of the pure lth component at unit molar volume and temperature T. Differentiation of Equation (7.53) with respect to the temperature at constant pressure and mole numbers yields

$$S[T, P, n] = -\sum_{l=1}^{C} n_l(R\ln(Py_l) - \tilde{S}_l^{\ominus}[T, 1\text{ bar}, y_l = 1]) \tag{7.55}$$

The expression for the enthalpy may be obtained from the relation $G = H - TS$, and is

$$H[T, n] = \sum_{l=1}^{C} n_l\tilde{H}_l^{\ominus}[T] \tag{7.56}$$

Similarly, the equation for the energy is

$$E[T, n] = \sum_{l=1}^{C} n_l\tilde{E}_l^{\ominus}[T] \tag{7.57}$$

In Equations (7.56) and (7.57) the indication of the pressure or volume is unnecessary because both the energy and enthalpy for an ideal gas are independent of these variables. Finally, the chemical potential of the kth component or species is

$$\mu_k[T, P, y] = RT\ln(Py_k) + \mu_k^{\ominus}[T, 1\text{ bar}, y_k = 1] \tag{7.58}$$

7.7 The changes of the thermodynamic functions on mixing of ideal gases

In this section we consider the changes of the values of the thermodynamic functions for the change of state in which an ideal gas mixture in a given state is the final state and the pure ideal gases in given states comprise the initial state. Of the large number of possible changes of state, two—both of which are isothermal—are probably the most important. The first change of state is the one used in Section 7.6, in which the pure gases each occupy the same volume and are mixed to give a mixture at the same volume. This change of state may be expressed as

$$n_1B_1[T, V] + n_2B_2[T, V] = (n_1 + n_2)M[T, V] \tag{7.59}$$

for a binary mixture, where B_1 and B_2 represent the two components and M the gas mixture. The fact that the Gibbs energy of the mixture is additive in the Gibbs energies of the pure components was used in Section 7.6 (Eq.

(7.51)). This is equivalent to the statement that the change of the Gibbs energy for this change of state is zero. The changes for all of the other thermodynamic functions may be shown to be zero by means of the relations between these functions and the Gibbs energy.

For the second change of state, the initial state consists of the individual pure gases all at the same pressure P and the same temperature T, and the final state is the gas mixture at the same pressure and temperature. This change of state may be expressed as

$$n_1B_1[T, P] + n_2B_2[T, P] = (n_1 + n_2)M[T, P] \tag{7.60}$$

for a binary mixture. The Gibbs energy for the gas mixture is given by Equation (7.53); from the same equation the expression for the Gibbs energy for a pure gas is

$$G_l[T, P, y = 1] = n_l(RT \ln P + \mu_l^{\ominus}[T, 1\text{ bar}, y_l = 1]) \tag{7.61}$$

Therefore, the change of the Gibbs energy, ΔG_{mix}, for this change of state is

$$\Delta G_{\text{mix}} = G - \sum_{l=1}^{C} G_l = \sum_{l=1}^{C} n_l RT \ln y_l \tag{7.62}$$

The change of the Helmholtz energy, ΔA_{mix}, is the same as that for the Gibbs energy. The change of the entropy, ΔS_{mix}, is given by

$$\Delta S_{\text{mix}} = S - \sum_{l=1}^{C} S_l = -\sum_{l=1}^{C} n_l R \ln y_l \tag{7.63}$$

The change of both the energy, ΔE_{mix}, and the enthalpy, ΔH_{mix}, is zero. Finally, the change in the chemical potential, $\Delta\mu_{k,\text{mix}}$, is

$$\Delta\mu_{k,\text{mix}} = \mu_k[T, P, y_k] - \mu_k[T, P, y_k = 1] = RT \ln y_k \tag{7.64}$$

7.8 The thermodynamic functions of real gases

The equation of state

$$V = \left(\sum_{l=1}^{C} n_l\right)\left(\frac{RT}{P} + \sum_{i=1}^{C}\sum_{j=1}^{C} y_i y_j \beta_{ij}\right)$$

is used to illustrate the method of obtaining expressions for the thermodynamic functions of real gases. In this equation the sums are taken over all components or species in the gas mixture for each index. As was done for the case of ideal gases, we assume for the present that no chemical reactions take place in the mixture and that the indices refer to the components of the mixture.

The Gibbs energy of the gas mixture is again obtained by integrating Equation (4.25) under conditions of constant temperature and constant mole numbers. Thus,

$$G[T, P, y] = \left(\sum_{l=1}^{C} n_l\right)\left(RT \ln P + \sum_{i=1}^{C}\sum_{j=1}^{C} y_i y_j \beta_{ij} P\right) + g(T, n) \quad (7.65)$$

where $g(T, n)$ again represents the integration constant and is a function of the temperature and the mole numbers. The same question of the dependence of this function on the mole numbers again arises. However, we know that the behavior of real gases approaches that of ideal gases as the pressure approaches zero. Consequently, Equation (7.65) must approach Equation (7.51) as the pressure approaches zero, and thus

$$G[T, P, y] = \sum_{l=1}^{C} n_l\left(RT \ln(Py_l) + \sum_{i=1}^{C}\sum_{j=1}^{C} y_i y_j \beta_{ij} P + g_l(T)\right) \quad (7.66)$$

The chemical potential of the kth component can be obtained by differentiating this equation with respect to n_k at constant temperature, pressure, and all other mole numbers, and is

$$\mu_k[T, P, y] = RT \ln(Py_k) + \sum_{i=1}^{C}\sum_{j=1}^{C} (2\beta_{ik} - \beta_{ij}) y_i y_j P + g_k(T) \quad (7.67)$$

The differentiation of the term containing the second virial coefficient may present some difficulty. In a specific case the summation could be expanded and then the differentiation carried out. In the general case we must evaluate

$$\left(\frac{\partial(\sum_{l=1}^{C} n_l)(\sum_{i=1}^{C}\sum_{j=1}^{C} y_i y_j \beta_{ij})}{\partial n_k}\right)_{T,P,n}$$

We can multiply through by the single summation and obtain

$$\left(\frac{\partial[\sum_{i=1}^{C}\sum_{j=1}^{C} (n_i n_j \beta_{ij}/\sum_{l=1}^{C} n_l)]}{\partial n_k}\right)_{T,P,n}$$

In the differentiation, three groups of similar terms may be considered. The first is obtained on setting both i and j not equal to k. Then

$$\left(\frac{\partial[\sum_{i\neq k}^{C-1}\sum_{j\neq k}^{C-1} (n_i n_j \beta_{ij}/\sum_{l=1}^{C} n_l)}{\partial n_k}\right)_{T,P,n} = -\sum_{\substack{i\neq k\\ j\neq k}}^{C} y_i y_j \beta_{ij}$$

The second group is obtained by letting $i = k$ but $j \neq k$, from which

$$\left(\frac{\partial[\sum_{j\neq k}^{C-1} (n_k n_j \beta_{kj}/\sum_{l=1}^{C} n_l)}{\partial n_k}\right)_{T,P,n} = \sum_{j\neq k}^{C} (y_j - y_k y_j)\beta_{kj}$$

The third group is obtained by letting $i \neq k$ but $j = k$. The result is similar to that of the second group:

$$\sum_{i=k}^{C-1} (y_i - y_i y_k)\beta_{ik}$$

One term for which both i and j are equal to k remains. In this case

$$\left(\frac{\partial(n_k^2\beta_{kk}/\sum_{l=1}^{C} n_l)}{\partial n_k}\right)_{T,P,n} = (2y_k - y_k^2)\beta_{kk}$$

These four groups can now be recombined to give

$$\sum_{i=1}^{C} 2y_i\beta_{ki} - \sum_{i=1}^{C}\sum_{j=1}^{C} y_i y_j \beta_{ij}$$

where the indices i and j again refer to all components of the system. Each term in the single summation may be multiplied by $\sum_{j=1}^{C} y_j$, which is equal to unity. Thus, we finally obtain

$$\sum_{i=1}^{C}\sum_{j=1}^{C} (2\beta_{ki} - \beta_{ij}) y_i y_j$$

The quantity $g_k(T)$ in Equation (7.67) is again a molar quantity, characteristic of the individual gas, and a function of the temperature. It can be related to the molar Gibbs energy of the kth substance by the use of Equation (7.67). The first two terms on the right-hand side of this equation are zero when the gas is pure and ideal and the pressure is 1 bar. Then $g_k(T)$ is the chemical potential or molar Gibbs energy for the pure kth substance in the ideal gas state at 1 bar pressure. We define this state to be the standard state of the kth substance and use the symbol $\mu_k^\ominus[T, 1\text{ bar}, y_k = 1]$ for the chemical potential of the substance in this state. Then Equation (7.66) may be written as

$$G[T, P, y] = \sum_{l=1}^{C} n_l\left(RT\ln(Py_l) + \sum_{i=1}^{C}\sum_{j=1}^{C} y_i y_j \beta_{ij} P + \mu_l^\ominus[T, 1\text{ bar}, y_l = 1]\right) \tag{7.68}$$

and Equation (7.67) as

$$\mu_k[T, P, y] = RT\ln(Py_k) + \sum_{i=1}^{C}\sum_{j=1}^{C} (2\beta_{ik} - \beta_{ij}) y_i y_j P + \mu_k^\ominus[T, 1\text{ bar}, y_k = 1] \tag{7.69}$$

The standard state defined in terms of the ideal gas presents no difficulty. In order to go from the ideal gas to the real gas at the same temperature, we may consider two changes of state. For the first change of state the pressure of the gas is changed from 1 bar to zero with the use of the ideal equation of state, and for the second the pressure of the gas is changed from zero to any finite pressure with the use of an equation of state for the real gas.

7.9 The changes of the thermodynamic functions on mixing of real gases

We consider the changes of various thermodynamic functions only for the change of state in which the initial state is that of the unmixed gases each at pressure P and temperature T and the final state in the gas mixture at the same pressure and temperature. The Gibbs energy of the final state is given by Equation (7.68), and that of the initial state is

$$\sum_{l=1}^{C} G_l = \sum_{l=1}^{C} n_l(RT\ln P + \beta_{ll}P + \mu_l^{\ominus}[T, 1\ \text{bar}, y_l = 1]) \tag{7.70}$$

according to the same equation. Then, the change of the Gibbs energy for this change of state is

$$\Delta G_{\text{mix}} = \sum_{l=1}^{C} n_l\left[RT\ln y_l + \left(\sum_{i=1}^{C}\sum_{j=1}^{C} y_i y_j \beta_{ij} - \beta_{ll}\right)P\right] \tag{7.71}$$

This equation may be written in the form

$$\Delta G_{\text{mix}} = \sum_{l=1}^{C} n_l\left[RT\ln y_l + \sum_{i=1}^{C}\sum_{j=1}^{C} y_i y_j(\beta_{ij} - \beta_{ll})P\right] \tag{7.72}$$

because $\sum_{i=1}^{C}\sum_{j=1}^{C} y_i y_j = 1$. When Equations (7.71) and (7.72) are expanded for specific cases, that part of the resultant expressions which pertain to the nonideality of the gases consists of terms of the general form $(2\beta_{ij} - \beta_{ii} - \beta_{jj})$. Consequently, if each β_{ij} is the algebraic mean of β_{ii} and β_{jj}, such terms become zero and the change of the Gibbs energy becomes that of an ideal gas. Such a mixture of gases might be called an *ideal mixture of real gases*.

The change of the entropy for the change of state is obtained by differentiating Equation (7.72) with respect to the temperature, and is

$$\Delta S_{\text{mix}} = -\sum_{l=1}^{C} n_l\left\{R\ln y_l + \sum_{i=1}^{C}\sum_{j=1}^{C} y_i y_j P\left[\left(\frac{d\beta_{ij}}{dT}\right) - \left(\frac{d\beta_{ll}}{dT}\right)\right]\right\} \tag{7.73}$$

From the definition of the Gibbs energy or appropriate differentiation of Equation (7.72), the change of enthalpy for the change of state is

$$\Delta H_{\text{mix}} = \sum_{l=1}^{C} n_l\left\{\sum_{i=1}^{C}\sum_{j=1}^{C} y_i y_j P\left[\left(\frac{d(\beta_{ij}/T)}{d(1/T)}\right) - \left(\frac{d(\beta_{ll}/T)}{d(1/T)}\right)\right]\right\} \tag{7.74}$$

Finally, the change of the chemical potential of the kth component is

$$\Delta\mu_{k,\text{mix}} = RT\ln y_k + \sum_{i=1}^{C}\sum_{j=1}^{C} y_i y_j P(2\beta_{ki} - \beta_{ij} - \beta_{kk}) \tag{7.75}$$

7.10 The equilibrium pressure

We have already defined Dalton's pressures and the partial pressures. There is still one more pressure of a gas in a gas mixture that must be considered. This is the equilibrium pressure, and was first discussed by Gibbs [15]. The equilibrium pressure of a component is the pressure exerted by the pure component when it is in equilibrium with a gas mixture through a rigid, diathermic membrane that is permeable only to that component. The temperature of the pure phase and that of the gas mixture must be the same, but the pressure of the two phases will be different. We may conceive of two phases at the same temperature separated by such a membrane, one phase being the pure kth component and the other the gas mixture. The condition of equilibrium is that the chemical potential of the kth component must be identical in each phase. If P_k^e is the equilibrium pressure, then by Equation (7.69)

$$RT \ln P_k^e + \beta_{kk} P_k^e = RT \ln(Py_k) + \sum_{i=1}^{C} \sum_{j=1}^{C} (2\beta_{ki} - \beta_{ij}) y_i y_j P \tag{7.76}$$

This equation is transcendental and the equilibrium pressure must be calculated by approximation methods. The first approximation could be obtained by letting the equilibrium pressure be the partial pressure and evaluating the terms corresponding to the nonideality of the gas by this approximation. The solution of the resultant equation for the equilibrium pressure in the logarithmic term would give a second approximation. By such successive approximations the actual value of the equilibrium pressure can be obtained easily by using computer calculations.

7.11 The fugacity and fugacity coefficient

Lewis and Randall [16] originated the term *fugacity* on considering the thermodynamics of real gases. It is a quantity that can be used in the same manner for real gases as the partial pressures are used for ideal gases. As such, it is a convenient quantity to use, and gives an alternate method of developing the thermodynamics of real gases instead of using the chemical potentials.

The fugacity is defined in terms of the chemical potential of a gas, either pure or in a mixture, as

$$\mu_k[T, P, y_k] = RT \ln f_k + g_k(T) \tag{7.77}$$

where $\mu_k[T, P, y_k]$ is the chemical potential of the kth component at the temperature T, the pressure P, and the mole fraction y_k. The symbol f_k represents the fugacity of the kth component under these conditions, and $g_k(T)$ is a function of the temperature and is characteristic of the substance. The similarity of this equation to that in terms of the partial pressure of an ideal gas as given by Equation (7.58) is obvious. The standard state of the gas in terms of the fugacity is defined in such a way that $g_k(T)$ is the chemical potential of the gas in its standard state. Thus, the *fugacity of the gas in its standard state is unity*. This standard state is again the state of the pure substance as an ideal gas at 1 bar pressure and the temperature T.

The value of the fugacity of a gas in a given state must be calculated by means of an equation of state, either algebraic or graphic; it is not determined directly by experimental means. An expression for the fugacity of the kth substance in a gas mixture can be obtained by comparison of Equations (7.67) and (7.77). These equations give two different ways of expressing the chemical potential, and consequently the two expressions must be equal. Thus,

$$RT \ln f_k = RT \ln(Py_k) + \sum_{i=1}^{C} \sum_{j=1}^{C} (2\beta_{ik} - \beta_{ij}) y_i y_j P \tag{7.78}$$

For a pure gas this equation becomes

$$RT \ln f_k = RT \ln P + \beta_{kk} P \tag{7.79}$$

The virial coefficients can be determined from studies of the pressure–volume–temperature relations of gases. A graphical method for determining the fugacity may be illustrated by the use of the equation of state

$$V = \left(\sum_{l=1}^{C} n_l\right)\left(\frac{RT}{P} + \alpha\right) \tag{7.80}$$

where α is a function of the temperature, pressure, and composition of the gas mixture. The quantity α represents the deviations of the real gas mixture from ideal behavior. The change of the chemical potential of the kth component with pressure at constant composition and temperature is given by

$$d\mu_k = \bar{V}_k \, dP$$

Therefore,

$$\mu_k[T, P, y_k] = RT \ln(Py_k) + \int \left[\alpha + \left(\sum_{l=1}^{C} n_l\right)\left(\frac{\partial \alpha}{\partial n_k}\right)_{T,P,n_l}\right] dP + g_k(T)$$

and

$$RT \ln f_k = RT \ln(Py_k) + \int_0^P \left[\alpha + \left(\sum_{l=1}^{C} n_l\right)\left(\frac{\partial \alpha}{\partial n_k}\right)_{T,P,n_l}\right] dP \tag{7.81}$$

For a pure gas Equation (7.81) reduces to

$$RT \ln f = RT \ln P + \int_0^P \alpha \, dP \tag{7.82}$$

The lower limit of zero pressure is introduced in Equations (7.81) and (7.82) because the behavior of a real gas approaches that of an ideal gas as the pressure goes to zero. The integral in these equations can be evaluated graphically when α is known as a function of the pressure at constant temperature and composition.

The fugacity, from its definition, is a function of the temperature, pressure, and composition of the system. It approaches the partial pressure of the particular substance as the total pressure approaches zero. Its temperature derivative is given by

$$\bar{S}_k[T, P, y] - \tilde{S}_k^{\ominus}[T, 1\ \text{bar}, y_k = 1] = -R \ln f_k - RT\left(\frac{\partial \ln f_k}{\partial T}\right)_{P,y} \tag{7.83}$$

where $\bar{S}_k^{\ominus}[T, P, y]$ is the partial molar entropy of the kth component in the given state and $\tilde{S}_k^{\ominus}[T, 1\ \text{bar}, y_k = 1]$ is the molar entropy of the same substance in its standard state. Similarly,

$$\bar{H}_k[T, P, y] - \tilde{H}_k^{\ominus}[T, 1\ \text{bar}, y_k = 1] = R\left(\frac{\partial \ln f_k}{\partial T}\right)_{P,y} \tag{7.84}$$

The *fugacity coefficient*, γ_k, of a substance in a gas mixture is defined as the ratio of its fugacity to its partial pressure, so $\gamma_k = f_k/Py_k$ and for a pure gas $\gamma = f/P$. Equations expressing the fugacity coefficients as a function of the pressure, temperature, and composition are readily derived from the equations in this section, which give the relations between the fugacity and the partial pressure. It is evident that the fugacity coefficient is related to the terms that express the nonideality of the pure gas or gas mixture. The temperature derivatives of the coefficients are derived easily from Equations (7.83) and (7.84).

A generalized method of presenting real gas behavior is due to Pitzer *et al.* [17]. They introduced an acentric factor, ω, defined as

$$\omega = -(\log_{10} P_r^s - 1) \qquad \text{at } T_r = 0.7 \tag{7.85}$$

where P_r^s is the reduced vapor pressure. The original tabulated data of Pitzer *et al.* are presented in graphical form in Figures 7.5 and 7.6, using as a correlation

$$Z = Z^{(0)} + \omega Z^{(1)} \tag{7.86}$$

In this equation $Z^{(0)}$ is $P\tilde{V}/RT$ for a simple fluid and the graph for $Z^{(0)}$ is Figure 7.5. $Z^{(1)}$ is the correction to obtain the real fluid compressibility factor and these values are presented in Figure 7.6. This correlation, based on the principle of corresponding states, is quite good for a wide range of substances. (See Edmister and Lee [18] for more details, including extensive tabulations and graphs, on this subject.)

7.12 The integration constant, $g_k(T)$

In the preceding work it was not necessary to consider the dependence of the integration constant on the temperature and in many cases it is not necessary to do so. At a given temperature it is a constant and, as long as only differences in the thermodynamic functions are considered at the same temperature, this constant cancels.

It is informative, however, to consider the dependence of this function on the temperature. Since we know it is characteristic of the pure gas, we consider only 1 mole of a pure gas. Moreover, we limit the discussion to an ideal gas. There are two possible methods, one concerning the energy and entropy and the other the enthalpy and entropy, but we use only the energy and entropy here. The differential of the energy of 1 mole of ideal gas is $d\tilde{E} = \tilde{C}_V\,dT$. On

Figure 7.5. Compressibility factor for simple fluids.

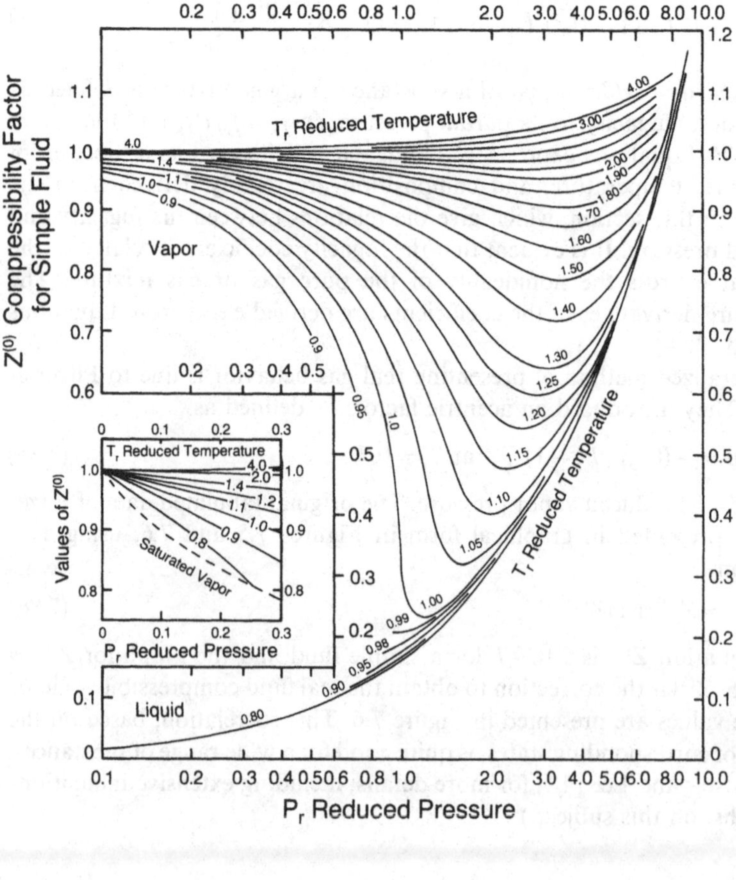

integration, the energy is given as

$$\tilde{E} = \int \tilde{C}_V \, \mathrm{d}T + \tilde{e} \tag{7.87}$$

where $\tilde{e}$ is a constant having units of any energy unit per mole, and is characteristic of the gas. Similarly, if the entropy is a function of the temperature and volume,

$$\mathrm{d}\tilde{S} = \frac{\tilde{C}_V}{T} \mathrm{d}T + \frac{R}{\tilde{V}} \mathrm{d}\tilde{V}$$

for an ideal gas. On integration, the entropy is given as

$$\tilde{S} = \int \frac{\tilde{C}_V}{T} \mathrm{d}T + R \ln \tilde{V} + \tilde{s}$$

because $\tilde{C}_V$ is independent of the volume for an ideal gas. In this equation, $\tilde{s}$ is a constant, characteristic of the gas and having molar entropy units. Since $\tilde{G} = \tilde{E} + P\tilde{V} - T\tilde{S}$ and $P\tilde{V} = RT$, we can obtain the Gibbs energy of 1 mole of an ideal gas as

$$\tilde{G} = \int \tilde{C}_V \, \mathrm{d}T - T \int \frac{\tilde{C}_V}{T} \mathrm{d}T + RT \ln P - RT \ln(RT) + RT + \tilde{e} - T\tilde{s} \tag{7.88}$$

Figure 7.6. Compressibility factor correction for deviation from simple fluids.

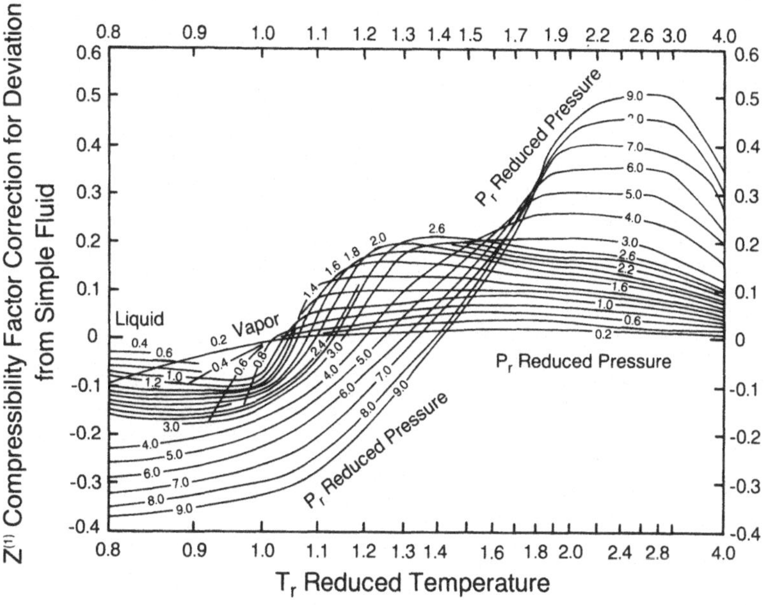

By comparisong with Equation (7.50), we find that

$$g(T) = \int \tilde{C}_V \, \mathrm{d}T - T \int \frac{\tilde{C}_V}{T} \, \mathrm{d}T - RT \ln(RT) + RT + \tilde{e} - T\tilde{s} \qquad (7.89)$$

Although the heat capacity is determinable, $\tilde{e}$ and $\tilde{s}$ are not known, and hence the value of $g(T)$ is not known. When we consider the difference, $g(T_2) - g(T_1)$, the integrals involving the heat capacities can be evaluated, but the term $(T_2 - T_1)\tilde{s}$ still remains. As long as $\tilde{s}$ is unknown, this term cannot be evaluated. The evaluation of $\tilde{s}$ is the province of the third law of thermodynamics.

8

Liquids and solids: reference and standard states

The thermodynamic equations for the Gibbs energy, enthalpy, entropy, and chemical potential of pure liquids and solids, and for liquid and solid solutions, are developed in this chapter. The methods used and the equations developed are identical for both pure liquids and solids, and for liquid and solid solutions; therefore, no distinction between these two states of aggregation is made. The basic concepts are the same as those for gases, but somewhat different methods are used between no single or common equation of state that is applicable to most liquids and solids has so far been developed. The thermodynamic relations for both single-component and multicomponent systems are developed.

The thermodynamic functions have been defined in terms of the energy and the entropy. These, in turn, have been defined in terms of differential quantities. The absolute values of these functions for systems in given states are not known.[1] However, differences in the values of the thermodynamic functions between two states of a system can be determined. We therefore may choose a certain state of a system as a standard state and consider the differences of the thermodynamic functions between any state of a system and the chosen standard state of the system. The choice of the standard state is arbitrary, and any state, physically realizable or not, may be chosen. The nature of the thermodynamic problem, experience, and convention dictate the choice. For gases the choice of standard state, defined in Chapter 7, is simple because equations of state are available and because, for mixtures, gases are generally miscible with each other. The question is more difficult for liquids and solids because, in addition to the lack of a common equation of state, limited ranges of solubility exist in many systems. The independent variables to which values must be assigned to fix the values of all of the

---

[1] The absolute value of the entropy has been determined for relatively few substances, and it is still advantageous to consider that the absolute values of the entropy are unknown. The knowledge of such values permits an alternate method of treating thermodynamic data and is useful in many applications.

intensive dependent variables are chosen to be the temperature, the pressure, and the composition variables of all but one of the components. For pure substances it is then sufficient to choose a given temperature and pressure for the standard state. For solutions we must also consider the composition variables.

Although the choice of standard states is arbitrary, two choices have been established by convention and international agreement. For some systems, when convenient, the pure component is chosen as the substance in the standard state. For other systems, particularly dilute solutions of one or more solutes in a solvent, another state that is not a standard state is chosen as a *reference state* [19]. This choice determines the standard state, which may or may not be a physically realizable state. The *reference state* of a component or species is that state to which all measurements are *referred*. The standard state is that state used to determine and report the differences in the values of the thermodynamic functions for the components or species between some state and the chosen standard state. When pure substances are used in the definition of a standard state, the standard state and the reference state are identical.

8.1 Single-phase, one-component systems

The state of a single-phase, one-component system may be defined in terms of the temperature, pressure, and the number of moles of the component as independent variables. The problem is to determine the difference between the values of the thermodynamic functions for any state of the system and those for the chosen standard state. Because the variables are not separable in the differential expressions for these functions, the integrations cannot be carried out directly to obtain general expressions for the thermodynamic functions without an adequate equation of state. However, each of the thermodynamic functions is a function of the state of the system, and the changes of these functions are independent of the path. The problem can be solved for specific cases by using the method outlined in Section 4.9 and illustrated in Figure 4.1.

The enthalpy is discussed first. The differential expression is

$$dH = C_P\, dT + \left[V - T\left(\frac{\partial V}{\partial T}\right)_{P,n}\right] dP + \tilde{H}\, dn \tag{8.1}$$

from Equation (4.86), where $\tilde{H}$ is the molar enthalpy of the substance. Each of the coefficients is a function of the independent variables T, P, and n. At constant pressure and mole number,

$$dH = C_P\, dT \tag{8.2}$$

where C_P is a function of the temperature and is to be evaluated at the particular constant pressure of interest. When Equation (8.2) is integrated with the use of an indefinite integral, we obtain

$$H[T,P] = \int C_P\,\mathrm{d}T + h[P,n] \tag{8.3}$$

where $h[P,n]$ is the integration constant. Its value depends upon the chosen pressure and mole number. The enthalpy is a homogenous function of the first degree in the mole number at constant temperature and pressure, and consequently $h[P,n]$ must also be. We can then write $h[P,n]$ as $n\tilde{h}[P]$, where $\tilde{h}[P]$ is a molar quantity. This quantity cannot be evaluated, and it is evident that standard states must be introduced. When Equation (8.2) is integrated with the use of a definite integral, we obtain

$$\Delta H = H[T_2,P] - H[T_1,P] = \int_{T_1}^{T_2} C_P\,\mathrm{d}T \tag{8.4}$$

This difference can be evaluated. If we choose to define a standard state of the substance to be the state of the substance at the pressure P and the temperature T_0, the difference in the value of the enthalpy at any temperature T and the given pressure P and that of the standard state is

$$\Delta H = H[T,P] - H^{\ominus}[T_0,P] = \int_{T_0}^{T} C_P\,\mathrm{d}T \tag{8.5}$$

where $H^{\ominus}[T_0,P]$ represents the enthalpy of the substance in the chosen standard state.

The differential of the enthalpy at a constant temperature and mole number is

$$\mathrm{d}H = \left[V - T\left(\frac{\partial V}{\partial T}\right)_{P,n}\right]\mathrm{d}P \tag{8.6}$$

where $[V - T(\partial V/\partial P)_{T,n}]$ is to be evaluated at the particular chosen temperature and is a function of the pressure. We again consider three different integrals. An indefinite integral might be used, so

$$H[T,P] = \int\left[V - T\left(\frac{\partial V}{\partial T}\right)_{P,n}\right]\mathrm{d}P + n\tilde{h}(T) \tag{8.7}$$

In this equation the concept that the enthalpy is a homogenous function of the first degree in the mole numbers at constant temperature and pressure has again been introduced. The quantity $\tilde{h}(T)$ is an undeterminable quantity, and again it is necessary to define a standard state. For the definite integral, we choose any two pressures, P_1 and P_2, and consider the integral between these two limits. The change of enthalpy of the system in going from the pressure P_1 to the pressure P_2 at the given temperature T is obtained from

$$\Delta H = H[T, P_2] - H[T, P_1] = \int_{P_1}^{P_2} \left[V - T\left(\frac{\partial V}{\partial T}\right)_{P,n}\right] \mathrm{d}P \tag{8.8}$$

We may then choose the standard state of the system at the temperature T to be at some arbitrary pressure P_0, so that the difference of the enthalpy of the system at the temperature T and pressure P and this standard state is given by

$$\Delta H = H[T, P] - H^{\ominus}[T, P_0] = \int_{P_0}^{P} \left[V - T\left(\frac{\partial V}{\partial T}\right)_{P,n}\right] \mathrm{d}P \tag{8.9}$$

where $H^{\ominus}[T, P_0]$ represents the enthalpy of the substance in this standard state.

It may be convenient to define the standard state of the system as the state at an arbitrary temperature, T_0, and an arbitrary pressure, P_0. The enthalpy of the system in any state defined by the temperature T and the pressure P may then be calculated by a combination of Equations (8.5) and (8.8). Two alternate equations, depending on the path we choose, are obtained. These are

$$H[T, P] - H^{\ominus}[T_0, P_0] = \int_{T_0}^{T} (C_P)_{P_0} \,\mathrm{d}T + \int_{P_0}^{P} \left[V - T\left(\frac{\partial V}{\partial T}\right)_{P,n}\right]_T \mathrm{d}P \tag{8.10}$$

and

$$H[T, P] - H^{\ominus}[T_0, P_0] = \int_{T_0}^{T} (C_P)_P \,\mathrm{d}T + \int_{P_0}^{P} \left[V - T\left(\frac{\partial V}{\partial T}\right)_{T,n}\right]_{T_0} \mathrm{d}P \tag{8.11}$$

Equation (8.10) is based on the two successive changes of state $B[T_0, P_0] = B[T, P_0] = B[T, P]$, where B represents the substance, whereas Equation (8.11) is based on the successive changes of state, $B[T_0, P_0] = B[T_0, P]$ and $B[T_0, P] = B[T, P]$. The subscripts to the integrands represent the value of the pressure or temperature at which the integrands must be evaluated. For the purposes of tabulating data, the usual standard state was taken at 298.15 K and 1 atm pressure (101.325 kPa). The new internationally recommended standard state is at 1 bar or 10^5 Pa [20–23].

Exactly the same methods as used above may be applied to the entropy. When the entropy is considered as a function of the temperature, pressure, and mole number, the differential of the entropy is

$$\mathrm{d}S = \frac{C_P}{T}\mathrm{d}T - \left(\frac{\partial V}{\partial T}\right)_{P,n} \mathrm{d}P + \tilde{S}\,\mathrm{d}n$$

where $\tilde{S}$ is the molar entropy of the substance and each coefficient is a function of the temperature, pressure, and mole number. Again, the variables are not separable and must be considered individually. At constant pressure and mole number

$$dS = (C_P/T)\, dT$$

from which we obtain the three alternate equations:

$$\Delta S = \int_{T_1}^{T_2} \frac{C_P}{T}\, dT \tag{8.12}$$

$$S[T, P] = \int \frac{C_P}{T}\, dT + n\tilde{s}(P) \tag{8.13}$$

$$S[T, P] = \int_{T_0}^{T} \frac{C_P}{T}\, dT + n\tilde{S}^{\ominus}[T_0, P] \tag{8.14}$$

Similarly, at constant temperature and mole number

$$dS = -\left(\frac{\partial V}{\partial T}\right)_{P,n} dP$$

from which

$$\Delta S = -\int_{P_1}^{P_2} \left(\frac{\partial V}{\partial T}\right)_{P,n} dP \tag{8.15}$$

$$S[T, P] = -\int \left(\frac{\partial V}{\partial T}\right)_{P,n} dP + n\tilde{s}(T) \tag{8.16}$$

and

$$S[T, P] = -\int_{P_0}^{P} \left(\frac{\partial V}{\partial T}\right)_{P} dP + n\tilde{S}^{\ominus}[T, P_0] \tag{8.17}$$

The standard state used in Equations (8.14) and (8.17) are defined in the same way as the standard states for the enthalpy in Equations (8.5) and (8.8). If we choose the standard state as the state of the substance at the temperature T_0 and the pressure P_0, the entropy of the system at the temperature T and the pressure P may be determined by either of the equations

$$S[T, P] - S^{\ominus}[T_0, P_0] = \int_{T_0}^{T} \left(\frac{C_P}{T}\right)_{P_0} dT - \int_{P_0}^{P} \left[\left(\frac{\partial V}{\partial T}\right)_{P,n}\right]_{T} dP \tag{8.18}$$

and

$$S[T, P] - S^{\ominus}[T_0, P_0] = \int_{T_0}^{T} \left(\frac{C_P}{T}\right)_P \mathrm{d}T - \int_{P_0}^{P} \left[\left(\frac{\partial V}{\partial T}\right)_{P,n}\right]_{T_0} \mathrm{d}P \tag{8.19}$$

The change of the Gibbs energy with temperature, pressure, and mole number is given by

$$\mathrm{d}G = -S\,\mathrm{d}T + V\,\mathrm{d}P + \mu\,\mathrm{d}n$$

Again, the variables are not separable and each term must be considered alone. No difficulty occurs when the pressure of a closed system is changed at constant temperature. Then

$$\mathrm{d}G = V\,\mathrm{d}P$$

and

$$\Delta G = \int_{P_1}^{P_2} V\,\mathrm{d}P \tag{8.20}$$

$$G[T, P] = \int V\,\mathrm{d}P + n\tilde{g}(T) \tag{8.21}$$

or

$$G[T, P] = \int_{P_0}^{P} V\,\mathrm{d}P + n\mu^{\ominus}[T, P_0] \tag{8.22}$$

In Equation (8.22) $\mu^{\ominus}[T, P_0]$ is the molar Gibbs energy of the substance in its standard state defined at the temperature T and the arbitrary pressure P_0. The change of the Gibbs energy with temperature cannot be determined, because the absolute value of the entropy is not known, as stated before.

For open systems the change in the mole number that corresponds to the addition or removal of material from the system would be done at constant temperature and pressure. The molar enthalpy, the molar entropy, and the chemical potential would then be constant and the change in the enthalpy, entropy, and Gibbs energy of the system would be the product of the molar quantities and the change in the number of moles. Again, such changes cannot be calculated, because the absolute values are not known. However, the concept of the operation is used in later chapters.

8.2 Changes of the state of aggregation in one-component systems

The changes of the temperature and pressure for one-component systems discussed in Section 8.1 are limited in range, because a change of phase might take place in the system with the change of temperature or pressure. Such changes of phase may be represented by

$$B'(T, P) = B''(T, P)$$

where B' represents the substance in the prime phase and B'' represents the substance in the double-prime phase. The change of phase takes place under equilibrium conditions, so the temperature and pressure of the two phases are the same. The change of enthalpy is simply the difference in the enthalpies of the two phases. Moreover, the molar enthalpies are constant at the given temperature and pressure. Thus, the change of enthalpy may be expressed as

$$\Delta H = H'' - H' = n(\tilde{H}'' - \tilde{H}') \tag{8.23}$$

where n is the number of moles of the substance that have undergone the change of phase. This quantity may be determined by calorimetric methods at constant pressure because ΔH is equal to the heat absorbed by the system for the constant-pressure processes. It may also be determined indirectly by means of the Clapeyron equation (Eq. 5.73)).

By use of similar arguments,

$$\Delta S = (S'' - S') = n(\tilde{S}'' - \tilde{S}') = \Delta H/T \tag{8.24}$$

is obtained for the change of the entropy. Finally, the chemical potentials of the substance in the two states of aggregation are equal at equilibrium, and consequently $\Delta G = 0$.

8.3 Two-phase, one-component systems

A one-component system that exists in two phases is univariant and hence indifferent. In order to define the state of the system, at least two of the independent variables must be extensive. We choose the temperature, volume, and number of moles as the independent variables, and consider the enthalpy, entropy, and Gibbs energy as functions of these variables.

The two condition equations

$$n = n' + n'' \tag{8.25}$$

and

$$V = n'\tilde{V}' + n''\tilde{V}'' \tag{8.26}$$

must always be satisfied; n' and n'' represent the number of moles of the substance in the two separate phases, and $\tilde{V}'$ and $\tilde{V}''$ represent the molar volumes of the substance in the two phases. The first of the two condition equations expresses the condition of mass balance, and the second expresses the condition that the total volume of the system must be the sum of the volumes of the two phases. The two equations can be combined with the elimination of either n' or n''. If we choose to eliminate n'', then

$$V = n'(\tilde{V}' - \tilde{V}'') + n\tilde{V}'' \tag{8.27}$$

This equation makes it possible to determine the number of moles of the component in each phase or to use n' as an independent variable rather than V, if we so choose. The molar volumes are properties of the separate phases, and consequently can be considered as functions of the temperature and pressure. However, the system is univariant, and consequently the pressure is a function of the temperature when the temperature is taken as the independent variable. The relation between the temperature and pressure may be determined experimentally or may be determined by means of the Clapeyron equation. The differential of each molar volume may then be expressed by

$$\mathrm{d}\tilde{V} = \left[\left(\frac{\partial \tilde{V}}{\partial T}\right)_P + \left(\frac{\partial \tilde{V}}{\partial P}\right)_T \left(\frac{\mathrm{d}P}{\mathrm{d}T}\right)_{\mathrm{sat}}\right] \mathrm{d}T \tag{8.28}$$

With this introduction, we consider the enthalpy of the system as a function of the temperature, volume, and number of moles. The differential of the enthalpy is given by

$$\mathrm{d}H = \left(\frac{\partial H}{\partial T}\right)_{V,n} \mathrm{d}T + \left(\frac{\partial H}{\partial V}\right)_{T,n} \mathrm{d}V + \left(\frac{\partial H}{\partial n}\right)_{T,V} \mathrm{d}n \tag{8.29}$$

and the problem reduces to obtaining expressions for the derivatives. The enthalpy of the system is the sum of the enthalpies of the two phases, so

$$H = n'\tilde{H}' + n''\tilde{H}'' \tag{8.30}$$

the differential of which is

$$\mathrm{d}H = (\tilde{H}' - \tilde{H}'')\,\mathrm{d}n' + n'\,\mathrm{d}\tilde{H}' + (n - n')\,\mathrm{d}\tilde{H}'' + \tilde{H}''\,\mathrm{d}n \tag{8.31}$$

From this equation we may write

$$\left(\frac{\partial H}{\partial T}\right)_{V,n} = (\tilde{H}' - \tilde{H}'')\left(\frac{\partial n'}{\partial T}\right)_{V,n} + n'\left(\frac{\partial \tilde{H}'}{\partial T}\right)_{V,n} + (n - n')\left(\frac{\partial \tilde{H}''}{\partial T}\right)_{V,n} \tag{8.32}$$

From Equation (8.27)

$$\left(\frac{\partial n'}{\partial T}\right)_{V,n} = \frac{(V - n\tilde{V}')(\partial \tilde{V}''/\partial T)_{\mathrm{sat}} - (V - nV'')(\partial \tilde{V}'/\partial T)_{\mathrm{sat}}}{(\tilde{V}' - \tilde{V}'')^2} \tag{8.33}$$

and each $(\partial \tilde{V}/\partial T)_{\mathrm{sat}}$ may be written as

$$\left(\frac{\partial \tilde{V}}{\partial T}\right)_{\mathrm{sat}} = \left(\frac{\partial \tilde{V}}{\partial T}\right)_P + \left(\frac{\partial \tilde{V}}{\partial P}\right)_T \left(\frac{\mathrm{d}P}{\mathrm{d}T}\right)_{\mathrm{sat}} \tag{8.34}$$

on the basis of Equation (8.28). The molar enthalpies of each of the phases which are saturated with respect to the other phase are functions of the temperature and pressure in general, but the pressure is a function of the temperature for the two-phase system. We then express $(\partial H/\partial T)_{V,n}$ as

$$\left(\frac{\partial \tilde{H}}{\partial T}\right)_{V,n} = \left(\frac{\partial \tilde{H}}{\partial T}\right)_{\text{sat}} = \tilde{C}_P + \left[\tilde{V} - T\left(\frac{\partial \tilde{V}}{\partial T}\right)_{P,n}\right]\left(\frac{\mathrm{d}P}{\mathrm{d}T}\right)_{\text{sat}} \tag{8.35}$$

for each of the phases. Equations (8.27), (8.33). (8.34), and (8.35) when substituted into Equation (8.32) are sufficient to evaluate $(\partial H/\partial T)_{V,n}$ at any permitted set of values of the independent variables.

The evaluation of $(\partial H/\partial V)_{T,n}$ is also based on Equation (8.31). At constant temperature both $\mathrm{d}H'$ and $\mathrm{d}H''$ are zero, and consequently

$$\mathrm{d}H = (\tilde{H}' - \tilde{H}'')\,\mathrm{d}n' \tag{8.36}$$

That is, the change of enthalpy of a closed system for a change in volume at constant temperature is simply the molar change of enthalpy for the change of phase multiplied by the number of moles that undergo the change of phase. The equation for the derivative may be written either as

$$\left(\frac{\partial H}{\partial V}\right)_{T,n} = (\tilde{H}' - \tilde{H}'')\left(\frac{\partial n'}{\partial V}\right)_{T,n} \tag{8.37}$$

or as

$$\left(\frac{\partial H}{\partial V}\right)_{T,n} = \frac{\tilde{H}' - \tilde{H}''}{\tilde{V}' - \tilde{V}''} \tag{8.38}$$

with the use of Equation (8.27).

The expression for $(\partial H/\partial n)_{T,V}$ is derived also from Equation (8.31), from which we obtain

$$\left(\frac{\partial H}{\partial n}\right)_{T,V} = (\tilde{H}' - \tilde{H}'')\left(\frac{\partial n'}{\partial n}\right)_{T,V} + \tilde{H}'' \tag{8.39}$$

From Equation (8.27) we have

$$\left(\frac{\partial n'}{\partial n}\right)_{T,V} = -\frac{\tilde{V}''}{\tilde{V}' - \tilde{V}''} \tag{8.40}$$

and therefore

$$\left(\frac{\partial H}{\partial n}\right)_{T,V} = \tilde{H}'' - \frac{(\tilde{H}' - \tilde{H}'')\tilde{V}''}{\tilde{V}' - \tilde{V}''} \tag{8.41}$$

The evaluation of Equation (8.41) presents some difficulty. The molar volumes, $\tilde{V}'$ and $\tilde{V}''$, can be determined experimentally, as can $(\tilde{H}' - \tilde{H}'')$. However, the equation contains the molar enthalpy of the double-primed phase, and the absolute value of this quantity is not known. It would then appear that the derivative cannot be evaluated. This difficulty is overcome by suitable choices of standard states for the system. We choose a standard state at each temperature and define it to be the double-primed phase; the value zero is assigned to the enthalpy of this phase at each temperature. Equation (8.41) then becomes

$$\left(\frac{\partial H}{\partial n}\right)_{T,V} = -\frac{\tilde{H}'\tilde{V}''}{\tilde{V}' - \tilde{V}''} \tag{8.42}$$

where $\tilde{H}'$ is actually the molar change of enthalpy for the change of state of aggregation at each temperature. The derivative can thus be evaluated, and the change of enthalpy for the addition of the component to the system at constant temperature can be calculated by integration of this equation at each temperature. All values of the enthalpy thus calculated are *relative* to these standard states, which are *different* for each temperature.

With these standard states, comparison between the values of $(\partial H/\partial n)_{T,V}$ at different temperatures cannot be made. In order to do so, we could choose the standard state to be the double-primed phase at some arbitrary temperature, T_0, with the condition that this phase is saturated with respect to the primed phase; the pressure of the standard state, P_0, is thus determined by the temperature. We then define the enthalpy of the double-primed phase to be zero at T_0. The molar enthalpy of the double-primed phase at any other temperature at which the two phases are at equilibrium may be calculated by the use of Equation (8.35), so

$$\tilde{H}''[T] = \int_{T_0}^{T} \left(\tilde{C}_P'' + \left[\tilde{V}'' - T\left(\frac{\partial \tilde{V}''}{\partial T}\right)_{P,n} \right] \left(\frac{\mathrm{d}P}{\mathrm{d}T}\right)_{\mathrm{sat}} \right) \mathrm{d}T \tag{8.43}$$

It must be remembered that $\tilde{C}_P''$, $\tilde{V}''$, and $(\partial \tilde{V}''/\partial T)_P$ are, in general, functions of the pressure as well as the temperature, but in this special case the pressure is a function of the temperature. The expression for $(\partial H/\partial n)_{T,V}$ finally becomes

$$\left(\frac{\partial H}{\partial n}\right)_{T,V} = \int_{T_0}^{T} \left(\tilde{C}_P'' + \left[\tilde{V}'' - T\left(\frac{\partial \tilde{V}''}{\partial T}\right)_{P,n} \right] \left(\frac{\mathrm{d}P}{\mathrm{d}T}\right)_{\mathrm{sat}} \right) \mathrm{d}T - \frac{(\tilde{H}' - \tilde{H}'')\tilde{V}''}{\tilde{V}' - \tilde{V}''} \tag{8.44}$$

with the use of this last standard state. The enthalpy of the system can thus be determined relative to the chosen standard state.

The integration of Equation (8.29) to determine the change of enthalpy of the system for a finite change of state would be rather a formidable task. If we represent the system by the symbol B and designate the initial state as $[T_1, V_1, n_1]$ and the final state as $[T_2, V_2, n_2]$, we wish to consider the change of state

$$B[T_1, V_1, n_1] = B[T_2, V_2, n_2]$$

For a specific case it would be more convenient to integrate along the path designated by the three successive changes of state

$$B[T_1, V_1, n_1] = B[T_2, V_1, n_1]$$

$$B[T_2, V_1, n_1] = B[T_2, V_2, n_1]$$

and

$$B[T_2, V_2, n_1] = B[T_2, V_2, n_2]$$

so that only one independent variable is changed at a time. Of course, other equivalent paths could be set up. The first of the three changes of state involves the integration of Equation (8.32). The integrand, represented by the right-hand expression of Equation (8.32) with the use of Equations (8.27) and (8.33)–(8.35), is a function of the temperature alone. The pressure–volume–temperature data for the two saturated phases must be known in order to carry out the integration. The integrals required for the other two changes of state present no difficulty because, according to Equations (8.38) and (8.41) or (8.44), only the molar quantities of the two saturated phases appear in the integrands, and these quantities are constant at constant temperature. One final comment concerning the molar heat capacity in Equations (8.35) and (8.43) needs to be made. The heat capacity may not be known as a function of the temperature and pressure. However, if it is known as a function of the temperature at some pressure P_0 and if the pressure–volume–temperature data are known, then we may use $(\partial C_P/\partial P)_{T,n}$ to determine the heat capacity at any pressure P. The derivative is given by

$$\left(\frac{\partial C_P}{\partial P}\right)_{T,n} = -T\left(\frac{\partial^2 V}{\partial T^2}\right)_{P,n} \tag{8.45}$$

and consequently, on integration between the limits P_0 and P,

$$\tilde{C}_P[T, P] = \tilde{C}_P[T, P_0] - \int_{P_0}^{P} T\left(\frac{\partial^2 V}{\partial T^2}\right)_{P,n} \mathrm{d}P \tag{8.46}$$

Equations, similar to those for the enthalpy, can be derived for the entropy and the problems related to standard states are identical. The relations for the Gibbs energy are

$$\left(\frac{\partial G}{\partial T}\right)_{V,n} = -n\tilde{S}' + n\tilde{V}'\left(\frac{\mathrm{d}P}{\mathrm{d}T}\right)_{\mathrm{sat}} = -n\tilde{S}'' + n\tilde{V}''\left(\frac{\mathrm{d}P}{\mathrm{d}T}\right)_{\mathrm{sat}} \tag{8.47}$$

$$\left(\frac{\partial G}{\partial V}\right)_{T,n} = 0 \tag{8.48}$$

and

$$\left(\frac{\partial G}{\partial n}\right)_{T,V} = \mu \tag{8.49}$$

8.4 Three-phase, one-component systems

A one-component system that exists in three phases is indifferent and has no degrees of freedom. In order to define the state of the system then, three extensive variables must be used. We choose for discussion the enthalpy, volume, and number of moles of the components. The enthalpy of the system is additive in the molar enthalpies of the three phases, as is the volume. We can then write three equations:

$$H = n'\tilde{H}' + n''\tilde{H}'' + n'''\tilde{H}''' \tag{8.50}$$

$$V = n'\tilde{V}' + n''\tilde{V}'' + n'''\tilde{V}'''$$

and

$$n = n' + n'' + n'''$$

For a given set of values for H, V, and n, values of n', n'', and n''' can be calculated if the molar enthalpies and molar volumes of the three phases can be determined. The molar volumes can be obtained experimentally, but the absolute values of the molar enthalpies are not known. In order to solve this problem, we make use of the concept of standard states. We choose one of the three phases and define the standard state to be the state of the system when all of the component exists in that phase at the temperature and pressure of the triple point. If we choose the triple-primed phase as the standard phase, we subtract $n\tilde{H}'''$ from each side of Equation (8.50) and obtain

$$(H - n\tilde{H}''') = n'(\tilde{H}' - \tilde{H}''') + n''(\tilde{H}'' - \tilde{H}''') \tag{8.51}$$

This equation gives the enthalpy of the system relative to the standard state, and the independent variable would now be $(H - n\tilde{H}''')$ rather than H itself. The quantities $(\tilde{H}' - \tilde{H}''')$ and $(\tilde{H}'' - \tilde{H}''')$ are the changes of enthalpy when the state of aggregation of 1 mole of the component is changed from the triple-primed state to the primed state and to the double-primed state, respectively, at the temperature and pressure of the triple point. These quantities can be determined experimentally or from the Clapeyron equation, as discussed in Section 8.2. The three simultaneous, independent equations can now be solved, provided values that permit a physically realizable solution have been given to $(H - n\tilde{H}''')$, V, and n. If such a solution is not obtained, the system cannot exist in three phases for the chosen set of independent variables. Actually, the standard state could be defined as one of the phases at any arbitrarily chosen temperature and pressure. The values of the enthalpy and entropy for the phase at the temperature and pressure of the triple point would then be calculated by the methods discussed in Section 8.1.

The values of the other thermodynamic functions are readily obtained for these independent variables. The chemical potential is identical in each phase, and consequently $G = n\mu$. If we choose the same standard state for the Gibbs energy as we did for the enthalpy, we have

$$G - n\mu''' = 0 \tag{8.52}$$

$$(H - n\tilde{H}''') = T(S - n\tilde{S}''') \tag{8.53}$$

and

$$S - n\tilde{S}''' = n'(\tilde{S}' - \tilde{S}''') + n''(\tilde{S}'' - \tilde{S}''') \tag{8.54}$$

8.5 Solutions

The equations developed in Section 8.1 for single-phase, one-component systems are all applicable to single-phase, multicomponent systems with the condition that the composition of the system is constant. The dependence of the thermodynamic functions on concentration are introduced through the chemical potentials because, for such a system,

$$G = \sum_{i=1}^{C} n_i\mu_i \tag{8.55}$$

and the other functions can be obtained from this equation by appropriate differentiation. In this equation the subscript i refers to a component or species and the sum is taken over all components or species. The dependence of the chemical potentials on composition is of special importance because the conditions of equilibrium are expressed in terms of the chemical potentials. *It is through these conditions of equilibrium and the dependence of the chemical potentials on composition that all of the relations involving phase equilibrium and chemical equilibrium are obtained.*

In addition, a change of phase in multicomponent systems does not take place, in general, under conditions of both constant temperature and constant pressure, or with constant composition of the individual phases. We consider, as an example, a change of state represented as

$$nM[\text{s}, T_1, P] = nM[\ell, T_2, P]$$

where M represents 1 mole of solution. Changes in the thermodynamic functions are desired for this change of state. We may separate it into three steps such as

$$nM[\text{s}, T_1, P] = nM[\ell, T', P]$$

$$nM[\text{s}, T', P] = nM[\ell, T'', P]$$

and

$$nM[\ell, T'', P] = nM[\ell, T_2, P]$$

where T' is the temperature at which the solid first begins to melt and T'' is that at which the last solid melts. The changes of the thermodynamic functions for the first and last changes of state can be calculated, but not that for the second. Although in some cases the overall change of enthalpy, entropy, or Gibbs energy including the effect of temperature may be known, it is not possible to calculate the changes directly. However, a different path can be set up in which the change of phase takes place when the components are pure. The changes of the thermodynamic functions are then calculated along this path. Consider a binary solution containing n_A moles of A and n_B moles of B, and the change of state

$$(n_A + n_B)M[\text{s}, T_1, P] = (n_A + n_B)M[\ell, T_2, P]$$

We use the path

$$(n_A + n_B)M[\text{s}, T_1, P] = n_A A[\text{s}, T_1, P] + n_B B[\text{s}, T_1, P]$$
$$n_A A[\text{s}, T_1, P] = n_A A[\text{s}, T_A, P]$$
$$n_A A[\text{s}, T_A, P] = n_A A[\ell, T_A, P]$$
$$n_A A[\ell, T_A, P] = n_A A[\ell, T_2, P]$$
$$n_B B[\text{s}, T_1, P] = n_B B[\text{s}, T_B, P]$$
$$n_B B[\text{s}, T_B, P] = n_B B[\ell, T_B, P]$$
$$n_B B[\ell, T_B, P] = n_B B[\ell, T_2, P]$$
$$n_A A[\ell, T_2, P] + n_B B[\ell, T_2, P] = (n_A + n_B)M[\ell, T_2, P]$$

where T_A and T_B are the melting points of A and B, respectively. Obviously, a similar path may be set up for a liquid–vapour or solid–solid transition. The problem is thus reduced to the determination of the change of the thermodynamic functions for the process of separating the solution into its components at constant temperature and pressure, and reforming the solution from the components against at constant temperature and pressure. This calculation requires a knowledge of the dependence of the chemical potentials on the composition.

In the following sections expressions are developed for the chemical potential of the components or species in solution in terms of the composition —first for an ideal solution and then for real solutions—with special emphasis on reference and standard states.

8.6 The ideal solution

The ideal solution is defined thermodynamically as one in which the dependence of the chemical potential of each component on the composition is expressed by

$$\mu_k[T, P, x] = RT \ln x_k + \mu_k^\ominus[T, P] \tag{8.56}$$

at all temperatures and pressures. The symbol x within the brackets refers to a set of $(C-1)$ values of the mole fractions of the C components. The quantity $\mu_k^\ominus[T, P]$ is first considered as an undetermined constant at each temperature and pressure. Meaning is given to this quantity by considering a system in which $x_k = 1$ and therefore the value of all other mole fractions is zero. Then

$$\mu_k[T, P, x_k = 1] = \mu_k^\ominus[T, P]$$

and we interpret this symbol to represent the chemical potential of the pure kth component at the same temperature and pressure of the solution and in the same state of aggregation as the solution. The complete expression of Equation (8.56) is then written as

$$\mu_k[T, P, x] = RT \ln x_k + \mu_k^\ominus[T, P, x_k = 1] \tag{8.57}$$

The *standard state* for the chemical potential of each component in the solution is thus defined as the pure component at the same temperature and pressure and in the same state of aggregation as the solution. We say that the chemical potential of the kth component in a solution at the temperature T, pressure P, and composition x referred to the pure kth component at the same T and P and in the same state of aggregation is equal to $RT \ln x_k$. This difference is also known as the *change of the chemical potential on mixing* at constant temperature and pressure, so

$$\Delta\mu_k^{\mathrm{M}} = \mu_k[T, P, x] - \mu_k^\ominus[T, P, x_k = 1] = RT \ln x_k \tag{8.58}$$

The Gibbs energy of the solution is given by

$$G[T, P, x] = \sum_{i=1}^{C} n_i RT \ln x_i + \sum_{i=1}^{C} n_i \mu_i^\ominus[T, P, x_i = 1] \tag{8.59}$$

according to Equation (8.55). The change of the Gibbs energy for the formation of the solution at the temperature T and pressure P from the components at the same temperature and pressure and in the same state of aggregation as the solution is given by

$$\Delta G^{\mathrm{M}}[T, P, x] = G[T, P, x] - \sum_{i=1}^{C} n_i \mu_i^\ominus[T, P, x_i = 1] = \sum_{i=1}^{C} n_i RT \ln x_i \tag{8.60}$$

The quantity ΔG^{M} is called the change of Gibbs energy on mixing at constant temperature and pressure or, more briefly, the Gibbs energy of mixing.

By appropriate differentiation of Equations (8.57), (8.59), and (8.60), we obtain the corresponding expressions for the entropy. Thus,

$$\bar{S}_k[T, P, x] = -R \ln x_k + \tilde{S}_k^{\ominus}[T, P, x_k = 1] \tag{8.61}$$

$$S[T, P, x] = -\sum_{i=1}^{C} n_i R \ln x_i + \sum_{i=1}^{C} n_i \tilde{S}_i^{\ominus}[T, P, x_i = 1] \tag{8.62}$$

and

$$\Delta S^{\mathrm{M}}[T, P, x] = -\sum_{i=1}^{C} n_i R \ln x_i \tag{8.63}$$

The quantity $\Delta S^{\mathrm{M}}[T, P, x]$ is referred to as the change of entropy on mixing at constant temperature and pressures or, more briefly, the entropy of mixing. Similarly, we obtain

$$\bar{H}_k[T, P, x] = \tilde{H}_k^{\ominus}[T, P, x_i = 1] \tag{8.64}$$

$$H[T, P, x] = \sum_{i=1}^{C} n_i \tilde{H}_i^{\ominus}[T, P, x_i = 1] \tag{8.65}$$

and

$$\Delta H^{\mathrm{M}}[T, P, x] = 0 \tag{8.66}$$

The change of enthalpy on mixing, $\Delta H^{\mathrm{M}}[T, P, x]$, at constant temperature and pressure is seen to be zero for an ideal solution. The change of the heat capacity on mixing at constant temperature and pressure is also zero for an ideal solution, as are all higher derivatives of ΔH^{M} with respect to both the temperature and pressure at constant composition. Differentiation of Equations (8.57), (8.59), and (8.60) with respect to the pressure yields

$$\bar{V}_k[T, P, x] = \tilde{V}_k^{\ominus}[T, P, x_k = 1] \tag{8.67}$$

$$V[T, P, x] = \sum_{i=1}^{C} n_i \tilde{V}_i^{\ominus}[T, P, x_i = 1] \tag{8.68}$$

and

$$\Delta V^{\mathrm{M}}[T, P, x] = 0 \tag{8.69}$$

Thus, the change of the volume on mixing at constant temperature and pressure, $\Delta V^{\mathrm{M}}[T, P, x]$, is also zero, as are all higher derivatives of ΔV^{M} with respect to both the temperature and pressure at constant composition.

In review, we see that, for an ideal solution, the change of the Gibbs energy on mixing and the change of the entropy on mixing both at constant temperature and pressure are given by Equations (8.60) and (8.63), respectively, at all temperatures and pressures; and that the enthalpy, the heat capacity, and the volume are all additive in the corresponding quantities of the components at all temperatures and pressures.

8.7 Real solutions: reference and standard states

The dependence on composition of the thermodynamic properties of a component in a real solution is developed in terms of the *deviations* of these properties from those of an ideal solution at the same temperature, pressure, and composition. We know that the chemical potential is a function of the state of the system, that it is an intensive variable, and that it is a homogenous function of zeroth degree in the mole numbers. Thus, we know that the chemical potential of a component in a single phase is a function of the composition and that its value, although unknown, is fixed when values have been assigned to the necessary number of intensive, independent variables. We choose these variables to be the temperature, pressure, and $(C-1)$ mole fractions, where C represents the number of components. We now introduce a quantity, $\mu_k^\ominus[T, P]$ that will refer eventually to the standard state but which for the present remains undefined. It is introduced to account for our lack of knowledge of the absolute value of the chemical potential. We need only to know that this quantity is a function of the temperature and pressure and is a constant, *independent of composition*, for any chosen set of values of the temperature and pressure. We then consider the difference between $\mu_k[T, P, x]$ and $\mu_k^\ominus[T, P]$. It is this difference, once $\mu_k^\ominus[T, P]$ has been defined, that is determined experimentally. The experimental determination of this difference is discussed in Chapter 10; throughout this chapter it must be remembered that the difference can be determined. We expect this difference for a real solution to approximate $RT \ln x_k$ (except for a constant) on the basis of the behavior of an ideal solution. We therefore write

$$\mu_k[T, P, x] - \mu_k^\ominus[T, P] = RT \ln x_k + \Delta\mu_{k,x}^{\mathrm{E}}[T, P, x] \tag{8.70}$$

or

$$\mu_k[T, P, x] = RT \ln x_k + \Delta\mu_{k,x}^{\mathrm{E}}[T, P, x] + \mu_k^\ominus[T, P] \tag{8.71}$$

where $\Delta\mu_{k,x}^{\mathrm{E}}$ represents the difference between the actual behaviour of $(\mu_k[T, P, x] - \mu_k^\ominus[T, P])$ and $RT \ln x_k$ (except for a constant) and is called the *excess change* of the chemical potential on mixing. The subscript x is used to designate the use of mole fractions as the composition variables.

Equation (8.71) relates two undefined quantities, $\Delta\mu_{k,x}^{\mathrm{E}}$ and $\mu_k^\ominus[T, P]$, in one equation. The first quantity is a function of the temperature, pressure, and $(C-1)$ mole fractions, but the zero of the function is not defined, and the second is a function of the temperature and pressure, but the state of the system to which it refers is not defined. Alternately, we may use

$$\Delta\mu_{k,x}^{\mathrm{M}}[T, P, x] = \mu_k[T, P, x] - \mu_k^{\ominus}[T, P] \tag{8.72}$$

for the second quantity. Then, $\Delta\mu_{k,x}^{\mathrm{M}}[T, P, x]$ is a function of the temperature, pressure, and $(C-1)$ mole fractions, but its zero is not defined. The arbitrary assignment of the zero value to one of the two quantities, $\Delta\mu_{k,x}^{\mathrm{E}}$ or $\Delta\mu_{k,x}^{\mathrm{M}}$, fixes the zero value for the other because the two are related through Equation (8.71). There are thus two ways to define these quantities. We may choose a state of the system at every temperature and pressure of interest and define the value of $\Delta\mu_{k,x}^{\mathrm{E}}$ to be zero for this state. It is convenient to call this state the *reference state* of the kth component, because all measurements are actually referred to this state. Alternately, we may define $\Delta\mu_{k,x}^{\mathrm{M}}[T, P, x]$ for the kth component to be zero for a given state of the system at every temperature and pressure of interest. This state is called the *standard state* of the kth component, and $\mu_k^{\ominus}[T, P]$ is equal to the chemical potential of the kth component in this state according to Equation (8.72). When the reference state of a component is defined at the same composition of the system for every temperature and pressure, the temperature and pressure derivatives of $\Delta\mu_{k,x}^{\mathrm{E}}$ at the reference state must be zero. Similarly, if the standard state of a component is defined rather than the reference state, and if the standard state is defined at the same composition for every temperature and pressure, then the temperature and pressure derivatives of $\Delta\mu_{k,x}^{\mathrm{M}}$ must be zero at the standard state.

The nomenclature and symbols used by Lewis and Randall [24] are closely related to those introduced here. The activity coefficient, γ_k, is related to $\Delta\mu_k^{\mathrm{E}}$ by

$$\Delta\mu_k^{\mathrm{E}} = RT \ln \gamma_k \tag{8.73}$$

and the activity, a_k, is related to $\Delta\mu_k^{\mathrm{M}}$ by

$$\Delta\mu_k^{\mathrm{M}} = RT \ln a_k \tag{8.74}$$

Thus, *the reference state of a component is the state of the system for which the activity coefficient of the component is defined to be unity and the standard state is the state of the system for which the activity of the component is unity or defined to be unity.*

In this discussion the temperature and pressure of the reference state and of the standard state have been taken to be those of the solution; this usage is consistent with the recommendations of the Commission on Symbols, Terminology, and Units of the Division of Physical Chemistry of the International Union of Pure and Applied Chemistry. For the standard state however, a fixed, arbitrary pressure (presumably 1 bar) might be chosen. If we define

$$\Delta\mu_k^{\mathrm{M}'} = \mu_k[T, P, x] - \mu_k^{\ominus}[T, P_0] \tag{8.75}$$

where P_0 is the fixed, arbitrary pressure, the difference between $\Delta\mu_k^{M'}$ and $\Delta\mu_k^{M}[T, P, x]$ is

$$\Delta\mu_{k,x}^{M'} - \Delta\mu_{k,x}^{M}[T, P, x] = \mu_k^{\ominus}[T, P] - \mu_k^{\ominus}[T, P_0] \tag{8.76}$$

according to Equation (8.72). This difference can be calculated because $(\partial\mu_k/\partial P)_{T,x} = \bar{V}_k$ and the integration can be performed when $\bar{V}_k$ is known as a function of the pressure for the kth component at the composition of the standard state. In this book the pressure of the standard state is always taken to be that of the solution, which may be variable, rather than a fixed arbitrary pressure. A fixed, arbitrary temperature cannot be used, in general, to define the standard state with respect to temperature, because the absolute values of the enthalpy or entropy are generally not known.

Similar arguments and definitions can be applied to the other partial molar thermodynamic functions and properties of the components in solution. By differentiation of Equation (8.71), the following expressions for the partial molar entropy, enthalpy, volume, and heat capacity of the kth component are obtained:

$$\bar{S}_k[T, P, x] = -R \ln x_k + \Delta\bar{S}_{k,x}^{E}[T, P, x] + \bar{S}_k^{\ominus}[T, P] \tag{8.77}$$

$$\bar{H}_k[T, P, x] = \Delta\bar{H}_{k,x}^{E}[T, P, x] + \bar{H}_k^{\ominus}[T, P] \tag{8.78}$$

$$\bar{V}_k[T, P, x] = \Delta\bar{V}_{k,x}^{E}[T, P, x] + \bar{V}_k^{\ominus}[T, P] \tag{8.79}$$

and

$$\bar{C}_{P,k}[T, P, x] = \Delta\bar{C}_{P,k,x}^{E}[T, P, x] + \bar{C}_{P,x}^{\ominus}[T, P] \tag{8.80}$$

The excess quantities in each equation represent the corresponding temperature and pressure derivatives of the excess chemical potential of mixing. When the values of the temperature and pressure derivatives of $\Delta\mu_{k,x}^{E}$ for the reference state or $\Delta\mu_{k,x}^{M}$ for the standard state are defined, then all quantities in Equations (8.77)–(8.80) are also defined.

In summary, a reference state or standard state must be defined for each component in the system. The definition may be quite arbitrary and may be defined for convenience for any thermodynamic system, but the two states cannot be defined independently. When the reference state is defined, the standard state is determined; conversely, when the standard state is defined, the reference state is determined. There are certain conventions that have been developed through experience but, for any particular problem, it is not necessary to hold to these conventions. These conventions are discussed in the following sections. *The general practice is to define the reference state.* This state is then a *physically realizable* state and is the one to which experimental measurements are referred. The standard state may or may not be physically realizable, and in some cases it is convenient to speak of the standard state for the chemical potential, for the enthalpy, for the entropy, for the volume, etc. However, the definition of the standard state of a component may be more convenient than the definition of the reference state in some instances. This question is discussed more fully in Chapters 10 and 11.

8.8 Definitions based on mole fractions

We may choose the reference state of any or all of the components to be the pure component or components in the same state of aggregation as the solution at all temperatures and pressures of interest. Then, according to Equation (8.71),

$$\mu_k^{\bullet}[T, P, x_k = 1] = \mu_k^{\ominus}[T, P] \tag{8.81}$$

that is, the chemical potential of the component in its standard state is also that of the pure component at all temperatures and pressures of interest. Similarly, we find that

$$\tilde{S}_k^{\bullet}[T, P, x_k = 1] = \tilde{S}_k^{\ominus}[T, P] \tag{8.82}$$

$$\tilde{H}_k^{\bullet}[T, P, x_k = 1] = \tilde{H}_k^{\ominus}[T, P] \tag{8.83}$$

$$\tilde{V}_k^{\bullet}[T, P, x_k = 1] = \tilde{V}_k^{\ominus}[T, P] \tag{8.84}$$

and

$$\tilde{C}_{P,k}^{\bullet}[T, P, x_k = 1] = \tilde{C}_{P,k}^{\ominus}[T, P] \tag{8.85}$$

Thus, the properties of the component in its standard state are identical to those of the component in its reference state. This reference state is primarily used for systems whose properties can be studied over the entire range of composition even though a region of partial immiscibility may exist.

The reference state of each component in a system may be defined in many other ways. As an example, we may choose the reference state of each component to be that at some composition with the condition that the composition of the reference state is the same at all temperatures and pressures of interest. For convenience and simplicity, we may choose a single solution of fixed composition to be the reference state for all components, and designate x_k^* to be the mole fraction of the kth component in this solution. If $(\Delta\mu_{k,x}^{\mathrm{E}})'$ represents the values of the excess chemical potential based on this reference state, then $(\Delta\mu_{k,x}^{\mathrm{E}}{}^*)'[T, P, x^*]$ is zero at all temperatures and pressures at the composition of the reference state. That this definition determines the standard state is seen from Equation (8.71), for then

$$\mu_k^{\ominus}[T, P] = \mu_k^*[T, P, x^*] - RT \ln x_k^* \tag{8.86}$$

where x^* in the brackets represents the $(C-1)$ values of the mole fractions in the reference state. The value of μ_k^* is the chemical potential of the kth component in the reference state, but is unknown, when values of the temperature, pressure, and mole fractions in the reference state are given. Therefore, the value of $\mu_k^{\ominus}[T, P]$ is fixed but is also unknown.

The question of what the mole fraction of a component is in its standard state for the chemical potential seldom arises; the important point is that the value of the chemical potential of the component in its standard state is fixed when the reference state of the component has been defined. It is of interest to discuss this question when the pure components are not the reference states, so that a better understanding of the standard state may be obtained. The mole fraction of the kth component in its standard state cannot be determined from Equation (8.86), because the absolute values of the chemical potential are unknown. However, according to Equations (8.71) and (8.72),

$$RT \ln x_k + (\Delta\mu_{k,x}^{\mathrm{E}})'[T, P, x] = 0 \tag{8.87}$$

for the standard state. The excess change of the chemical potential is a function of $(C-1)$ mole fractions, and we assume that this function has been determined experimentally. We can then calculate the value of x_k that satisfies Equation (8.87) after assigning values to $(C-2)$ other mole fractions. The value of x_k obviously depends upon the arbitrary values that are assigned to the other mole fractions. We thus observe that, except in a binary system, the value of the mole fraction of a component in its standard state in a multicomponent system does not have a fixed value. Moreover, the values so obtained may be greater than unity, or possibly negative, and therefore do not correspond to a physically realizable state. We may consider that $\Delta\mu_{k,x}^{\mathrm{M}}$ represents a surface in a multidimensional space of $(C-1)$ coordinates at a fixed temperature and pressure. All points on this surface that satisfy Equation (8.87) represent standard states of the kth component at the given temperature and pressure. A different surface is obtained for each temperature and each pressure; thus, the standard state is a function of the temperature and pressure.

The determination of the mole fraction of the kth component in its standard state for the entropy follows the same argument that was used for the chemical potential. By the use of Equation (8.77) and the condition that $\tilde{S}[T, P, x]$ must equal $\tilde{S}^{\ominus}[T, P]$ for the standard state, we find that

$$-R \ln x_k + (\Delta\tilde{S}_{k,x}^{\mathrm{E}})'[T, P, x] = 0 \tag{8.88}$$

must be satisfied. The arguments are the same, and we can calculate a value for x_k when arbitrary values have been assigned to $(C-2)$ other mole fractions. This value is generally a different value from that obtained from Equation (8.87) for the chemical potential for the same set of values assigned to the $(C-2)$ independent mole fractions. Thus, the value of the mole fraction of the kth component in its standard state for the entropy depends upon the temperature, pressure, and the arbitrary values assigned to the $(C-2)$ mole fractions; it need not have physical reality.

A different result is obtained when we consider the partial molar enthalpy, the partial molar volume, the partial molar heat capacity, and all other higher derivatives taken with respect to the temperature or pressure. At the composition of the reference state, $\Delta\bar{H}^{E}_{k,x}{}^{*}$, $\Delta\bar{V}^{E}_{k,x}{}^{*}$, and $\Delta\bar{C}^{E}_{P,k,x}{}^{*}$ are all equal to zero. Then we have, from Equations (8.78)–(8.80),

$$\bar{H}^{*}_{k}[T, P, x^{*}] = \bar{H}^{\ominus}_{k,x}[T, P] \tag{8.89}$$

$$\bar{V}^{*}_{k}[T, P, x^{*}] = \bar{V}^{\ominus}_{k,x}[T, P] \tag{8.90}$$

and

$$\bar{C}^{*}_{P,k}[T, P, x^{*}] = \bar{C}^{\ominus}_{P,x}[T, P] \tag{8.91}$$

where x^{*} represents the set of values of the mole fractions at the reference state. We find that the partial molar enthalpy, volume, and heat capacity of the kth component in its standard state is equal to the same quantities at the reference state. This result is also true of the higher derivatives with respect to the temperature and pressure.

We find from this discussion that, when the reference state of a component in a multicomponent system is taken to be the pure component at all temperatures and pressures of interest, the properties of the standard state of the component are also those of the pure component. When the reference state of a component in a multicomponent system is taken at some fixed concentration of the system at all temperatures and pressures of interest, the system or systems that represent the standard state of the component are different for the chemical potential, the partial molar entropy, and for the partial molar enthalpy, volume, and heat capacity. There is no real state of the system whose properties are those of the standard state of a component. In such cases it may be better to speak of the standard state of a component for each of the thermodynamic quantities.

The usual choice of a reference state other than the pure components is the infinitely dilute solution for which the mole fractions of all solutes are infinitesimally small and the mole fraction of the solvent approaches unity; that is, the values of the thermodynamic properties of the system in the reference state are the limiting values as the mole fractions of all the solutes approach zero. However, this is not the only choice, and care must be taken in defining a reference state for multicomponent systems other than binary systems. We use a ternary system for purposes of illustration (Fig. 8.1). If we choose the component A to be the solvent, we may define the reference state to be the infinitely dilute solution of both B and C in A. Such a reference state would be useful for all possible compositions of the ternary systems. In other cases it may be advantageous to take a solution of A and B of fixed composition, such as that at M, as the solvent. The ratio of the moles of A to the moles of B in any solution whose composition lies on the line CM is a constant and is equal to that at the point M. We could then choose a reference state as the infinitely dilute solution of C in the mixed solvent M. The values of the thermodynamic properties of the reference state are those obtained in the limit as the mole fraction of C approaches zero along the line CM. A different reference state is obtained for each arbitrarily chosen value of M.

The problem of reference and standard states is illustrated in Figures 8.2–8.8 for a binary solution, for which there is only one independent mole fraction. In the first case we choose the reference state of the first component to be the pure component at all temperatures, and assume that $\Delta\mu_1^E = 500x_2^2$ at 298 K and $200x_2^2$ at 373 K, in J mol^{-1}. The curves for these two equations (curve A for 298 K and curve B for 373 K) are given in Figure 8.2, where the mole fraction scale is continued beyond unity for illustrative purposes. The corresponding curves for $\Delta\mu_1^M$ are given in Figure 8.3. The curve for $\Delta\bar{S}_1^E$ is given in Figure 8.4, that for $\Delta\bar{S}_1^M$ is given in the upper curve of Figure 8.5, and that for $\Delta\bar{H}_1^E$ is given in Figure 8.6. These last curves are determined on the assumption that $\Delta\mu_1^E$ is a linear function of the temperature, and consequently $\Delta\bar{S}_1^E$, $\Delta\bar{S}_1^M$, and $\Delta\bar{H}_1^E$ are independent of the temperature. As a consequence of the chosen reference state, the value of each of the functions is zero for the pure component. If, however, we choose the reference state for the first component to be the system in which the mole fraction of the first component is 0.3, then $\Delta\mu_1^E$ is zero at this mole fraction at all temperatures and pressures. This definition simply shifts the zero on the axis for $\Delta\mu_1$ and the two equations become $\Delta\mu_1^E = 500x_2^2 - 245$ at 298 K and $200x_2^2 - 98$ at 373 K. These curves are given in Figure 8.7. The corresponding curves for $\Delta\mu_1^M$ are given in Figure 8.8. The zero value of this function now occurs at a hypothetical mole fraction of 1.37 at 298 K at 1.14 at 373 K. The curves

Figure 8.1. A ternary system.

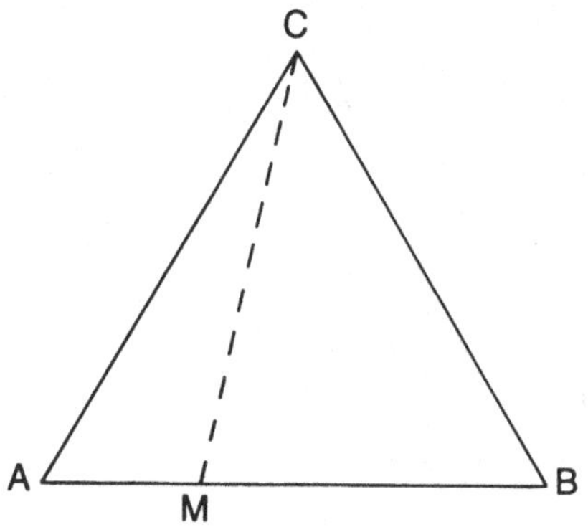

for $\Delta\bar{S}_1^E$ and $\Delta\bar{H}_1^M$ remain the same but the zero value is shifted as shown by the broken horizontal line and the coordinate scale on the right-hand side of Figures 8.4 and 8.6. The new values for $\Delta\bar{S}_1^M$ are given by the lower curve in Figure 8.5. The zero value of this function occurs at a mole fraction of 0.55. Thus, the standard states of the first component occurs at a mole fraction of 1.37 for the chemical potential at 298 K, at 0.55 for the entropy, and at 0.3 for the enthalpy.

8.9 Definitions based on molalities

When the concentration of a multicomponent system is expressed in terms of the molalities of the solutes, the expression for the chemical potential of the individual solutes and for the solvent are somewhat different. For dilute solutions the molality of a solute is approximately proportional to its mole fraction. (The molality, m, is the number of moles of solute per kilogram of solvent. When two or more substances, pure or mixed, may be considered as solvents, a choice of solvent must be clearly stated.) In conformity with Equation (8.68), we then express the chemical potential of a solute in a solution at a given temperature and pressure as

$$\mu_k[T, P, m] = RT \ln m_k + \Delta\mu_{k,m}^E[T, P, m] + \mu_k^\ominus[T, P] \tag{8.92}$$

where $\mu_k^\ominus[T, P]$ is a function of the temperature and pressure and for the

Figure 8.2. Excess chemical potentials in a binary system for the reference state as the pure component.

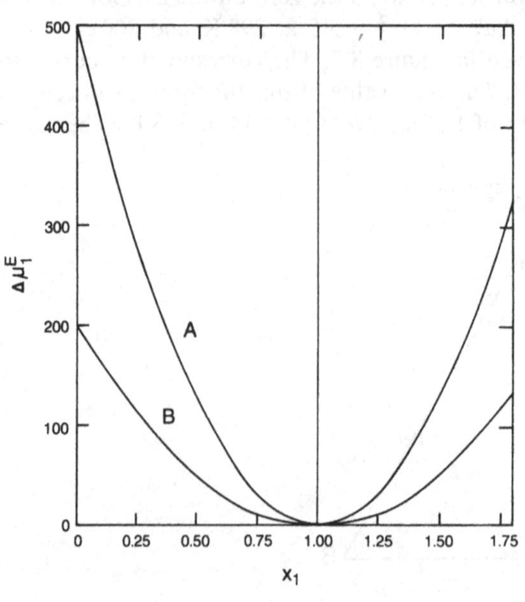

present is undetermined otherwise. The excess quantity, $\Delta\mu^{E}_{k,m}$, is a function of the molalities of all the solutes as well as the temperature and pressure. It may alternatively be written as $RT \ln \gamma_{k,m}$. The quantiy $\gamma_{k,m}$ is called the activity coefficient of the kth component based on molalities. The subscript m in both $\Delta\mu^{E}_{k,m}$ and $\gamma_{k,m}$ is used to indicate that the quantities are expressed in terms of the molality rather than the mole fraction. The symbol m in the brackets represents the molalities of all of the solutes. The reference state when molalities are used is usually taken to be the infinitely dilute solution with respect to all solutes at every temperature and pressure of interest. With this definition $\Delta\mu^{E}_{k,m}$ is defined to go to zero as all molalities go to zero at a constant temperature and pressure. Alternately, the value of the activity coefficient must go to unity as all molalities go to zero. In addition, the values of all the temperature and pressure derivatives of $\Delta\mu^{E}_{k,m}$, such as $\Delta\bar{S}^{E}_{k,m}$, $\Delta\bar{H}^{E}_{k,m}$, and $\Delta\bar{V}^{E}_{k,m}$, $\Delta\bar{C}^{E}_{P,k,m}$, must go to zero as all molalities go to zero. The standard states for the chemical potential, the entropy, and all the other thermodynamic properties of the kth solute are then determined. Thus, from

Figure 8.3. Chemical potentials of mixing in a binary system for reference state as the pure component.

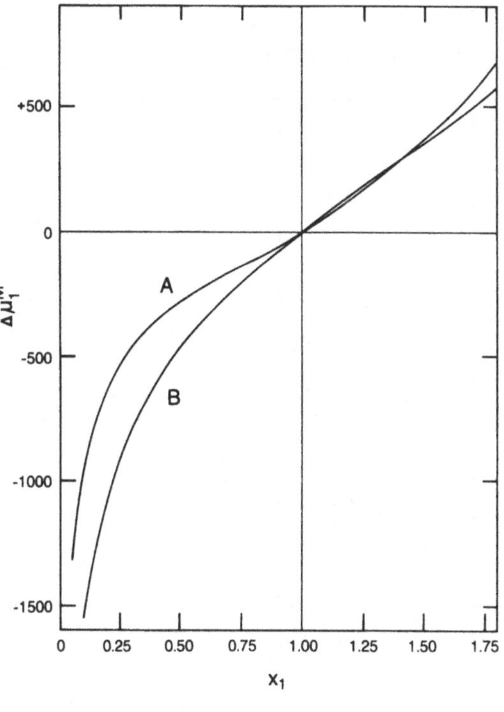

Equation (8.92)

$$\mu_k^{\ominus}[T, P] = \mu_k^{\infty}[T, P, m^{\infty}] - RT \ln m_k^{\infty} \tag{8.93}$$

where m^{∞} refers to the very small molalities of the solutes in the reference state and m_k^{∞} to the particular molality of the kth solute in the reference state. Similarly, for the entropy,

$$\bar{S}_k^{\ominus}[T, P] = \bar{S}_k^{\infty}[T, P, m^{\infty}] + R \ln m_k^{\infty} \tag{8.94}$$

These two equations are useful only to change from one reference state to another. As described in Section 8.8 for mole fractions, the compositions of the standard states of the kth solute for the chemical potential and for the entropy can only be determined by the solution of such equations as

$$RT \ln m_k + \Delta\mu_{k,m}^{\mathrm{E}}[T, P, m] = 0 \tag{8.95}$$

and

$$-R \ln m_k + \Delta\bar{S}_{k,m}^{\mathrm{E}}[T, P, m] = 0 \tag{8.96}$$

respectively. The standard states of the kth solute for the other functions are identical to the reference states as shown by the equations

$$\bar{H}_k^{\ominus}[T, P] = \bar{H}_k^{\infty}[T, P, m^{\infty}] \tag{8.97}$$

$$\bar{V}_k^{\ominus}[T, P] = \bar{V}_k^{\infty}[T, P, m^{\infty}] \tag{8.98}$$

Figure 8.4. Excess entropy of mixing in a binary system.

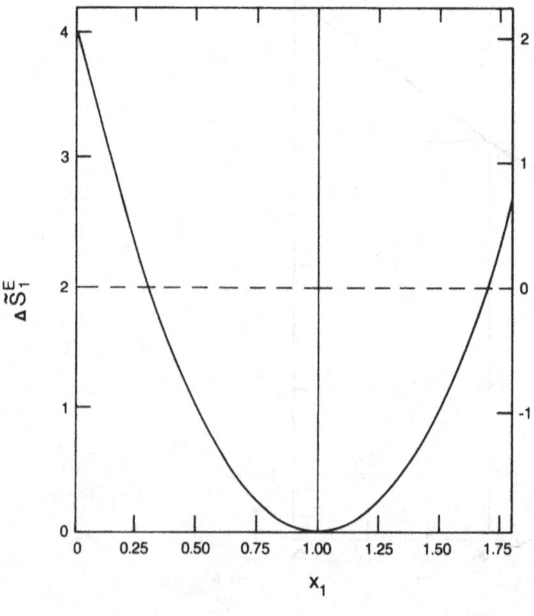

and

$$\bar{C}^{\ominus}_{P,k}[T, P] = \bar{C}^{\infty}_{P,k}[T, P, m^{\infty}] \tag{8.99}$$

which are derived from Equation (8.92) and the definition of the reference state. There is thus no one state of the system that is the standard state for a solute, and if reference is made to such a state, that state is 'hypothetical.'

The chemical potential of the solvent is a function of the molalities of all the solutes at each temperature and pressure, and, for the present, is written as

$$\mu_1[T, P, m] = f(m) + \Delta\mu^{E}_{1,m} + \mu^{\ominus}_1[T, P] \tag{8.100}$$

where $f(m)$ is an undetermined function of all the molalities and $\mu^{\ominus}_1[T, P]$ represents the chemical potential of the standard state of the solvent, also undefined for the present. The subscript 1 is used to represent the solvent. The function $f(m)$ may be determined by use of the Gibbs–Duhem equation at constant temperature and pressure, so that $\sum_{i=1}^{C} n_i \, \mathrm{d}\mu_i = 0$, the sum being taken over all components. However, it is convenient for the purposes of calculation to make the excess chemical potentials obey the Gibbs–Duhem

Figure 8.5. Entropy of mixing in a binary system; reference states are the pure component for the upper curve and $x_1 = 0.3$ for the lower curve.

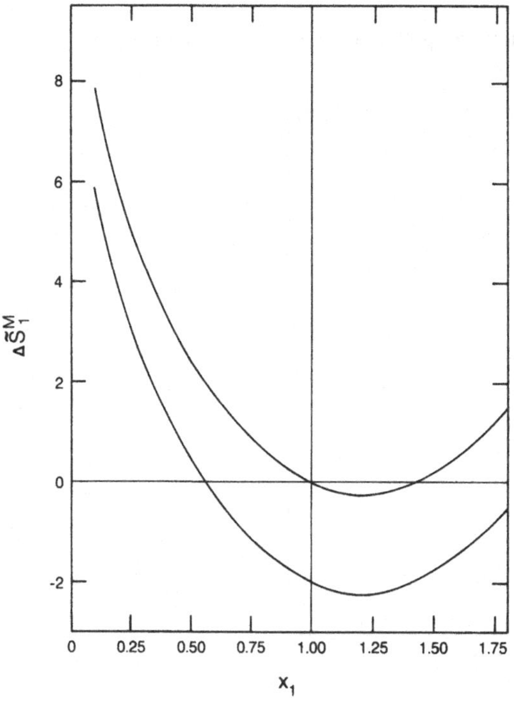

Figure 8.6. Excess enthalpy of mixing in a binary system.

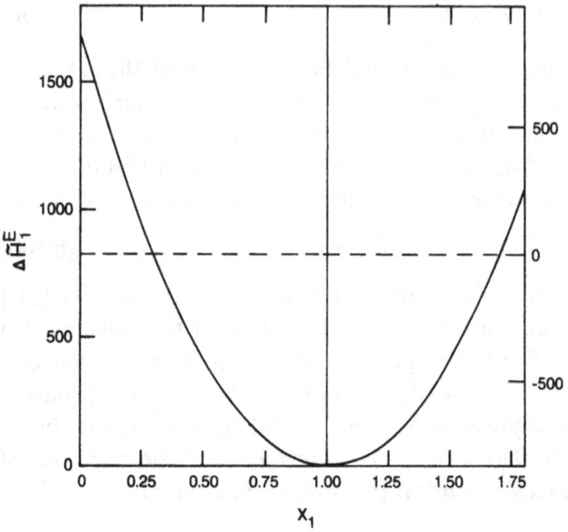

Figure 8.7. Excess chemical potential of mixing in a binary system for reference state at $x_1 = 0.3$.

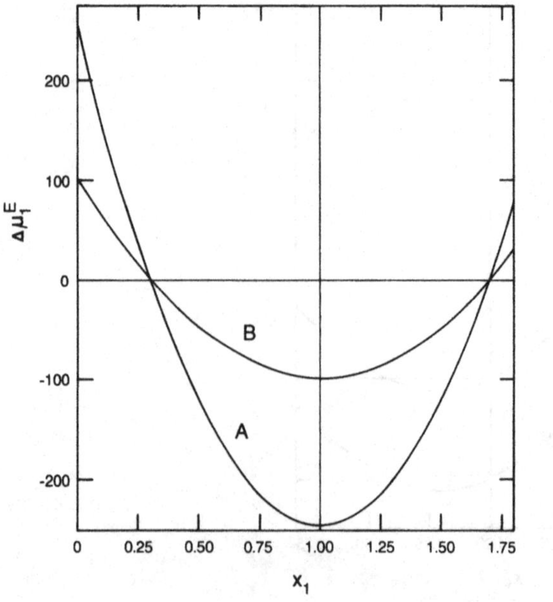

equation, so that $\sum_{i=1}^{C} n_i \, \mathrm{d}\Delta\mu_i^{\mathrm{E}} = 0$ at constant temperature and pressure. When this condition is applied, we find that

$$n_1 \, \mathrm{d}f(m) + \sum_{i=2}^{C} n_i \frac{RT}{m_i} \, \mathrm{d}m_i = 0$$

with the use of Equation (8.92), so that

$$\mathrm{d}f(m) = -\sum_{i=2}^{C} \frac{n_i}{n_1} \frac{RT}{m_i} \, \mathrm{d}m_i \tag{8.101}$$

However,

$$n_i/n_1 = m_i M_1/1000$$

and, therefore,

$$\mathrm{d}f(m) = -\sum_{i=2}^{C} \frac{RTM_1}{1000} \, \mathrm{d}m_i$$

Figure 8.8. Chemical potential of mixing in a binary system for reference state at $x_1 = 0.3$.

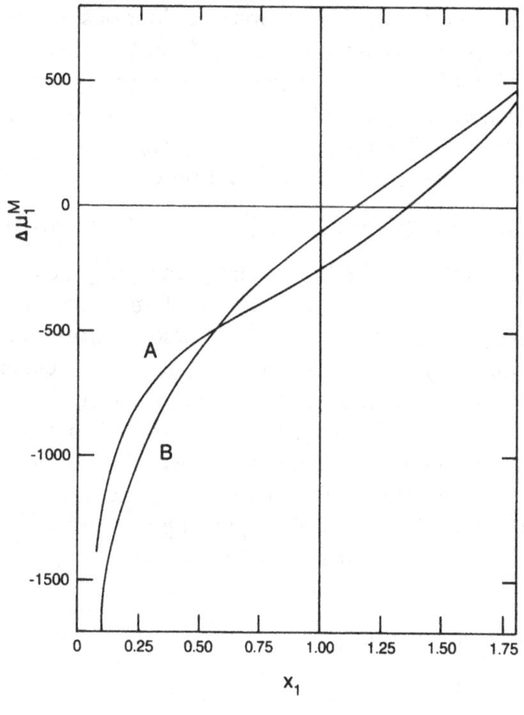

Then, on integration, we find that

$$f(m) = -\sum_{i=2}^{C} \frac{RTM_1}{1000} m_i + f(T, P)$$

where $f(T, P)$ is a function of the temperature and pressure. The complete equation for the chemical potential of the solvent becomes

$$\mu_1[T, P, m] = -\sum_{i=2}^{C} \frac{RTM_1}{1000} m_i + \Delta\mu_{1,m}^{E} + \mu_1^{\ominus}[T, P] \tag{8.102}$$

When the infinitely dilute solution with respect to all solutes is chosen as the reference state for the solvent, the first two terms on the right-hand side of Equation (8.102) become zero at the reference state and

$$\mu_1^{\ominus}[T, P] = \mu^{\bullet}[T, P, m^{\infty}] \tag{8.103}$$

Thus, the standard state of the solvent is the pure solvent and is identical to the reference state for the solvent in all of its thermodynamic properties.

8.10 Definitions based on molarities

The definitions based on molarities, c, are very similar to those based on molalities, and again the solvent must be considered separately from the solutes. (The molarity is defined as the number of moles per liter of solution, and is dependent on the pressure and temperature.) Molarities, like the molalities, are used primarily for solutions for which the concentration ranges are limited. For dilute solutions the molarities of the solutes are approximately proportional to their mole fractions. We thus express the chemical potential of the kth solute in solution at a given temperature and pressure as

$$\mu_k[T, P, c] = RT \ln c_k + \Delta\mu_{k,c}^{E}[T, P, c] + \mu_k^{\ominus}[T, P] \tag{8.104}$$

where $\Delta\mu_{k,c}^{E}$ may be written alternately as $RT \ln \gamma_{k,c}$. The quantity $\gamma_{k,c}$ is called the activity coefficient of the kth component in terms of molarity. The subscript c in both $\Delta\mu_{k,c}^{E}$ and $\gamma_{k,c}$ is used to indicate that these quantities are expressed in terms of the molarity. In this equation c, in the brackets, represents the set of values of the molarities of all the solutes and $\mu_k^{\ominus}[T, P]$ is again undefined at present. The reference state is usually defined to be the infinitely dilute solution for all solutes at all temperatures and pressures. Thus, each $\Delta\mu_{k,c}^{E}$ is defined to go to zero as all molarities go to zero, and all the temperature and pressure derivatives of each $\Delta\mu_{k,c}^{E}$ must go to zero under these conditions. For the standard state for each solute we have the set of equations

$$\mu_k^{\ominus}[T, P] = \mu_k^{\infty}[T, P, c^{\infty}] - RT \ln c_k^{\infty} \tag{8.105}$$

$$\bar{S}_k^{\ominus}[T, P] = \bar{S}_k^{\infty}[T, P, c^{\infty}] + R \ln c_k^{\infty} \tag{8.106}$$

$$\bar{H}_k^{\ominus}[T, P] = \bar{H}_k^{\infty}[T, P, c^{\infty}] \tag{8.107}$$

$$\bar{V}_k^{\ominus}[T, P] = \bar{V}_k^{\infty}[T, P, c^{\infty}] \tag{8.108}$$

and

$$\bar{C}_{P,k}^{\ominus}[T, P] = \bar{C}_{P,k}^{\infty}[T, P, c^{\infty}] \tag{8.109}$$

The first two of these equations are useful, in addition to showing that the standard state of the kth solute for the chemical potential and the entropy are determined, only in converting from one reference state to another. If it is ever necessary, the composition of the standard state for the chemical potential and the entropy would have to be determined by solution of equations such as

$$RT \ln c_k + \Delta\mu_{k,c}^{\mathrm{E}} = 0 \tag{8.110}$$

and

$$-R \ln c_k + \Delta\bar{S}_{k,c}^{\mathrm{E}} = 0 \tag{8.111}$$

when $\Delta\mu_{k,c}^{\mathrm{E}}$ and $\Delta\bar{S}_{k,c}^{\mathrm{E}}$ are known as functions of the molarities of all of the solutes. We see from Equations (8.107)–(8.109) that the standard states for the volume, enthalpy, and heat capacity are identical to the reference state. Thus, again it is better to speak of the standard state of a solute for the various functions, rather than a single standard for the solute. Indeed, such a single state is hypothetical.

For the solvent, we must again make use of the Gibbs–Duhem equation to find a suitable function of the molarities of the solutes. We set

$$\mu_1[T, P, c] = f(c) + \Delta\mu_{1,c}^{\mathrm{E}}[T, P, c] + \mu^{\ominus}[T, P] \tag{8.112}$$

where $f(c)$ represents a function of the molarities of all of the solutes. Then, according to the Gibbs–Duhem equation at constant temperature and pressure,

$$n_1 \, \mathrm{d}f(c) + RT \sum_{i=2}^{C} \frac{n_i \, \mathrm{d}c_i}{c_i} + \sum_{i=1}^{C} n_i \, \mathrm{d}\Delta\mu_{i,c}^{\mathrm{E}} = 0 \tag{8.113}$$

where the solvent is excluded in the first sum but is included in the second sum. Then

$$df(c) = -RT \sum_{i=2}^{C} \frac{n_i}{n_1} \frac{\mathrm{d}c_i}{c_i} \tag{8.114}$$

when $\sum_{i=1}^{C} n_i \, \mathrm{d}\Delta\mu_{i,c}^{\mathrm{E}}$ is set equal to zero. The number of moles of a given solute in 1 liter of solution is c_i and the number of moles of solvent is $(1000\rho - \sum_{i=2}^{C} c_i M_i)/M_1$, where ρ is the density of the solution in units of g ml^{-1} and M_i and M_1 are the molecular masses of the solute and solvent, respectively. Then

$$\frac{n_i}{n_1} = \frac{c_i M_1}{1000\rho - \sum_{i=2}^{C} c_i M_i} \tag{8.115}$$

Integration of Equation (8.114) after substitution of Equation (8.115) cannot be performed easily, because ρ is a function of the concentrations of the solutes and is generally different for each system. However, for dilute solutions 1000ρ is large with respect to $\sum_{i=2}^{C} c_i M_i$ and ρ approximates ρ_1, the density of the pure solvent. If these approximations are made, then

$$\mathrm{d}f(c) = -\frac{RTM_1}{1000\rho_1} \sum_{i=2}^{C} \mathrm{d}c_i \tag{8.116}$$

and

$$f(c) = -\frac{RTM_1}{1000\rho_1} \sum_{i=2}^{C} c_i + f[T, P] \tag{8.117}$$

The chemical potential of the solvent is then written as

$$\mu_1[T, P, c] = -\frac{RTM_1}{1000\rho_1} \sum_{i=2}^{C} c_i + \Delta\mu_{1,c}^{\mathrm{E}}[T, P, c] + \mu^{\ominus}[T, P] \tag{8.118}$$

The difference between the integral of Equation (8.114) and that of Equation (8.116) is contained in the excess chemical potential, and consequently $\sum_{i=1}^{C} n_i \, \mathrm{d}\Delta\mu_{i,c}^{\mathrm{E}}$ cannot strictly be zero and can only approximate zero for dilute solutions. Thus, Equation (8.113), after substitution of Equation (8.116), must always be used for the Gibbs–Duhem equation when molarities are used for the composition variable.

When the infinitely dilute solution, with respect to all solutes, is used as the reference state of the solution at all temperatures and pressures, $\Delta\mu_{1c}^{\mathrm{E}}$ approaches zero as all c_is approach zero. Thus, the standard state of the solvent is the pure solvent at all temperature and pressures and is identical to the reference state of the solvent for all thermodynamic functions.

8.11 The osmotic coefficient

Another function, the *osmotic coefficient*, has been used in place of the excess chemical potential or the activity coefficient. It is a multiplicative factor rather than additive, and is defined in terms of the chemical potential of the solvent. Two such functions are used, one based on molalities and the other on molarities. The first is defined, except for its absolute value, by

$$\mu_1[T, P, m] = -\frac{RTM_1\phi_m}{1000} \sum_{i=2}^{C} m_i + \mu_1^{\ominus}[T, P] \tag{8.119}$$

and the other, except for its absolute value, by

$$\mu_1[T,P,c] = -\frac{RTM_1\phi_c}{1000\rho_1}\sum_{i=2}^{C} c_i + \mu_1^{\ominus}[T,P] \tag{8.120}$$

The definition is completed by assigning a value to ϕ_m and ϕ_c in some reference state. To conform with the definitions made in Sections 8.9 and 8.10, the infinitely dilute solution with respect to all molalities or molarities is usually used as the reference state at all temperatures and pressures, and both ϕ_m and ϕ_c are made to approach unity as the sum of the molalities or molarities of the solutes approaches zero. The standard state of the solvent is again the pure solvent, and is identical to its reference state in all of its thermodynamic functions.

8.12 The dependence of the activity and the activity coefficient on temperature

In this chapter we have introduced the four quantities, $\Delta\mu_k^{\mathrm{M}}$, $\Delta\mu_k^{\mathrm{E}}$, a_k, and γ_k. The temperature dependence of these quantities are easily obtained from their definitions. Thus,

$$\left(\frac{\partial \Delta\mu_k^{\mathrm{M}}}{\partial T}\right)_{P,n} = -(\bar{S}_k - \bar{S}_k^{\ominus}) = -\Delta\bar{S}_k^{\mathrm{M}} \tag{8.121}$$

and with the relation that

$$\Delta\mu_k^{\mathrm{M}} = RT\ln a_k$$

we have

$$\Delta\bar{S}_k^{\mathrm{M}} = -R\ln a_k - RT\left(\frac{\partial \ln a_k}{\partial T}\right)_{P,n} \tag{8.122}$$

Moreover,

$$\left(\frac{\partial(\Delta\mu_k^{\mathrm{M}}/T)}{\partial(1/T)}\right)_{P,n} = \bar{H}_k - \bar{H}_k^{\ominus} = \Delta\bar{H}_k^{\mathrm{M}} \tag{8.123}$$

and

$$\left(\frac{\partial \ln a_k}{\partial T}\right)_{P,n} = -\frac{\bar{H}_k - \bar{H}_k^{\ominus}}{RT^2} \tag{8.124}$$

Similarly, when we differentiate $\Delta\mu_k^{\mathrm{E}}$ with respect to the temperature, we obtain

$$\Delta\bar{S}_k^{\mathrm{E}} = -R\ln\gamma_k - RT\left(\frac{\partial \ln\gamma_k}{\partial T}\right)_{P,n} \tag{8.125}$$

If we use the alternate differentiation with respect to temperature, we obtain

$$\left(\frac{\partial \ln \gamma_k}{\partial T}\right)_{P,n} = -\frac{\bar{H}_k - \bar{H}_k^\ominus}{RT^2} \tag{8.126}$$

Equations (8.125) and (8.126) are valid for the case in which either mole fractions or molalities are used to express the concentrations. When molarities are used, we must include the temperature derivative of the molarity and of the density or molar volume of the solvent when necessary. Thus, for a solute

$$\left(\frac{\partial \ln \gamma_{k,c}}{\partial T}\right)_{P,n} = -\frac{\bar{H}_k - \bar{H}_k^\ominus}{RT^2} + \frac{1}{V}\left(\frac{\partial V}{\partial T}\right)_{P,n} \tag{8.127}$$

and for the solvent

$$\left(\frac{\partial \ln \gamma_{1,c}}{\partial T}\right)_{P,n} = -\frac{\bar{H}_1 - \bar{H}_1^\ominus}{RT^2} - \frac{\bar{V}_1^\ominus \sum_{i=2}^{C} c_i}{1000V}\left(\frac{\partial V}{\partial T}\right)_{P,n} + \frac{\sum_{i=2}^{C} c_1}{1000}\left(\frac{\partial \bar{V}_1^\ominus}{\partial T}\right)_{P,n} \tag{8.128}$$

Similar equations in terms of the excess entropy may also be obtained.

Constant mole numbers have been used in deriving all of the equations in this section. No changes need be made when we consider the derivatives at constant mole fractions or constant molalities, because these concentration variables are independent of the temperature and pressure. The equations obtained when molarities are used are more complicated and quite inconvenient. If molarities are used, then the volume should be used as an independent variable rather than the pressure.

8.13 The dependence of the activity and the activity coefficient on pressure

Identical methods are used for determining the dependence of the activity and activity coefficient on the pressure as were used for their dependence on the temperature. Again, however, when molarities are used, their dependence on pressure must also be considered. Since the terms, $RT \ln x$, $RT \ln m$, or $(RTM_1 \sum_{i=2}^{C} m_i)/1000$ are not dependent upon the pressure,

$$\left(\frac{\partial \ln a_k}{\partial P}\right)_{T,n} = \left(\frac{\partial \ln \gamma_k}{\partial P}\right)_{T,n} = \frac{\bar{V}_k - \bar{V}_k^\ominus}{RT} \tag{8.129}$$

when mole fractions or molalities are used. When molarities are used

$$\left(\frac{\partial \ln a_{k,c}}{\partial P}\right)_{T,n} = \frac{\bar{V}_k - \bar{V}_k^\ominus}{RT} \tag{8.130}$$

for both the solvent and solute, but

$$\left(\frac{\partial \ln \gamma_{k,c}}{\partial P}\right)_{T,n} = \frac{\bar{V}_k - \bar{V}_k^{\ominus}}{RT} + \frac{1}{V}\left(\frac{\partial V}{\partial P}\right)_{T,n} \tag{8.131}$$

for a solute and

$$\left(\frac{\partial \ln \gamma_{1,c}}{\partial P}\right)_{T,n} = \frac{\bar{V}_1 - \bar{V}_1^{\ominus}}{RT} - \frac{\bar{V}_1 \sum_{i=2}^{C} c_i}{1000}\left(\frac{\partial V}{\partial P}\right)_{T,n} + \frac{\sum_{i=2}^{C} c_i}{1000}\left(\frac{\partial \bar{V}_1^{\ominus}}{\partial P}\right)_{T,n} \tag{8.132}$$

for the solvent. The same comments concerning the constancy of the concentration variables made in Section 8.12 are applicable here.

8.14 Conversion from one reference state to another

Occasionally the problem arises of converting values of various thermodynamic functions of the components of a solution that have been determined on the basis of one reference state to values based on another reference state. To do so we equate the two relations of the thermodynamic function of interest obtained for the two reference states, because the value of the function at a given temperature, pressure, and composition must be the same irrespective of the reference state. We also equate the relation for the thermodynamic function for the component in the new reference state expressed in terms of the new reference state to that for the same state expressed in terms of the old reference state. The desired relation is obtained when the chemical potentials of the component in the different standard states are eliminated from the two equations. For examples, we use only the chemical potentials and discuss three cases.

(A) Consider that the reference state of a component of a solution has been chosen to be the pure component and it is desired to change to a solution of a specific composition, the temperature and the pressure being the same. Let mole fractions be used to express the composition. For the first reference state

$$\mu_k[T, P, x] = RT \ln x_k + \Delta\mu_{k,x}^{\mathrm{E}}[T, P, x] + \mu_k^{\ominus}[T, P, x_k = 1] \tag{8.133}$$

and for the second

$$\mu_k[T, P, x] = RT \ln x_k + (\Delta\mu_{k,x}^{\mathrm{E}})'[T, P, x] + \mu_k^{\ominus}[T, P] \tag{8.134}$$

for any solution, where the prime indicates the values of $\Delta\mu_{k,x}^{\mathrm{E}}$ based on the new reference state. These two equations must be identical, and we have

$$\Delta\mu_{k,x}^{\mathrm{E}}[T, P, x] + \mu_k^{\ominus}[T, P, x_k = 1] = (\Delta\mu_{k,x}^{\mathrm{E}})'[T, P, x] + \mu_k^{\ominus}[T, P] \tag{8.135}$$

A solution in which the mole fractions are given by a set of values, x^*, is chosen as the new reference state. Then the chemical potential of the kth component is expressed as

$$\mu_k^*[T, P, x^*] = RT \ln x_k^* + \mu_k^{\ominus}[T, P] \tag{8.136}$$

at the composition of the new reference state whereas, based on the old reference state, it is expressed as

$$\mu_k^*[T, P, x^*] = RT \ln x_k^* + \Delta\mu_{k,x}^{\mathrm{E}}[T, P, x^*] + \mu_k^{\ominus}[T, P, x_k = 1] \tag{8.137}$$

where $\Delta\mu_{k,x}^{\mathrm{E}}[T, P, x^*]$ is the excess chemical potential of the kth component in a solution having mole fractions x^* when the reference state is the pure substance. These two equations must be identical, therefore

$$\Delta\mu_{k,x}^{\mathrm{E}}[T, P, x^*] + \mu_k^{\ominus}[T, P, x_k = 1] = \mu_k^{\ominus}[T, P] \tag{8.138}$$

On substitution of this equation into Equation (8.135), we obtain

$$\Delta\mu_{k,x}^{\mathrm{E}}[T, P, x] = (\Delta\mu_{k,x}^{\mathrm{E}})'[T, P, x] + \Delta\mu_{k,x}^{\mathrm{E}}[T, P, x^*] \tag{8.139}$$

or

$$(\Delta\mu_{k,x}^{\mathrm{E}})'[T, P, x] = \Delta\mu_{k,x}^{\mathrm{E}}[T, P, x] - \Delta\mu_{k,x}^{\mathrm{E}}[T, P, x^*] \tag{8.140}$$

This becomes

$$\gamma'_{k,x} = \gamma_{k,x}/\gamma_{k,x}^* \tag{8.141}$$

in terms of the activity coefficients (Eq. (8.73)). The relation for the activities is

$$a'_k = a_k x_k^*/a_k^* \tag{8.142}$$

The particular conversion is discussed in Section 11.8 and illustrated in Figures 8.1 and 8.6. It is seen from Equation (8.140) that the conversion in terms of the excess chemical potentials is simply the shifting of the zero point as illustrated in the two figures.

(B) As a second example, suppose that the original reference state of the kth component, considered as a solute, is the pure substance and that mole fractions are used as the composition variable. It is then desired to make the infinitely dilute solution the reference state and to use the molality for the composition variable. Here, again, we express the chemical potential of the kth component in the two equivalent ways:

$$\mu_k[T, P, x] = RT \ln x_k + \Delta\mu_{k,x}^{\mathrm{E}}[T, P, x] + \mu_k^{\ominus}[T, P, x_k = 1] \tag{8.143}$$

and

$$\mu_k[T, P, m] = RT \ln m_k + \Delta\mu_{k,m}^{\mathrm{E}}[T, P, m] + \mu_k^{\ominus}[T, P] \tag{8.144}$$

so

$$RT \ln x_k + \Delta\mu_{k,x}^{\mathrm{E}}[T, P, x] + \mu_k^{\ominus}[T, P, x_k = 1] = RT \ln m_k + \Delta\mu_{k,m}^{\mathrm{E}} + \mu_k^{\ominus}[T, P] \qquad (8.145)$$

The second reference state yields the two equations

$$\mu_k^*[T, P, m^*] = RT \ln m_k^* + \mu_k^{\ominus}[T, P] \qquad (8.146)$$

and

$$\mu_k^*[T, P, x^*] = RT \ln x_k^* + \Delta\mu_{k,x}^{\mathrm{E}}[T, P, x^*] + \mu_k^{\ominus}[T, P, x_k = 1] \qquad (8.147)$$

where x_k^* is the mole fraction of the component corresponding to the molality m_k^*. Therefore

$$RT \ln x_k^* + \Delta\mu_{k,x}^{\mathrm{E}}{}^*[T, P, x^*] + \mu_k^{\ominus}[T, P, x_k = 1] = RT \ln m_k^* + \mu_k^{\ominus}[T, P] \qquad (8.148)$$

When the constant terms are eliminated from Equations (8.145) and (8.148), the resultant equation is

$$RT \ln \frac{x_k}{x_k^*} + \Delta\mu_{k,x}^{\mathrm{E}}[T, P, x] - \Delta\mu_{k,x}^{\mathrm{E}}{}^*[T, P, x^*] = RT \ln \frac{m_k}{m_k^*} + \Delta\mu_{k,m}^{\mathrm{E}}[T, P, m] \qquad (8.149)$$

or

$$\Delta\mu_{k,m}^{\mathrm{E}}[T, P, m] = \Delta\mu_{k,x}^{\mathrm{E}}[T, P, x] - \Delta\mu_{k,x}^{\mathrm{E}}[T, P, x^*] + RT \ln \frac{x_k m_k^*}{m_k x_k^*} \qquad (8.150)$$

However,

$$x_k = m_k \Big/ \left(\frac{1000}{M_1} + \sum_{i=2}^{C} m_i \right) \qquad (8.151)$$

and, because each m_i^* is very small with respect to $1000/M_1$,

$$x_k^* = \frac{m_k^*}{1000/M_1} \qquad (8.152)$$

Furthermore,

$$x_1 = \frac{1000}{M_1} \Big/ \left(\frac{1000}{M_1} + \sum_{i=2}^{C} m_i \right) \qquad (8.153)$$

Then, Equation (8.149) becomes

$$\Delta\mu^{E}_{k,m}[T, P, m] = \Delta\mu^{E}_{k,x}[T, P, x] - \Delta\mu^{E}_{k,x}{}^{*}[T, P, x^{*}] + RT \ln x_1 \tag{8.154}$$

and, in terms of activity coefficients,

$$\gamma_{k,m} = \frac{\gamma_{k,x} x_1}{\gamma^{*}_{k,x}} \tag{8.155}$$

Here, again, we see that the change is one of displacing the zero point of the excess chemical potential when mole fractions are used with an additional term related to the change of the composition variable.

As an alternative, consider the same change of reference state when the kth component is the solvent. Then

$$\begin{aligned} &RT \ln x_1 + \Delta\mu^{E}_{1,x}[T, P, x] + \mu^{\ominus}[T, P, x_1 = 1] \\ &= -\frac{RTM_1}{1000} \sum_{i=2}^{C} m_i + \Delta\mu^{E}_{1,m}[T, P, m] + \mu^{\ominus}[T, P] \end{aligned} \tag{8.156}$$

However, in this case

$$\mu^{\ominus}[T, P, x_1 = 1] = \mu^{\ominus}[T, P]$$

because the standard state of the solvent is the same for both cases. Therefore

$$\Delta\mu^{E}_{1,m}[T, P, m] = \Delta\mu^{E}_{1,x}[T, P, x] + RT \ln x_1 + \frac{RTM_1}{1000} \sum_{i=2}^{C} m_i \tag{8.157}$$

(C) In this example let the original reference state of the kth component, considered as a solute, be the infinitely dilute solution, and let the molalities be used to express the composition. Let the new reference state of the kth component, again considered as a solute, be the infinitely dilute solution, but with the molarities being used as the composition variable. Then we have

$$\begin{aligned} RT \ln m_k + \Delta\mu^{E}_{k,m}[T, P, m] + \mu^{\ominus}_{k,m}[T, P] = RT \ln c_k + \Delta\mu^{E}_{k,c}[T, P, c] \\ + \mu^{\ominus}_{k,c}[T, P] \end{aligned} \tag{8.158}$$

In this equation the subscripts m and c have been used in the terms for the chemical potential in the standard state in order to emphasize that these two terms are not necessarily identical. The equality of the chemical potential in the two reference states yields

$$RT \ln m^{*}_k + \mu^{\ominus}_{k,m}[T, P] = RT \ln c^{*}_k + \mu^{\ominus}_{k,c}[T, P] \tag{8.159}$$

On elimination of the constant terms,

$$\Delta\mu^{E}_{k,c}[T, P, c] = \Delta\mu^{E}_{k,m}[T, P, m] + RT\ln(m_k c_k^*/m_k^* c_k) \tag{8.160}$$

However,

$$c_k = \frac{1000\rho m_k}{1000 + \sum_{i=2}^{C} M_i m_i} \tag{8.161}$$

and

$$c_k^* = m_k^* \rho_1 \tag{8.162}$$

Consequently,

$$\Delta\mu^{E}_{k,c}[T, P, c] = \Delta\mu^{E}_{k,m}[T, P, m] + RT\ln\frac{(1000 + \sum_{i=2}^{C} M_i m_i)}{1000\rho} \tag{8.163}$$

This equation in terms of the activity coefficients is

$$\gamma_{k,c} = \frac{\gamma_{k,m}\rho_1}{\rho}\left(1 + \frac{\sum_{i=2}^{C} M_i m_i}{1000}\right) \tag{8.164}$$

8.15 Reference and standard states for species

In the previous sections concerning reference and standard states we have developed expressions for the thermodynamic functions in terms of the components of the solution. The equations derived and the definitions of the reference and standard states for components are the same in terms of species when reactions take place in the system so that other species, in addition to the components, are present. Experimental studies of such systems and the thermodynamic treatment of the data in terms of the components yield the values of the excess thermodynamic quantities as functions of the temperature, pressure, and composition variables. However, no information is obtained concerning the equilibrium constants for the chemical reactions, and no correlations of the observed quantities with theoretical concepts are possible. Such information can be obtained and correlations made when the thermodynamic functions are expressed in terms of the species actually present or assumed to be present. The methods that are used are discussed in Chapter 11. Here, general relations concerning the expressions for the thermodynamic functions in terms of species and certain problems concerning the reference states are discussed.

When the state of a system is defined by assigning values to the necessary independent variables, the values of all of the thermodynamic functions are fixed. For a single-phase, multicomponent system the independent variables are usually the temperature, pressure, and mole numbers of the components. The Gibbs energy of such a system at a given temperature and pressure is additive in the chemical potentials of the components by Equation (5.62), but, by the same arguments used to obtain Equation (5.62), we can show that the Gibbs energy is also additive in the chemical potentials of the species. Thus, for a given state of the system we can write

$$G = \sum_{i=1}^{C} n_i\mu_i = \sum_{j=1}^{S} n'_j\mu'_j \tag{8.165}$$

where the unprimed quantities refer to components and the primed quantities refer to species.[2] The sums are to be carried over all of the components i and all of the species j. We can show from this equation that the chemical potential of a given compound is the same whether we consider it as a component or as a species. As an example, we consider a four-component system in which three of the components react to form a new species, resulting in a system containing five species. Equation (8.165) becomes

$$G = n_1\mu_1 + n_2\mu_2 + n_3\mu_3 + n_5\mu_5 \tag{8.166}$$

$$= n'_1\mu'_1 + n'_2\mu'_2 + n'_3\mu'_3 + n'_4\mu'_4 + n'_5\mu'_5 \tag{8.167}$$

where the subscript 4 represents the new species.

If we let the chemical reaction be represented as

$$\nu_1 B + \nu_2 B_2 = \nu_3 B_3 + \nu_4 B_4$$

then, by mass balance

$$n'_1 = n_1 - (\nu_1/\nu_4)n'_4$$

$$n'_2 = n_2 - (\nu_2/\nu_4)n'_4$$

and

$$n'_3 = n_3 + (\nu_3/\nu_4)n'_4$$

When this set of equations is substituted into Equation (8.167), we obtain

$$G = n_1\mu'_1 + n_2\mu'_2 + n_3\mu'_3 + n'_4\left(\mu'_4 + \frac{\nu_3}{\nu_4}\mu'_3 - \frac{\nu_2}{\nu_4}\mu'_2 - \frac{\nu_1}{\nu_4}\mu'_1\right) + n'_5\mu'_5$$

but

$$\mu'_4 + \frac{\nu_3}{\nu_4}\mu'_3 - \frac{\nu_2}{\nu_4}\mu'_2 - \frac{\nu_1}{\nu_4}\mu'_1 = 0$$

according to the condition of chemical equilibrium. Also, n_5 must equal n'_5, so

$$G = n_1\mu'_1 + n_2\mu'_2 + n_3\mu'_3 + n_5\mu'_5$$

[2] The molar Gibbs energy of a system based on components need not be the same as that based on species, because the total number of moles in the system based on components need not be the same as that based on species.

For this equation to equal Equation (8.166), the chemical potential of a substance considered as a species must be identical to the chemical potentials of the same substance considered as a component. We must emphasize that the chemical potential of the component and that of the species must refer to the same mass of the substance.

When the same substance may be taken either as a component or as a species, we can write, in terms of mole fractions,

$$\mu_k = RT \ln x_k + \Delta\mu_{k,x}^{E} + \mu_k^{\ominus} = RT \ln x_k' + (\Delta\mu_{k,x}^{E})' + (\mu_k^{\ominus})' \qquad (8.168)$$

There are four undetermined quantities $\Delta\mu_{k,x}^{E}$, $(\Delta\mu_{k,x}^{E})'$, $\mu_k^{\ominus}$, and $(\mu_k^{\ominus})'$ and two equations. We must, therefore, define two of the four quantities, which in turn determines the other two quantities and the relationship between them. We can define the reference states for the component and the species. The difference between the standard chemical potential of the component and that of the species is then expressed in terms of the mole fractions in the reference state. The problem is the determination of this difference. The different species may be known from our knowledge of the chemical system, or they may be assumed. However, a definite decision must be made concerning the species, and all calculations must be carried out based upon this decision. Several examples concerning reference and standard states are discussed here and in the following sections.

The simplest case is one in which we can use the pure substance as the reference state for both the component and the species. If, then, the pure substance consists entirely of the same molecular entity and this entity is taken as the component and the species, the standard states for the component and the species are identical. Equation (8.168) becomes

$$RT \ln x_k + \Delta\mu_k^{E} = RT \ln x_k' + (\Delta\mu_k^{E})' \qquad (8.169)$$

This case is applicable when the components of the system are simple substances that react to form new species in solution.

For the second case the component may actually be a solution of polymeric entities of some monomer. As a simple example we consider that only monomers and dimers exist, so that we have the equilibrium, $2B = B_2$. When we choose the unit mass for the chemical potential of the component to be equal to the molecular mass of the monomer, we have

$$\mu_k = \mu_{\mathrm{M}}' = \tfrac{1}{2}\mu_{\mathrm{D}}' \qquad (8.170)$$

where the subscripts M and D refer to the monomer and the dimer, respectively. The second equality arises from the condition of chemical equilibrium. When we use only the first equality, Equation (8.168) becomes

$$RT \ln x_k + \Delta\mu_k^{E} + \mu_k^{\ominus} = RT \ln x_{\mathrm{M}}' + (\Delta\mu_{\mathrm{M}}^{E})' + \mu_{\mathrm{M}}^{\ominus} \qquad (8.171)$$

We choose the reference state for the component to be the pure component and the reference state for the monomer to be the pure monomer. The standard state of the component is then the pure component in the same state of aggregation as the solution, whereas that of the monomer is a fictitious substance in the same state of aggregation as the solution but containing only monomeric entities. The problem is to find a relationship between the standard chemical potentials of the component and of the species. We do so by recognizing that the pure component is also a solution of monomers and dimers. Again, the chemical potential of the pure component based on the molecular mass of the monomer must be equal to the chemical potential of the monomer in the pure component. Thus,

$$\mu_k^{\ominus}[T, P, x_k = 1] = \mu_{\mathrm{M}}^{*}[T, P, x^{*}] = RT \ln x_{\mathrm{M}}^{*} + (\Delta\mu_{\mathrm{M}}^{\mathrm{E}})^{*} + \mu_{\mathrm{M}}^{\ominus} \quad (8.172)$$

where the asterisk is used here to designate quantities applicable to the monomer in the pure component. When Equation (8.172) is substituted into Equation (8.171), we obtain

$$RT \ln x_k + \Delta\mu_k^{\mathrm{E}} = RT \ln(x_{\mathrm{M}}/x_{\mathrm{M}}^{*}) + [\Delta\mu_{\mathrm{M}}^{\mathrm{E}} - (\Delta\mu_{\mathrm{M}}^{\mathrm{E}})^{*}] \quad (8.173)$$

where x_{M}^{*} is the mole fraction of the monomer in the pure component and $(\Delta\mu_{\mathrm{M}}^{\mathrm{E}})^{*}$ is the excess chemical potential of monomer in the pure component. In general $(\Delta\mu_{\mathrm{M}}^{\mathrm{E}})^{*}$ cannot be evaluated, but it is only the difference $[\Delta\mu_{\mathrm{M}}^{\mathrm{E}} - (\Delta\mu_{\mathrm{M}}^{\mathrm{E}})^{*}]$ with which we are concerned. The arguments are similar if we use the second equality in Equation (8.170) so that $\mu_k = \frac{1}{2}\mu_{\mathrm{D}}'$ or if we use the molecular mass of the dimer as the mass unit for the chemical potential of the component so that $\mu_k = \mu_{\mathrm{D}}' = 2\mu_{\mathrm{M}}'$.

The problems of the reference and standard states are slightly different when we use the infinitely dilute solution of all solutes as the reference states. When we take the reference state of both the component and the species as the infinitely dilute solution, Equation (8.168) becomes

$$\mu_k^{\ominus} - {}_{k}^{\ominus\prime} = RT \ln(x_k')^{*} - RT \ln x_k^{*} \quad (8.174)$$

where the asterisks again refer to the reference states. The problem is the evaluation of the difference of the two standard chemical potentials. When there is only one molecular entity that can be associated with the component, the ratio of the mole fraction of the species to that of the component must approach unity as the mole fraction of the component approaches zero. In such a case the two standard chemical potentials are equal and Equation (8.168) becomes

$$RT \ln x_k + \Delta\mu_k^{\mathrm{E}} = RT \ln x_k' + (\Delta\mu_k^{\mathrm{E}})' \quad (8.175)$$

This case applies to the systems in which the components react together to form new species.

A different problem is met when the component is polymerized. Again, we take the case in which the component exists in solution in both monomeric and dimeric forms so that the equilibrium $2B = B_2$ exists. As the mole fraction of the component is decreased, the equilibrium shifts to the monomeric form. If we choose the unit of mass for the chemical potential of the component to be the molecular mass of the dimer, the ratio of $(x'_D)^*/x_k^*$ approaches zero as x_k^* goes to zero; that is, no dimer exists in the infinitely dilute solution. Under such conditions the difference between the values of the two standard chemical potentials (Eq. (8.174)) goes to $-\infty$. To avoid such an infinity, we choose the unit of mass for the chemical potential of the component to be the molecular mass of the monomer. The ratio $(x'_M)^*/x_k^*$ then approaches unity as x_k^* approaches zero and the two standard chemical potentials are identical. Equation (8.168) reduces to Equation (8.175) for this case. This result is quite general and, whenever a component can undergo an association or dissociation in solution, the basis for the chemical potential of the component must be the simplest molecular entity that the component exhibits in solution.

8.16 Solutions of strong electrolytes

Solutions of electrolytes form a class of thermodynamic systems for which the concept of species is all-important. In this section we discuss the problems of reference and standard states for strong electrolytes as solutes dissolved in some solvent.

We consider an electrolyte having the general formula $M_{\nu_+}A_{\nu_-}$, where ν_+ and ν_- represent the number of positive and negative ions, respectively, formed from one molecule of the electrolyte. The only species present in solution, based on the concepts of strong electrolytes, are stipulated to be the ions M_+ and A_-. The symbols M_+ and A_- are used to represent the ions without reference to the actual charge. Under this condition we cannot equate the chemical potential of the component to the chemical potential of the nonionized species. The required relation is obtained by the same methods used in Section 8.15. If we consider a binary system having n_1 moles of solvent and n_2 moles of solute, the Gibbs energy is given by

$$G = n_1\mu_1 + n_2\mu_2 \tag{8.176}$$

We also have

$$G = n_1\mu_1 + n_+\mu_+ + n_-\mu_- \tag{8.177}$$

in terms of the species. For the general electrolyte that we are considering, $n_+ = \nu_+ n_2$ and $n_- = \nu_- n_2$. On substituting these relations into Equation (8.177) and equating Equations (8.176) and (8.177), we find that

$$\mu_2 = \nu_+\mu_+ + \nu_-\mu_- \tag{8.178}$$

Equation (8.178) is the basic equation that is needed, and is applicable to all solutes that are assumed to be strong electrolytes.

We recognize that we cannot determine experimentally the thermodynamic properties of a single type of ion in solution, because both positive and negative ions must be present to satisfy the condition of electrical neutrality. However, we can use equations based on those previously derived, and express the chemical potential of a single type of ion in terms of the concentration variables at a given temperature and pressure. We follow convention here and use molalities and activity coefficients. Then we have

$$\mu_+[T, P, m] = RT \ln m_+ + RT \ln \gamma_+[T, P, m] + \mu_+^\ominus[T, P] \quad (8.179)$$

and

$$\mu_-[T, P, m] = RT \ln m_- + RT \ln \gamma_-[T, P, m] + \mu_-^\ominus[T, P] \quad (8.180)$$

When we substitute these two equations into Equation (8.178), we obtain

$$\mu_2[T, P, m] = RT \ln m_+^{\nu_+} m_-^{\nu_-} + RT \ln \gamma_+^{\nu_+}\gamma_-^{\nu_-}[T, P, m] + \nu_+\mu_+^\ominus[T, P] + \nu_-\mu_-^\ominus[T, P] \quad (8.181)$$

The mean activity coefficient is defined as

$$\nu_\pm^\nu = \gamma_+^{\nu_+}\gamma_-^{\nu_-} \quad (8.182)$$

and the mean molality as

$$m_\pm^\nu = m_+^{\nu_+} m_-^{\nu_-} \quad (8.183)$$

where $\nu = \nu_+ + \nu_-$. Equation (8.181) can then be written as

$$\mu_2[T, P, m] = \nu RT \ln m_\pm + \nu RT \ln \gamma_\pm[T, P, m] + \nu_+\mu_+^\ominus[T, P] + \nu_-\mu_-^\ominus[T, P] \quad (8.184)$$

The reference state of the electrolyte can now be defined in terms of this equation. We use the infinitely dilute solution of the component in the solvent and let the mean activity coefficient go to unity as the molality or mean molality goes to zero. This definition fixes the standard state of the solute on the basis of Equation (8.184). We find later in this section that it is neither profitable nor convenient to express the chemical potential of the component in terms of its molality and activity. Moreover, we are not able to separate the individual quantities, $\mu_+^\ominus$ and $\mu_-^\ominus$. Consequently, we arbitrarily define the standard chemical potential of the component by

$$\mu^\ominus[T, P] = \nu_+\mu_+^\ominus[T, P] + \nu_-\mu_-^\ominus[T, P] \quad (8.185)$$

With this definition, Equations (8.181) and (8.184) can be written as

$$\mu_2[T, P, m] = RT \ln m_+^{\nu_+} m_-^{\nu_-} + \nu RT \ln \gamma_\pm[T, P, m] + \mu_2^\ominus[T, P] \quad (8.186)$$

or

$$\mu_2[T, P, m] = \nu RT \ln m_\pm + \nu RT \ln \gamma_\pm[T, P, m] + \mu^\ominus[T, P] \quad (8.187)$$

These equations are used whenever we need an expression for the chemical potential of a strong electrolyte in solution. We have based the development only on a binary system. The equations are exactly the same when several strong electrolytes are present as solutes. In such cases the chemical potential of a given solute is a function of the molalities of all solutes through the mean activity coefficients. In general the reference state is defined as the solution in which the molality of all solutes is infinitesimally small. In special cases a mixed solvent consisting of the pure solvent and one or more solutes at a fixed molality may be used. The reference state in such cases is the infinitely dilute solution of all solutes except those whose concentrations are kept constant. Again, when two or more substances, pure or mixed, may be considered as solvents, a choice of solvent must be made and clearly stated.

It is useful to review the problems when we express the chemical potential of the solute in terms of its molality and activity coefficient as well as by Equation (8.181). When we do so, we obtain

$$\mu_2 = RT \ln m_2 + RT \ln \gamma_2 + \mu_2^\ominus = RT \ln m_+^{\nu_+} m_-^{\nu_-} + \nu RT \ln \gamma_\pm + \nu_+ \mu_+^\ominus + \nu_- \mu_-^\ominus$$

If we define the reference state for both γ_2 and $\gamma_\pm$ as the infinitely dilute solution, we have

$$\mu_2^\ominus - (\nu_+ \mu_+^\ominus + \nu_- \mu_-^\ominus) = RT \ln \left(\frac{m_+^{\nu_+} m_-^{\nu_-}}{m_2} \right)^*$$

for the reference state. The ratio of the molalities in the reference state is zero, and consequently the difference between $\mu^\ominus$ and $(\nu_+ \mu_+^\ominus + \nu_- \mu_-^\ominus)$ is $-\infty$. When we define the reference state as the infinitely dilute solution for $\gamma_\pm$ and use Equation (8.185), we obtain

$$\gamma_2^* = \left(\frac{m_+^{\nu_+} m_-^{\nu_-}}{m_2} \right)^*$$

so that γ_2^* is zero rather than unity in the reference state. It is for these reasons that we do not use γ_2, the activity coefficient of the component when the component is a strong electrolyte. Thus, the activity coefficient of a strong electrolyte taken as a component in solution remains undefined.

The situation with the activities is different. When we use Equation (8.74) for the component and the separate ions, we have

$$\mu_2 = RT \ln a_2 + \mu_2^\ominus = RT \ln a_+^{\nu_+} a_-^{\nu_-} + \nu_+ \mu_+^\ominus + \nu_- \mu_-^\ominus$$

We can define the mean activity as

$$a_{\pm}^{\nu} = a_{+}^{\nu_+} a_{-}^{\nu_-} \tag{8.188}$$

Then, with the definition of the standard chemical potentials given in Equation (8.185),

$$a_2 = a_{\pm}^{\nu} = a_{+}^{\nu_+} a_{-}^{\nu_-} \tag{8.189}$$

8.17 Solutions of weak electrolytes

The methods for obtaining expressions for the chemical potential of a component that is a weak electrolyte in solution are the same as those used for strong electrolutes. For illustration we choose a binary system whose components are a weak electrolyte represented by the formula M_2A and the solvent. We assume that the species are M^+, MA^-, A^{2-}, and M_2A. We further assume that the species are in equilibrium with each other according to

$$M_2A = M^+ + MA^- \tag{8.190}$$

and

$$MA^- = M^+ + A^{2-} \tag{8.191}$$

The first problem is to obtain relations between the chemical potential of the component and those of the species. The Gibbs energy of a solution containing n_1 moles of solvent and n_2 moles of solute is given by

$$G = n_1\mu_1 + n_2\mu_2 \tag{8.192}$$

In terms of the species, we have

$$G = n_1\mu_1 + n_{M^+}\mu_{M^+} + n_{MA^-}\mu_{MA^-} + n_{A^{2-}}\mu_{A^{2-}} + n_{M_2A}\mu_{M_2A} \tag{8.193}$$

Neither the mole numbers of the species nor their chemical potentials are all independent, but are subject to the condition of mass balance

$$n_2 = n_{MA^-} + n_{A^{2-}} + n_{M_2A}$$

the condition of electroneutrality

$$n_{M^+} = n_{MA^-} + 2n_{A^{2-}}$$

and the two conditions of chemical equilibrium

$$\mu_{M^+} + \mu_{MA^-} - \mu_{M_2A} = 0 \tag{8.194}$$

and

$$\mu_{M^+} + \mu_{A^{2-}} - \mu_{MA^-} = 0 \tag{8.195}$$

When these four condition equations are used in Equation (8.193) to make all of the remaining mole numbers and chemical potentials independent, we obtain one of the following three equations:

$$\mu_2 = \mu_{M_2A} \tag{8.196}$$

$$\mu_2 = \mu_{M^+} + \mu_{MA^-} \tag{8.197}$$

or

$$\mu_2 = 2\mu_{M^+} + \mu_{A^{2-}} \tag{8.198}$$

depending on which variables are eliminated. Given one of Equations (8.196)–(8.198), the other two can be obtained by use of Equations (8.194) and (8.195). Thus, the three equations are equivalent. When we express the chemical potential of each of the species in terms of its molality and activity coefficient, we obtain

$$\mu_2 = RT \ln m_{M_2A} + RT \ln \gamma_{M_2A} + \mu^{\ominus}_{M_2A} \tag{8.199}$$

$$\mu_2 = RT \ln m_{M^+} m_{MA^-} + RT \ln \gamma_{M^+}\gamma_{MA^-} + \mu^{\ominus}_{M^+} + \mu^{\ominus}_{MA^-} \tag{8.200}$$

and

$$\mu_2 = RT \ln m^2_{M^+} m_{A^{2-}} + RT \ln \gamma^2_{M^+}\gamma_{A^{2-}} + 2\mu_{M^+} + \mu^{\ominus}_{M^{2-}} \tag{8.201}$$

We define the reference state as the infinitely dilute solution of the component in the solvent, so that γ_{M_2A}, $\gamma_{M^+}\gamma_{MA^-}$, and $\gamma^2_{M^+}\gamma_{A^{2-}}$ all go to unity as m_2 goes to zero. This definition fixes the values of the quantities $\mu^{\ominus}_{M_2A}$, $(\mu^{\ominus}_{M^+} + \mu^{\ominus}_{MA^-})$, and $(2\mu^{\ominus}_{M^+} + \mu^{\ominus}_{A^{2-}})$ according to Equations (8.199), (8.200), and (8.201), respectively.

The quantity, $\mu^{\ominus}_{M_2A}$, represents the chemical potential of the undissociated species M_2A in its standard state, and not that of the component. The other standard-state quantities represent the chemical potential of the designated species in their standard states, but, for the present, we cannot separate the two standard chemical potentials from the sums, neither is it important to do so. The standard-state quantities appearing in Equations (8.199)–(8.201) are not all independent, because the three equations are equivalent. If we equate Equations (8.199) and (8.200), we obtain an expression that can be evaluated experimentally for the quantity $(\mu^{\ominus}_{M^+} + \mu^{\ominus}_{MA^-} - \mu^{\ominus}_{M_2A})$. Similarly, we obtain an expression that can be evaluated experimentally for the quantity $(2\mu^{\ominus}_{M^+} + \mu^{\ominus}_{A^{2-}} - \mu^{\ominus}_{M_2A})$ when we equate Equations (81.99) and (8.201). These last two quantities are related to the equilibrium constants for the chemical reactions. This relation is developed in Chappter 11 and the basic experimental methods are discussed in Chapters 10 and 11.

8.18 Mixtures of molten salts

Salts, when in the liquid state, continue to exhibit the ionic properties of the solid state. It is therefore advantageous to consider the pure molten salts and mixtures of molten salts as systems composed of ionic species, and to express the thermodynamic functions in terms of the concentrations of the ionic species and their excess chemical potentials. The methods are very similar to those used in the previous few sections. We consider separately systems that are completely dissociated into the simplest ions and those that may have complex ions.

For the first case we consider a salt having the general formula $M_{\nu_+}A_{\nu_-}$ and assume that the only species are M_+ ions and A_- ions. The chemical potential of the component can be expressed as

$$\mu_k = \nu_+\mu_+ + \nu_-\mu_- \tag{8.202}$$

in terms of the chemical potentials of the ions. This relation is obtained by the use of the methods used previously. Before proceeding, however, we must choose the composition variables that we wish to use. We choose mole fractions, because of the lack of a solvent and because many molten salt mixtures can be studied over the whole range of concentration. In studying thermodynamic properties of mixtures, we wish to interpret the excess chemical potentials as deviations from some ideal or theoretical behavior. Such considerations, which are beyond the scope of this book, indicate that positive-ion fractions and negative-ion fractions are appropriate for the mixtures discussed in this section. (Other composition variables can be used, and the decision concerning which variables to use is an individual choice.) The *positive-ion fraction* of a given positive ionic species is defined as the number of moles of the given species divided by the total number of moles of positively charged ionic species contained in the system. The *negative-ion fraction* is defined in the same way for the negatively charged species. We then write for the chemical potential of the M_+ species

$$\mu_+ = RT\ln x_+ + \Delta\mu_+^{\mathrm{E}} + \mu_+^{\ominus} \tag{8.203}$$

and a similar expression for the A_- species:

$$\mu_k = \nu_+RT\ln x_+ + \nu_-RT\ln x_- + \nu_+\Delta\mu_+^{\mathrm{E}} + \nu_-\Delta\mu_-^{\mathrm{E}} + \nu_+\mu_+^{\ominus} + \nu_-\mu_-^{\ominus} \tag{8.204}$$

We choose the pure component to be the reference state for the compound, and therefore also the standard state. We choose the reference state of the M_+ species to be a fictitious system that contains only M_+ molecular entities and the reference state of A_- species to be a fictitious system containing only A_- molecular entities according to Equation (8.203). The symbols, $\mu_{M_+}^{\ominus}$ and $\mu_{A_-}^{\ominus}$ then represent the chemical potentials of the M_+ and A_- species, respectively, in their standard state—the fictitious systems. However, the pure component is also a mixture of the two ions and, according to Equations (8.202) and (8.203), we have

$$\mu_k^{\ominus} = \nu_+\mu_+^* + \nu_-\mu_-^* \tag{8.205}$$

$$\mu_k^{\ominus} = \nu_+ RT \ln x_+^* + \nu_- RT \ln x_-^* + \nu_+ (\Delta\mu_+^{\mathrm{E}})^* + \nu_- (\Delta\mu_-^{\mathrm{E}})^* + \nu_+ \mu_+^{\ominus} + \nu_- \mu_-^{\ominus} \quad (8.206)$$

where the asterisks refer to the indicated quantities in the pure liquid component. Actually both $x_{\mathrm{M}_+}^*$ and $x_{\mathrm{A}_-}^*$ are unity because only one positive ionic species and one negative ionic species are present in the pure component. Elimination of $(\nu_+ \mu_{\mathrm{M}_+}^{\ominus} + \nu_- \mu_{\mathrm{A}_-}^{\ominus})$ from Equations (8.204) and (8.206) yields

$$\mu_k = \nu_+ RT \ln x_+ + \nu_- RT \ln x_- + \{\nu_+[\Delta\mu_+^{\mathrm{E}} - (\Delta\mu_+^{\mathrm{E}})^*] + \nu_-[\Delta\mu_-^{\mathrm{E}} - (\Delta\mu_-^{\mathrm{E}})^*]\} + \mu_k^{\ominus} \quad (8.207)$$

None of the individual quantities within the brackets can be evaluated, neither can the two differences. It is therefore convenient to define the excess chemical potential of the component as the quantity within the brackets, so that finally

$$\mu_k = \nu_+ RT \ln x_+ + \nu_- RT \ln x_- + \Delta\mu_k^{\mathrm{E}} + \mu_k^{\ominus} \quad (8.208)$$

Some salts may not dissociate completely into the simplest ion, so some more complex ions are present in solution. For illustrative purposes we choose a salt having the formula M_2A, and assume that the species are M^+, MA^-, and A^{2-}. The derivation of the equations for the chemical potential of the component follows the methods used before. We first have

$$\mu_k = \mu_{\mathrm{M}^+} + \mu_{\mathrm{MA}^-} \quad (8.209)$$

or

$$\mu_k = 2\mu_{\mathrm{M}^+} + \mu_{\mathrm{A}^{2-}} \quad (8.210)$$

Equations similar to Equation (8.203) are written for the chemical potentials of the ionic species. We take the pure liquid component as the reference state, and hence the standard state for the component, and express the chemical potential of the component in terms of the chemical potentials of the ionic species to obtain equations similar to Equation (8.206). In this particular case $x_{\mathrm{M}^+}^*$ is unity but both $x_{\mathrm{MA}^-}^*$ and $x_{\mathrm{A}^{2-}}^*$ have fixed values whose sum must be unity. Finally, with appropriate definitions of the excess chemical potentials, we obtain either

$$\mu_k = RT \ln x_{\mathrm{M}^+} + RT \ln(x_{\mathrm{MA}^-}/x_{\mathrm{MA}^-}^*) + \Delta\mu_k^{\mathrm{E}} + \mu_k^{\ominus} \quad (8.211)$$

or

$$\mu_k = 2RT \ln x_{\mathrm{M}^+} + RT \ln(x_{\mathrm{A}^{2-}}/x_{\mathrm{A}^{2-}}^*) + (\Delta\mu_k^{\mathrm{E}})' + \mu_k^{\ominus} \quad (8.212)$$

The two quantities $\Delta\mu_k^{\mathrm{E}}$ and $(\Delta\mu_k^{\mathrm{E}})'$ are not equal, but are related by the two equations.

Throughout the discussions in Sections 8.15–8.18, we have emphasized methods for obtaining expressions for the chemical potential of a component when we choose to treat the thermodynamic systems in terms of the species that may be present in solution. A complete presentation of all possible types of systems containing charged or neutral molecular entities is not possible. However, no matter how complicated the system is, the pertinent equations can always be developed by the use of the methods developed here, together with the careful definition of reference states or standard states. We should also recall at this point that it is the quantity $(\mu_k - \mu_k^{\ominus})$ that is determined directly or indirectly from experiment.

9

Thermochemistry

Thermochemistry is concerned with the determination of the heat absorbed by a system when some process occurs within the system. The quantity of heat absorbed may be determined experimentally by the use of calorimeters or by calculation from prior knowledge of the thermodynamic properties of the system. The equations relating the heat absorbed by a system for a given process to the change of energy or enthalpy of the system for the change of state that occurs during the process are the mathematical statements of the first law of thermodynamics. They are Equations (2.26) and (2.30), written here as

$$đQ = \mathrm{d}E + P\,\mathrm{d}V - đW' \qquad (9.1)$$

and

$$đQ = \mathrm{d}H - V\,\mathrm{d}P - đW' \qquad (9.2)$$

In the use of these equations, care must be taken to define the system and its surroundings in order to assure the proper interpretation of both the heat absorbed and the work done by the system. The equations in their integral form may be interpreted in two ways: the quantity of heat absorbed by the system may be measured calorimetrically, from which either the change of energy or the change of enthalpy for the change of state that takes place is calculated by use of Equations (9.1) or (9.2) or, conversely, the quantity of heat absorbed by the system for a given process may be calculated from prior knowledge of the change of energy or enthalpy for the change of state that takes place.

Because of the close relationship between the quantity of heat absorbed by a system for a given process and the accompanying change of energy or enthalpy for the change of state that takes place, the concept of thermochemistry has been extended to include studies of the changes of enthalpy or of energy for changes of state of thermodynamic systems without reference to heat. This chapter is concerned primarily with this extended concept of thermochemistry. Problems associated with the change of enthalpy for various types of changes of state are discussed. The enthalpy is used in preference to the energy, because of the convenience of using the pressure rather than the volume as an experimental variable. The methods developed for the enthalpy are also applicable to the energy.

Several subjects that might be considered under the title of thermochemistry are discussed in previous chapters. Such subjects are the heat capacities of a single-phase system, the dependence of the enthalpy of a single-phase system on temperature and pressure, and the dependence of the enthalpy of a one-component, multiphase system on the temperature, volume, and mole numbers. Here we are concerned with heat capacities of multiphase systems, with changes of enthalpy for the formation of a solution and for a change of concentration of the solution, and with changes of enthalpy of systems in which chemical reactions occur. First the basic concepts of calorimetry are reviewed.

9.1 Basic concepts of calorimetry

We may classify all calorimeters into two groups when we limit the processes to those that involve only the work of expansion or compression: those that operate at constant volume and those that operate at constant pressure. The application of Equation (9.1) to constant-volume calorimeters shows that the heat absorbed by the system equals the change of energy of the system for the change of state that takes place in the system. Similarly, the heat absorbed by the system in constant-pressure calorimeters is equal to the change of enthalpy for the change of state that takes place in the system according to Equation (9.2).

We must define the system and determine the initial and final states of the system very carefully. Although our interest may be the change of energy or enthalpy of a system comprising only the chemical substances undergoing a given change of state, even then the thermodynamic system may be defined in at least two ways. We may consider the chemical substances whose properties we wish to determine as the thermodynamic system. Then the container and all other parts of the calorimeter are considered as the surroundings. Alternately, we may consider the chemical substances and all the other parts of the calorimeter as the thermodynamic system. In either case the heat absorbed by or the work done on the thermodynamic system across the boundary is the quantity that is measured experimentally. This heat or work is equal to the change of energy or the change of enthalpy, as the case may be, of the thermodynamic system. If the chemical system is identical to the thermodynamic system, of course the change of energy or enthalpy of the chemical system must be equal to that of the thermodynamic system. When the container and other parts of the calorimeter are considered to be parts of the thermodynamic system in addition to the chemical system, the change of energy or enthalpy of the thermodynamic system is equal to the sum of the changes of the individual parts of the system. In order to determine the changes of energy or enthalpy of the chemical system, we must determine the changes of the other parts of the thermodynamic system. Finally, one small correction may arise in constant-volume calorimeters when the thermodynamic system includes parts of the calorimeter as well as the chemical system. In this case the volume of the thermodynamic system is kept constant, but the volume of the various parts of the system might change. Then corrections have to be made according to Equation (9.1). Such corrections would not occur in constant-pressure calorimeters unless the parts of the thermodynamic system were separated by a membrane that would sustain a pressure difference across it. Of course, the end-result of determining the change of energy or enthalpy for a given change of state must be the same; the requirement of *carefully defining the system* leads to correct reasoning and results.

The calorimetric determination of changes of energy or enthalpy for changes of states that involve chemical reactions needs some elaboration. We have seen from the above discussion that we obtain the change of energy or enthalpy for the change of state that occurs in the calorimeter for the given process. However, this change of state is not usually the change of state that is desired. In order to determine the change of energy or enthalpy for the desired change of state from that obtained calorimetrically, we must first know or determine the change of state that occurs in the calorimeter; that is, we must know or determine the initial and final states of the chemical system in the calorimeter. We then devise paths connecting the two changes of state and calculate the changes of energy or enthalpy along these paths. If we consider the two changes of state and the connecting paths as a cyclic change of state, then the net change of energy or enthalpy must be zero. This relation then enables us to calculate the change of energy or enthalpy for the desired change of state. As an example, we may want to determine the change of energy or enthalpy for the change of state

$$\begin{aligned}&CH_4[\text{g}, 298\text{ K}, 1\text{ bar, ideal}] + 2O_2[\text{g}, 298\text{ K}, 1\text{ bar, ideal}]\\&\quad = CO_2[\text{g}, 298\text{ K}, 1\text{ bar, ideal}] + 2H_2O[\ell, 298\text{ K}, 1\text{ bar}]\end{aligned}$$

For the experimental determination, we use a bomb calorimeter and assume that the process takes place at constant volume, so that the volume of the chemical system remains constant. An excess of oxygen must be used to ensure complete combustion. The initial state of the chemical system is then a mixture of CH_4 and O_2 with known mole numbers of CH_4 and O_2 at a temperature equal or close to 298 K and at a measured total pressure. The individual gases as well as the gas mixture are not ideal gases. The final state is a two-phase system at a known temperature and measured (or calculated) pressure, one phase being a real gas mixture of CO_2, H_2O, and O_2 and the second phase being a liquid phase containing CO_2, H_2O, and O_2 as components. The temperature of this final state may or may not be equal to the initial temperature. The devising of the connecting paths between these two changes of state is straightforward.

9.2 Molar heat capacities of saturated phases

The heat capacities that have been discussed previously refer to closed, single-phase systems. In such cases the variables that define the state of the system are either the temperature and pressure or the temperature and volume, and we are concerned with the heat capacities at constant pressure or constant volume. In this section and Section 9.3 we are concerned with a more general concept of heat capacity, particularly the molar heat capacity of a phase that is in equilibrium with other phases and the heat capacity of a thermodynamic system as a whole. Equation (2.5), $C = đQ/\mathrm{d}T$, is the basic equation for the definition of the heat capacity which, when combined with Equation (9.1) or (9.2), gives the relations by which the more general heat capacities can be calculated. Actually $đQ/\mathrm{d}T$ is a ratio of differentials and has no value until a path is defined. The general problem becomes the determination of the variables to be used in each case and of the restrictions that must be placed on these variables so that only the temperature is independent.

As a first example for saturated phases, we consider one phase of a two-phase, single-component system that is closed. The molar enthalpy, and hence the molar heat capacity, of a phase is a function of the temperature and pressure. However, the pressure of the saturated phase is a function of the temperature because, in the two-phase system, there is only one degree of freedom. The differential of the molar enthalpy is given by

$$\mathrm{d}\tilde{H} = \left\{\tilde{C}_P + \left[\tilde{V} - T\left(\frac{\partial \tilde{V}}{\partial T}\right)_P\right]\left(\frac{\mathrm{d}P}{\mathrm{d}T}\right)_{\text{sat}}\right\}\mathrm{d}T \tag{9.3}$$

for these conditions, and by use of Equation (9.2) becomes

$$\mathrm{d}Q = \tilde{C}_{\text{sat}}\,\mathrm{d}T = \mathrm{d}\tilde{H} - \tilde{V}\,\mathrm{d}P \tag{9.4}$$

The molar heat capacity of a saturated phase is thus determined to be

$$\tilde{C}_{\text{sat}} = \tilde{C}_P - T\left(\frac{\partial \tilde{V}}{\partial T}\right)_P\left(\frac{\mathrm{d}P}{\mathrm{d}T}\right)_{\text{sat}} \tag{9.5}$$

This equation may be written as

$$\tilde{C}_{\text{sat}} = \tilde{C}_P - \left(\frac{\partial \tilde{V}}{\partial T}\right)_P\frac{\Delta \tilde{H}}{\Delta \tilde{V}} \tag{9.6}$$

according to the Clapeyron equation. In Equation (9.6) $\Delta\tilde{H}$ is the molar change of enthalpy for the change of phase and $\Delta\tilde{V}$ is the corresponding change in molar volume. Equations (9.5) and (9.6) are applicable to all phases, but the results are particularly interesting for gases. If we neglect the molar volume of the liquid or solid phase with respect to the gas phase and make use of the ideal gas law, Equation (9.6) becomes

$$\tilde{C}_{\text{sat}} = \tilde{C}_P - \frac{\Delta\tilde{H}}{T} \tag{9.7}$$

The molar heat capacity of a gas is of the order of 20–32 J K^{-1} for many gases and $\Delta\tilde{H}/T$ is approximately 84 J K^{-1} for many normal liquids. Thus, $\tilde{C}_{\text{sat}}$ is approximately -50 to -63 J K^{-1}; that is, approximately 50–63 J of heat must be removed from 1 mole of gas, which is saturated with respect to a liquid or solid, in order to increase its temperature by 1 K. The negative values arise because the molar volume of the saturated gas decreases with increasing temperature.

The determination of the molar heat capacity of a phase saturated with respect to other phases in a multicomponent system requires the application of sufficient conditions to define the heat capacity. Although expressions are developed here for the molar heat capacity of a saturated phase in general, the expressions can be evaluated only if the phase is pure. The molar enthalpy of a phase is a function of the temperature, pressure, and $(C-1)$ mole fractions, where C represents the number of components. Thus,

$$d\tilde{H} = \left(\frac{\partial\tilde{H}}{\partial T}\right)_{P,x} dT + \left(\frac{\partial\tilde{H}}{\partial P}\right)_{T,x} dP + \sum_{i=1}^{C-1}\left(\frac{\partial\tilde{H}}{\partial x_i}\right)_{T,P,x} dx_i \tag{9.8}$$

If the system is univariant, the pressure and each mole fraction are functions of the temperature and the derivative $d\tilde{H}/dT$ becomes

$$\frac{d\tilde{H}}{dT} = \tilde{C}_P + \left[\tilde{V} - T\left(\frac{\partial\tilde{V}}{\partial T}\right)_{P,x}\right]\left(\frac{dP}{dT}\right)_{\text{sat}} + \sum_{i=1}^{C-1}\left(\frac{\partial\tilde{H}}{\partial x_i}\right)_{T,P,x}\left(\frac{\partial x_i}{\partial T}\right)_{\text{sat}} \tag{9.9}$$

The derivatives $(\partial P/\partial T)_{\text{sat}}$ and $(\partial x_i/\partial T)_{\text{sat}}$ may be determined experimentally or by solution of the set of Gibbs–Duhem equations applicable to each phase, provided we have sufficient knowledge of the system. If the system is multivariant, a sufficient number of intensive variables—the pressure or mole fractions of the components in one or more phases—must be held constant to make the system univariant. Thus, for a divariant system either the pressure or one mole fraction of one of the phases must be held constant. When the pressure is constant, Equation (9.9) becomes

$$\frac{d\tilde{H}}{dT} = \tilde{C}_P + \sum_{i=1}^{C-1}\left(\frac{\partial\tilde{H}}{\partial x_i}\right)_{T,P,x}\left(\frac{\partial x_i}{\partial T}\right)_P \tag{9.10}$$

Two equations are possible when one of the mole fractions is held constant. If the mole fraction is one of the mole fractions of the saturated phase of interest, then Equation (9.9) becomes

$$\frac{d\tilde{H}}{dT} = \tilde{C}_P + \left[\tilde{V} - T\left(\frac{\partial \tilde{V}}{\partial T}\right)_{P,x}\right]\left(\frac{\partial P}{\partial T}\right)_{x_j} + \sum_{i \neq j}^{C-2}\left(\frac{\partial \tilde{H}}{\partial x_i}\right)_{T,P,x}\left(\frac{\partial x_i}{\partial T}\right)_{x_j} \tag{9.11}$$

where x_j indicates the mole fraction that is held constant and the sum is taken over $(C-2)$ mole fractions. If the constant mole fraction is one of the mole fractions of another phase, then Equation (9.9) becomes

$$\frac{d\tilde{H}}{dT} = \tilde{C}_P + \left[\tilde{V} - T\left(\frac{\partial \tilde{V}}{\partial T}\right)_{P,x}\right]\left(\frac{\partial P}{\partial T}\right)_{x'_j} + \sum_{i=1}^{C-1}\left(\frac{\partial \tilde{H}}{\partial x_i}\right)_{T,P,x}\left(\frac{\partial x_i}{\partial T}\right)_{x'_j} \tag{9.12}$$

where x'_j represents the mole fraction that is held constant. The derivatives $(\partial x_i/\partial T)_P$, $(\partial P/\partial T)_{xj}$, and $(\partial x_i/\partial T)_{x'_j}$ may be determined experimentally or by use of the Gibbs–Duhem equations subject to the indicated conditions. The extension to systems having a larger number of degrees of freedom is obvious.

Equations (9.9), (9.10), (9.11), or (9.12) in conjunction with Equation (9.2) give expressions for the molar heat capacity of a saturated phase. However, each equation contains the quantity $(\partial \tilde{H}/\partial x_i)_{T,P,x}$, which in turn contains terms such as $(\bar{H}_i - \bar{H}_k)$ when x_k is taken to be the dependent mole fraction. Evaluation of such quantities requires the knowledge of the absolute values of the enthalpies. Therefore, such terms cannot be evaluated, and the values of the molar heat capacities cannot be calculated. The necessity of knowing the absolute values of the enthalpies arises from the fact that a number of moles of some components must be added to, and the same number of moles of other components must be removed from the 1 mole of saturated phase in order to change the mole fractions of the phase. However, if the saturated phase is pure, even though it is in equilibrium with other phases that are solutions, the molar enthalpy of the phase is not a function of the mole fractions and Equations (9.9)–(9.12) reduce to Equation (9.3).

9.3 Heat capacities of multiphase, closed systems

In this section we consider the heat capacity of a complete system rather than that of a single phase. Equation (9.2) continues to be the basic equation with the condition that $đW' = 0$. The development of the appropriate equations requires expressions for the differential of the enthalpy with a sufficient number of conditions that the heat capacity is defined completely. There are two general cases: univariant systems and multivariant systems.

The state of a closed, univariant system is defined by assigning values to the temperature, the volume, and the mole numbers of the components. For a closed system the mole numbers are constant. Then, to define the heat capacity, we must make the volume constant or an arbitrary function of the temperature. The simplest case is to make the volume constant. The heat capacity of the system is then given by

$$C_V = \left(\frac{\partial H}{\partial T}\right)_{V,n} - V\left(\frac{\mathrm{d}P}{\mathrm{d}T}\right)_{\mathrm{sat}} \tag{9.13}$$

where H represents the enthalpy of the system, V the volume of the system, and $(\partial P/\partial T)_{\mathrm{sat}}$ the rate of change of the pressure of the univariant system with the temperature. The enthalpy of the system is the sum of the enthalpies of the phases, so

$$H = \sum^{\prime} n'\tilde{H}' \tag{9.14}$$

where the primes refer to the phases and the sum is taken over all of the phases. When it is possible to make the standard state of a component the same for all phases, it may be convenient to write Equation (9.14) as

$$H = \sum^{\prime} n'\left(\tilde{H}' - \sum_{i=1}^{C} x_i'\tilde{H}_i^{\ominus}\right) + \sum_{i=1}^{C}\left(\sum^{\prime} x_i'n'\tilde{H}_i^{\ominus}\right) \tag{9.15}$$

where the enthalpies of the components in the standard states have been added and subtracted. The advantage is that the first term on the right-hand side of Equation (9.15) can be written as

$$\sum^{\prime} n'\left(\tilde{H}' - \sum_{i=1}^{C} x_i\tilde{H}_i^{\ominus}\right) = \sum^{\prime} n'\left(\sum_{i=1}^{C} x_i'(\bar{H}_1' - \tilde{H}_i^{\ominus})\right) \tag{9.16}$$

Equation (9.14) is subject to the condition equations

$$V = \sum^{\prime} n'\tilde{V}' = \sum^{\prime} n'\left(\sum_{i=1}^{C} x_i'\bar{V}_i'\right) \tag{9.17}$$

and one equation of the form

$$n_i^0 = \sum^{\prime} x_i'n' \tag{9.18}$$

for each component. Equation (9.17) expresses the volume of the system, whose value has been assigned, in terms of the molar volumes and number of moles of each phase, and Equation (9.18) expresses the mass balance relation for each component. The additional condition equations are $(\mathrm{d}V/\mathrm{d}T) = 0$ and $(\mathrm{d}n_i^0/\mathrm{d}T) = 0$ for each component; these relations express the condition of constant volume and constant mole numbers. The molar enthalpy of a phase, the partial molar enthalpies, the molar volume of a phase, and the partial molar volumes are all functions of the temperature, pressure, and $(C-1)$ mole fractions of a phase. The derivatives of these quantities with respect to the temperature under the given conditions can be expressed in terms of the derivatives $(\partial P/\partial T)_{\text{sat}}$ and $(\partial x_i/\partial T)_{\text{sat}}$ (refer to Eq. (9.9) as an example). These derivatives are evaluated by the solution of the appropriate set of Gibbs–Duhem equations. These equations are then sufficient to obtain the expressions for C_V.

The simplest univariant system is a one-component, two-phase system. The development of the expression for $(\partial H/\partial T)_{V,n}$ for such a system is discussed in Section 8.3. The next-simplest system is a two-component, three-phase system. The appropriate equations for this system are

$$H[T, V, n] = n'\tilde{H}'[T, P, x'] + n''\tilde{H}''[T, P, x''] + n'''\tilde{H}'''[T, P, x'''] \tag{9.19}$$

or

$$\begin{aligned} H[T, V, n] = {} & n'[x_1'(\bar{H}_1' - \tilde{H}_1^{\ominus}) + x_2'(\bar{H}_2' - \tilde{H}_2^{\ominus})][T, P, x'] \\ & + n''[x_1''(\bar{H}_1'' - \tilde{H}_1^{\ominus}) + x_2''(\bar{H}_2'' - \tilde{H}_2^{\ominus})][T, P, x''] \\ & + n'''[x_1'''(\bar{H}_1''' - \tilde{H}_1^{\ominus}) + x_2'''(\bar{H}_2''' - \tilde{H}_2^{\ominus})][T, P, x'''] \\ & + n_1^0\tilde{H}_1^{\ominus}[T, P] + n_2^0\tilde{H}_2^{\ominus}[T, P] \end{aligned} \tag{9.20}$$

$$V = n'\tilde{V}'[T, P, x'] + n''\tilde{V}''[T, P, x''] + n'''\tilde{V}'''[T, P, x'''] \tag{9.21}$$

$$n_1^0 = x_1'n' + x_1''n'' + x_1'''n''' \tag{9.22}$$

and

$$n_2^0 = x_2'n' + x_2''n'' + x_2'''n''' \tag{9.23}$$

The quantities n'' and n''' can be expressed in terms of n', n_1^0, n_2^0, and the indicated mole fractions with the use of Equations (9.22) and (9.23). Equation (9.21) can then be used to evaluate n' in terms of the volume of the system, the molar volumes of the phases, and the mole fractions. The derivative $(\partial n'/\partial T)_{V,n}$ can also be evaluated from the resultant equations. The derivative $(\partial H/\partial T)_{V,n}$ is obtained by appropriate differentiation of Equation (9.19) or (9.20), and can be evaluated with the use of expressions obtained by the methods discussed below.

The state of a multivariant system is defined by assigning values to either the temperature, volume, and mole numbers of the components or the temperature, pressure, and mole numbers. Thus, we define heat capacities at constant volume or heat capacities at constant pressure for such closed systems. The equations and method of calculation are exactly the same as those outlined for univariant systems when the heat capacity at constant volume is desired. For the heat capacity at constant pressure, Equation (9.14) or (9.15) and the set of equations, one for each component, illustrated by Equation (9.18) are still applicable. The method of calculation is the same, with the exception that the volume of the system is a dependent variable and does not enter into the calculations, and that the derivatives of the various enthalpy functions are simpler because of the constancy of the pressure.

9.4 Changes of enthalpy for changes of state involving solutions

Changes of enthalpy for changes of state involving solutions are important because of their close relation to the heat absorbed for a process associated with such a change of state, and because of the relationship of the enthalpy to the temperature dependence of the Gibbs energy and the chemical potential. In this section we discuss the change of enthalpy for the formation of a solution from the components, the change of enthalpy for the change of composition of a solution by the addition of one of the components, and the change of enthalpy when two solutions that contain the same components but have different compositions are mixed to form a third solution. The relations between the actual change of enthalpy and the partial molar enthalpies are developed and the methods by which the partial molar enthalpies relative to suitable standard states may be determined are outlined. It is assumed that no chemical reactions take place. Moreover, all changes of state are assumed to be isothermal and isopiestic.

The change of state for the formation of a solution from the components for a binary system is

$$n_1B_1[\ell, T, P] + n_2B_2[\ell, T, P] = (n_1 + n_2)M[\ell, T, P]$$

with the condition that the temperature and pressure of the final state (the solution) and of the initial state (the separate components) are the same. In this change of state, B_1 and B_2 refer to the components and M refers to the solution; the subscripts 1 and 2 refer to the components B_1 and B_2, respectively. The state of aggregation of all substances has been indicated in this case as liquid, but this is not a limitation of the general considerations. The change of state could involve all gases or all solids, or some of the components may be in different states of aggregation. The change of enthalpy is always the enthalpy of the final state minus that of the initial state, so

$$\Delta H = H - n_1\tilde{H}_1^{\cdot} - n_2\tilde{H}_2^{\cdot} \tag{9.24}$$

where the standard states of the components are the pure substances in the state of aggregation indicated in the equation for the change of state. The change of enthalpy for this change of state has been called the integral change of enthalpy of solution (the integral heat of solution). By Euler's theorem,

$$H = n_1\bar{H}_1 + n_2\bar{H}_2 \tag{9.25}$$

at constant temperature and pressure, and consequently ΔH may be written as

$$\Delta H = n_1(\bar{H}_1 - \tilde{H}_1^{\cdot}) + n_2(\bar{H}_2 - \tilde{H}_2^{\cdot}) \tag{9.26}$$

This equation gives the relation between the change of enthalpy for this change of state and the partial molar enthalpies of the components relative to the chosen standard states.[1] Equation (9.26) can easily be generalized to

$$\Delta H = \sum_{i=1}^{C} n_i(\bar{H}_i - \tilde{H}_i^\bullet) \tag{9.27}$$

for the change of state in which a solution is formed from several components.

A change of standard state of one of the components can easily be accomplished when the pure component at the given temperature and pressure is chosen as the standard state. As an example we assume that ΔH given by Equation (9.26) for the change of state

$$n_1B_1[\text{s}, T, P] + n_2B_2[\ell, T, P] = (n_1 + n_2)M[\ell, T, P]$$

has been determined by calorimetric methods, but that it is desirable to change the standard state of the component B_1 from the solid state to the liquid state. The change of standard state can be accomplished by determining ΔH for the change of state

$$B_1[\ell, T, P] = B_1[\text{s}, T, P]$$

for which $\Delta\tilde{H}$ is $(\tilde{H}_1^\bullet(\text{s}) - \tilde{H}_1^\bullet(\ell))$ and adding this quantity to the original change of enthalpy for the formation of the solution. Thus,

$$\Delta H' = \Delta H + n_1(\tilde{H}_1^\bullet(\text{s}) - \tilde{H}_1^\bullet(\ell)) = n_1(\bar{H}_1 - \tilde{H}_1^\bullet(\ell)) + n_2(\bar{H}_2 - \tilde{H}_2^\bullet(\ell)) \tag{9.28}$$

for the change of state

$$n_1B_1[\ell, T, P] + n_2B_2[\ell, T, P] = (n_1 + n_2)M[\ell, T, P]$$

If the solid is the stable phase of component B_1 at the given temperature and pressure, the liquid state is unstable and the value of $(\tilde{H}_1^\bullet(\text{s}) - \tilde{H}_1^\bullet(\ell))$ must be calculated rather than measured. This calculation requires the calculation of the change of enthalpy for each of the changes of state

$$B_1[\ell, T, P] = B_1[\ell, T_{\text{mp}}, P]$$

$$B_1[\ell, T_{\text{mp}}, P] = B_1[\text{s}, T_{\text{mp}}, P]$$

and

$$B_1[\text{s}, T_{\text{mp}}, P] = B_1[\text{s}, T, P]$$

where T_{mp} represents the melting point of B_1 at the pressure P. The calculation of the change of enthalpy for the first change of state of this series requires

---

[1] Lewis used the symbol L_i for the quantities $(\bar{H}_i - \tilde{H}_i^\ominus)$ and named them the relative partial molar enthalpies. Because of our emphasis on standard states, we prefer the symbol $(\bar{H}_i - \tilde{H}_i^\ominus)$, which is much more explicit.

that the equation for the molar heat capacity of the liquid phase may be extrapolated into the unstable region from T_{mp} to T without serious error.

Equation (9.27) gives the relation needed to determine the partial molar enthalpy of a component relative to a standard state. We write this equation as

$$\Delta H = H(\text{soln}) - \sum_{i=1}^{C} n_i \tilde{H}_i^{\ominus}$$

and, by the definition of partial molar quantities, we find that

$$\left(\frac{\partial \Delta H}{\partial n_k}\right)_{T,P,n} = \bar{H}_k - \tilde{H}_k^{\ominus} \tag{9.29}$$

This equation implies an experimental determination of ΔH as a function of the mole numbers. The experimental data that are required may be published in many different ways. For example, the change of enthalpy per mole of solution may be given as a function of the mole fractions, molalities, or molarities. The methods required to calculate $(\bar{H}_k - \tilde{H}_k^{\ominus})$ for the components from such data are exactly those discussed in Sections 6.3 and 6.5.

Special consideration must be given to systems involving liquid solutions of at least one solid component, for which the choice of either the pure solid or pure supercooled liquid as the standard state is not convenient. This case is encountered for all solutions in which the pure solute is not chosen as the reference state. As an example, we consider an aqueous solution of a solid B and choose the reference state to be the infinitely dilute solution. Then a general change of state for the formation of the solution from the components is written as

$$n_2 B[\text{s}, T, P] + n_1 \text{H}_2\text{O}[\ell, T, P] = n_2\left(B \cdot \frac{n_1}{n_2} \text{H}_2\text{O}\right)[T, P]$$

or

$$n_2 B[\text{s}, T, P] + n_1 \text{H}_2\text{O}[\ell, T, P] = (n_1 + n_2) M[T, P, m]$$

where in the last change of state the letter M refers to the solution and m to the molality. The change of enthalpy for this change of state in terms of the partial molar enthalpies is

$$\Delta H = n_1(\bar{H}_1 - \tilde{H}_1^{\bullet}) + n_2(\bar{H}_2 - \tilde{H}_2^{\bullet}(\text{s})) \tag{9.30}$$

When the reference state is the infinitely dilute solution, the standard state for the enthalpy is also the infinitely dilute solution. We then change the standard state of component B from the pure solid to the infinitely dilute solution by adding to and subtracting from Equation (9.30) the quantity $n_2 \bar{H}_2^{\infty}$, where $\bar{H}_2^{\infty}$ is the partial molar enthalpy of the component in the infinitely dilute solution. Then we have

$$\Delta H = n_1(\bar{H}_1 - \tilde{H}_1^{\bullet}) + n_2(\bar{H}_2 - \bar{H}_2^{\infty}) - n_2(\tilde{H}_2^{\bullet}(\mathrm{s}) - \bar{H}_2^{\infty}) \tag{9.31}$$

This equation gives the change of enthalpy in terms of the partial molar enthalpies of the solvent relative to the molar enthalpy of the pure solvent and of the partial molar enthalpy of the solute and the molar enthalpy of the pure solid, both relative to its partial molar enthalpy in the infinitely dilute solution.

Experimental measurements give ΔH for a given change of state as a function of n_1 and n_2, from which the separate quantities, $(\bar{H}_1 - \tilde{H}_1^{\bullet})$, $(\bar{H}_2 - \bar{H}_2^{\infty})$, and $(\tilde{H}_2^{\bullet}(\mathrm{s}) - \bar{H}_2^{\infty})$, can be determined. If Equation (9.31) is expressed as

$$\Delta H = H - n_1\tilde{H}_1^{\bullet} - n_2\bar{H}_2^{\infty} - n_2(\tilde{H}_2^{\bullet}(\mathrm{s}) - \bar{H}_2^{\infty}) \tag{9.32}$$

we find that

$$\left(\frac{\partial\, \Delta H}{\partial n_1}\right)_{T,P,n_2} = \bar{H}_1 - \tilde{H}_1^{\bullet} \tag{9.33}$$

Thus, if ΔH is plotted as a function of n_1 at constant n_2 or $\Delta H/n_2$ as a function of n_1/n_2, the slope of the curve so obtained gives values of $(\bar{H}_1 - \tilde{H}_1^{\bullet})$. Similarly, we find that

$$\left(\frac{\partial\, \Delta H}{\partial n_2}\right)_{T,P,n_1} = (\bar{H}_2 - \bar{H}_2^{\infty}) - (\tilde{H}_2^{\bullet}(\mathrm{s}) - \bar{H}_2^{\infty}) \tag{9.34}$$

However, in order to determine the values of $(\bar{H}_2 - \bar{H}_2^{\infty})$, the quantity $(\tilde{H}_2^{\bullet}(\mathrm{s}) - \bar{H}_2^{\infty})$ must be determined separately. If ΔH is plotted as a function of n_2 at constant n_1 or $\Delta H/n_1$ as a function of n_2/n_1, the slope of the curve gives the difference between $(\bar{H}_2 - \bar{H}_2^{\infty})$ and $(\tilde{H}_2^{\bullet}(\mathrm{s}) - \bar{H}_2^{\infty})$. The limiting slope at $n_2 = 0$ gives the value $-(\tilde{H}_2^{\bullet}(\mathrm{s}) - \bar{H}_2^{\infty})$ because in the limit $(\bar{H}_2 - \bar{H}_2^{\infty})$ goes to zero. Alternately, we can write Equation (9.31) as

$$\frac{\Delta H}{n_2} = \frac{n_1}{n_2}(\bar{H}_1 - \tilde{H}_1^{\bullet}) + (\bar{H}_2 - \bar{H}_2^{\infty}) - (\tilde{H}_2^{\bullet}(\mathrm{s}) - \bar{H}_2^{\infty}) \tag{9.35}$$

As n_1 becomes large, $(\bar{H}_1 - \tilde{H}_1^{\bullet})$ and $(\bar{H}_2 - \bar{H}_2^{\infty})$ both approach zero, so

$$\lim_{n_1 \to \infty} \frac{\Delta H}{n_2} = -(\tilde{H}_2^{\bullet}(\mathrm{s}) - \bar{H}_2^{\infty}) \tag{9.36}$$

Thus, $-(\tilde{H}_2^{\bullet}(\mathrm{s}) - \bar{H}_2^{\infty})$ is obtained as the limiting value of the curve as n_1 becomes large when $\Delta H/n_2$ is plotted as a function of n_1/n_2. Experimental values of ΔH are reported in the literature as the change of enthalpy per mole of solute, per mole of solvent, or per 1000 g of solvent and as a function of the molality or the mole ratios, n_1/n_2 or n_2/n_1. The methods discussed in

Section 6.3, in addition to those discussed here, can be used to determine the quantities $(\bar{H}_1 - \tilde{H}_1^\bullet)$ and $(\bar{H}_2 - \bar{H}_2^\infty)$ coupled with the determination of $(\tilde{H}_2^\bullet(\mathrm{s}) - \bar{H}_2^\infty)$ from the limiting values of appropriate quantities.

For another change of state, we consider the change of composition of a solution from one composition to another represented by the equation

$$n_2\left(B\cdot\frac{n_1}{n_2}\,\mathrm{H_2O}\right)[T, P] + n_1'\mathrm{H_2O}[\ell, T, P]$$
$$= n_2\left(B\cdot\frac{n_1 + n_1'}{n_2}\,\mathrm{H_2O}\right)[T, P]$$

The change of enthalpy is written as the difference between the enthalpies of the system in the final and initial states, so

$$\Delta H = H' - H - n_1'\tilde{H}_1^\bullet \tag{9.37}$$

where H' is the enthalpy of the final solution and H is that of the initial solution. This equation can be written as

$$\begin{aligned}\Delta H = {} & (n_1 + n_1')(\bar{H}_1' - \tilde{H}_1^\bullet) + n_2(\bar{H}_2' - \bar{H}_2^\infty) \\ & - n_1(\bar{H}_1 - \tilde{H}_1^\bullet) - n_2(\bar{H}_2 - \bar{H}_2^\infty)\end{aligned} \tag{9.38}$$

when Equation (9.25) is introduced. We observe that ΔH is the difference between the changes of enthalpy for the formation of the two solutions from the components. This is the expected result, because the change of state can be obtained as the difference of the two changes of state

$$n_2B[\mathrm{s}, T, P] + (n_1 + n_1')\mathrm{H_2O}[\ell, T, P] = n_2\left(B\cdot\frac{n_1 + n_1'}{n_2}\,\mathrm{H_2O}\right)[T, P]$$

and

$$n_2B[\mathrm{s}, T, P] + n_1\mathrm{H_2O}[\ell, T, P] = n_2\left(B\cdot\frac{n_1}{n_2}\,\mathrm{H_2O}\right)[T, P]$$

Similarly, if we mix two solutions to form a third solution according to the change of state

$$n_2\left(B\cdot\frac{n_1}{n_2}\,\mathrm{H_2O}\right)[T, P] + n_2'\left(B\cdot\frac{n_1'}{n_2'}\,\mathrm{H_2O}\right)[T, P]$$
$$= (n_2 + n_2')\left(B\cdot\frac{n_1 + n_1'}{n_2 + n_2'}\,\mathrm{H_2O}\right)[T, P]$$

the change of enthalpy is the algebraic sum of the changes of enthalpy for the three changes of state:

$$n_2B[T, P] + n_1H_2O[\ell, T, P] = n_2\left(B\cdot\frac{n_1}{n_2}H_2O\right)[T, P]$$

$$n_2'B[T, P] + n_1'H_2O[\ell, T, P] = n_2'\left(B\cdot\frac{n_1'}{n_2'}H_2O\right)[T, P]$$

and

$$(n_2 + n_2')B[T, P] + (n_1 + n_1')H_2O[\ell, T, P]$$
$$= (n_2 + n_2')\left(B\cdot\frac{n_1 + n_1'}{n_2 + n_2'}H_2O\right)[T, P]$$

Instead of diluting a solution by the addition of solvent, we might increase the composition of the solution by the addition of the solute. The corresponding change of state is

$$n_2'B[s, T, P] + n_2\left(B\cdot\frac{n_1}{n_2}H_2O\right)[T, P] = (n_2 + n_2')\left(B\cdot\frac{n_1}{n_2 + n_2'}\right)[T, P]$$

for which the change of enthalpy is given by

$$\Delta H = n_1(\bar{H}_1' - \bar{H}_1) + n_2(\bar{H}_2' - \bar{H}_2) + n_2'(\bar{H}_2' - \tilde{H}_2^{\bullet}(s)) \tag{9.39}$$

or

$$\Delta H = n_1(\bar{H}_1' - \tilde{H}_1^{\bullet}) + (n_2 + n_2')(\bar{H}_2' - \bar{H}_2^{\infty}) - n_1(\bar{H}_1 - \tilde{H}_1^{\bullet})$$
$$- n_2(\bar{H}_2 - \bar{H}_2^{\infty}) - n_2'(\tilde{H}_2^{\bullet}(s) - \bar{H}_2^{\infty}) \tag{9.40}$$

Again, the change of enthalpy is the difference between the changes of enthalpy for the formation of the two solutions from the components.

Two further concepts need to be defined. First we consider the addition of a very small quantity of solvent to a solution. The change of enthalpy, based on 1 mole of solvent, for the change of state is called the *differential or partial change of enthalpy on dilution*. If we designate Δn_1 as the small quantity of solvent added to the solution, Equation (9.38) can be written as

$$\Delta H = \Delta n_1(\bar{H}_1' - \tilde{H}_1^{\bullet}) + n_1(\bar{H}_1' - \bar{H}_1) + n_2(\bar{H}_2' - \bar{H}_2) \tag{9.41}$$

Then, in the limit as Δn_1 goes to zero, we find that

$$\lim_{\Delta n_1 \to 0} \frac{\Delta H}{\Delta n_1} = \bar{H}_1 - \tilde{H}_1^{\bullet} \tag{9.42}$$

because both $(\bar{H}_1' - \bar{H}_1)$ and $(\bar{H}_2' - \bar{H}_2)$ go to zero in the limit. Thus, the differential change of enthalpy on dilution is equal to $(\bar{H}_1 - \tilde{H}_1^{\bullet})$. We note that Equation (9.42) is equivalent to Equation (9.33).

Second, we consider the addition of a very small quantity of solute to a solution. The change of enthalpy, based on 1 mole of solute, for the change of state is called the *differential or partial change of enthalpy of solution.* If the solute is a solid and the solution a liquid, and if we designate Δn_2 as the small quantity of solute added to the solution, then Equation (9.39) can be written as

$$\Delta H = \Delta n_2[(\bar{H}'_2 - \bar{H}^{\infty}_2) - (\tilde{H}^{\bullet}_2(\text{s}) - \bar{H}^{\infty}_2)] + n_1(\bar{H}'_1 - \bar{H}_1) + n_2(\bar{H}'_2 - \bar{H}_2) \tag{9.43}$$

In the limit as Δn_2 goes to zero, we find that

$$\lim_{\Delta n_2 \to 0} \frac{\Delta H}{\Delta n_2} = (\bar{H}_2 - \bar{H}^{\infty}_2) - (\tilde{H}^{\bullet}_2(\text{s}) - \bar{H}^{\infty}_2) \tag{9.44}$$

because both $(\bar{H}'_1 - \bar{H}_1)$ and $(\bar{H}'_2 - \bar{H}_2)$ go to zero in the limit. Thus, the differential change of enthalpy of solution is equal to

$$[(\bar{H}_2 - \bar{H}^{\infty}_2) - (\tilde{H}^{\bullet}_2(\text{s}) - \bar{H}^{\infty}_2)]$$

We again note that Equation (9.44) is equivalent to Equation (9.34). Although we have primarily discussed aqueous solutions, the principles and methods applied in this section are applicable to any solvent, pure or mixed at constant concentration.

9.5 Changes of enthalpy for changes of state involving chemical reactions

The change of any thermodynamic function for any change of state is always the difference between the value of the function for the final state and that for the initial state. Therefore, for a change of state represented by

$$aB_1[\text{g}, T, P] + bB_2[\ell, T, P] = mB_3[\text{g}, T, P] + nB_4[\text{s}, T, P]$$

the change of enthalpy is

$$\Delta H = m\tilde{H}_{B_4} + n\tilde{H}_{B_3} - a\tilde{H}_{B_1} - b\tilde{H}_{B_2} \tag{9.45}$$

In these equations the capital letters represent the substance, the lower case letters represent the moles of each substance taking part in the change of state, and the $\tilde{H}$s represent the molar enthalpies of the substances in the state given in the expression for the change of state. The state of aggregation of each substance is generally the stable phase at the given temperature and pressure. The temperature and pressure of each substance need not be the same. In general terms we represent the change of state for any chemical reaction as

$$\sum_i \nu_i B_i = 0$$

where the v_is are the stoichiometric coefficients for the chemical reaction and are positive for the products of the reaction and negative for the reactants. The state of each substance, B_i, must be stated explicitly. For such a generalized change of state, the change of enthalpy is given by

$$\Delta H = \sum_i v_i \tilde{H}_i \tag{9.46}$$

This equation indicates a relation between the change of enthalpy for the change of state and the absolute enthalpies of the substances taking part in the change of state. In order to avoid the problem of determining experimentally the change of enthalpy for every possible chemical reaction, we seek a method to replace the absolute enthalpies in Equation (9.46) by quantities whose values can be obtained. We know that every chemical reaction can be written as a sum of the reactions for the formation of the compounds that take part in the reaction from the elements. Then, the change of state for a chemical reaction can be expressed as the sum of the changes of state for the formation of the compounds from the elements, provided the state of an element is the same in each change of state. The change of enthalpy for the given change of state involving a chemical reaction can then be written as the algebraic sum of the changes of enthalpies for the formation of the compounds. This is true because the change of enthalpy in going from some initial state to a final state is independent of the path.[2]

Consequently, the changes of enthalpy for the formation of a compound from the elements, $\Delta\tilde{H}_{f,i}$, may be substituted for each $\tilde{H}_i$ in Equation (9.46), so

$$\Delta H = \sum_i v_i \Delta\tilde{H}_{f,i} \tag{9.47}$$

As an example we consider the change of state

$$NH_3[g, T, P] + HCl[g, T, P] = NH_4Cl[s, T, P]$$

The changes of state for the formation of each compound from the elements are

$$\tfrac{1}{2}N_2[g, T, P] + 2H_2[g, T, P] + \tfrac{1}{2}Cl_2[g, T, P] = NH_4Cl[s, T, P]$$
$$\tfrac{1}{2}N_2[g, T, P] + \tfrac{3}{2}H_2[g, T, P] = NH_3[g, T, P]$$

[2] The additivity of changes of enthalpy on the addition of changes of state is identical to a modification of the law that Hess announced in 1840. This law stated that the heat associated with a chemical reaction was independent of the path used in going from the initial to the final state, and thus the heats of chemical reactions were additive. This statement, of course, is not exactly true, because the heat does depend on the path. It is the changes of the state functions such as the enthalpy that are independent of the path.

and

$$\tfrac{1}{2}H_2[g, T, P] + \tfrac{1}{2}Cl_2[g, T, P] = HCl[g, T, P]$$

from which we obtain

$$\Delta H = \Delta H_f(NH_4Cl) - \Delta H_f(NH_3) - \Delta H_f(HCl) \quad (9.48)$$

for the change of enthalpy. We emphasize here that the state of each element in the set of changes of state of formation must be the same in each change of state; otherwise the cancellation of the element in the algebraic addition of the changes of state and the absolute enthalpy of the element implied in Equation (9.48) would not cancel.

9.6 Standard changes of enthalpy (standard enthalpies) of formation[3]

Equation (9.47) relates the change of enthalpy for a change of state involving a chemical reaction and the changes of enthalpy of formation of the substances involved in the change of state. Thus, ΔH may be calculated when each $\Delta \tilde{H}_{f,i}$ is known. It is advantageous then to compile tables of values of the change of enthalpy on the formation of a compound from the elements. Only the values of the change of enthalpy on the formation of a compound in one standard state from the elements in their standard states need be listed. The change of enthalpy of formation when the compound and elements are not in their standard states can be calculated from the standard changes of enthalpy of formation.

The standard states of both pure compounds and elements are usually the natural state of aggregation at a pressure of 1 bar and some arbitrarily chosen temperature. In actual practice the two temperatures 18°C and 25°C have generally been used, with the former being found in the older literature. It is important in using any table to determine the particular temperature used to define the standard state. For a specific purpose, any other standard state may be defined. (See the appendices for tables of data.)

Several comments need to be made concerning the state of aggregation of the substances. For gases, the standard state is the ideal gas at a pressure of 1 bar; this definition is consistent with the standard state developed in Chapter 7. When a substance may exist in two allotropic solid states, one state must be chosen as the standard state; for example, graphite is usually chosen as the standard form of carbon, rather than diamond. If the chemical reaction takes place in a solution, there is no added complication when the standard state of the components of the solution can be taken as the pure components, because the change of enthalpy on the formation of a compound in its standard state is identical whether we are concerned with the pure

[3] We use the expression standard change of enthalpy, because we wish to emphasize a standard change of state. The alternate expression, enthalpy of formation relative to a standard state, could also be used. These expressions are frequently shortened to 'standard enthalpies.'

compound or the compound in a solution. The situation is somewhat different when the reference state of a solute is taken to be the infinitely dilute solution. In such a case we are concerned with the change of enthalpy for the formation of the compound in an infinitely dilute solution. If we use an aqueous solution of sodium chloride as an example, the change of state is written as

$$\begin{aligned}&\mathrm{Na[s, 298\ K, 1\ bar]} + \tfrac{1}{2}\mathrm{Cl_2[g, ideal, 298\ K, 1\ bar]}\\&\quad + \infty \mathrm{H_2O}[\ell, \mathrm{298\ K, 1\ bar}] = \mathrm{NaCl\cdot\infty H_2O[298\ K, 1\ bar]}\end{aligned}$$

This change of state can be considered as the sum of two changes of state:

$$\begin{aligned}&\mathrm{Na[s, 298\ K, 1\ bar]} + \tfrac{1}{2}\mathrm{Cl[g, ideal, 298\ K, 1\ bar]}\\&\quad = \mathrm{NaCl[s, 298\ K, 1\ bar]}\end{aligned}$$

and

$$\begin{aligned}&\mathrm{NaCl[s, 298\ K, 1\ bar]} + \infty \mathrm{H_2O}[\ell, \mathrm{298\ K, 1\ bar}]\\&\quad = \mathrm{NaCl\cdot\infty H_2O[298\ K, 1\ bar]}\end{aligned}$$

The standard change of enthalpy for the formation of sodium chloride in an infinitely dilute aqueous solution is thus given by the sum of the standard changes of enthalpy for the last two changes of state. Therefore,

$$\Delta H_f^\ominus(\mathrm{NaCl\cdot\infty H_2O}) = \Delta H_f^\ominus(\mathrm{NaCl(s)}) + (\bar{H}_2^\infty - \tilde{H}_2^\ominus(\mathrm{s}))(\mathrm{NaCl}) \tag{9.49}$$

We observe that the standard change of enthalpy for the formation of a solute in an infinitely dilute solution is the sum of the standard change of enthalpy for the formation of the pure compound and the difference between its partial molar enthalpy in the infinitely dilute solution and its molar enthalpy when pure.

Many tables of values of standard changes of enthalpy of formation list values for individual ions, particularly in aqueous solutions. In order to do so, an arbitrary definition must be introduced because the properties of individual ions in solution cannot be determined. We consider an electrolyte $M_{\nu_+}A_{\nu_-}$ which is completely ionized in the infinitely dilute solution. We choose this solution to be the standard state for the enthalpy. The enthalpy of this solution per mole of solute, H', is given by

$$H' = \nu_+\bar{H}^\infty(\mathrm{M}_+) + \nu_-\bar{H}^\infty(\mathrm{A}_-) + \infty\bar{H}^\infty(\mathrm{H_2O}) \tag{9.50}$$

where $\nu_+\bar{H}^\infty(\mathrm{M}_+) + \nu_-\bar{H}^\infty(\mathrm{A}_-)$ is set equal to $\bar{H}_2^\infty$, the enthalpy of the solute, by Equation (8.97). The standard change of enthalpy for the formation of the solution from the elements is then

$$\begin{aligned}\Delta H' = {} & \nu_+\bar{H}^\infty(\mathrm{M}_+) + \nu_-\bar{H}^\infty(\mathrm{A}_-) + \infty\bar{H}^\infty(\mathrm{H_2O}) - \nu_+\tilde{H}^\ominus(\mathrm{M})\\ & - \nu_-\tilde{H}^\ominus(\mathrm{A}) - \infty\tilde{H}^\bullet(\mathrm{H_2O})\end{aligned} \tag{9.51}$$

or

$$\Delta H' = \nu_+[\bar{H}^\infty(M_+) - \tilde{H}^\ominus(M)] + \nu_-[\bar{H}^\infty(A_-) - \tilde{H}^\ominus(A)] \qquad (9.52)$$

because $\tilde{H}^\bullet(H_2O) = \bar{H}^\infty(H_2O)$. The two quantities $\tilde{H}^\ominus(M)$ and $\tilde{H}^\ominus(A)$ in Equations (9.51) and (9.52) represent the sum of the enthalpies of the elements that are contained in the ion; for a monatomic metal ion such as zinc or silver, $\tilde{H}^\ominus(M)$ represents the molar standard enthalpy of the metal; for the ammonium ion, $\tilde{H}^\ominus(M)$ represents $\frac{1}{2}\tilde{H}^\ominus(N_2) + 2\tilde{H}^\ominus(H_2)$; for the chloride ion, $\tilde{H}^\ominus(A)$ represents $\frac{1}{2}\tilde{H}^\ominus(Cl_2)$; and for the sulfate ion, $\tilde{H}^\ominus(A)$ represents $\tilde{H}^\ominus(S) + 2\tilde{H}^\ominus(O_2)$. The two quantities in the brackets in Equation (9.52) are the standard changes of enthalpy for the formation of the ion from the elements, represented by the changes of state

$$M[T, P] + \infty H_2O = (M_+ \cdot \infty H_2O)[T, P] + z_+ e^-$$

and

$$A[T, P] + \infty H_2O + z_- e^- = (A_- \cdot \infty H_2O)[T, P]$$

where z_+ and z_- represents the absolute values of the number of electronic charges on the respective ions and e^- represents the electron. The enthalpies associated with the electrons in these changes of state are not included in the change of enthalpy. In the addition of these changes of state to give the change of state for the formation of the solution from the elements, the electrons must cancel and, in the corresponding addition of the change of enthalpies, the enthalpies of the electrons also cancel.

If we define the change of enthalpy for one of these changes of state to be zero, then the standard of enthalpy for the other change of state must equal $\Delta H'$. Standard changes of enthalpy for the formation of ions can thus be defined relative to that for some arbitrarily chosen ion. In practice, *this ion is the hydrogen ion and the change of enthalpy for the change of state*

$$\tfrac{1}{2}H_2[\text{g, ideal}, T, P] + \infty H_2O[\ell, T, P] = (H^+ \cdot \infty H_2O)[T, P] + e^-$$

is defined to be zero at all temperatures and pressures.

9.7 The dependence of the change of enthalpy for a chemical reaction on the temperature and pressure

We may know the change of enthalpy for a given change of state and need to determine the change of enthalpy for a change of state in which the same substances are used but in which the states of the substances are different. We may consider the initial change of state to be given by $\sum_i \nu_i B_i[T_i, P_i] = 0$, where T_i and P_i represent the temperature and pressure of each substance. These temperatures and pressures need not be the same. The state of aggregation of each substance—gas, liquid, or solid—or the concentrations of a substance in solution would have to be given to define the change of state completely. For this change of state, the change of enthalpy is given by

$$\Delta H = \sum_i \nu_i \tilde{H}_i[T_i, P_i] \tag{9.53}$$

In this equation and those immediately following, the tilde represents, as usual, the molar quantity, but a partial molar quantity symbolized as $\bar{H}_i$ may be substituted when necessary. The new change of state may be represented by

$$\sum_i \nu_i B_i[T'_i, P'_i] = 0$$

where T'_i and P'_i represent the new temperatures and pressures of the substances. Again, the states of aggregation or concentrations would have to be specified. The change of enthalpy for the new change of state is given by

$$\Delta H' = \sum_i \nu_i \tilde{H}_i[T'_i, P'_i] \tag{9.54}$$

The substraction of Equation (9.53) from (9.54) yields

$$\Delta H' = \Delta H + \sum_i \nu_i (\tilde{H}_i[T'_i, P'_i] - \tilde{H}_i[T_i, P_i]) \tag{9.55}$$

The determination of $\Delta H'$ when ΔH is known requires then the evaluation of each quantity $(\tilde{H}_i[T'_i, P'_i] - \tilde{H}_i[T_i, P_i])$. If the substance is in solution, evaluation of the quantity $(\bar{H}_i[T'_i, P'_i, x_i] - \bar{H}_i[T_i, P_i, x_i])$ is required with the condition of constant composition. No further distinction is made in this section because the calculations are the same whether the substance is pure or in solution.

Two cases must be considered: one in which the state of aggregation is the same in the initial and final state, and the other in which the state of aggregation is different in the two states. In the first case the enthalpy is a continuous function of the temperature and pressure in the interval between (T_i, P_i) and (T'_i, P'_i). Equation (4.86) can be used for a closed system and the integration of this equation is discussed in Section 8.1, where the emphasis is on standard states of pure substances. The result of the integration is valid in the present instance with change of the limits of integration and limitation to molar quantities. Equations (8.10) and (8.11) then become

$$\tilde{H}_i[T'_i, P'_i] - \tilde{H}_i[T_i, P_i] = \int_{T_i}^{T'_i} (\tilde{C}_{P_i})_{P_i}\, dT + \int_{P_i}^{P'_i} \left[\tilde{V}_i - T\left(\frac{\partial \tilde{V}_i}{\partial T}\right)_P\right]_{T'_i} dP \tag{9.56}$$

and

$$\tilde{H}_i[T'_i, P'_i] - \tilde{H}_i[T_i, P_i] = \int_{T_i}^{T'_i} (\tilde{C}_{P_i})_{P_i}\, \mathrm{d}T + \int_{P_i}^{P'_i} \left[\tilde{V}_i - T\left(\frac{\partial \tilde{V}_i}{\partial T}\right)_{P_i}\right]_{T'_i} \mathrm{d}P \tag{9.57}$$

respectively. We use only Equation (9.56) for further discussion because the two equations are equivalent. Substitution of Equation (9.56) into Equation (9.55) yields

$$\Delta H' = \Delta H + \sum_i \int_{T_i}^{T'_i} \nu_i(\tilde{C}_{P_i})_{P_i}\, \mathrm{d}T + \sum_i \int_{P_i}^{P'_i} \nu_i\left[\tilde{V}_i - T\left(\frac{\partial \tilde{V}_i}{\partial T}\right)_{P_i}\right]_{T'_i} \mathrm{d}P \tag{9.58}$$

The calculation of the value of $\Delta H'$, knowing the value of ΔH, thus requires knowledge of the molar heat capacity of each substance as a function of the temperature at the pressure P_i and knowledge of the equation of state of each substance as a function of the pressure at the temperature T'_i. The last term in Equations (9.56)–(9.58) is usually small unless the difference in the pressures is greater than several hundred bars.

In the special case that the limits of integration for the temperature and for the pressure are the same for each substance, the summation signs can be placed inside the integral signs. We can also define a quantity $(\Delta C_P)_{P_i}$ as

$$(\Delta C_P)_{P_i} = \sum_i \nu_i(\tilde{C}_{P_i})_{P_i} \tag{9.59}$$

and a quantity $[\Delta V_i - T(\partial\, \Delta\tilde{V}_i/\partial T)_{P_i}]_{T_i}$ as

$$\begin{aligned}\left[\Delta\tilde{V}_i - T\left(\frac{\partial\, \Delta\tilde{V}_i}{\partial T}\right)_{P_i}\right]_{T'_i} &= \Delta\left[\tilde{V}_i - T\left(\frac{\partial \tilde{V}_i}{\partial T}\right)_{P_i}\right]_{T'_i} \\ &= \sum_i \nu_i\left[\tilde{V}_i - T\left(\frac{\partial \tilde{V}_i}{\partial T}\right)_{P_i}\right]_{T'_i}\end{aligned} \tag{9.60}$$

Equation (9.58) can then be written as

$$\Delta H' = \Delta H + \int_{T_i}^{T'_i} (\Delta\tilde{C}_P)_{P_i}\, \mathrm{d}T + \int_{P_i}^{P'_i} \left[\Delta\tilde{V}_i - T\left(\frac{\partial\, \Delta\tilde{V}_i}{\partial T}\right)_{P_i}\right]_{T'_i} \mathrm{d}P \tag{9.61}$$

subject to the above condition and definitions. The numerical calculations required by this equation are considerably less than those required by Equation (9.58). Further simplification is obtained when the pressure of each substance is the same in the initial and final changes of state, so Equation

(9.61) reduces to

$$\Delta H' = \Delta H + \int_{T_i}^{T'_i} (\Delta C_P)_{P_i}\, \mathrm{d}T \tag{9.62}$$

An alternate equation may be obtained by differentiation of this equation to give

$$\left(\frac{\partial\, \Delta H}{\partial T}\right)_{P_i} = (\Delta C_P)_{P_i} \tag{9.63}$$

with the concept that the upper limit of integration is a variable rather than a fixed value. It is important to note that for Equation (9.62) it is the pressure of each substance that is kept constant in both equations but the condition that each pressure be the same is no longer necessary. Similarly, when the temperature of a substance is the same in both the initial and final changes of state, Equation (9.61) reduces to

$$\Delta H' = \Delta H + \int_{P_i}^{P'_i} \left[\Delta \tilde{V}_i - T\left(\frac{\partial\, \Delta \tilde{V}_i}{\partial T}\right)_{P_i}\right]_{T_i} \mathrm{d}P \tag{9.64}$$

which, on differentiation with respect to the pressure, yields

$$\left(\frac{\partial\, \Delta H}{\partial P}\right)_{T_i} = \left[\Delta \tilde{V}_i - T\left(\frac{\partial \Delta \tilde{V}_i}{\partial T}\right)_{P_i}\right]_{T_i} \tag{9.65}$$

Here, in this case, the temperature of each substance need not be the same, but the differentiation of Equation (9.64) or the integration of Equation (9.65) requires the temperature of each substance to be constant.

The determination of $(\tilde{H}_i[T'_i, P'_i] - \tilde{H}_i[T_i, P_i])$ for a pure substance is only a little more complex when the state of aggregation is different in the two states. The molar enthalpy of a pure substance considered as a function of the temperature and pressure is discontinuous at all points where a change of phase occurs. The molar heat capacity and the equation of state is different on either side of the discontinuity. It is therefore necessary to carry out the integrations discussed in the previous paragraph in parts, so that any single integration is carried out only over those regions that are continuous. The change of enthalpy for the change of phase must be included in determining the overall change of enthalpy. As an example we may be required to determine the change of enthalpy for the change of state

$$H_2O[\ell, 298\ \text{K}, 1\ \text{bar}] = H_2O[\text{g}, 500, \text{K}, 3\ \text{bar}]$$

This change of state may be considered to be the sum of the following changes of state:

$$H_2O[\ell, 298\text{ K}, 1\text{ bar}] = H_2O[\ell, 373\text{ K}, 1\text{ bar}]$$

$$H_2O[\ell, 373\text{ K}, 1\text{ bar}] = H_2O[g, 373\text{ K}, 1\text{ bar}]$$

$$H_2O[g, 373, \text{K}, 1\text{ bar}] = H_2O[g, 500\text{ K}, 1\text{ bar}]$$

and

$$H_2O[g, 500\text{ K}, 1\text{ bar}] = H_2O[g, 500\text{ K}, 3\text{ bar}]$$

The total change of enthalpy is the sum of the changes of enthalpy for the four changes of state. Methods for determining the change of enthalpy for each step have been given previously.

For a solution it would be necessary to consider the changes of enthalpy for the separation of the solution into its components, the change of enthalpy for the change of temperature, pressure, and state of aggregation of the components and, finally, the change of enthalpy for the reformation of the solution from the components under the new condition, as outlined in Section 8.5.

In concluding this chapter, we briefly review the general concepts related to the changes of enthalpy associated with changes of state involving chemical reactions. The enthalpy of a system is a point function of the variables used to define the state of the system. A change of enthalpy for a change of state then depends only upon the initial and final states, and not at all upon the path. Therefore, when it is necessary to calculate the change of enthalpy for a change of state under one set of conditions, having knowledge of the change of enthalpy for a change of state involving the same substances but under another set of conditions, it is only necessary to set up a path between the two changes of state along which the change of enthalpy can be calculated. The path, in general, consists of simple changes of state in which only one variable is changed at a time.

10

Phase equilibrium

The study of phase equilibrium affords methods by which the excess chemical potentials of the components in a solution may be determined. It is concerned with the relations that exist between the intensive variables—temperature, pressure, and composition of the phases—when two or more phases are in equilibrium. In this respect, such studies also yield information on the colligative properties of a solution. For single-component systems without restrictions, the temperature and pressure are the only variables of interest. When restrictions are applied, an additional number of temperature or pressure variables must be introduced. In experimental studies, conditions are set so that any given system is univariant and it is possible to determine the dependence of one variable, a temperature or pressure, on one other variable, again a temperature or a pressure. Composition variables must be used for multicomponent systems, in addition to the temperature and pressure, and are introduced here through the chemical potentials. The total number of intensive variables and the number of independent intensive variables for a given system can be determined from the appropriate set of Gibbs–Duhem equations, consistent with any restrictions that may or may not be placed upon the system. The experimental and thermodynamic study of such a system then requires that a sufficient number of limitations be placed on the system so that it is univariant. Thus, we determine the dependence of one dependent variable on one independent variable, all other independent variables being held constant.

Two methods may be used, in general, to obtain the thermodynamic relations that yield the values of the excess chemical potentials or the values of the derivative of one intensive variable. One method, which may be called an *integral method*, is based on the condition that the chemical potential of a component is the same in any phase in which the component is present. The second method, which may be called a *differential method*, is based on the solution of the set of Gibbs–Duhem equations applicable to the particular system under study. The results obtained by the integral method must yield on differentiation those obtained by the differential method, and the result obtained by the differential method must give on integration those obtained by the integral method. The choice between the two methods for any problem depends upon the nature of the problem. The integral method may be used to advantage when evaluation of the excess chemical potentials is desired. When a certain derivative is desired or is of interest, the differential method may be advantageous. Also, the differential method may be most convenient for multiphase, multicomponent systems. Both methods are illustrated in this chapter.

One-component, two-phase systems are discussed in the first part of this chapter. The major part of the chapter deals with two-component systems with emphasis on the colligative properties of solutions and on the determination of the excess chemical potentials of the components in the solution. In the last part of the chapter three-component systems are discussed briefly.

ONE-COMPONENT SYSTEMS

10.1 Phase transitions

The common characteristics of phase transitions are that the Gibbs energy is continuous. Although the conditions of equilibrium and the continuity of the Gibbs energy demand that the chemical potential must be the same in the two phases at a transition point, the molar entropies and the molar volumes are not. If, then, we have two such phases in equilibrium, we have a set of two Gibbs–Duhem equations, the solution of which gives the Clapeyron equation (Eq. (5.73))

$$\frac{\mathrm{d}P}{\mathrm{d}T} = \frac{\tilde{S}'' - \tilde{S}'}{\tilde{V}'' - \tilde{V}'} = \frac{\Delta\tilde{S}}{\Delta\tilde{V}} \tag{10.1}$$

The primes distinguish the separate phases. This equation may also be written as

$$\frac{\mathrm{d}P}{\mathrm{d}T} = \frac{\tilde{H}'' - \tilde{H}'}{T(\tilde{V}'' - \tilde{V}')} = \frac{\Delta\tilde{H}}{T\,\Delta\tilde{V}} \tag{10.2}$$

because the chemical potentials are equal at equilibrium and therefore $(\tilde{H}'' - \tilde{H}') = T(\tilde{S}'' - \tilde{S}')$. These equations relate the slope of the saturation curve with the change of molar entropy or enthalpy and the change of molar volume for the phase transition at an equilibrium point. A knowledge of any two of these quantities permits the calculation of the third. The integration of either equation is not possible in the general case, because the entropies or enthalpies and the volumes are functions of the temperature and pressure and the variables are not separable.

The entropy, enthalpy, and volume quantities appearing in Equations (10.1) and (10.2) are molar quantities, but care must be taken that such quantities refer to the same mass of material. This requirement is readily proved. Consider one phase, the primed phase, to contain $(n_1^0)'$ moles of component based on a specified chemical formula, and assume that no chemical reaction takes place in this phase. Assume that the double-primed phase contains $(n_1^0)''$ moles of component but that a reaction, possibly polymerization or decomposition, takes place in this phase. The chemical reaction may be expressed as $\sum_{i=1}^{S} \nu_i B_i = 0$, where the sum is taken over all reacting species. The two Gibbs–Duhem equations may then be written as

$$S'\,dT - V'\,dP + (n_1^0)'\,d\mu_1 = 0 \tag{10.3}$$

and

$$S''\,dT - V''\,dP + \sum_{i=1}^{S} n_i''\,d\mu_i'' = 0 \tag{10.4}$$

If α is the fraction of the original number of moles of component 1 which react, the number of moles of species 1 at equilibrium is $(n_1^0)''(1-\alpha)$ and the number of moles of each new species is $\alpha(n_1^0)''\nu_i/\nu_1$. Then Equation (10.4) may be written as

$$S''\,dT - V''\,dP + (n_1^0)''\,d\mu_1 + \frac{\alpha(n_1^0)''}{\nu_1}\sum_{i=1}^{S} \nu_i\,d\mu_i'' = 0$$

However, $\sum_{i=1}^{S} \nu_i\,d\mu_i'' = 0$ by the condition of equilibrium, and thus

$$S''\,dT - V''\,dP + (n_i^0)''\,d\mu_1 = 0 \tag{10.5}$$

The comparison of Equations (10.3) and (10.5) shows that the molar quantities obtained by dividing Equation (10.3) by $(n_1^0)'$ and Equation (10.5) by $(n_1^0)''$ must refer to the same mass of component.

10.2 Gas and condensed phase equilibrium: the Clausius–Clapeyron equation

The Clapeyron equation can be simplified to some extent for the case in which a condensed phase (liquid or solid) is in equilibrium with a gas phase. At temperatures removed from the critical temperature, the molar volume of the gas phase is very much larger than the molar volume of the condensed phase. In such cases the molar volume of the condensed phase may be neglected. An equation of state is then used to express the molar volume of the gas as a function of the temperature and pressure. When the virial equation of state (accurate to the second virial coefficient) is used,

Equation (10.2) may be written as

$$\frac{dP}{dT} = \frac{\Delta\tilde{H}}{(RT^2/P) + \beta T} \tag{10.6}$$

or

$$\frac{d \ln P}{dT} = \frac{\Delta\tilde{H}}{RT^2 + \beta TP} \tag{10.7}$$

When the vapor pressure is sufficiently small that the ideal gas equation of state may be used, Equation (10.7) becomes

$$\frac{d \ln P}{dT} = \frac{\Delta\tilde{H}}{RT^2} \tag{10.8}$$

This equation is called the *Clausius–Clapeyron equation*

The integration of Equation (10.8) still presents a problem. For integration, the equation may be written as

$$\int_{P_1}^{P_2} d \ln P = \int_{T_1}^{T_2} \frac{\Delta\tilde{H}}{RT^2} \, dT \tag{10.9}$$

However, $\Delta\tilde{H}$, the difference between the molar enthalpy of the gas and the condensed phase, depends in general on both the temperature and the pressure. The enthalpy for an ideal gas is independent of pressure and, fortunately, the enthalpy for the condensed phase is only a slowly varying function of the pressure. It is therefore possible to assume that $\Delta\tilde{H}$ is independent of the pressure and a function of the temperature alone, provided that the limits of integration do not cover too large an interval. With this final assumption, the integration can be carried out. When the molar heat capacities of the two phases are known as functions of the temperature, $\Delta\tilde{H}$ is obtained by integration. If $\Delta\tilde{C}_P$, the difference in the molar heat capacities of the two phases, is expressed as

$$\Delta\tilde{C}_P = A + BT + CT^2 \tag{10.10}$$

then

$$\Delta\tilde{H} = \Delta\tilde{H}_0 + AT + \tfrac{1}{2}BT^2 + \tfrac{1}{3}CT^3 \tag{10.11}$$

where $\Delta\tilde{H}_0$ is the integration constant. Substitution of Equation (10.11) into Equation (10.9) would permit the integration to be carried out.

10.3 The dependence of the boiling point on pressure

The reciprocal of dP/dT in Equation (10.2) gives the basic relation for the change of the boiling point of a liquid with pressure, so that dT/dP may be calculated from a knowledge of $\Delta\tilde{H}$ and $\Delta\tilde{V}$. However, the integration cannot be performed directly. Of course, any equation giving the vapor pressure of a liquid as a function of the temperature or the integrated form of the Clausius–Clapeyron equation may be used to calculate the boiling point at various pressures within the range of the applicability of the equation.

A simple but approximate equation is obtained by writing the reciprocal of Equation (10.8) as

$$\frac{\mathrm{d}T}{\mathrm{d}P} = \frac{RT^2}{P\,\Delta\tilde{H}} \tag{10.12}$$

If P_0 is any arbitrary pressure and the change in pressure from P_0 is small, the right-hand side of Equation (10.12) can be considered a constant. Then, on integration,

$$(T - T_0) = \frac{RT_0^2}{P_0\,\Delta\tilde{H}}(P - P_0) \tag{10.13}$$

where T_0 is the boiling point at P_0 and T is the boiling point at P. It must be emphasized again that this equation is only approximate and cannot be used for work of precision.

10.4 Dependence of the vapor pressure of a condensed phase on total pressure

Whenever the total pressure is increased on a condensed phase, the chemical potential of the phase is increased. As a consequence, the pressure of the vapor in equilibrium with the condensed phase must also increase. The discussion here is limited to the liquid phase, but the basic equations that are developed are applicable also to a solid phase.

Two cases arise. The simpler case is one in which we imagine that the liquid is confined in a piston-and-cylinder arrangement with a rigid membrane that is permeable to the vapor but not to the liquid, as indicated in Figure 10.1. Pressure may then be exerted on the liquid independently of the pressure of the vapor. The temperatures of the two phases are equal and are held constant. The Gibbs–Duhem equation for the vapor phase is

$$-\tilde{V}(\mathrm{g})\,\mathrm{d}P(\mathrm{g}) + \mathrm{d}\mu = 0 \tag{10.14}$$

and that for the liquid is

$$-\tilde{V}(\ell)\,\mathrm{d}P(\ell) + \mathrm{d}\mu = 0 \tag{10.15}$$

Because of the condition of equilibrium, the chemical potentials can be eliminated in the two equations to yield

$$\frac{\mathrm{d}P(\mathrm{g})}{\mathrm{d}P(\ell)} = \frac{\tilde{V}(\ell)}{\tilde{V}(\mathrm{g})} \tag{10.16}$$

At temperatures far from the critical temperature, $\tilde{V}(\text{g})$ is very large compared with $\tilde{V}(\ell)$, and consequently the effect is quite small. However, it cannot be neglected in precise work.

In the second case the liquid and vapor are at equilibrium in a closed vessel without restrictions. An inert gas is pumped into the vessel at constant temperature in order to increase the total pressure. For the present we assume that the inert gas is not soluble in the liquid. (The system is actually a two-component system, but it is preferable to consider the problem in this section.) The Gibbs–Duhem equations are now

$$-\tilde{V}(\text{g})\,\mathrm{d}P + y_1\,\mathrm{d}\mu_1 + y_2\,\mathrm{d}\mu_2 = 0 \tag{10.17}$$

and

$$-\tilde{V}(\ell)\,\mathrm{d}P + \mathrm{d}\mu_1 = 0 \tag{10.18}$$

where y_2 is the mole fraction of the inert gas. By elimination of $\mathrm{d}\mu_1$ we obtain

$$[y_1\tilde{V}(\ell) - \tilde{V}(\text{g})]\,\mathrm{d}P + y_2\,\mathrm{d}\mu_2 = 0 \tag{10.19}$$

Now μ_2 is a function of P and y_2, and therefore

$$\mathrm{d}\mu_2 = \bar{V}_2\,\mathrm{d}P + \left(\frac{\partial\mu_2}{\partial y_2}\right)_{T,P}\mathrm{d}y_2 \tag{10.20}$$

Also,

$$\tilde{V}(\text{g}) = y_1\bar{V}_1(\text{g}) + y_2\bar{V}_2(\text{g}) \tag{10.21}$$

Substitution of Equations (10.20) and (10.21) into Equation (10.19) yields

$$y_1[\tilde{V}(\ell) - \bar{V}_1(\text{g})]\,\mathrm{d}P + y_2\left(\frac{\partial\mu_2}{\partial y_2}\right)_{T,P}\mathrm{d}y_2 = 0 \tag{10.22}$$

The expressions for $\bar{V}_1(\text{g})$ and $(\partial\mu_2/\partial y_2)_{T,P}$ in terms of the temperature,

Figure 10.1. Equilibrium between a pure liquid and gas across a rigid membrane.

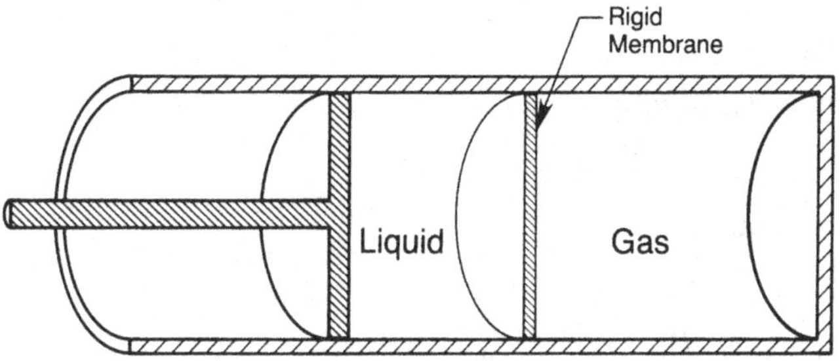

pressure, and mole fraction depend upon the particular equation of state used for the gas phase. For an ideal gas, Equation (10.22) becomes

$$y_1\left(\tilde{V}(\ell)-\frac{RT}{P}\right)\mathrm{d}P - RT\,\mathrm{d}y_1 = 0 \tag{10.23}$$

where the relation $\mathrm{d}y_2 = -\mathrm{d}y_1$ is used. The partial pressure P_1 can be introduced by its definition, $P_1 = Py_1$. After substitution and rearrangement

$$\frac{\mathrm{d}P_1}{\mathrm{d}P} = \frac{\tilde{V}(\ell)}{(RT/Py_1)} = \frac{\tilde{V}(\ell)}{(RT/P_1)} \tag{10.24}$$

is obtained. This equation is the same as Equation (10.16) when it is recognized that RT/P_1 is the molar volume of the pure vapor at the partial pressure P_1.

10.5 Second-order transitions

The transitions between phases discussed in Section 10.1 are classed as first-order transitions. Ehrenfest [25] pointed out the possibility of higher-order transitions, so that second-order transitions would be those transitions for which both the Gibbs energy and its first partial derivatives would be continuous at a transition point, but the second partial derivatives would be discontinuous. Under such conditions the entropy and volume would be continuous. However, the heat capacity at constant pressure, the coefficient of expansion, and the coefficient of compressibility would be discontinuous. If we consider two systems, on either side of the transition point but infinitesimally close to it, then the molar entropies of the two systems must be equal. Also, the change of the molar entropies must be the same for a change of temperature or pressure. If we designate the two systems by a prime and a double prime, we have

$$\mathrm{d}\tilde{S}' = \mathrm{d}\tilde{S}''$$

or

$$\left(\frac{\tilde{C}_P}{T}\right)'\mathrm{d}T + \left(\frac{\partial\tilde{V}}{\partial T}\right)_P'\mathrm{d}P = \left(\frac{\tilde{C}_P}{T}\right)''\mathrm{d}T + \left(\frac{\partial\tilde{V}_P}{\partial T}\right)_P''\mathrm{d}P \tag{10.25}$$

and, therefore,

$$\frac{\mathrm{d}T}{\mathrm{d}P} = \frac{T[(\partial\tilde{V}/\partial T)_P' - (\partial\tilde{V}/\partial T)_P'']}{(\tilde{C}_P)' - (\tilde{C}_P)''} \tag{10.26}$$

Similarly, we obtain

$$\frac{\mathrm{d}T}{\mathrm{d}P} = \frac{(\partial\tilde{V}/\partial P)_T' - (\partial\tilde{V}/\partial P)_T''}{(\partial\tilde{V}/\partial T)_P' - (\partial\tilde{V}/\partial T)_P''} \tag{10.27}$$

when we consider the continuity of the volume at the transition point. Equations (10.26) and (10.27) are called *Ehrenfest's equations.*

Ehrenfest's concept of the discontinuities at the transition point was that the discontinuities were finite, similar to the discontinuities in the entropy and volume for first-order transitions. Only one second-order transition, that of superconductors in zero magnetic field, has been found which is of this type. The others, such as the transition between liquid helium-I and liquid helium-II, the Curie point, the order–disorder transition in some alloys, and transition in certain crystals due to rotational phenomena all have discontinuities that are large and may be infinite. Such discontinuities are particularly evident in the behavior of the heat capacity at constant pressure in the region of the transition temperature. The curve of the heat capacity as a function of the temperature has the general form of the Greek letter *lambda* and, hence, the points are called *lambda points.* Except for liquid helium, the effect of pressure on the transition temperature is very small. The behavior of systems at these second-order transitions is not completely known, and further thermodynamic treatment must be based on molecular and statistical concepts. These concepts are beyond the scope of this book, and no further discussion of second-order transitions is given.

TWO-COMPONENT SYSTEMS

Except for the gas phase, the determination of the value of the thermodynamic functions of a single-phase binary system and those for the components of the system require the study of the equilibrium between at least two phases. When the chemical potential of a component in one of the phases is known as a function of the temperature, pressure, and composition, then the chemical potential of the same component in the other phase can be determined by means of the conditions of equilibrium. In this part of the chapter we first develop the methods for the evaluation of the excess chemical potentials of the components in a binary condensed phase, and hence of the molar excess Gibbs energy of the phase, by the study of various two-phase equilibria. In so doing we include the colligative properties of the binary solution. The methods of treating a liquid phase or a solid phase are identical. Consequently, although emphasis is placed on the liquid, no real distinction need be made except when the phase equilibria include both a liquid and a solid phase. After these considerations, applications of the Gibbs–Duhem equations to phase equilibria leading to the relations between the various intensive variables in terms of derivatives are developed. Finally, we outline methods for the converse problem of determining phase diagrams from known thermodynamic data.

10.6 Liquid–vapor equilibria at constant temperature

The condition of equilibrium, in addition to the equality of the temperature and pressure of the two phases, demands that the chemical potential of a component in each phase in which it exists must be the same. *The problem is to obtain expressions for the chemical potential of a component in terms of quantities that are experimentally observable.*

The particular equations developed depend upon the experimental methods that are used and on the particular reference states that are chosen. The methods are the same for either component when both components are volatile and, consequently, we consider only one component and use the subscript 1 to indicate this component.

The condition of equilibrium in terms of the chemical potential is

$$\mu_1[T, P, x] = \mu_1[T, P, y] \tag{10.28}$$

where y is used to represent the mole fraction of the component in the gas phase. For the purposes of illustration, we choose to use the virial equation of state accurate through the second virial coefficient for the gas phase. When both components are volatile, we can express $\mu_1[T, P, y]$ as

$$\begin{aligned}\mu_1[T, P, y] = RT \ln(Py_1) + \beta_{11}P + \delta_{12}Py_2^2 \\ + \mu_1^{\ominus}[T, 1 \text{ bar, ideal}, y_1 = 1]\end{aligned} \tag{10.29}$$

where $\delta_{12} = 2\beta_{12} - \beta_{11} - \beta_{22}$. If only one component were volatile, then y_2 would be zero. We choose to express the composition of the liquid phase in terms of the mole fraction, so

$$\mu_1[T, P, x] = RT \ln x_1 + \Delta\mu_1^{\mathrm{E}}[T, P, x] + \mu_1^{\ominus}[T, P] \tag{10.30}$$

according to Equation (8.71). A relationship between the two standard states can be obtained if the pure component exists as a liquid at the experimental temperature and is volatile. Under such conditions we choose the pure component at the temperature and pressure of the experiment to be the reference state and, thus, the standard state for the component in the liquid phase. We then measure the vapor pressure, $P_1^{\bullet}$, of the pure liquid, and by the condition of equilibrium we have

$$\mu_1^{\bullet}[T, P_1^{\bullet} x_1 = 1] = RT \ln P_1^{\bullet} + \beta_{11}P_1^{\bullet} + \mu_1^{\ominus}[T, 1 \text{ bar, ideal}, y_1 = 1] \tag{10.31}$$

where we have assumed that the vapor phase is also pure. The combination of Equations (10.28)–(10.31) yields

$$\begin{aligned}RT \ln x_1 + \Delta\mu_1^{\mathrm{E}}[T, P, x] = RT \ln(Py_1/P_1^{\bullet}) + \beta_{11}(P - P_1^{\bullet}) + \delta_{12}Py_2^2 \\ + \mu_1^{\bullet}[T, P_1^{\bullet}, x_1 = 1] - \mu_1^{\ominus}[T, P, x_1 = 1]\end{aligned} \tag{10.32}$$

where $\mu_1^\ominus[T, P, x_1 = 1]$ represents the standard state of the component in the liquid phase. The standard state for the gas phase has been eliminated, but instead we have the difference of the chemical potential of the pure liquid at two pressures. This difference is evaluated by the integral

$$\mu_1^{\bullet}[T, P_1^{\bullet}\, x_1 = 1] - \mu_1^\ominus[T, P, x_1 = 1] = \int_P^{P_1^{\bullet}} \tilde{V}_1^{\bullet}\, \mathrm{d}P \qquad (10.33)$$

where $\tilde{V}_1^{\bullet}$ is the molar volume of the pure liquid component at the experimental temperature.[1] While the combination of Equations (10.32) and (10.33) eliminates the standard states, so that values of $\Delta\mu_1^{\mathrm{E}}[T, P, x]$ can be determined from the experimental methods suggested, we are still faced with one problem. A two-phase, binary system at constant temperature is univariant, and thus the pressure is a function of the mole fraction. The values of $\Delta\mu_1^{\mathrm{E}}[T, P, x]$ so determined depend only on the mole fraction, but the pressure is not constant. In order to obtain values of the excess chemical potential at some arbitrary pressure, P_0, the correction

$$\Delta\mu_1^{\mathrm{E}}[T, P_0, x] - \Delta\mu_1^{\mathrm{E}}[T, P, x] = \int_P^{P_0} (\bar{V}_1 - \tilde{V}_1^{\bullet})\, \mathrm{d}P \qquad (10.34)$$

must be made. The final equation for the excess chemical potential is then

$$\Delta\mu_1^{\mathrm{E}}[T, P_0, x] = RT \ln \frac{Py_1}{P_1^{\bullet} x_1} + \beta_{11}(P - P_1^{\bullet}) + \delta_{12} P y_2^2 + \int_P^{P_1^{\bullet}} \tilde{V}_1^{\bullet}\, \mathrm{d}P - \int_P^{P_0} (\bar{V}_1 - \tilde{V}_1^{\bullet})\, \mathrm{d}P \qquad (10.35)$$

The values of the excess chemical potential obtained by the use of Equation (10.32) after substitution of Equation (10.33) refer to the liquid phase in equilibrium with a vapor phase. In contrast, the values obtained by the use of Equation (10.35) refer to the liquid phase alone at the temperature T and the pressure P_0. The concept of an equilibrium between the liquid phase and a gas phase has been removed by Equation (10.34).

The dynamic method of studying vapor–liquid equilibria requires the use of an inert gas that is passed over the liquid phase under conditions that equilibrium is attained. Under such conditions the total pressure is controlled and can be made the same in each experiment. The system is actually a three-component system in which the solubility of the inert gas in the liquid phase is extremely small and its effect is neglected. The chemical potential of the first component in the gas phase in equilibrium with the solution is

[1] We note that it is the quantity $\Delta\mu_1^{\mathrm{M}} = RT \ln x_1 + \Delta\mu_1^{\mathrm{E}}$ that is obtained on elimination of the standard states. This point was simply accepted in Chapter 8.

now given by

$$\mu_1[T, P, y] = RT\ln(Py_1) + \beta_{11}P + \delta_{12}Py_2^2 + \delta_{13}Py_3^2 + (\delta_{12} + \delta_{13} - \delta_{23})Py_2y_3 + \mu_1^{\ominus}[T, 1\text{ bar, ideal}, y_1 = 1] \qquad (10.36)$$

where the subscript 3 refers to the inert gas, δ_{13} is $(2\beta_{13} - \beta_{11} - \beta_{33})$, and δ_{23} is $(2\beta_{23} - \beta_{22} - \beta_{33})$. When the second component is not volatile, y_2 is, of course, zero. The chemical potential of the component in the gas phase in equilibrium with the pure component is given by

$$\mu_1[T, P, y'] = RT\ln(Py_1') + \beta_{11}P + \delta_{13}P(y_3')^2 + \mu_1^{\ominus}[T, 1\text{ bar, ideal}, y_1 = 1] \qquad (10.37)$$

The conditions of equilibrium are for the solution

$$\mu_1[T, P, x] = \mu_1[T, P, y] \qquad (10.38)$$

and for the pure component

$$\mu_1^{\ominus}[T, P, x_1 = 1] = \mu_1[T, P, y'] \qquad (10.39)$$

Combination of these four equations yields

$$\Delta\mu_1^{\mathrm{E}}[T, P, x] = RT\ln\frac{y_1}{y_1'x_1} + \delta_{12}P(y_2)^2 + \delta_{13}[(y_3)^2 - (y_3')^2]P + (\delta_{12} + \delta_{13} - \delta_{23})Py_2y_3 \qquad (10.40)$$

Because of the experimental conditions, the excess chemical potential is obtained at the constant pressure P, and no corrections need be made.

In many systems the pure liquid phase of a component is not attainable under the experimental conditions and thus cannot be used as the reference state for the component. It is then necessary to choose some other state, usually the infinitely dilute solution of the component in the liquid, as a reference state. We choose to illustrate the development under such circumstances by the use of Equations (10.28)–(10.30) under the appropriate experimental conditions. The combination of these equations yields

$$RT\ln x_1 + \Delta\mu_1^{\mathrm{E}}[T, P, x] + \mu_1^{\ominus}[T, P] = RT\ln(Py_1) + \beta_{11}P + \delta_{12}Py_2^2 + \mu_1^{\ominus}[T, 1\text{ bar, ideal}, y_1 = 1] \qquad (10.41)$$

We define the reference state of the component to be the infinitely dilute solution of the component in the second component, so that $\Delta\mu_1^{\mathrm{E}}$ approaches zero as x_1 approaches zero. Then, in the limit,

$$\mu_1^\ominus[T, P] - \mu_1^\ominus[T, \text{ideal}, 1\text{ bar}, y_1 = 1]$$

$$= \lim_{x_1 \to 0} [RT \ln(Py_1/x_1) + \beta_{11}P + \delta_{12}Py_2^2]$$

$$= RT \ln(Py_1^\infty/x_1^\infty) + \beta_{11}P^\infty + \delta_{12}P^\infty(y_2^\infty)^2 \tag{10.42}$$

where the superscript ∞ is used to indicate the limiting values at infinite dilution. The elimination of the chemical potentials for the standard states in Equation (10.41) by the use of Equations (10.42) and

$$\mu_1^\ominus[T, P^\infty] - \mu_1^\ominus[T, P] = \int_P^{P^\infty} \bar{V}_1^\infty \, dP = \int_P^{P^\infty} \bar{V}_1^\ominus \, dP \tag{10.43}$$

yields

$$\Delta\mu_1^E[T, P, x] = \left(RT \ln \frac{Py_1}{x_1} + \beta_{11}P + \delta_{12}Py_2^2\right) - RT \ln \frac{P^\infty y_1^\infty}{x_1^\infty}$$

$$+ \beta_{11}P^\infty + \delta_{12}P^\infty(y_2^\infty)^2 + \int_P^{P^\infty} \bar{V}_1^\ominus \, dP \tag{10.44}$$

In Equations (10.43) and (10.44) $\bar{V}_1^\infty$ represents the partial molar volume of the component in the infinitely dilute solution, which is *also* the partial molar volume of the component in the standard state. The right-hand side of Equation (10.44) contains only quantities that can be determined experimentally, and thus $\Delta\mu_1^E[T, P, x]$ can be determined. However, just as in the previous case, the pressure is a function of the mole fraction. Therefore, if we require values of $\Delta\mu_1^E$ at some arbitrary constant pressure, the correction expressed in Equation (10.34) must be made with the substitution of $\bar{V}_1^\infty$ for $\tilde{V}_1^\ominus$.

10.7 Raoult's law

The relation between Raoult's law and the definition of an ideal solution given by Equation (8.57) is obtained by a study of Equation (10.35) or (10.40). If a solution is ideal, then $\Delta\mu_1^E$ must be zero and the right-hand side of both equations must be zero. If we write Py_1 in both equations as P_1, the partial pressure of the component, and Py_1' in Equation (10.40) as $P_1^\bullet$, then the logarithmic term becomes $\ln(P_1/P_1^\bullet x_1)$, which is zero when Raoult's law, given in the form $P_1 = P_1^\bullet x_1$, is obeyed. We then see that to define an ideal solution in terms of Raoult's law and still be consistent with Equation (10.57) requires that the experimental measurements be made at the same total pressure and that the vapor behaves as an ideal gas.

10.8 Henry's law

The simplest approach to Henry's law is by the use of equations (10.38), (10.30), and (10.36) under the experimental condition of constant total pressure. Combination of these equations results in

$$\Delta\mu_1^E[T, P, x] = RT\ln(Py_1/x_1) + \beta_{11}P + \delta_{12}Py_2^2 + \delta_{13}Py_3^2 + (\delta_{12} + \delta_{13} - \delta_{23})Py_2y_3 + \mu_1^\ominus[T, 1\text{ bar, ideal } y_1 = 1] - \mu_1^\ominus[T, P] \tag{10.45}$$

We choose the reference state to be the infinitely dilute solution, so

$$\mu_1^\ominus[T, 1\text{ bar, ideal, } y = 1] - \mu_1^\ominus[T, P] = -\left(RT\ln\frac{Py_1^\infty}{x_1^\infty} + \beta_{11}P + \delta_{12}P(y_2^\infty)^2 + \delta_{13}P(y_3^\infty)^2 + (\delta_{12} + \delta_{13} - \delta_{23})Py_2^\infty y_3^\infty\right) \tag{10.46}$$

We now define the Henry's law constant (some authors call it the Henry coefficient or the Henry constant), k, by

$$\mu_1^\ominus[T, 1\text{ bar, ideal, } y_1 = 1] - \mu_1^\ominus[T, P] = -RT\ln k \tag{10.47}$$

Equation (10.47) indicates that k is a function of the temperature and pressure. For a given series of experiments at the constant pressure P, k is a function of the temperature alone, and not a function of the composition. Substitution of Equation (10.47) into Equation (10.46) gives

$$RT\ln(Py_1^\infty/kx_1^\infty) = -[\beta_{11}P + \delta_{12}(Py_2^\infty)^2 + \delta_{13}(Py_3^\infty)^2 + (\delta_{12} + \delta_{13} - \delta_{23})Py_2^\infty y_3^\infty] \tag{10.48}$$

Henry's law in the form $Py_1^\infty/kx_1^\infty = 1$ can be true only with the assumption that the gas phase behaves as an ideal gas or that the corrections for nonideal behavior are negligible.

When the Henry's law constant has been evaluated according to Equations (10.46) and (10.47) for a series of experiments at constant temperature and pressure, Equation (10.45) may be written as

$$RT\ln(Py_1/kx_1) = \Delta\mu_1^E[T, P, x] - [\beta_{11}P + \delta_{12}Py_2^2 + \delta_{13}Py_3^2 + (\delta_{12} + \delta_{13} - \delta_{23})Py_2y_3] \tag{10.49}$$

Henry's law in the form $Py_1/kx_1 = 1$ can be valid over a finite range of composition only when the right-hand side of Equation (10.49) is zero. The corrections for the nonideality of the gas must be negligible and $\Delta\mu_1^E$ must remain zero over the range of composition starting from the infinitely dilute solution. In the strict sense, then, *Henry's law is only a limiting law*, being obeyed in the limit as x_1 approaches zero with the assumption of ideal gas behavior.

The development of Henry's law is only a little more complicated when the experimental measurements are made without the use of an inert gas. The Henry's law constant is defined in terms of Equation (10.42) so that

$$\mu_1^{\ominus}[T, P^{\infty}] - \mu_1^{\ominus}[T, 1\text{ bar, ideal}, y_1 = 1] = RT \ln k \tag{10.50}$$

We see from this definition that k is not a function of the composition, but it is a function of the temperature and the pressure. For a given series of experiments in which the mole fraction is varied at constant temperature, k is evaluated at the limiting pressure P^{∞} whereas the total pressure is different for each experiment. If we wished to evaluate k at any given pressure, then

$$RT\left(\frac{\partial \ln k}{\partial P}\right)_T = \bar{V}_1^{\infty} \tag{10.51}$$

would be used. In most cases the corrections are small and are usually omitted. Moreover, when Henry's law is used in the form $P_1 = kx_1$, the deviations from ideal behavior in the gas phase must be negligible.

Another approach to Henry's law based on Equation (10.35) is of interest. When the deviations from ideal gas behavior and the corrections for the pressure on the liquid phase are negligible, this equation may be written in the form

$$P_1 = P_1^{\bullet} x_1 \exp(\Delta\mu_1^{\mathrm{E}}/RT) \tag{10.52}$$

Here we have chosen the pure liquid to be the reference state. If now we assume that measurements can be made over the entire range of composition, so that the *limiting* value of $\Delta\mu_1^{\mathrm{E}}$ can be obtained as x_1 approaches zero, we have

$$P_1^{\infty} = P_1^{\bullet} \exp(\Delta\mu_1^{\mathrm{E}}/RT)x_1^{\infty} = kx_1^{\infty} \tag{10.53}$$

This definition involves indirectly the change of the reference state from the pure liquid to the infinitely dilute solution.

Henry's law is also used in some cases as a limiting expression of the behavior of a component in a single condensed phase at a given temperature and an arbitrary constant pressure P_0 with no concept of an equilibrium between the condensed phase and a vapor phase. In order to obtain the concept of Henry's law under these conditions, two different expressions are required for the chemical potential of the component with the use of two different standards states. In terms of mole fractions, the usual practice is to use Equation (10.30) for one expression and

$$\mu_1[T, P_0, x] = RT \ln f_1 + \mu_1^\ominus[T, 1 \text{ bar, ideal}, y_1 = 1] \tag{10.54}$$

for the other expression. This equation defines the fugacity of the component in the phase with the ideal gas at 1 bar as the standard state. When the two expressions are equated,

$$\begin{aligned} &RT \ln x_1 + \Delta\mu_1^E[T, P_0, x] + \mu_1^\ominus[T, P_0] \\ &\quad = RT \ln f_1 + \mu_1^\ominus[T, 1 \text{ bar, ideal}, y_1 = 1] \end{aligned} \tag{10.55}$$

We choose the reference state for the left-hand side of this equation as the infinitely dilute solution, so

$$\mu_1^\ominus[T, P_0] - \mu_1^\ominus[T, 1 \text{ bar, ideal } y_1 = 1] = RT \ln(f_1^\infty/x_1^\infty) \tag{10.56}$$

We then define the Henry's law constant in terms of the difference between the chemical potentials of the component in the two standard states by

$$\mu_1^\ominus[T, P_0] - \mu_1^\ominus[T, 1 \text{ bar, ideal}, y_1 = 1] = RT \ln k \tag{10.57}$$

and

$$k = f_1^\infty/x_1^\infty \tag{10.58}$$

Here k is determined at the temperature T and the arbitrary pressure P_0. When the ratio of f_1/x_1 has the same value over a finite range of concentration as that in the limit, we have

$$f_1 = kx_1 \tag{10.59}$$

The experimental determination of the fugacity, and hence the value of k, requires the study of the equilibrium between a vapor phase and the condensed phase. The problem becomes identical to those already discussed.

10.9 Limited uses of the Gibbs–Duhem equation

Three different uses of the Gibbs–Duhem equation associated with the integral method are discussed in this section: (A) the calculation of the excess chemical potential of one component when that of the other component is known; (B) the determination of the minimum number of intensive variables that must be measured in a study of isothermal vapor–liquid equilibria and the calculation of the values of other variables; and (C) the study of the thermodynamic consistency of the data when the data are redundant.

(A) When only one of the two components in a binary solution is volatile, the excess chemical potential of the volatile component can be determined by the methods that have been discussed. However, we require the valucs of the excess chemical potential of the other component or of the molar

Gibbs energy. Let us assume that $\Delta\mu_1^E[T, P, x]$ has been determined over a range of composition at constant temperature and pressure. Then, by the Gibbs–Duhem equation for the single phase at constant temperature and pressure,

$$d(\Delta\mu_2^E) = -(x_1/x_2)\, d(\Delta\mu_1^E) \tag{10.60}$$

or

$$\Delta\mu_2^E[T, P, x''] - \Delta\mu_2^E[T, P, x'] = -\int_{x_1'}^{x_1''} (x_1/x_2) d(\Delta\mu_1^E) \tag{10.61}$$

where x_1' and x_1'' represent the limits of integration. In general, the upper limit may be variable rather than a fixed value. However, in order to obtain $\Delta\mu_2^E$ as a function of the mole fraction, $\Delta\mu_2^E[T, P, x']$ must be evaluated by choosing an appropriate reference state. When $\Delta\mu_1^E$ has been determined so that its value may be extrapolated to $x_1 = 0$, the pure liquid may be used as the reference state of the second component. Otherwise the limit of x_1 is usually $x_1 = 1$ and the reference state of the second component is the infinitely dilute solution.

An alternative calculation, particularly applicable when the experimental studies can be carried out over the entire range of composition, involves the direct calculation of the molar excess Gibbs energy. We have

$$\Delta\tilde{G}^E = x_1\, \Delta\mu_1^E + x_2\, \Delta\mu_2^E \tag{10.62}$$

and

$$\Delta\mu_2^E = -\int_{x_1=0}^{x_1} (x_1/x_2)\, d(\Delta\mu_1^E) \tag{10.63}$$

so that

$$\Delta\tilde{G}^E = x_1\, \Delta\mu_1^E - x_2 \int_{x_1=0}^{x_1} (x_1/x_2)\, d(\Delta\mu_1^E) \tag{10.64}$$

where we have chosen the reference state of the second component to be the pure component. Integration by parts yields

$$\Delta\tilde{G}^E = -x_2 \int_{x_1=0}^{x_1} (\Delta\mu_1^E/x_2^2)\, dx_1 \tag{10.65}$$

(B) We have pointed out that experimental studies are usually arranged so that the system is univariant. The experimental measurements then involve the determination of the values of the dependent intensive variables for chosen values of the one independent variable. Actually, the values of only one dependent variable need be determined, because of the condition that the Gibbs–Duhem equations, applicable to the system at equilibrium, must be satisfied. In two of the type of systems that we have discussed, another dependent variable exists but its values can be determined by use of the Gibbs–Duhem equations, with appropriate choice of variables.

First we consider the binary systems when no inert gas is used. When only one of the components is volatile, the intensive variables of the system are the temperature, the pressure, and the mole fraction of one of the components in the liquid phase. When the temperature has been chosen, the pressure must be determined as a function of the mole fraction. When both components are volatile, the mole fraction of one of the components in the gas phase is an additional variable. At constant temperature the relation between two of the three variables P_1, x_1, and y_1 must be determined experimentally; the values of the third variable might then be calculated by use of the Gibbs–Duhem equations. The particular equations for this case are

$$-\tilde{V}(\mathrm{g})\,\mathrm{d}P + y_1\,\mathrm{d}\mu_1 + y_2\,\mathrm{d}\mu_2 = 0 \tag{10.66}$$

and

$$-\tilde{V}(\ell)\,\mathrm{d}P + x_1\,\mathrm{d}\mu_1 + x_2\,\mathrm{d}\mu_2 = 0 \tag{10.67}$$

When $\mathrm{d}\mu_2$ is eliminated from these two equations, Equation (10.29) is used to express μ_1 as a function of P and y_1 with the relation

$$\mathrm{d}\mu_1 = \left(\frac{\partial \mu_1}{\partial P}\right)_{T,y} \mathrm{d}P + \left(\frac{\partial \mu_1}{\partial y_1}\right)_{T,P} \mathrm{d}y_1 \tag{10.68}$$

and the equation of state

$$\tilde{V}(\mathrm{g}) = (RT/P) + \beta_{11}y_1 + \beta_{22}y_2 + \delta_{12}y_1y_2 \tag{10.69}$$

is used for $\tilde{V}(\mathrm{g})$, the relation

$$\left(\frac{\partial \ln P}{\partial y_1}\right)_T = \frac{(y_1 - x_1)[RT - 2\delta_{12}Py_1y_2]}{y_1y_2[RT + x_1\beta_{11}P + x_2\beta_{22}P + \delta_{12}P(y_1^2x_2 + y_2^2x_1) - P\tilde{V}(\ell)]} \tag{10.70}$$

is obtained. When P is measured as a function of y_1, x_1 can be calculated by a series of approximations. When y_1 is measured as a function of x_1, P can be calculated by integration, again using a series of approximations. However, y_1 cannot be calculated when P is measured as a function of x_1.

Similar conditions prevail when an inert gas is used. In this case both the temperature and pressure are fixed and made constant. When only one of the components in the liquid phase is volatile, the two variables are the mole fraction of one of the components in the liquid phase and the mole fraction of either the volatile component or the inert gas in the gas phase. Experimentally one of these variables must be determined as a function of the other. An additional variable, a second mole fraction in the gas phase, must be considered when both components in the liquid phase are volatile. Then, of the three variables, say x_1, y_1, and y_3, one variable must be measured as a function of another variable, properly chosen, and the third may be calculated. However, it should be noted that, according to Equation (10.40), y_3 needs to be known only to correct for the nonideality of the gas phase. The two Gibbs–Duhem equations for this case are

$$x_1\, d\mu_1 + x_2\, d\mu_2 = 0 \tag{10.71}$$

and

$$y_1\, d\mu_1 + y_2\, d\mu_2 + y_3\, d\mu_3 = 0 \tag{10.72}$$

With the elimination of $d\mu_2$ and the evaluation of $d\mu_1$ and $d\mu_3$ in terms of dy_1 and dy_3,

$$\left(\frac{\partial y_3}{\partial y_1}\right)_{T,P} = -\frac{[y_1 - y_2(x_1/x_2)](\partial\mu_1/\partial y_1)_{T,P,y_3} + y_3(\partial\mu_3/\partial y_1)_{T,P,y_3}}{[y_1 - y_2(x_1/x_2)](\partial\mu_1/\partial y_3)_{T,P,y_1} + y_3(\partial\mu_3/\partial y_3)_{T,P,y_1}} \tag{10.73}$$

is obtained. The partial derivatives can be expressed in terms of T, P, y_1, and y_3 by means of a suitable equation of state. When the virial equation of state is used as before, only terms containing the δ quantities appear and no term appears containing a β alone. Then, for ideal mixtures of real gases, Equation (10.73) becomes

$$\left(\frac{\partial y_3}{\partial y_1}\right)_{T,P} = -\frac{y_1 - x_1 + x_1 y_3}{y_1(1 - x_1)} \tag{10.74}$$

We observe that when y_3 and y_1 are determined, x_1 can be calculated and, when x_1 and y_1 are determined, y_3 may be calculated. However, y_1 cannot be obtained when y_3 and x_1 are determined.

(C) In many experimental studies, all of the intensive variables are determined, giving a redundancy of experimental data. However, Equations (10.70) and (10.73) afford a means of checking the thermodynamic consistency of the data at each experimental point for the separate cases. Thus, for Equation (10.70), the required slope of the curve of P versus y_1, consistent with the thermodynamic requirements of the Gibbs–Duhem equations, can be calculated at each experimental point from the measured values of P, x_1, and y_1 at the experimental temperature. This slope must agree within the experimental error with the slope, at the same composition, of the best curve drawn through all of the experimental points. When this agreement is not achieved for a given set of data, that set is thermodynamically inconsistent.

Many other tests, too numerous to discuss individually, have been devised, *all* of which are based on the Gibbs–Duhem equation. Only one such test, given by Redlich, is discussed here and is applicable to the case in which both components are volatile and in which experimental studies can be made over the entire range of composition. The reference states are chosen to be the pure liquid at the experimental temperature and a constant arbitrary pressure P_0. The values of $\Delta\mu_1^E[T, P_0, x]$ and $\Delta\mu_2^E[T, P_0, x]$ will have been calculated from the experimental data. The molar excess Gibbs energy is given by Equation (10.62), from which we conclude that $\Delta\tilde{G}^E = 0$ when $x_1 = 0$ and when $x_1 = 1$. Therefore,

$$\int_0^1 \left(\frac{\partial\, \Delta\tilde{G}^E}{\partial x_1}\right)_{T,P} dx_1 = 0 \tag{10.75}$$

Differentiation of Equation (10.62) with respect to x_1 at constant temperature and pressure gives

$$\left(\frac{\partial\, \Delta\tilde{G}^E}{\partial x_1}\right)_{T,P} = \Delta\mu_1^E - \Delta\mu_2^E \tag{10.76}$$

because

$$x_1\left(\frac{\partial\, \Delta\mu_1^E}{\partial x_1}\right)_{T,P} + x_2\left(\frac{\partial\, \Delta\mu_2^E}{\partial x_1}\right)_{T,P} = 0 \tag{10.77}$$

according to the Gibbs–Duhem equation. Therefore,

$$\int_0^1 (\Delta\mu_1^E - \Delta\mu_2^E)\, dx_1 = 0 \tag{10.78}$$

Thus, when values of $(\Delta\mu_1^E - \Delta\mu_2^E)$ are plotted as a function of x_1, the area between the best smooth curve drawn through the points and the composition axis must be zero within the experimental accuracy. This test concerns the thermodynamic consistency of the data as a whole rather than that of each individual set of experimental values. It also applies strictly to the liquid solution at the arbitrary pressure P_0 and only to the two-phase system at equilibrium through the calculation of $\Delta\mu_1^E$ and $\Delta\mu_2^E$ from the experimental data.

10.10 Liquid–vapor equilibria at constant pressure

The thermodynamic treatment of data obtained from the study of the liquid–vapor equilibria at constant pressure differs from that used in constant-temperature experiments in the treatment of the standard states. This difference arises because we cannot calculate the change in the chemical potential of a single-phase system with temperature in the general case. We consider the case for which the pure liquid state of a volatile component can be used as the reference state, and thus the standard state. No inert gas is used as a third component in the system. The condition of equilibrium is again the equality of the chemical potentials of the component in the two phases. When we use the mole fraction as the composition variable in the liquid phase, we have

$$RT \ln x_1 + \Delta\mu_1^{\mathrm{E}}[T, P, x] + \mu_1^{\bullet}[T, P, x_1 = 1] = \mu_1[T, P, y] \qquad (10.79)$$

We can subtract from each side, $\mu_1^{\bullet}[T, P, y = 1]$, the chemical potential of the pure, real gas at the same temperature and pressure as the experiment. We then have

$$\begin{aligned} RT \ln x_1 &+ \Delta\mu_1^{\mathrm{E}}[T, P, x] + \mu_1^{\bullet}[T, P, x_1 = 1] - \mu_1^{\bullet}[T, P, y_1 = 1] \\ &= \mu_1[T, P, y] - \mu_1^{\bullet}[T, P, y_1 = 1] \end{aligned} \qquad (10.80)$$

By Equation (10.29)

$$\mu_1[T, P, y] - \mu_1^{\bullet}[T, P, y_1 = 1] = RT \ln y_1 + \delta_{12} P y_2^2 \qquad (10.81)$$

The difference between the chemical potentials of the pure component in the liquid and gas phases is calculated, with the knowledge that the chemical potentials must be equal at the boiling point of the pure substance, $T^{\bullet}$, at the pressure P. We choose the series of changes of state

$$B[T, P, y_1 = 1] \rightarrow B[T^{\bullet}, P, y_1 = 1]$$

$$B[T^{\bullet}, P, y_1 = 1] \rightarrow B[T^{\bullet}, P, x_1 = 1]$$

$$B[T^{\bullet}, P, x_1 = 1] \rightarrow B[T, P, x_1 = 1]$$

where B represents the component, and then make use of the relation (Eq. 4.41))

$$\left(\frac{\partial(\mu/T)}{\partial T}\right)_{P,x} = -\frac{\bar{H}}{T^2}$$

For the first change of state

$$\frac{\mu_1^{\bullet}[T^{\bullet}, P, y_1 = 1]}{T^{\bullet}} - \frac{\mu_1^{\bullet}[T, P, y_1 = 1]}{T} = -\int_T^{T^{\bullet}} \frac{\tilde{H}_1^{\bullet}(\mathrm{g})}{T^2}\,\mathrm{d}T \qquad (10.82)$$

and for the third change of state

$$\frac{\mu_1^{\bullet}[T, P, x_1 = 1]}{T} - \frac{\mu_1^{\bullet}[T^{\bullet}, P, x_1 = 1]}{T^{\bullet}} = -\int_{T^{\bullet}}^{T} \frac{\tilde{H}_1^{\bullet}(\ell)}{T^2}\,\mathrm{d}T \qquad (10.83)$$

so

$$\mu_1^{\bullet}[T, P, x_1 = 1] - \mu_1^{\bullet}[T, P, y_1 = 1] = T \int_{T^{\bullet}}^{T} \frac{\tilde{H}_1^{\bullet}(\mathrm{g}) - \tilde{H}_1^{\bullet}(\ell)}{T^2} \, \mathrm{d}T \tag{10.84}$$

with the condition that

$$\mu_1^{\bullet}[T^{\bullet}, P, x_1 = 1] = \mu_1^{\ominus}[T^{\bullet}, P, y_1 = 1] \tag{10.85}$$

Substitution of Equations (10.81) and (10.84) into Equation (10.80) yields

$$\Delta\mu_1^{\mathrm{E}}[T, P, x] = RT \ln \frac{y_1}{x_1} + \delta_{12} P y_2^2 - T \int_{T^{\bullet}}^{T} \frac{\tilde{H}_1^{\bullet}(\mathrm{g}) - \tilde{H}_1^{\bullet}(\ell)}{T^2} \, \mathrm{d}T \tag{10.86}$$

Thus, values of $\Delta\mu_1^{\mathrm{E}}[T, P, x]$ can be calculated from the measured values of the equilibrium temperature T and the two mole fractions, x_1 and y_1. The same equation with change of subscripts is valid for the second component when its standard state can be taken as the pure liquid component.

Some comments need to be made concerning the integral in Equations (10.84) and (10.86). Certainly, when the molar heat capacities at the constant pressure P are known as a function of the temperature for both phases, the integral can be evaluated. However, the integration requires the extrapolation of the heat capacity of one of the phases into a region where the phase is metastable with respect to the other phase. Figures 10.2 and 10.3 illustrate this point (refer also to Fig. 5.7). The curve in Figure 10.2 represents the vapor pressure curve of the component and $T^{\bullet}$, the boiling point of the component at the chosen pressure. When the equilibrium temperature of the solution is greater than $T^{\bullet}$, the integration is carried out along the path *a* into the region of the gas phase; the liquid phase is then metastable. When the equilibrium temperature of the solution is less than $T^{\bullet}$, the integration is carried out along the path *b* into the liquid region; the gas phase is then metastable. The same problem is illustrated in Figure 10.3 in terms of the chemical potentials. The curves in this figure represent the chemical potential of the pure component in the two different phases as a function of the temperature at constant pressure. The curve labeled *ll′* represents the liquid and the curve labeled *gg′* represents the gas. When the integration is carried to temperatures greater than $T^{\bullet}$, the chemical potential of the gas phase follows the solid portion of the gas curve whereas that for the liquid phase follows the broken portion of the liquid curve. The liquid phase is then metastable. When the integration is carried out to temperatures lower than $T^{\bullet}$, the chemical potential of the gas phase becomes greater than that of the liquid, and hence is metastable. The difference between the values of the chemical potentials on the two curves at a given temperature T is the difference required in Equation (10.84). We must recognize, then, that the

Figure 10.2. Integrations into metastable regions.

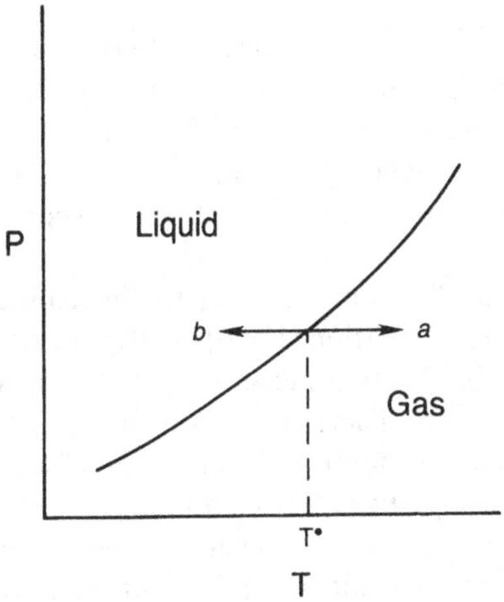

Figure 10.3. Integrations into metastable regions.

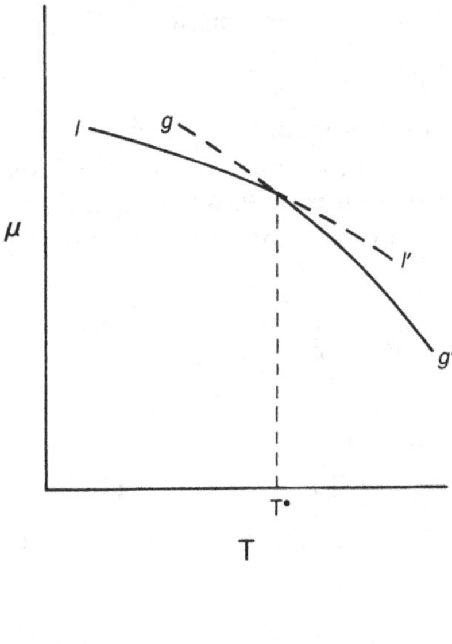

value of the integral used in the calculation of $\Delta\mu_1^E$ is uncertain to some unknown amount.

The values of $\Delta\mu_1^E[T, P, x]$ at different compositions obtained by the case of Equation (10.86) are not isothermal, because in the univariant system the temperature is a function of the mole fractions. Corrections must be made to obtain isothermal values of $\Delta\mu_1^E$. We choose some arbitrary temperature T_0. Then, according to Equation (11.120),

$$\frac{\Delta\mu_1^E[T_0, P, x]}{T_0} - \frac{\Delta\mu_1^E[T, P, x]}{T} = -\int_T^{T_0} \frac{\bar{H}_1 - \tilde{H}_1^\bullet(\ell)}{T^2}\,\mathrm{d}T \tag{10.87}$$

In order to evaluate this integral, $(\bar{H}_1 - \tilde{H}_1^\bullet(\ell))$ must be known at each composition over the indicated range of temperature, or the value must be known at one temperature together with the values of $(\bar{C}_{P_1} - \tilde{C}_{P_1}^\bullet(\ell))$. Just as $\Delta\mu_1^E[T, P_0, x]$ in the isothermal case refers to the liquid phase alone, so here $\Delta\mu_1^E[T_0, P, x]$ refers to the liquid phase alone, with no concept of an equilibrium between the two phases.

The experimental determination of T, y_1, and x_1 is actually redundant. However, the equation corresponding to Equation (10.70) involves the partial molar entropies of the components in the two phases. In many cases the values of these quantities are not known. Therefore, a test of the thermodynamic consistency of each experimental point generally cannot be made, neither can the value of one of the three quantities T, y_1, and x_1 be calculated when the other two are measured. The test for the consistency of the overall data according to Equation (10.78) can be made when the isothermal values of both $\Delta\mu_1^E[T_0, P, x]$ and $\Delta\mu_2^E[T_0, P, x]$ have been determined.

The use of the infinitely dilute solution as the reference state of a volatile component is not convenient for the experimental study of isopiestic vapor–liquid equilibria. Equation (10.79) may be written as

$$RT^\infty \ln x_1 + \mu_1^\ominus[T^\infty, P] = \mu_1^\ominus[T^\infty, P, y_1^\infty] \tag{10.88}$$

for the infinitely dilute solution with the condition that $\Delta\mu_1^E[T^\infty, P, x_1^\infty]$ is zero. Here T^∞ is the temperature at which the infinitely dilute solution is in equilibrium with the vapor at the constant pressure P. When Equation (10.88) is subtracted from Equation (10.79) and use is made of Equation (10.29), we obtain

$$\begin{aligned}\Delta\mu_1^E[T, P, x] = {} & RT \ln \frac{y_1 x_1^\infty}{x_1 y_1^\infty} + (\beta_{11} - \beta_{11}^\infty)P \\ & + [\delta_{12} y_2^2 - \delta_{12}(y_2^\infty)^2]P \\ & + (\mu_1^\ominus[T^\infty, P] - \mu_1^\ominus[T^\infty, 1\ \text{bar}, y_1 = 1]) \\ & - (\mu_1^\ominus[T, P] - \mu_1^\ominus[T, 1\ \text{bar}, y_1 = 1])\end{aligned} \tag{10.89}$$

The differences between the standard states of the component in the condensed and gaseous states at the two temperatures cannot be evaluated without knowledge of the composition of standard states in the condensed phases. However, this requires prior knowledge of the excess chemical potentials. The experimental study of the isothermal vapor–liquid equilibria is therefore more convenient and yields the same information.

10.11 The boiling point elevation

The study of the elevation of the boiling point of a solvent by the addition of a nonvolatile solute is a special case of isopiestic vapor–liquid equilibria. The solvent is the only volatile component, and the standard state of the solvent is chosen as the pure solvent. Under these conditions Equation (10.81) may be written as

$$RT \ln x_1 + \Delta\mu_1^{\mathrm{E}}[T, P, x] = -T \int_{T^{\cdot}}^{T} \frac{\tilde{H}_1^{\cdot}(\mathrm{g}) - \tilde{H}_1^{\cdot}(\ell)}{T^2} \, \mathrm{d}T \qquad (10.90)$$

This equation can be interpreted as giving the temperature of the equilibrium system as a function of the mole fraction of the liquid phase when $\Delta\mu_1^{\mathrm{E}}[T, P, x]$ is known as a function of the temperature and mole fraction. For values of x_1 very close to unity, $\Delta\mu_1^{\mathrm{E}}$ may be taken as zero, and $(\tilde{H}_1^{\cdot}(\mathrm{g}) - \tilde{H}_1^{\cdot}(\ell))$ may be considered to be independent of the temperature and equal to the molar change of enthalpy on evaporation of the pure liquid at $T_1^{\cdot}$. Then we obtain on integration

$$\ln x_1 = -\frac{\Delta\tilde{H}_{\mathrm{v}}^{\cdot}(T - T_1^{\cdot})}{RTT_1^{\cdot}} \qquad (10.91)$$

When we write $(T - T_1^{\cdot}) = \Delta T$ and let T be approximately $T_1^{\cdot}$, we have

$$\Delta T = -\frac{R(T_1^{\cdot})^2}{\Delta\tilde{H}_{\mathrm{v}}^{\cdot}} \ln x_1 \qquad (10.92)$$

Here $[R(T_1^{\cdot})^2/\Delta\tilde{H}_{\mathrm{v}}^{\cdot}]$ is a property of the pure solvent. Equation (10.92) is the basic equation for the simpler expressions involving the mole fraction or molality of the solute of the equation for the boiling point elevation, but it must be emphasized that *it is only an approximate equation, valid in the limit as x_1 approaches unity*. Even for the approximation of ideal solutions, Equation (10.90) should be used for the calculation of the boiling point when x_1 is removed from unity.

10.12 Liquid–pure solid equilibria at constant pressure

The study of the equilibria between a liquid solution and a pure solid phase of one of the components is generally made under conditions of constant pressure. The temperature is then a function of the composition of the liquid, or the composition of the liquid is a function of the temperature. In the first case we usually speak of the freezing point of the solution, whereas in the second case we speak of the solubility of the solid. The thermodynamic treatment is identical in either case, and the difference is simply one of interpretation. The condition of equilibrium is the equality of the chemical potential of the component in the two phases, so we write

$$\mu_1[T, P, x_1] = \mu_1^{\cdot}[T, P, s] \tag{10.93}$$

We again choose to use Equation (10.30) to express the chemical potential of the component in the liquid phase in terms of the mole fraction and choose the pure liquid to be the standard state. Then we have

$$RT \ln x_1 + \Delta\mu_1^{\mathrm{E}}[T, P, x] = \mu_1^{\cdot}[T, P, s] - \mu_1^{\cdot}[T, P, x_1 = 1] \tag{10.94}$$

The difference in the chemical potential of the pure component in the liquid and solid phases can be evaluated in the same manner as was done in Section 10.10, because of the equality of the two chemical potentials at the melting point for the pressure P. The result is

$$\mu_1^{\cdot}[T, P, s] - \mu_1^{\cdot}[T, P, x_1 = 1] = -T \int_T^{T_1^{\cdot}} \frac{\tilde{H}_1^{\cdot}(\ell) - \tilde{H}_1^{\cdot}(\mathrm{s})}{T^2}\, \mathrm{d}T \tag{10.95}$$

where $T_1^{\cdot}$ is the melting point of the pure component. Substitution of Equation (10.95) into Equation (10.94) yields

$$\Delta\mu_1^{\mathrm{E}}[T, P, x_1] = -RT \ln x_1 - T \int_T^{T_1^{\cdot}} \frac{\tilde{H}_1^{\cdot}(\ell) - \tilde{H}_1^{\cdot}(\mathrm{s})}{T^2}\, \mathrm{d}T \tag{10.96}$$

Thus, values of $\Delta\mu_1^{\mathrm{E}}[T, P, x]$ can be calculated when the equilibrium values of T and x_1 are determined.

The evaluation of the integral requires, as in Section 13.10, a knowledge of the molar heat capacities of the pure solid and liquid phases as a function of the temperature. In this case the equilibrium temperature for the solution is always less than the melting point of the pure substance. Then the heat capacity of the liquid phase is extrapolated into the region in which the liquid is metastable with respect to the solid phase. Because of this, there is always some uncertainty in the evaluation of the integral and the range of temperature over which the integration is carried out should not be too large.

The values of $\Delta\mu_1^{\mathrm{E}}[T, P, x]$ obtained from Equation (10.96) are not isothermal. When isothermal values are desired at some arbitrary temperature T_0, the same correction as that given in Equation (10.87) must be applied.

When the excess chemical potential of the solute in the liquid phase is required as a function of the mole fraction at the constant temperature T_0 and pressure P, an integration of the Gibbs–Duhem equation must be used. For this the infinitely dilute solution of the solute in the solvent must be used as the reference state for the solute, because the experimental measurements can be made only over a limited range of composition starting with the pure solvent.

10.13 Freezing point lowering and solubility

The comments concerning the lowering of the freezing point of a solvent by a solute, based on Equation (10.96), are the same as those made concerning the boiling point elevation. Equation (10.96) gives indirectly the temperature as a function of the composition when $\Delta\mu_1^{\mathrm{E}}$ is known as a function of these two variables. In the limit as x_1 approaches unity, $\Delta\mu_1^{\mathrm{E}}$ approaches zero and $(\tilde{H}_1^{\bullet}(\ell) - \tilde{H}_1^{\bullet}(\mathrm{s}))$ approaches the molar change of enthalpy on fusion, $\Delta\tilde{H}_{\mathrm{f}}^{\bullet}$, of the component at its melting point at the pressure P. We may then write Equation (10.96) as

$$\Delta T = -\frac{R(T_1^{\bullet})^2}{\Delta\tilde{H}_{\mathrm{f}}^{\bullet}}\ln x_1 \tag{10.97}$$

where $\Delta T = (T_1^{\bullet} - T)$ and T is set approximately to $T_1^{\bullet}$. This equation, of the same form as Equation (10.92), is the basic equation for the elementary formula for the freezing point lowering in terms of the mole fraction or molality of the solute, but again we emphasize that *it is an approximation, valid only in the limit.* Equation (10.96) should be used for an ideal solution, with $\Delta\mu_1^{\mathrm{E}}$ set equal to zero, to calculate the freezing point as a function of the mole fraction for values of x_1 not close to unity.

When we consider the solubility of a solid component in a solvent, the emphasis is placed on obtaining the mole fraction or other composition variable as a function of the temperature. Thus, Equation (10.96) gives the solubility as a function of the temperature in this interpretation. The solubility in an ideal solution is given by

$$\ln x_1 = -\frac{1}{R}\int_{T}^{T_1^{\bullet}} \frac{\tilde{H}_1^{\bullet}(\ell) - \tilde{H}_1^{\bullet}(\mathrm{s})}{T^2}\,\mathrm{d}T \tag{10.98}$$

10.14 Liquid–solid solution equilibria at constant pressure

When a liquid solution and a solid solution are in equilibrium, both components exist in both phases. We again develop the pertinent equations for only one component, because the equations for the second component are identical except for the change of subscripts. The condition of equilibrium is the equality of the chemical potential of the component in the two phases, so

$$\mu_1[T, P, x] = \mu_1[T, P, z] \tag{10.99}$$

where z represents the mole fraction in the solid phase. With the use of Equation (10.30) to express the chemical potentials in terms of the mole fraction, Equation (10.99) becomes

$$RT \ln x_1 + \Delta\mu_1^E[T, P, x] + \mu_1^{\bullet}[T, P, x_1 = 1]$$
$$= RT \ln z_1 + \Delta\mu_1^E[T, P, z] + \mu_1^{\bullet}[T, P, z = 1] \quad (10.100)$$

where we have chosen the pure liquid and the pure solid to be the standard states. This choice is made because we can determine the difference between the chemical potentials of the component in the two states by knowing the melting point of the pure component at the experimental pressure according to Equation (10.95). We then have

$$\Delta\mu_1^E[T, P, x] - \Delta\mu_1^E[T, P, z] = RT \ln \frac{z_1}{x_1} - \int_T^{T_1^{\bullet}} \frac{\tilde{H}_1^{\bullet}(\ell) - \tilde{H}_1^{\bullet}(s)}{T^2}\, dT \quad (10.101)$$

It is apparent from this equation that we can determine only the difference between the two excess chemical potentials at the experimental conditions. If one is known from other studies, then the other can be determined. It must be pointed out that the difference is not isothermal and, if isothermal quantities are desired, the same corrections as discussed in Section 10.12 must be made by the use of Equation (10.87).

The equilibrium temperature may be intermediate between the freezing points of the two components, or may be above or below both freezing points. Two special cases are worthy of additional comment. The mole fractions are equal at the composition of a maximum or minimum in the temperature–composition curve. Then, at such an extremum,

$$\Delta\mu_1^E[T, P, x] - \Delta\mu_1^E[T, P, z] = -T \int_T^{T_1^{\bullet}} \frac{\tilde{H}_1^{\bullet}(\ell) - \tilde{H}_1^{\bullet}(g)}{T^2}\, dT \quad (10.102)$$

For a maximum, $T > T_1^{\bullet}$, the right-hand side of Equation (10.102) is positive, and therefore $(\Delta\mu_1^E[T, P, x] - \Delta\mu_1^E[T, P, z])$ must be positive. For a minimum, $(\Delta\mu_1^E[T, P, x] - \Delta\mu_1^E[T, P, z])$ must be negative. The same conditions must apply to the excess chemical potential for the second component.

10.15 Liquid–liquid equilibria at constant pressure

When a system exhibits partial immiscibility, we encounter an equilibrium between two phases in the same state of aggregation. In this case the same standard state of a component can be used for both phases. With the use of the condition of equilibrium in terms of the chemical potentials and Equation (10.30), we have

$$RT \ln x_1' + \Delta\mu_1^E[T, P, x_1'] = RT \ln x_1'' + \Delta\mu_1^E[T, P, x''] \quad (10.103)$$

where the primes and double primes are used to distinguish the two phases. The measurement of the composition of the two saturated phases at a given temperature permits the calculation of only the difference in the excess chemical potentials of each component in the saturated phases. It is not possible to obtain directly the values of the excess chemical potentials in either phase as a function of the composition. Approximate values can be obtained with certain assumptions. Because the two phases are in the same state of aggregation, we may assume that the isothermal molar excess Gibbs energy of both homogenous phases is given by the same function of the composition. When the form of this function is known from experience or on a theoretical basis, the excess chemical potentials of both components can be obtained. Then, the two equations, Equation (10.103) and the same equation for the second component, give two simultaneous, independent equations which may be used to evaluate two coefficients in the equation for $\Delta\tilde{G}^{\mathrm{E}}$. The calculations would be repeated at each experimental temperature. As an example, for nonelectrolytic solutions, $\Delta\tilde{G}^{\mathrm{E}}$ might be expressed by

$$\Delta\tilde{G}^{\mathrm{E}}[T, P, x] = x_1 x_2(a + bx_1) \tag{10.104}$$

from which we obtain

$$\Delta\mu_1^{\mathrm{E}} = x_2^2(a + 2x_1 b) \tag{10.105}$$

and

$$\Delta\mu_2^{\mathrm{E}} = x_1^2[a + (x_1 - x_2)b] \tag{10.106}$$

When Equation (10.105) is substituted in Equation (10.103) and Equation (10.106) in the similar equation for the second component, with appropriate values of the mole fraction, the two equations can be solved to give the values of a and b. We must emphasize that the values so obtained are valid only for the saturated phases, and the use of such values to obtain values of the excess chemical potentials in the homogenous regions of composition depend upon the validity of Equation (10.104).

10.16 Osmotic pressure

The measurement of osmotic pressure and the determination of the excess chemical potential of a component by means of such measurements is representative of a system in which certain restrictions are applied. In this case the system is separated into two parts by means of a diathermic, rigid membrane that is permeable to only one of the components. For the purpose of discussion we consider the case in which the pure solvent is one phase and a binary solution is the other phase. The membrane is permeable only to the solvent. When a solute is added to a solvent at constant temperature and pressure, the chemical potential of the solvent is decreased. The pure solvent would then diffuse into such a solution when the two phases are separated by the semipermeable membrane but are at the same temperature and pressure. The chemical potential of the solvent in the solution can be increased by an increase of pressure on the solution until equilibrium is achieved. *The excess pressure is the osmotic pressure.*

A binary system consisting of two parts separated by a diathermic, rigid membrane has three degrees of freedom. The particular system under discussion can be made univariant by fixing the temperature and pressure of the pure solvent; the equilibrium pressure on the solution is then a function of the composition of the solution. The condition of equilibrium is

$$\mu_1[T, P', x] = \mu_1^{\bullet}[T, P, x_1 = 1] \tag{10.107}$$

and when Equation (10.30) is used, the equation becomes

$$RT \ln x_1 + \Delta\mu_1^{\mathrm{E}}[T, P', x] + \mu_1^{\bullet}[T, P', x_1 = 1] = \mu_1^{\bullet}[T, P, x_1 = 1] \tag{10.108}$$

The difference between the chemical potential for the two standard states is

$$\mu_1^{\bullet}[T, P, x_1 = 1] - \mu_1^{\bullet}[T, P', x_1 = 1] = \int_{P'}^{P} \tilde{V}_1^{\bullet}\, dP \tag{10.109}$$

so

$$\Delta\mu_1^{\mathrm{E}}[T, P', x] = -RT \ln x_1 + \int_{P'}^{P} \tilde{V}_1^{\bullet}\, dP \tag{10.110}$$

A measurement of P' for a given mole fraction, x_1, permits the calculation of the value of $\Delta\mu_1^{\mathrm{E}}$ at P' and x_1. Such values are not at a constant pressure. When values of $\Delta\mu_1^{\mathrm{E}}$ are desired at the pressure of the solvent, P, the correction

$$\Delta\mu_1^{\mathrm{E}}[T, P, x] - \Delta\mu_1^{\mathrm{E}}[T, P', x] = \int_{P'}^{P} (\bar{V}_1 - \tilde{V}_1^{\bullet})\, dP \tag{10.111}$$

must be made. With this correction we have

$$\Delta\mu_1^{\mathrm{E}}[T, P, x] = -RT \ln x_1 + \int_{P'}^{P} \bar{V}_1\, dP \tag{10.112}$$

For an ideal solution, Equation (10.112) becomes

$$RT \ln x_1 = \int_{P'}^{P} \tilde{V}_1^{\bullet}\, dP \tag{10.113}$$

where the partial molar volume is now equal to the molar volume of the pure component. If $\tilde{V}_1^{\bullet}$ is assumed to be constant over the range of pressure, we can write Equation (10.113) as

$$RT \ln x_1 = -\pi \tilde{V}_1^{\bullet} \tag{10.114}$$

where $\pi = P' - P$ and is the *osmotic pressure*. This equation is the basic equation for the more-elementary expressions for the dependence of the osmotic pressure on the mole fraction or molality of the solute. It is also valid for solutions that are not ideal in the limit as x_1 approaches unity because, in these cases, $\Delta\mu_1^E$ approaches zero and $\bar{V}_1$ approaches $\tilde{V}_1^{\bullet}$.

10.17 Other concentration variables

We have used the mole fraction as the composition variable throughout the previous discussion. Where appropriate, we could have used the molality or molarity instead of the mole fraction. For the molality, Equation (8.92) would be used for the solute and either Equation (8.102) or Equation (8.119) for the solvent. For the molarity, Equation (8.104) would be used for the solute and Equation (8.118) or Equation (8.120) for the solvent. The methods used to develop the corresponding equations are identical to those that have been used for the mole fractions.

10.18 Solutions of strong electrolytes

The thermodynamic treatment of systems in which at least one component is an electrolyte needs special comment. Such systems present the first case where we must choose between treating the system in terms of components or in terms of species. No decision can be based on thermodynamics alone. If we choose to work in terms of components, any effect of the presence of new species that are different from the components, would appear in the excess chemical potentials. No error would be involved, and the thermodynamic properties of the system expressed in terms of the excess chemical potentials and based on the components would be valid. It is only when we wish to explain the observed behavior of a system, to treat the system on the basis of some theoretical concept or, possibly, to obtain additional information concerning the molecular properties of the system, that we turn to the concept of species. For example, we can study the equilibrium between a dilute aqueous solution of sodium chloride and ice in terms of the components: water and sodium chloride. However, we know that the observed effect of the lowering of the freezing point of water is approximately twice that expected for a nondissociable solute. This effect is explained in terms of the ionization. In any given case the choice of the species is dictated largely by our knowledge of the system obtained outside of the field of thermodynamics and, indeed, may be quite arbitrary.

In the discussion presented here, we consider the electrolytic component to be the solute dissolved in some nonelectrolytic liquid solvent. We express the chemical formula of the solute as $M_{\nu_+}A_{\nu_-}$ where ν_+ and ν_- represent the number of positive ions and negative ions, respectively, present in one molecule of the component. We choose the species in solution to be only the positive and negative ions represented as M_+ and A_- and solvent. We introduce by this means the concept of strong electrolytes; the thermodynamic treatment concerning solutions of weak electrolytes is discussed in Chapter 11. The reference state of the ions is chosen as the infinitely dilute solution of both ions in the solvent; the choice is the usual one dictated by the limited concentration range of most electrolytic solutions. The chemical potential of the solute is

$$\mu_2[T, P, m] = \nu_+\mu_+[T, P, m_+] + \nu_-\mu_-[T, P, m_-] \tag{10.115}$$

according to Equation (8.178). With the use of Equations (8.179) and (8.180), we have Equation (8.181)

$$\mu_2[T, P, m] = RT \ln m_+^{\nu_+} m_-^{\nu_-} + RT \ln \gamma_+^{\nu_+}\gamma_-^{\nu_-} + \nu_+\mu_+^{\ominus}[T, P] + \nu_-\mu_-^{\ominus}[T, P] \tag{10.116}$$

If m represents the molality of the solute, $m_+ = \nu_+ m$ and $m_- = \nu_- m$. With the introduction of the mean activity coefficient, Equation (10.116) becomes

$$\mu_2[T, P, m] = \nu RT \ln m + \nu RT \ln \gamma_\pm + RT \ln \nu_+^{\nu_+}\nu_-^{\nu_-} + \nu_+\mu_+^{\ominus}[T, P] + \nu_-\mu_-^{\ominus}[T, P] \tag{10.117}$$

where $\nu = \nu_+ + \nu_-$. This equation is the basic equation for the chemical potential of a strong electrolyte considered as a solute. The factor ν present in the first two terms on the right-hand side of the equation is to be noted; it is this factor that distinguishes the behavior of solutions of strong electrolytes from that of nonelectrolytic solutions. The third term is a new term, and is independent of the molality and pressure.

When the solute is present in both phases, the condition of equilibrium is again the equality of the chemical potential of the component in the two phases. The same methods used in the previous sections of this part of the chapter are used to develop the equations for the specific type of equilibrium. As an example, when the solute is volatile, the conditions of equilibrium lead to

$$\nu RT \ln m + \nu RT \ln \gamma_\pm + RT \ln \nu_+^{\nu_+}\nu_-^{\nu_-} + \nu_+\mu_+^{\ominus}[T, P] + \nu_-\mu_-^{\ominus}[T, P] = RT \ln Py_2 + f(\beta)P + \mu_2^{\ominus}[T, 1 \text{ bar, ideal}, y_2 = 1] \tag{10.118}$$

where $f(\beta)$ is used to express the deviations of the vapor phase from ideality in a general form. With due consideration of the pressure as discussed in Section 10.8, Henry's law constant may be defined as

$$\mu_2^{\ominus}[T, 1 \text{ bar, ideal}, y_2 = 1] - (\nu_+\mu_+^{\ominus}[T, P] + \nu_-\mu_-^{\ominus}[T, P]) = RT \ln k \tag{10.119}$$

In many cases the solute will not be present in the second phase. Then values of the excess chemical potential, the activity coefficient of the solvent, or the osmotic coefficient will be obtained from the study of the phase equilibria. The values of the mean activity or mean activity coefficient of the solute would then be obtained by use of the Gibbs–Duhem equation. The experimental studies must be made over a range of compositions starting at $m=0$, when the infinitely dilute solution is used as the reference state of the ions. The appropriate expression for the chemical potential of the solvent must be used. The correct expressions based on Equation (8.102) or (8.120) are

$$\mu_1[T, P, m] = -\frac{RTM_1}{1000} \nu m + \Delta\mu_{1m}^{\mathrm{E}} + \mu_1^{\ominus}[T, P] \tag{10.120}$$

or

$$\mu_1[T, P, m] = -\frac{RTM_1\phi}{1000} \nu m + \mu_1^{\ominus}[T, P] \tag{10.121}$$

10.19 Solutions of fused salts

The same choice of using component or species is also met in the study of binary systems of molten salts. We could choose to deal solely with components, but because of the concept of the ionic nature of the melts we choose to express the chemical potential of the component in terms of the chemical potential of the species. Equation (8.202) gives the relation for the chemical potential of the general salt $M_{\nu_+}A_{\nu_-}$ in terms of the chemical potentials of the ions. The discussion given in Section 8.18 results in the more convenient equation for the chemical potential of the salt (Eq. (8.208))

$$\mu_1[T, P, x] = RT \ln x_+^{\nu_+} x_-^{\nu_-} + \Delta\mu_1^{\mathrm{E}}[T, P, x] + \mu_1^{\ominus}[T, P, x_1 = 1] \tag{10.122}$$

where the subscript 1 is used to indicate the component without distinction between a solvent and solute. However, at times it may be advantageous to define an average of the excess chemical potentials of the ions based on Equations (8.207) and (8.208) such that

$$\Delta\mu_1^{\mathrm{E}} = \nu\overline{\Delta\mu_1^{\mathrm{E}}} = \nu_+[\Delta\mu_+^{\mathrm{E}} - (\Delta\mu_+^{\mathrm{E}})^{\infty}] + \nu_-[\Delta\mu_-^{\mathrm{E}} - (\Delta\mu_-^{\mathrm{E}})^{\infty}] \tag{10.123}$$

Equation (10.122) is the basic equation that would be used in obtaining expressions for the excess chemical potential by means of the condition for phase equilibrium.

10.20 Applications of the Gibbs–Duhem equations

We discuss in the next few sections the applications of the Gibbs–Duhem equations to various phase equilibria. In so doing we obtain expressions for the derivatives of one intensive variable with respect to another under the condition that the system is univariant. In this section we discuss the general problem, and then we consider a few cases of special interest in the following sections.

Consider a binary, two-phase system without specifying the specific phases. The two Gibbs–Duhem equations on a molar basis are

$$\tilde{S}'\,\mathrm{d}T - \tilde{V}'\,\mathrm{d}P + x_1'\,\mathrm{d}\mu_1 + x_2'\,\mathrm{d}\mu_2 = 0 \tag{10.124}$$

and

$$\tilde{S}''\,\mathrm{d}T - \tilde{V}''\,\mathrm{d}P + x_1''\,\mathrm{d}\mu_1 + x_2''\,\mathrm{d}\mu_2 = 0 \tag{10.125}$$

where the single and double primes are used to distinguish the two phases. The problem is to solve these two equations under the particular conditions of a given system, with the provision that one of the variables must be kept constant in order to make the system univariant. There are 24 such derivatives when the chemical potentials are used as variables. The number is reduced to seven if we consider a derivative and its reciprocal to be the same, and if we exclude the interchange of μ_1 and μ_2 as giving a new derivative. Although some of the derivatives may be useful in special cases, such derivatives in general are not convenient, because of the use of the chemical potentials as variables and because, in some of the derivatives, values of the absolute entropies are required.

More-useful derivatives are obtained when composition variables are used, rather than the chemical potentials. We use the mole fraction throughout, although the molality or the molarities could be used. The four intensive variables that are used are T, P, x_1', and x_1''. There are again seven derivatives between the four variables under the same conditions stated in the previous paragraph except that, rather than excluding the interchange of μ_1 and μ_2 as giving a new derivative, we exclude the interchange of x_1' and x_1''; the seven derivatives are $(\partial T/\partial P)_{x_1'}$, $(\partial T/\partial x_1')_P$, $(\partial T/\partial x_1')_{x_1''}$, $(\partial P/\partial x_1')_T$, $(\partial P/\partial x_1')_{x_1''}$, $(\partial x_1'/\partial x_1'')_T$, and $(\partial x_1'/\partial x_1'')_P$.

Four examples are used to illustrate the methods that can be used. First we consider the derivative $(\partial T/\partial x_1')_P$. At constant pressure, Equations (10.124) and (10.125) become

$$\tilde{S}'\,\mathrm{d}T + x_1'\,\mathrm{d}\mu_1 + x_2'\,\mathrm{d}\mu_2 = 0 \tag{10.126}$$

$$\tilde{S}''\,\mathrm{d}T + x_1''\,\mathrm{d}\mu_1 + x_2''\,\mathrm{d}\mu_2 = 0 \tag{10.127}$$

We eliminate $\mathrm{d}\mu_2$ to obtain

$$\left(\tilde{S}'' - \frac{x_2''\tilde{S}'}{x_2'}\right)\mathrm{d}T + \left(x_1'' - \frac{x_2''x_1'}{x_2'}\right)\mathrm{d}\mu_1 = 0 \tag{10.128}$$

Then $\mathrm{d}\mu_1$ can be written as

$$\mathrm{d}\mu_1' = -\bar{S}_1'\,\mathrm{d}T + \left(\frac{\partial \mu_1'}{\partial x_1'}\right)_{T,P}\mathrm{d}x_1' \tag{10.129}$$

where μ_1' is used to introduce x_1' as the independent variable. After substitution and simplification, we have

$$[x_1''(\bar{S}_1''-\bar{S}_1')+x_2''(\bar{S}_2''-\bar{S}_2')]\,\mathrm{d}T+\left(\frac{x_1''-x_1'}{x_2'}\right)\left(\frac{\partial\mu_1'}{\partial x_1'}\right)_{T,P}\mathrm{d}x_1'=0 \tag{10.130}$$

or

$$\left(\frac{\mathrm{d}T}{\mathrm{d}x_1'}\right)_P=-\frac{(x_1''-x_1')(\partial\mu_1'/\partial x_1')_{T,P}}{x_2'[x_1''(\bar{S}_1''-\bar{S}_1')+x_2''(\bar{S}_2''-\bar{S}_2')]} \tag{10.131}$$

We note here that the value of this derivative is zero when the compositions of the two phases are equal in accordance with one of the Gibbs–Konovalov theorems. The second derivative which we choose to evaluate is $(\partial T/\partial P)_{x_1'}$. We choose to work again with μ_1' in order to introduce x_1' as a variable. Therefore, we first eliminate $\mathrm{d}\mu_2$ from Equation (10.124) and (10.125) to obtain

$$\left(\tilde{S}''-\frac{x_2''}{x_2'}\tilde{S}'\right)\mathrm{d}T-\left(\tilde{V}''-\frac{x_2''\tilde{V}'}{x_2'}\right)\mathrm{d}P+\left(x_1''-\frac{x_2''x_1'}{x_2'}\right)\mathrm{d}\mu_1=0 \tag{10.132}$$

Then μ_1 is a function of the temperature and pressure at constant x_1', so

$$\mathrm{d}\mu_1=-\bar{S}_1'\,\mathrm{d}T+\bar{V}_1'\,\mathrm{d}P \tag{10.133}$$

After substituion and simplification, we obtain

$$\left(\frac{\mathrm{d}T}{\mathrm{d}P}\right)_{x_1'}=\frac{x_1''(\bar{V}_1''-\bar{V}_1')+x_2''(\bar{V}_2''-\bar{V}_2')}{x_1''(\bar{S}_1''-\bar{S}_1')+x_2''(\bar{S}_2''-\bar{S}_2')} \tag{10.134}$$

Some care must be used in evaluating the two derivatives $(\partial T/\partial x_1')_{x_1''}$ and $(\partial x_1'/\partial x_1'')_P$. In these derivatives both μ_1 and μ_2 must be retained and not eliminated as such, in order to introduce both mole fractions. If we choose to introduce x_1' by means of μ_1', then we must use μ_2'' to introduce x_1''. At constant x_1'', μ_2'' is a function only of the temperature and pressure, so

$$\mathrm{d}\mu_2=\mathrm{d}\mu_2''=-\bar{S}_2''\,\mathrm{d}T+\bar{V}_2''\,\mathrm{d}P \tag{10.135}$$

Also, $\mathrm{d}\mu_1$ may be written as

$$\mathrm{d}\mu_1=\mathrm{d}\mu_1'=-\bar{S}_1'\,\mathrm{d}T+\bar{V}_1'\,\mathrm{d}P+\left(\frac{\partial\mu_1'}{\partial x_1'}\right)_{T,P}\mathrm{d}x_1' \tag{10.136}$$

Substitution of these two equations into Equations (10.127) and (10.128) and elimination of $\mathrm{d}P$ yields

$$\left(\frac{\partial T}{\partial x_1'}\right)_{x_1''}=\frac{[x_1'(\bar{V}_1''-\bar{V}_1')-x_2'(\bar{V}_2'-\bar{V}_2'')](\partial\mu_1'/\partial x_1')_{T,P}}{x_2'[(\bar{S}_2'-\bar{S}_2'')(\bar{V}_1''-\bar{V}_1')-(\bar{S}_1''-\bar{S}_1')(\bar{V}_2'-\bar{V}_2'']} \tag{10.137}$$

Similar methods are used for the last derivative, $(\partial x'_1/\partial x''_1)_P$. We introduce x'_1 by means of μ'_1 and x''_1 by means of μ''_2. Then, we have

$$d\mu_1 = d\mu'_1 = -\bar{S}'_1\, dT + \left(\frac{\partial \mu'_1}{\partial x'_1}\right)_{T,P} dx'_1 \tag{10.138}$$

and

$$d\mu_2 = d\mu''_2 = -\bar{S}''_2\, dT + \left(\frac{\partial \mu''_2}{\partial x''_1}\right)_{T,P} dx''_1 \tag{10.139}$$

Substitution of these equations into Equations (10.124) and (10.125) and elimination of dT gives

$$\left(\frac{\partial x'_1}{\partial x''_1}\right)_P = -\frac{x'_2[x''_2(\bar{S}''_2 - \bar{S}'_2) + x''_1(\bar{S}''_1 - \bar{S}'_1)](\partial \mu''_2/\partial x''_1)_{T,P}}{x''_1[x'_2(\bar{S}''_2 - \bar{S}'_2) + x'_1(\bar{S}''_1 - \bar{S}'_1)](\partial \mu'_1/\partial x'_1)_{T,P}} \tag{10.140}$$

or

$$\left(\frac{\partial x'_1}{\partial x''_1}\right)_P = \frac{x'_2[x''_2(\bar{S}''_2 - \bar{S}'_2) + x''_1(\bar{S}''_1 - \bar{S}'_1)](\partial \mu''_1/\partial x''_1)_{T,P}}{x''_2[x'_2(\bar{S}''_2 - \bar{S}'_2) + x'_1(\bar{S}''_1 - \bar{S}'_1)](\partial \mu'_1/\partial x'_1)_{T,P}} \tag{10.141}$$

In order to evaluate each of the derivatives, such quantities as $(\bar{V}''_i - \bar{V}'_i)$, $(\bar{S}''_i - \bar{S}'_i)$, and $(\partial \mu'_i/x'_i)_{T,P}$ need to be evaluated. The difference in the partial molar volumes of a component between the two phases presents no problem; the dependence of the molar volume of a phase on the mole fraction must be known from experiment or from an equation of state for a gas phase. In order to determine the difference in the partial molar entropies, not only must the dependence of the molar entropy of a phase on the mole fraction be known, but also the difference in the molar entropy of the component in the two standard states must be known or calculable. If the two standard states are the same, there is no problem. If the two standard states are the pure component in the two phases at the temperature and pressure at which the derivative is to be evaluated, the difference can be calculated by methods similar to that discussed in Sections 10.10 and 10.12. In the case of vapor–liquid equilibria in which the reference state of a solute is taken as the infinitely dilute solution, the difference between the molar entropy of the solute in its two standard states may be determined from the temperature dependence of the Henry's law constant. Finally, the expression used for μ_i in evaluating $(\partial x'_i/\partial x''_i)_{T,P}$ must be appropriate for the particular phase of interest. This phase is dictated by the particular choice of the mole fraction variables.

Alternate forms of the equations for the various derivatives which contain the partial molar entropies are obtained by substituting $\bar{H}_i/T$ for each partial molar entropy. The substitution is possible because the chemical potential of a component is

$$\mu_i = \bar{H}_i - T\bar{S}_i \tag{10.142}$$

and under the conditions of equilibrium $\mu_i' = \mu_i''$.

The expressions for the derivatives become somewhat simpler when one of the components is not present in one of the phases. For the purposes of discussion we assume that the second component is not present in the double-primed phase. Then the two Gibbs–Duhem equations become

$$\tilde{S}'\,\mathrm{d}T - \tilde{V}'\,\mathrm{d}P + x_1'\,\mathrm{d}\mu_1 + x_2'\,\mathrm{d}\mu_2 = 0 \tag{10.143}$$

and

$$(\tilde{S}_1^{\bullet})''\,\mathrm{d}T - (\tilde{V}_1^{\bullet})''\,\mathrm{d}P + \mathrm{d}\mu_1 = 0 \tag{10.144}$$

We can immediately eliminate μ_1 from the two equations, to obtain

$$[\tilde{S}' - x_1'(\tilde{S}_1^{\bullet})'']\,\mathrm{d}T - [\tilde{V}' - x_1'(\tilde{V}_1^{\bullet})'']\,\mathrm{d}P + x_2'\,\mathrm{d}\mu_2 = 0 \tag{10.145}$$

where $(\tilde{S}_1^{\bullet})''$ and $(\tilde{V}_1^{\bullet})''$ are the molar entropy and volume of the pure component in the double-primed phase at the temperature and pressure at which the derivative is to be evaluated. There are three derivatives involving the three variables T, P, and either x_1' or x_2' when a derivative and its reciprocal are considered to be equivalent. The three derivatives are $(\partial T/\partial P)_{x_2'}$, $(\partial T/\partial x_2')_P$, and $(\partial P/\partial x_2')_T$. For the first of these derivatives μ_1' is a function of the temperature and pressure, so

$$\mathrm{d}\mu_2' = -\bar{S}_2'\,\mathrm{d}T + \bar{V}_2'\,\mathrm{d}P \tag{10.146}$$

Then, on substitution and simplification, we obtain

$$\left(\frac{\partial T}{\partial P}\right)_{x_2'} = \frac{(\bar{V}_1)' - (\tilde{V}_1^{\bullet})''}{(\bar{S}_1)' - (\tilde{S}_1^{\bullet})''} \tag{10.147}$$

For the second and third derivatives we write $\mathrm{d}\mu_2$ as

$$\mathrm{d}\mu_2' = -\bar{S}_2\,\mathrm{d}T + \bar{V}_2'\,\mathrm{d}P + (\partial\mu_2'/\partial x_2')_{T,P}\,\mathrm{d}x_2' \tag{10.148}$$

Then, on substitution and simplification under the appropriate conditions, we obtain

$$\left(\frac{\partial T}{\partial x_2'}\right)_P = -\frac{x_2'(\partial\mu_2'/\partial x_2')_{T,P}}{x_1'[\bar{S}_1' - (\tilde{S}_1^{\bullet})'']} \tag{10.149}$$

and

$$\left(\frac{\partial P}{\partial x_2'}\right)_T = \frac{x_2'(\partial\mu_2'/\partial x_2')_{T,P}}{x_1'[\bar{V}_1' - (\tilde{V}_1^{\bullet})'']} \tag{10.150}$$

Here, as in the other derivatives, each entropy quantity can be substituted by the corresponding enthalpy term divided by the temperature. The evaluation of the differences between the entropy quantities or the enthalpy quantities requires, as before, the knowledge of or the ability to calculate the difference of the molar entropy or enthalpy in the two standard states.

10.21 Raoult's law

In order to illustrate Raoult's law most simply, we choose a system in which the solvent, indicated by the subscript 1, is the only volatile component, and make use of Equation (10.150). We change the independent variable from x_2 to x_1, so that the equation is written as

$$\left(\frac{\partial P}{\partial x_1}\right)_T = \frac{x_2(\partial \mu_2/\partial x_2)_{T,P}}{x_1 \tilde{V}_1^{\bullet}(\mathrm{g})\left[1 - \dfrac{\tilde{V}_1(\ell)}{\tilde{V}_1^{\bullet}(\mathrm{g})}\right]} \tag{10.151}$$

where we now indicate specific phases. The derivative $(\partial \mu_2/\partial x_2)_{T,P}$ is

$$\left(\frac{\partial \mu_2}{\partial x_2}\right)_{T,P} = \frac{RT}{x_2} + \left(\frac{\partial\, \Delta\mu_2^{\mathrm{E}}}{\partial x_2}\right)_{T,P} \tag{10.152}$$

so

$$\left(\frac{\partial P}{\partial x_1}\right)_T = \frac{RT[1 + x_2(\partial\, \Delta\mu_2^{\mathrm{E}}/\partial x_2)_{T,P}/RT]}{x_1 \tilde{V}_1^{\bullet}(\mathrm{g})\{1 - [\tilde{V}_1(\ell)/\tilde{V}_1^{\bullet}(\mathrm{g})]\}} \tag{10.153}$$

In the limit of x_1 going to unity and with the use of the ideal gas equation of state, the equation becomes

$$\lim_{x_1 \to 1}\left(\frac{\partial P}{\partial x_1}\right)_T = \frac{P_1^{\bullet}}{\{1 - [\tilde{V}_1(\ell)/\tilde{V}_1^{\bullet}(\mathrm{g})]\}} \tag{10.154}$$

This equation approximates Raoult's law and becomes equal to it when the molar volume of the pure liquid solvent is negligibly small with respect to the molar volume of the gas. We see, once again, that Raoult's law is a limiting law, strictly valid only when the equation of state for the ideal gas is applicable to the gas phase, and that the molar volume of the pure liquid is negligibly small with respect to that of the gas phase.

10.22 Henry's law

In order to discuss Henry's law in the simplest manner, we assume that the solute indicated by the subscript 2 is the only volatile component. Equation (10.150) may be written as

$$\left(\frac{\partial P}{\partial x_1}\right)_T = \frac{x_1(\partial \mu_1/\partial x_1)_{T,P}}{x_2[\tilde{V}_2(\ell) - \tilde{V}_2^{\bullet}(\mathrm{g})]} \tag{10.155}$$

Again, we change the independent variable to x_2, so

$$\left(\frac{\partial P}{\partial x_2}\right)_T = \frac{x_1(\partial \mu_1/\partial x_1)_{T,P}}{x_2\tilde{V}_2^\circ(\text{g})\{1 - [\bar{V}_2(\ell)/\tilde{V}_2^\circ(\text{g})]\}} \tag{10.156}$$

With the relations that

$$\frac{x_1}{x_2}\left(\frac{\partial \mu_1}{\partial x_1}\right)_{T,P} = -\left(\frac{\partial \mu_2}{\partial x_1}\right)_{T,P} \tag{10.157}$$

and

$$\left(\frac{\partial \mu_2}{\partial x_1}\right)_{T,P} = -\frac{RT}{x_2} + \left(\frac{\partial\, \Delta\mu_2^{\text{E}}}{\partial x_1}\right)_{T,P} \tag{10.158}$$

we have

$$\left(\frac{\partial P}{\partial x_2}\right)_T = \frac{RT\left[1 - \dfrac{x_2}{RT}\left(\dfrac{\partial\, \Delta\mu_2^{\text{E}}}{\partial x_1}\right)_{T,P}\right]}{x_2\tilde{V}_2^\circ(\text{g})\left[1 - \dfrac{\bar{V}_2(\ell)}{\tilde{V}_2^\circ(\text{g})}\right]} \tag{10.159}$$

Then, in the limit as x_2 goes to zero, with the use of the equation of state for an ideal gas and with the assumption that the partial molar volume of the solute in solution is negligibly small with respect to the molar volume of the gas, Equation (10.159) becomes

$$\lim_{x_2\to 0}\left(\frac{\partial P}{\partial x_2}\right)_T = \frac{P}{x_2} \tag{10.160}$$

Such a relation requires P/x_2 to be a constant, and consequently

$$\lim_{x_2\to 0}\left(\frac{\partial P}{\partial x_2}\right)_T = k \tag{10.161}$$

We see again that *Henry's law is a limiting law* valid when the molar volume of the gas is large with respect to the partial molar volume of the solute. When the pressure is used as the dependent variable, the applicability of the ideal gas equation of state is also required.

10.23 Boiling point elevation

For the boiling point elevation, we require the solvent, indicated by the subscript 1, to be the only volatile component. The applicable equation is

$$\left(\frac{\partial T}{\partial x_2}\right)_P = -\frac{RT + x_2(\partial\, \Delta\mu_2^{\text{E}}/\partial x_2)_{T,P}}{x_1[\bar{S}_1(\ell) - \tilde{S}_1^\circ(\text{g})]} \tag{10.162}$$

from Equation (10.149), where we have used the relation given by Equation (10.152). In the limit as x_2 approaches zero, $x_2(\partial\, \Delta\mu_2^{\mathrm{E}}/\partial x_2)_{T,P}$ approaches zero, T approaches $T^{\bullet}$, the boiling point of the pure solvent at the pressure P, and $\bar{S}_1(\ell)$ approaches $\tilde{S}_1^{\bullet}(\ell)$. Then

$$\lim_{x_2 \to 0}\left(\frac{\partial T}{\partial x_2}\right)_P = \frac{RT_1^{\bullet}}{\tilde{S}_1^{\bullet}(\mathrm{g}) - \tilde{S}_1^{\bullet}(\ell)} \tag{10.163}$$

where the difference between the molar entropies are determined at the boiling point of the pure solvent. An alternate expression is

$$\lim_{x_2 \to 0}\left(\frac{\partial T}{\partial x_2}\right)_P = \frac{R(T_1^{\bullet})^2}{\tilde{H}_1^{\bullet}(\mathrm{g}) - \tilde{H}_1^{\bullet}(\ell)} \tag{10.164}$$

because $\tilde{H}_1^{\bullet}(\mathrm{g}) - \tilde{H}_1^{\bullet}(\ell) = T_1^{\bullet}[\tilde{S}_1^{\bullet}(\mathrm{g}) - \tilde{S}_1^{\bullet}(\ell)]$ at the boiling point of the pure solvent. The quantities given in Equations (10.162) and (10.164) are the usual constants for the boiling point elevation in terms of the mole fraction.

More information can be obtained for real systems by the use of Equation (10.162). This equation can be written, alternately, in two forms:

$$\left(\frac{\partial \ln x_1}{\partial \ln T}\right)_P = -\frac{\tilde{S}_1^{\bullet}(\mathrm{g}) - \bar{S}_1(\ell)}{R\left[1 + \dfrac{x_2}{RT}\left(\dfrac{\partial\, \Delta\mu_2^{\mathrm{E}}}{\partial x_2}\right)_{T,P}\right]} \tag{10.165}$$

and

$$\left(\frac{\partial \ln x_1}{\partial (1/T)}\right)_P = \frac{\tilde{H}_1^{\bullet}(\mathrm{g}) - \bar{H}_1(\ell)}{R\left[1 + \dfrac{x_2}{RT}\left(\dfrac{\partial\, \Delta\mu_2^{\mathrm{E}}}{\partial x_2}\right)_{T,P}\right]} \tag{10.166}$$

Thus, the value of $(\tilde{S}_1^{\bullet}(\mathrm{g}) - \bar{S}_1(\ell))$ at a given temperature may be determined from the slope of the curve of $\ln x_1$ plotted as a function of $\ln T$ at the given temperature, provided that $(\partial\, \Delta\mu_2^{\mathrm{E}}/\partial x_2)_{T,P}$ can be evaluated from experiment or theory. Similarly, $(\tilde{H}_1^{\bullet}(\mathrm{g}) - \bar{H}_1(\ell))$ can be calculated at a given temperature from the slope of the curve of $\ln x_1$ plotted as a function of $1/T$ at the given temperature with the same provision. The values so determined are not isothermal; when isothermal values are desired, then a knowledge of the partial molar heat capacity of the solvent in the liquid phase and the molar heat capacity of the component in the gas phase would be required.

10.24 Solubility and freezing point lowering

In the study of the solubility of a solid component in a solution or the lowering of the freezing point of a solvent by the presence of a solute, the solid phase must be considered as the pure component. We indicate this component, the one that is present in both phases, by the subscript 1. The applicable equations are identical to those developed in Section 10.23 when the molar entropy or enthalpy of the pure phase is changed from that of the gas to that of the solid. However, it is convenient to change the order of the entropies or enthalpies with an accompanying change of sign in order to make the differences between the entropies or enthalpies positive. The pertinent equations then are:

$$\lim_{x_2 \to 0} \left(\frac{\partial T}{\partial x_2} \right)_P = -\frac{RT_1^\bullet}{\tilde{S}_1^\bullet(\ell) - \tilde{S}_1^\bullet(\mathrm{s})} \tag{10.167}$$

$$\lim_{x_2 \to 0} \left(\frac{\partial T}{\partial x_2} \right)_P = -\frac{R(T_1^\bullet)^2}{\tilde{H}_1^\bullet(\ell) - \tilde{H}_1^\bullet(\mathrm{s})} \tag{10.168}$$

$$\left(\frac{\partial \ln x_1}{\partial \ln T} \right)_P = \frac{\bar{S}_1(\ell) - \tilde{S}_1^\bullet(\mathrm{s})}{R\left[1 + \frac{x_2}{RT}\left(\frac{\partial \Delta\mu_2^{\mathrm{E}}}{\partial x_2}\right)_{T,P}\right]} \tag{10.169}$$

and

$$\left(\frac{\partial \ln x_1}{\partial (1/T)} \right)_P = -\frac{\bar{H}_1(\ell) - \tilde{H}_1^\bullet(\mathrm{s})}{R\left[1 + \frac{x_2}{RT}\left(\frac{\partial \Delta\mu_2^{\mathrm{E}}}{\partial x_2}\right)_{T,P}\right]} \tag{10.170}$$

The first two of these equations give alternate forms of the constant for the freezing point lowering of the solvent on the addition of a solute, with the use of mole fractions as the composition variable. Equations (10.169) and (10.170) can be used to determine $(\bar{S}_1(\ell) - \tilde{S}_1^\bullet(\mathrm{s}))$ or $(\bar{H}(\ell) - \tilde{H}_1^\bullet(\mathrm{s}))$ as discussed in Section 10.23.

The various curves of $\ln x_1$ considered as a function of $1/T$ are illustrated in Figure 10.4. At the melting point, indicated by $T_1^\bullet$, the mole fraction must be equal to unity and $\ln x_1$ must be zero. All curves showing the solubility of a component in some liquid phase must terminate at this point. We assume here that the melting point of the pure component is not too far removed from the experimental temperatures of interest. The slope for an ideal solution is given by

$$\left(\frac{\partial \ln x_1}{\partial (1/T)} \right)_P = \frac{\bar{H}_1(\ell) - \tilde{H}_1^\bullet(\mathrm{s})}{R} \tag{10.171}$$

based on Equation (10.170). The solubility of a component in an ideal liquid solution is thus dependent only on the properties of that component, and not at all on the properties of the second component. The curve for ideal solubility is illustrated by curve A in the figure. When the deviations from ideal behavior are positive; that is, when the values of $\Delta\mu_1^{\mathrm{E}}$ are positive, the solubility of the component is less than ideal according to Equation (10.96) and the slope is given by Equation (10.170). Curve *B* illustrates the solubility for a system in which the deviations from ideal behavior are not too large. When the deviations are large and positive, the liquid phase exhibits partial immiscibility, as illustrated by curves *C* and *F*. Curve *C* illustrates the solid–liquid equilibria. The broken portion of the continuous curve represents the states of the system which are metastable or unstable. Curve *F* represents the liquid–liquid equilibria. The vertical broken line is the tie-line that connects the points of intersection of curve *F* with curve *C*. These points of intersection give the composition of the two liquid phases when they are in equilibrium with the solid phase. Curve *D* represents the solubility curve that passes through the upper critical solution temperature and critical solution composition of a given system. At the critical solution temperature, the curve has an infinite point of inflection. This behavior is consistent with Equation (5.165), because at the critical solution point the denominator must be zero based on the condition that $(\partial\mu_1/\partial x_1)_{T,P}$ must be zero at this point. The broken curve which is just tangent at the point of inflection represents the unstable liquid–liquid equilibria. When the deviations from ideality are negative, the solubility is greater than ideal according to Equation (10.96), whereas the slope of the curve is still given by Equation (10.193). Curve *E* illustrates the solubility when negative deviations occur. The limiting value of the slope of each curve as x_1 approaches unity is the slope of the ideal curve.

When the solute is a strong electrolyte, the expression for the dependence of the temperature on the molality is

$$\left(\frac{\partial T}{\partial m}\right)_P = -\frac{\nu RTM[1 + m(\partial \ln \gamma_\pm/\partial m)_{T,P}]}{1000[\bar{S}_1(\ell) - \bar{S}_1^\circ(\mathrm{s})]} \qquad (10.172)$$

Figure 10.4. Solubility relations as a function of temperature.

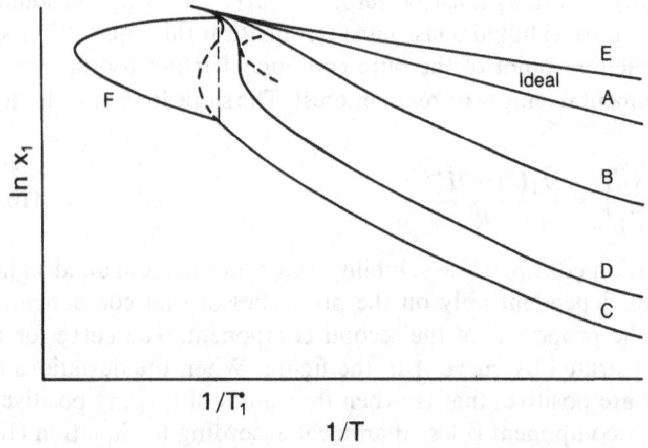

In the limit as m goes to zero, $m(\partial \ln \gamma_{\pm}/m)$ goes to zero and

$$\lim_{m\to 0}\left(\frac{\partial T}{\partial m}\right)_P = -\frac{\nu R T_1^{*} M}{1000[\tilde{S}_1^{*}(\ell) - \tilde{S}_1^{*}(\mathrm{s})]} \tag{10.173}$$

Thus, the limiting value of the slope of the curve of the temperature taken as a function of the molality is ν times the value for a nonionic solute.

When the solute is completely polymerized in solution, the change in the temperature with the concentration is less than that for an unpolymerized solute. As an example we assume that the solute is completely dimerized at all concentrations including the infinitely dilute solution. Then,

$$\mu_2 = \tfrac{1}{2}\mu_2' \tag{10.174}$$

and

$$\mu_2' yz = RT \ln x_2' + \Delta\mu_2^{\mathrm{E}'} + (\mu_2')^{\ominus} \tag{10.175}$$

where μ_2 is the chemical potential of the monomeric solute and μ_2' is that of the dimer and x_2' is the mole fraction of the dimer. For these two equations we find that

$$x_2\left(\frac{\partial \mu_2}{\partial x_2}\right)_{T,P} = \frac{RT}{2-x_2}\left[1 + \frac{x_2'}{RT}\left(\frac{\partial\, \Delta\mu_2^{\mathrm{E}'}}{\partial x_2'}\right)_{T,P}\right] \tag{10.176}$$

with the relation that $x_2 = 2x_2'(1 + x_2')$. The relation

$$\left(\frac{\partial \ln x_1}{\partial(1/T)}\right)_P = -\frac{(2-x_2)[\bar{H}_1(\ell) - \tilde{H}_1^{*}(\mathrm{s})}{R\left[1 + \dfrac{x_2}{RT}\left(\dfrac{\partial\, \Delta\mu_2^{\mathrm{E}'}}{\partial x_2'}\right)_{T,P}\right]} \tag{10.177}$$

is obtained by the same methods that have been used previously. We observe that in the limit at x_2 goes to zero, the value of $[\partial \ln x_1/\partial(1/T)]_P$ is twice that for a monomeric solute or, inversely, the dependence of the temperature on the mole fraction is half that for a monomeric solute. The problems related to incomplete ionization or polymerization are discussed in Chapter 11.

Some systems exhibit the phenomena known as *retrograde solubility*. In such cases the solubility decreases in the normal way as the temperature is decreased, passes through a minimum, and then increases as the temperature is continually decreased. The curve of $\ln x_1$ plotted as a function of $1/T$ then passes through a minimum and, at the minimum, $(\bar{H}_1(\ell) - \tilde{H}_1^{*}(\mathrm{s}))$ must be zero according to Equation (10.170). The quantity $(\bar{H}_1(\ell) - \tilde{H}_1^{*}(\mathrm{s}))$ may be written as

$$\begin{aligned}\bar{H}_1(\ell) - \tilde{H}_1^{*}(\mathrm{s}) &= [\bar{H}_1(\ell) - \tilde{H}_1^{*}(\ell)] + [\tilde{H}_1^{*}(\ell) - \tilde{H}_1^{*}(\mathrm{s})] \\ &= \Delta\bar{H}_1^{\mathrm{M}} + [\tilde{H}_1^{*}(\ell) - \tilde{H}_1^{*}(\mathrm{s})]\end{aligned} \tag{10.178}$$

Then, at the minimum solubility, the partial molar enthalpy of mixing of the component must be equal to the negative of the difference between the molar enthalpies of the pure component in the liquid and solid phases.

10.25 Liquid solution–solid solution equilibria

The equations for $(\partial T/\partial x_1)_P$ and $(\partial P/\partial x_1)_T$ for systems in which both the liquid and the solid phases are solutions are somewhat more complex than those discussed in Section 10.24, and little new information can be obtained. A comparison of the slopes of the curves of $\ln x$ considered as a function of $1/T$ when the solid phase is a solution and when it is pure, however, is of some interest. The equation

$$\left(\frac{\partial \ln x_1}{\partial(1/T)}\right)_P = -\frac{x_2\{z_1[\bar{H}_1(\ell) - \bar{H}_1(\mathrm{s})] + z_2[\bar{H}_2(\ell) - \bar{H}_2(\mathrm{s})]\}}{(z_1 - x_1)R\left[1 + \dfrac{x_1}{RT}\left(\dfrac{\partial\,\Delta\mu_1^{\mathrm{E}}}{\partial x_1}\right)_{T,P}\right]} \tag{10.179}$$

is based on Equation (10.131), where z has been used to represent the mole fraction in the solid phase, the primed phase has been chosen as the liquid phase, and the double-primed phase has been chosen as the solid phase. The entire term on the right-hand side of the equation, neglecting the negative sign, is generally positive except for the factor $(z_1 - x_1)$. The sign of $[\partial \ln x_1/\partial(1/T)]_P$ then must be consistent with the sign of $(z_1 - x_1)$ and can be positive or negative. For the purposes of discussion, which can only be qualitative, we consider the limiting value of $[\partial \ln x_1/\partial(1/T)]_P$ as x_1 approaches unity. In the limit, Equation (10.179) becomes

$$\lim_{x_1 \to 1}\left(\frac{\partial \ln x_1}{\partial(1/T)}\right)_P = -\frac{\tilde{H}_1^{\bullet}(\ell) - \tilde{H}_1^{\bullet}(\mathrm{s})}{R}\lim_{x_1 \to 1}\frac{1}{1 - (z_2/x_2)} \tag{10.180}$$

The limiting value of x_2/z_2 is determined by

$$RT\ln\frac{x_2}{z_2} = \Delta\mu_2^{\mathrm{E}}(\mathrm{s}) - \Delta\mu_2^{\mathrm{E}}(\ell) - T\int_T^{T_2^{\bullet}}\frac{\tilde{H}_2^{\bullet}(\ell) - \tilde{H}_2^{\bullet}(\mathrm{s})}{T^2}\,\mathrm{d}T \tag{10.181}$$

In the limit as x_1 approaches unity, T becomes $T_1^{\bullet}$ and both of the excess chemical potentials remain finite. Therefore, x_2/z_2 is finite and may have the entire range of positive values from approximately zero to extremely large values. Figures 10.5 and 10.6 illustrate this behavior. In Figure 10.5 the limiting slopes of curve of $\ln x_1$ taken as a function of $1/T$ are illustrated. The terminal point of all of the lines is at $T_1^{\bullet}$, the melting point of the pure first component, and $\ln x_1 = 0$. Curve A illustrates an extreme case in which z_2 is very large with respect to x_2 and the slope approaches zero. Curve B illustrates a more normal case in which $z_2 > x_2$. Curve C is a special case for which the ratio of z_2/x_2 is one in the limit. Curve D represents the case for which $z_2 < x_2$. The ideal curve represents the limit of this family of curves. A negative limiting slope more positive than the ideal would require the ratio of z_2/x_2 to be negative, which is impossible. Figure 10.6 illustrates the liquidus and solidus curves of the temperature taken as a function of x_2 or z_2 in the region of large values of x_1 for the same cases given in Figure 10.5. For the set of curves A, the liquid curve is almost vertical, and the solid curve has a very small positive slope in order to make z_2/x_2 extremely large.

Figure 10.5. Limiting curves for liquid solution–solid solution equilibria.

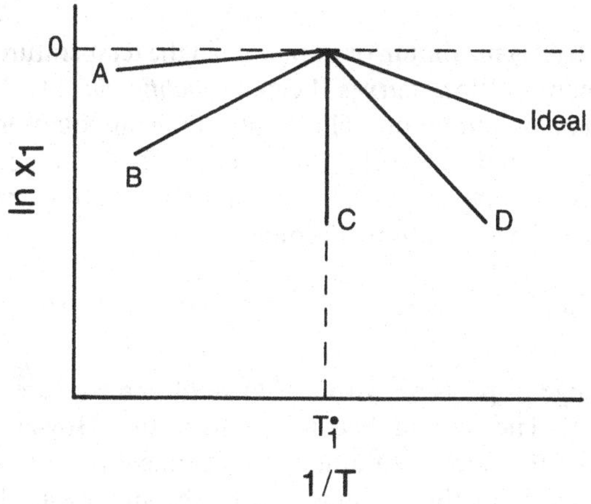

Figure 10.6. Various liquid solution–solid solution equilibria.

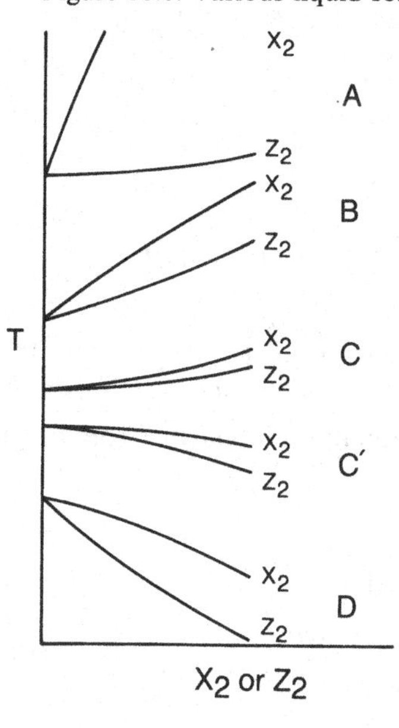

The set of curves B represents the general behavior when the temperature increases with increasing x_2. Curves C and C' represent a special case in which a minimum (curve C) or a maximum (curve C') in the temperature occurs just at the ordinate axis ($x_1 = 1$). Curve D illustrates the general behavior when the temperature decreases with increasing values of x_2.

10.26 Indifferent systems

Binary systems that have a maximum or minimum in the temperature–composition or pressure–composition curves become *indifferent* at the composition of the maximum or minimum. The Gibbs–Duhem equations applicable to each phase are Equations (10.124) and (10.125). The compositions of the two phases are equal at the maximum or minimum and, therefore, the solution of these two equations becomes

$$\frac{\mathrm{d}P}{\mathrm{d}T} = \frac{\tilde{S}'' - \tilde{S}'}{\tilde{V}'' - \tilde{V}'} \tag{10.182}$$

This equation gives the change of pressure with a change of temperature at the maximum or minimum. The system becomes univariant. However, because of the cancellation of the terms containing the chemical potentials, the determination of the change of the composition of the phases at the maximum or minimum with change of temperature or pressure cannot be determined from the Gibbs–Duhem equations alone. In order to do so we introduce the equality of the differentials of the chemical potentials in the two phases for one of the components, so

$$\mathrm{d}\mu_1' = \mathrm{d}\mu_1'' \tag{10.183}$$

or

$$-\bar{S}_1' \,\mathrm{d}T + \bar{V}_1' \,\mathrm{d}P + \left(\frac{\partial \mu_1'}{\partial x_1'}\right)_{T,P} \mathrm{d}x_1' = -\bar{S}_1'' \,\mathrm{d}T + \bar{V}_1'' \,\mathrm{d}P + \left(\frac{\partial \mu_1''}{\partial x_1'}\right)_{T,P} \mathrm{d}x_1' \tag{10.184}$$

where we use x_1' to indicate the mole fraction in both phases. From Equation (10.184) we have either

$$\frac{\mathrm{d}x_1'}{\mathrm{d}T} = \frac{\bar{S}_1'' - \bar{S}_1'}{[(\partial \mu_1''/\partial x_1')_{T,P} - (\partial \mu_1'/\partial x_1')_{T,P}]} - \frac{\bar{V}_1'' - \bar{V}_1'}{[(\partial \mu_1''/\partial x_1')_{T,P} - (\partial \mu_1'/\partial x_1')_{T,P}]} \frac{\mathrm{d}P}{\mathrm{d}T} \tag{10.185}$$

or

$$\frac{dx'_1}{dP} = \frac{\bar{S}''_1 - \bar{S}'_1}{[(\partial\mu''_1/\partial x'_1)_{T,P} - (\partial\mu'_1/\partial x'_1)_{T,P}]}\frac{dT}{dP} - \frac{\bar{V}''_1 - \bar{V}'_1}{[(\partial\mu''_1/\partial x'_1)_{T,P} - (\partial\mu'_1/\partial x'_1)_{T,P}]} \qquad (10.186)$$

The two equations obtained by substituting Equation (10.182) for dP/dT in Equations (10.185) and (10.186) then give the dependence of the composition at the maximum or minimum with temperature or pressure, respectively.

10.27 Three-phase systems

The three Gibbs–Duhem equations may be written as

$$\tilde{S}'\,dT - \tilde{V}'\,dP + x'_1\,d\mu_1 + x'_2\,d\mu_2 = 0$$

$$\tilde{S}''\,dT - \tilde{V}''\,dP + x''_1\,d\mu_1 + x''_2\,d\mu_2 = 0 \qquad (10.187)$$

and

$$\tilde{S}'''\,dT - \tilde{V}'''\,dP + x'''_1\,d\mu_1 + x'''_2\,d\mu_2 = 0$$

where the primes are used to indicate the phases. It is evident that the system is univariant. The elimination of the two chemical potentials yields

$$\frac{dP}{dT} = \frac{\tilde{S}'(x''_1 - x'''_1) + \tilde{S}''(x'''_1 - x'_1) + \tilde{S}'''(x'_1 - x''_1)}{\tilde{V}'(x''_1 - x'''_1) + \tilde{V}''(x'''_1 - x'_1) + \tilde{V}'''(x'_1 - x''_1)} \qquad (10.188)$$

which gives the change of the pressure with a change of temperature for the three-phase system. This equation could be written as

$$\frac{dP}{dT} = \frac{\Delta S}{\Delta V} \qquad (10.189)$$

but we must determine the change of state to which both ΔS and ΔV refer. In order to do so, let us imagine that, when heat is removed from the system at constant temperature and pressure, 1 mole of the triple-primed phase is formed from n' moles of the primed phase and $(1 - n')$ moles of the double-primed phase. By mass balance on the first component,

$$n'x'_1 + (1 - n')x''_1 = x'''_1 \qquad (10.190)$$

so

$$n' = \frac{x'''_1 - x''_1}{x'_1 - x''_1} \qquad (10.191)$$

and

$$1 - n' = \frac{x'_1 - x'''_1}{x'_1 - x''_1} \qquad (10.192)$$

Then, the change of entropy per mole of the triple-primed phase is

$$\Delta\tilde{S} = \tilde{S}''' - \frac{x_1''' - x_1''}{x_1' - x_1''}\,\tilde{S}' - \frac{x_1' - x_1'''}{x_1' - x_1''}\,\tilde{S}'' \tag{10.193}$$

We see by inspection that ΔS and ΔV refer to the formation of $(x_1' - x_1'')$ moles of the triple-primed phase from $(x_1''' - x_1'')$ moles of the primed phase and $(x_1' - x_1''')$ moles of the double-primed phase.

The three Gibbs–Duhem equations can also be used to determine the change in the mole fraction of one of the phases with temperature or with pressure. We choose here to determine $\mathrm{d}x_1'/\mathrm{d}T$. We first eliminate $\mathrm{d}\mu_2$ from the three equations, to obtain

$$(x_2'''\tilde{S}' - x_2'\tilde{S}''')\,\mathrm{d}T - (x_2'''\tilde{V}' - x_2'\tilde{V}''')\,\mathrm{d}P + (x_1' - x_1''')\,\mathrm{d}\mu_1 = 0 \tag{10.194}$$

and

$$(x_2'''\tilde{S}'' - x_2''\tilde{S}''')\,\mathrm{d}T - (x_2'''\tilde{V}'' - x_2''\tilde{V}''')\,\mathrm{d}P + (x_1'' - x_1''')\,\mathrm{d}\mu_1 = 0 \tag{10.195}$$

We then choose μ_1 to be μ_1', so

$$\mathrm{d}\mu_1 = \mathrm{d}\mu_1' = -\bar{S}_1'\,\mathrm{d}T + \bar{V}_1'\,\mathrm{d}P + \left(\frac{\partial\mu_1'}{\partial x_1'}\right)_{T,P}\mathrm{d}x_1' \tag{10.196}$$

Substitution of Equation (10.196) into Equations (10.194) and (10.195) with the subsequent elimination of $\mathrm{d}P$ yields

$$\frac{\mathrm{d}x_1'}{\mathrm{d}T} = -\frac{\begin{aligned}&[(x_1' - x_2''')[x_2'''\tilde{V}'' - x_2''\tilde{V}''' - (x_1'' - x_1''')\bar{V}_1']\\ &\quad - (x_1'' - x_1''')[x_2'''\tilde{V}' - x_2'\tilde{V}''' - (x_1' - x_1''')\bar{V}_1']](\partial\mu_1'/\partial x_1'')_{T,P}\end{aligned}}{\begin{aligned}&[x_2'''\tilde{S}' - x_2'\tilde{S}''' - (x_1' - x_1''')\bar{S}_1'][x_2'''\tilde{V}'' - x_2''\tilde{V}''' - (x_1'' - x_1''')\bar{V}_1']\\ &- [x_2'''\tilde{S}'' - x_2''\tilde{S}''' - (x_1'' - x_1''')\bar{S}_1'][x_2'''\tilde{V}' - x_2'\tilde{V}''' - (x_1' - x_1''')\bar{V}_1']\end{aligned}} \tag{10.197}$$

The evaluation of the derivatives thus requires a knowledge of the entropies and volumes of the three phases as well as a knowledge of the excess chemical potential of the first component in the primed phase.

10.28 Calculation of phase equilibria from the chemical potentials

In the foregoing we have discussed the determination of the chemical potentials as functions of the temperature, pressure, and composition by means of experimental studies of phase equilibria. The converse problem of determining the phase equilibria from a knowledge of the chemical potentials is of some importance. For any given phase equilibrium the required equations are the same as those developed for the integral method. The solution of the equation or equations requires that a sufficient number of conditions be set that the number of variables equals the number of equations. Thus, when both components are present in both phases for two-phase systems, two equations, one for each of the components, are available. However, there are *four* variables: the temperature, the pressure, and the mole fraction of one component in each phase. Therefore, two of the four variables must be fixed and the equations solved for the other two variables. The usual practice would be to fix the values of the temperature and pressure, and solve for the two mole fractions. In the case that one of the components is not present in one of the phases, only one equation is available, but there is also one less variable. The situation is then the same. The solution of the equations requires knowledge of the excess chemical potentials as a function of the mole fraction at the equilibrium temperature and pressure that have been chosen. When isothermal and isobaric values are not available, appropriate corrections must be made to obtain values under the equilibrium conditions. These corrections are identical to those used to obtain isothermal values when nonisothermal values were obtained experimentally and isobaric values when nonisobaric values were obtained.

The equations are transcendental equations, which must be solved by approximation methods. This presents no problem with the use of modern computers. However, it is still appropriate to discuss graphical aids to the solution of the equations. We discuss only one example. Let us consider the equilibrium between a solid solution and a liquid solution at a constant pressure. For the present we choose the pure solid phases and the pure liquid phases as the standard states of the two components for each phase. The equation for equilibrium for the first component is

$$RT\ln x_1 + \Delta\mu_1^{\mathrm{E}}[T, P, x_1] = RT\ln z_1 + \Delta\mu_1^{\mathrm{E}}[T, P, z_1] + \mu_1^{\ominus}[T, P, z_1 = 1] - \mu_1^{\ominus}[T, P, x_1 = 1] \tag{10.198}$$

based on Equation (10.30). The difference between the chemical potentials in the two standard states is

$$T\int_{T_1^{\cdot}}^{T} \{[\tilde{H}_1^{\cdot}(\ell) - \tilde{H}_1^{\cdot}(\mathrm{s})]/T^{\cdot}\}\,\mathrm{d}T$$

The equation for the second component is the same with change of subscripts. The problem is to determine values of x_1 and z_1 at specified temperatures by the use of the two equations. We define the quantity $\Delta\mu_1^{\mathrm{M}}[T, P]$ as equal to both sides of Equation (10.198). This quantity is the change of the chemical potential on mixing for the liquid phase that has been defined previously; however, for the solid phase the standard state is now the pure *liquid* component. A similar definition is made for $\Delta\mu_2^{\mathrm{M}}[T, P]$. The conditions of equilibrium then become

$$\Delta\mu_1^{\mathrm{M}}[T, P, \ell] = \Delta\mu_1^{\mathrm{M}}[T, P, \mathrm{s}] \tag{10.199}$$

and

$$\Delta\mu_2^M[T, P, \ell] = \Delta\mu_2^M[T, P, s] \tag{10.200}$$

We must emphasize that these conditions are valid only when the standard state of a component is the same for both phases. We now choose a temperature, calculate values of $\Delta\mu_1^M$ and $\Delta\mu_2^M$ for each phase for various mole fractions and plot $\Delta\mu_1^M$ against $\Delta\mu_2^M$ for each phase. Two curves are obtained, one for each phase. The point of intersection of the curves gives the equilibrium values of $\Delta\mu_1^M$ and $\Delta\mu_2^M$ according to Equations (10.199) and (10.200). The calculation of x_1 by the use of the left-hand side of Equation (10.198) and z_1 by the use of the right-hand side of Equation (10.198) is then relatively simple. If no intersection occurs, then no equilibrium between the two phases exists at the chosen temperature and pressure. The calculation must be repeated for each temperature. Activities could be used in place of $\Delta\mu^M$ provided the same standard state for a component is used for both phases.

For other examples, the reader is referred to the original literature [26].

THREE-COMPONENT SYSTEMS

The experimental studies of three-component systems based on phase equilibria follow the same principles and methods discussed for two-component systems. The integral form of the equations remains the same. The added complexity is the additional composition variable; the excess chemical potentials become functions of two composition variables, rather than one. Because of the similarity, only those topics that are pertinent to ternary systems are discussed in this section of the chapter. We introduce pseudobinary systems, discuss methods of determining the excess chemical potentials of two of the components from the experimental determination of the excess chemical potential of the third component, apply the set of Gibbs–Duhem equations to only one type of phase equilibria in order to illustrate additional problems that occur in the use of these equations, and finally discuss one additional type of phase equilibria.

10.29 Pseudobinary systems

In some cases we are concerned with systems in which the mole ratio of two of the components is kept constant. Such a mixture can be considered as a solvent of fixed composition with the third component acting as a solute and is, therefore, a *pseudobinary system.*

The thermodynamic relations for such pseudobinary systems are readily developed from those of a ternary system. The molar Gibbs energy of a ternary system is expressed by

$$\tilde{G} = x_1\mu_1 + x_2\mu_2 + x_3\mu_3 \tag{10.201}$$

If we take the second and third components as those that form the mixed solvent and express the ratio x_2/x_3 as k, a constant, then we can eliminate x_2 from Equation (10.201), so

$$\tilde{G} = x_1\mu_1 + x_3(k\mu_2 + \mu_3) \tag{10.202}$$

Furthermore, if we choose to express the composition of the system in terms of the mole fraction of the solute, x_1, and a mole fraction of the solvent, x_s, we have

$$x_s = x_2 + x_3 = x_3(k+1) \tag{10.203}$$

Then the molar Gibbs energy may be written as

$$\tilde{G} = x_1\mu_1 + x_s\left(\frac{k\mu_2 + \mu_3}{k+1}\right) \tag{10.204}$$

or

$$\tilde{G} = x_1\mu_1 + x_s\mu_s \tag{10.205}$$

where

$$\mu_s = \frac{k\mu_2 + \mu_3}{k+1} \tag{10.206}$$

The quantity μ_s can be considered as the chemical potential of the mixed solvent. The Gibbs–Duhem equation may be written as

$$\tilde{S}\,dT - \tilde{V}\,dP + x_1\,d\mu_1 + x_s\,d\mu_s = 0 \tag{10.207}$$

after the use of the same arguments.

Equations (10.205) and (10.207), together with Equation (10.206), comprise the basic equations to use in considering a ternary system that has a constant ratio of x_2/x_3 as a pseudobinary system.

10.30 Determination of excess chemical potentials

The study of a ternary system over the entire range of composition presents a formidable experimental problem. Several methods have been developed by which the excess chemical potentials of two of the components can be calculated from known values of only one component. Only two of these methods are discussed here. Throughout the discussion we assume that the values of the excess chemical potential of the first component are measured or known at constant temperature and pressure.

Darken [27] has made use of the concept of pseudobinary systems and has developed equations by which the molar excess Gibbs energy can be calculated. Equation (10.205) is written in terms of the excess quantities, so

$$\Delta\tilde{G}^{E} = x_1 \Delta\mu_1^{E} + x_s \Delta\mu_s^{E} \tag{10.208}$$

where $\Delta\mu_s^E$ is defined by Equation (10.206). Differentiation of this equation with respect to x_1 at constant temperature, constant pressure, and constant ratio of x_2/x_3, equal to k, yields

$$\left(\frac{\partial\, \Delta\tilde{G}^{E}}{\partial x_1}\right)_{T,P,k} = \Delta\mu_1^{E} - \Delta\mu_s^{E} \tag{10.209}$$

On elimination of $\Delta\mu_s^E$ by Equation (10.208), we obtain

$$\Delta\tilde{G}^{E} + (1 - x_1)\left(\frac{\partial\, \Delta\tilde{G}^{E}}{\partial x_1}\right)_{T,P,k} = \Delta\mu_1^{E} \tag{10.210}$$

where $(1 - x_1)$ has been substituted for x_s. After dividing by $(1 - x_1)^2$, Equation (10.210) is readily transformed to

$$\left(\frac{\partial\Delta\tilde{G}^{E}/(1 - x_1)}{\partial x_1}\right)_{T,P,k} = \frac{\Delta\mu_1^{E}}{(1 - x_1)^2} \tag{10.211}$$

Thus, when $\Delta\mu_1^E$ is determined as a function of x_1 at a fixed k, $\Delta\tilde{G}^E$ can be obtained for the pseudobinary systems by integration. The most convenient reference state of component 1 is the pure component at the chosen temperature and pressure, because the reference state chosen as the infinitely dilute solution of the component in a pseudobinary solvent is different for every ratio of x_2/x_3.

With the appropriate choice of the reference state, values of the molar excess Gibbs energy can be determined over the entire range of composition of the ternary system by making studies on a set of pseudobinary systems. Values of $\Delta\mu_2^E$ and $\Delta\mu_3^E$ can then be calculated by means of the usual methods.

Three cases concerning the integration must be discussed. In the first case the integral is determined over the range from $x_1 = 0$ to variable values of x_1, so

$$\left(\frac{\Delta\tilde{G}^{E}}{1 - x_1}\right)_{x_1} - [\Delta\tilde{G}^{E}]_{x_1 = 0} = \int_0^{x_1} \frac{\Delta\mu_1^{E}}{(1 - x_1)^2}\, dx_1 \tag{10.212}$$

The value of $\Delta\tilde{G}^E$ when $x_1 = 0$ is that of the binary system composed of the second and third components at the composition where $x_2/x_3 = k$. This value is dependent, of course, on the reference states chosen for the two components. In the other two cases the integration is carried out between the limits $x_1 = 1$ and variable values of x_1. The integral is then written as

$$\left[\frac{\Delta\tilde{G}^{E}}{1 - x_1}\right] - \lim_{x_1 \to 1}\left[\frac{\Delta\tilde{G}^{E}}{1 - x_1}\right] = \int_{x_1 = 1}^{x_1} \frac{\Delta\mu_1^{E}}{(1 - x_1)^2}\, dx_1 \tag{10.213}$$

The integrand must be finite as x_1 approaches zero. In the case of a dissociating substance, a proper choice of the components or species is required. The value of $\Delta\tilde{G}^{\mathrm{E}}/(1-x_1)$ becomes indeterminate in the limit as x_1 goes to unity, but can be determined by the use of l'Hôpital's theorem. We have

$$\lim_{x_1\to 1}\left[\frac{\Delta\tilde{G}^{\mathrm{E}}}{1-x_1}\right] = -\lim_{x_1\to 1}\left[\frac{\partial\,\Delta\tilde{G}^{\mathrm{E}}}{\partial x_1}\right]_{T,P,k} \tag{10.214}$$

With the use of Equations (10.208) and (10.209), together with Equation (10.206) expressed in terms of the excess chemical potentials, we obtain

$$-\lim_{x_1\to 1}\left[\frac{\partial\,\Delta\tilde{G}^{\mathrm{E}}}{\partial x_1}\right]_{T,P,k} = \lim_{x_1\to 1}\left[\Delta\mu_1^{\mathrm{E}} - \frac{k\,\Delta\mu_2^{\mathrm{E}} + \Delta\mu_3^{\mathrm{E}}}{k+1}\right] \tag{10.215}$$

When the reference state of the first component is chosen as the pure component and that of each of the other two components as the infinitely dilute solution of the component in the first component, each of the excess chemical potentials become zero in the limit and Equation (10.214) can be written as

$$\Delta\tilde{G}^{\mathrm{E}} = (1-x_1)\int_{x_1=1}^{x_1}\frac{\Delta\mu_1^{\mathrm{E}}}{(1-x_1)^2}\,\mathrm{d}x_1 \tag{10.216}$$

The third case occurs when we choose the reference state of each of the components as the pure components. In this case $\Delta\mu_1^{\mathrm{E}}$ becomes zero in the limit, and the values of $\Delta\mu_2^{\mathrm{E}}$ and $\Delta\mu_3^{\mathrm{E}}$ are limiting values in the binary systems composed of the first and second components and of the first and third components, respectively. We then have

$$\left[\frac{\Delta\tilde{G}^{\mathrm{E}}}{1-x_1}\right]_{x_1} - \frac{k}{k+1}[\Delta\mu_2^{\mathrm{E}}]_{x_1=1} - \frac{1}{k+1}[\Delta\mu_3^{\mathrm{E}}]_{x_1=1}$$
$$= \int_{x_1=1}^{x_1}\frac{\Delta\mu_1^{\mathrm{E}}}{(1-x_1)^2}\,\mathrm{d}x_1 \tag{10.217}$$

The second method for studying ternary systems, which has been developed by Gokcen [28], makes use of such equations expressed in general form as

$$\left(\frac{\partial\mu_i}{\partial n_j}\right)_{T,P,n} = \left(\frac{\partial\mu_j}{\partial n_i}\right)_{T,P,n} \tag{10.218}$$

The equations are derived from the differential of the Gibbs energy for a one-phase, multicomponent system Equation (2.33). The differential is exact, and therefore the condition of exactness must be satisfied. Two equations

$$\left(\frac{\partial \mu_2}{\partial n_1}\right)_{T,P,n_2,n_3} = \left(\frac{\partial \mu_1}{\partial n_2}\right)_{T,P,n_1,n_3} \tag{10.219}$$

and

$$\left(\frac{\partial \mu_3}{\partial n_1}\right)_{T,P,n_2,n_3} = \left(\frac{\partial \mu_1}{\partial n_3}\right)_{T,P,n_1,n_2} \tag{10.220}$$

are obtained for ternary systems. When values of μ_1 are known for various values of two of the composition variables, values of μ_2 and μ_3 can be obtained by the use of Equations (10.219) and (10.220), respectively.

We limit our discussion here to Equation (10.219) and make use of the excess chemical potentials rather than the chemical potentials themselves. We then start with

$$\left(\frac{\partial\, \Delta\mu_2^{\mathrm{E}}}{\partial n_1}\right)_{T,P,n_2,n_3} = \left(\frac{\partial\, \Delta\mu_1^{\mathrm{E}}}{\partial n_2}\right)_{T,P,n_1,n_3} \tag{10.221}$$

New composition variables are defined as

$$p = \frac{n_3}{n_1 + n_3} = \frac{x_3}{x_1 + x_3} = \frac{x_3}{1 - x_2} \tag{10.222}$$

and

$$q = \frac{n_3}{n_2 + n_3} = \frac{x_3}{x_2 + x_3} = \frac{x_3}{1 - x_1} \tag{10.223}$$

We then have

$$\left(\frac{\partial\, \Delta\mu_2^{\mathrm{E}}}{\partial n_1}\right)_{T,P,n_2,n_3} = \left(\frac{\partial\, \Delta\mu_2^{\mathrm{E}}}{\partial p}\right)_{T,P,q} \left(\frac{\partial p}{\partial n_1}\right)_{n_2,n_3} \tag{10.224}$$

and

$$\left(\frac{\partial\, \Delta\mu_1^{\mathrm{E}}}{\partial n_2}\right)_{T,P,n_1,n_3} = \left(\frac{\partial\, \Delta\mu_1^{\mathrm{E}}}{\partial q}\right)_{T,P,p} \left(\frac{\partial q}{\partial n_2}\right)_{n_1,n_3} \tag{10.225}$$

and Equation (10.221) becomes

$$\left(\frac{\partial\, \Delta\mu_2^{\mathrm{E}}}{\partial p}\right)_{T,P,q} = \frac{q^2}{p^2} \left(\frac{\partial\, \Delta\mu_1^{\mathrm{E}}}{\partial q}\right)_{T,P,p} \tag{10.226}$$

Values of $\Delta\mu_2^{\mathrm{E}}$ can be obtained from known values of $(\partial\, \Delta\mu_1^{\mathrm{E}}/\partial q)_{T,P,p}$ on integration.

The actual calculations associated with the integrations are outlined with reference to Figure 10.7, where the equilateral triangle gives the usual representation of the composition of a ternary system. Each apex of the triangle represents a pure component as given by the numbers 1, 2, and 3. The solid straight line from apex 2 to the 1–3 edge represents all systems that have a given, constant value of p. Similarly, the solid straight line drawn from apex 1 to the 2–3 edge represents all systems that have a given, constant value of q. The values of q along any line of constant p range from unity at the 1–3 edge to zero at the apex, and the values of p along any line of constant q range from unity at the 2–3 edge to zero at the apex. We assume for the purposes of discussion that the integration is carried out along the line drawn from 1 to q. We need to know values of $\Delta\mu_1^E$ as a function of q for various constant values of p; e.g., along the line drawn from 2 to p. With this knowledge, values of $(\partial\,\Delta\mu_1^E/\partial q)_{T,P,p}$ are obtained at the chosen value of q for various values of p. The integration is then carried out after multiplication by p^2/q^2.

Two choices of the lower limit for the integration are apparent. In one case the lower limit of p is chosen as unity, for which case $x_1 = 0$. Then the integral of Equation (10.226) becomes

$$\Delta\mu_2^E[p, q] - \Delta\mu_2^E[p = 1, q] = \int_{p=1}^{p} \frac{q^2}{p^2}\left(\frac{\partial\,\Delta\mu_1^E}{\partial q}\right)_{T,P,p} dp \qquad (10.227)$$

Here $\Delta\mu_2^E[p = 1, q]$ is the excess chemical potential of the second component in the binary system composed of the second and third components at the composition equal to q. The reference state of the second component may be either the pure component or the infinitely dilute solution of the second component in the third component. In the limit of $p = 1$, q/p becomes $1/(k + 1)$, where $k = x_2/x_3$ and, with the proper choice of components or

Figure 10.7. Ternary systems.

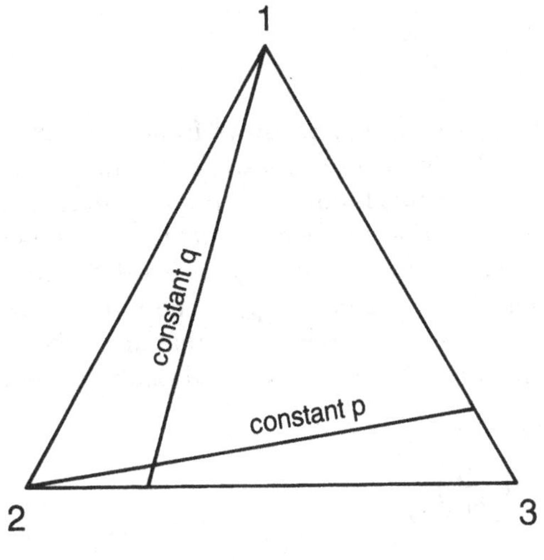

species, $(\partial\,\Delta\mu_1^{\mathrm{E}}/\partial q)_{T,P,p}$ remains finite in the limit. Hence, the integrand is finite in the limit.

The second choice of the lower limit is to set $p=0$. The integral then becomes

$$\Delta\mu_2^{\mathrm{E}}[p,q]-\Delta\mu_2^{\mathrm{E}}[p=1,q]=\int_{p=0}^{p}\frac{q^2}{p^2}\left(\frac{\partial\,\Delta\mu_1^{\mathrm{E}}}{\partial q}\right)_{T,P,p}\mathrm{d}p \tag{10.228}$$

The line for which $p=0$ is the 1–2 edge of the triangle. Therefore, the limiting value of $\Delta\mu_2^{\mathrm{E}}$ as p goes to zero along any line of constant q is that of the second component in the infinitely dilute solution of the component in the first component. Whereas a knowledge of the 2–3 binary system is required when the lower limit is at $p=1$, here a knowledge of the 1–2 binary system is required. From the definitions of p and q given in Equations (10.222) and (10.223), we observe that the value of q/p becomes large as p goes to zero. At the same time the value of $(\partial\,\Delta\mu_1^{\mathrm{E}}/\partial q)_{T,P,p}$ becomes small. Experimental work has shown that the product of the two quantities generally remains finite and the integral is well behaved. The proper choice of components or species may be required in some cases, particularly for electrolytic solutions. In any case, studies should be made to determine that the integrand does, indeed, remain finite in the limit. Values of $\Delta\mu_2^{\mathrm{E}}$ are determined over the entire range of composition by integrating along various lines of constant q.

Values of $\Delta\mu_3^{\mathrm{E}}$ are determined similarly by the use of Equation (10.220) expressed in terms of the excess chemical potentials. In this case the new variables would be

$$s=\frac{n_2}{n_1+n_2}=\frac{x_2}{x_1+x_2}=\frac{x_2}{1-x_3} \tag{10.229}$$

and

$$t=\frac{n_2}{n_2+n_3}=\frac{x_2}{x_2+x_3}=\frac{x_2}{1-x_1} \tag{10.230}$$

Lines of constant s in Figure 10.7 would be drawn from the apex marked 3 to the 1–2 edge of the triangle, and lines of constant t are the same as those at constant p. Gokcen has pointed out a method of determining the compositions to be studied in a ternary system in order to achieve a minimum of experimental work. Each angle of the triangle is divided into an equal number of parts by drawing lines of constant p, q, and s. The intersections of the three lines then determine the compositions that are to be used.

Wagner [29] and Schuhmann [30] have developed similar methods based on

$$\left(\frac{\partial\mu_2}{\partial\mu_1}\right)_{T,P,n_2,n_3}=-\left(\frac{\partial n_1}{\partial n_2}\right)_{T,P,\mu_2,n_3} \tag{10.231}$$

This equation may be derived from the differential of the function $(G - \mu_2 n_2)$, so

$$d(G - \mu_2 n_2) = -S\,dT + V\,dP + \mu_1\,dn_1 - n_2\,d\mu_2 + \mu_3\,dn_3 \quad (10.232)$$

The differential is exact, and Equation (10.231) follows directly from the condition of exactness. The reader is referred to the original work for detailed discussion involving the use of this equation.

The methods of Darken and Gokcen discussed here can be applied to quaternary systems. New composition variables are defined in order to convert the quaternary system into pseudobinary systems. The paths of integration can be illustrated by the use of an equilateral tetrahedron.

The methods discussed in this section can be extended to systems that have more than three components. The problem is to convert each system to a pseudobinary system. For a quaternary system, the properties of an equilateral tetrahedron may be used to depict the composition of the system. The composition axes would be four lines drawn from the four apexes perpendicular to the opposite faces. Planes cutting the tetrahedron parallel to the bases would represent pseudoternary systems for which one composition variable would be constant. Pseudobinary systems would be depicted by the intersections of two of the pseudoternary planes. Indeed, the experimental measurements and calculations would be extensive.

10.31 The use of the Gibbs–Duhem equations

The development of derivatives of one intensive variable with respect to another in ternary systems follows the same general methods used for binary systems. A difference does arise because the chemical potentials are functions of two composition variables rather than one. Therefore, we cannot eliminate all but one chemical potential from the applicable set of Gibbs–Duhem equations as was done for binary systems, without introducing two composition variables. As an illustration we consider a three-phase system at constant pressure. The phases are assumed to be a liquid in which all three components occur and two pure, solid phases. Such a system is univariant. We wish to determine the change of the temperature with a change of the mole fraction of the first component in the liquid phase. We designate the pure, solid phases as being the pure second and third components. The set of Gibbs–Duhem equations for this case are

$$\tilde{S}\,dT + x_1\,d\mu_1 + x_2\,d\mu_2 + x_3\,d\mu_3 = 0 \quad (10.233)$$

and

$$\tilde{S}_2^{\circ}(s)\,dT + d\mu_2 = 0 \quad (10.234)$$

$$\tilde{S}_3^{\circ}(s)\,dT + d\mu_3 = 0 \quad (10.235)$$

We can eliminate $d\mu_3$ in Equation (10.233) to obtain

$$(\tilde{S} - x_3\tilde{S}_3^*)\,dT + x_1\,d\mu_1 + x_2\,d\mu_2 = 0 \tag{10.236}$$

However, if we eliminate $d\mu_2$ from Equation (10.236) by the use of Equation (10.234) and then substitute for $d\mu_1$ by an equation giving $d\mu_1$ as a function of the temperature and two mole fractions, we obtain one equation involving three variables for a univariant system. This difficulty can be avoided if we first express $d\mu_2$ as

$$d\mu_2 = -\bar{S}_2\,dT + \left(\frac{\partial \mu_2}{\partial x_1}\right)_{T,P,x_2} dx_1 + \left(\frac{\partial \mu_2}{\partial x_2}\right)_{T,P,x_1} dx_2 \tag{10.237}$$

and $d\mu_1$ as

$$d\mu_1 = -\bar{S}_1\,dT + \left(\frac{\partial \mu_1}{\partial x_1}\right)_{T,P,x_2} dx_1 + \left(\frac{\partial \mu_1}{\partial x_2}\right)_{T,P,x_1} dx_2 \tag{10.238}$$

for the liquid phase. Equations (10.236) and (10.234) then become

$$(\tilde{S} - x_1\bar{S}_1 - x_2\bar{S}_2 - x_3\tilde{S}_3^*)\,dT + x_1\left(\frac{\partial \mu_1}{\partial x_1}\right)_{T,P,x_2} dx_1 + x_1\left(\frac{\partial \mu_1}{\partial x_2}\right)_{T,P,x_1} dx_2$$
$$+ x_2\left(\frac{\partial \mu_2}{\partial x_1}\right)_{T,P,x_2} dx_1 + x_2\left(\frac{\partial \mu_2}{\partial x_2}\right)_{T,P,x_1} dx_2 = 0 \tag{10.239}$$

and

$$[\bar{S}_2^*(s) - \bar{S}_2]\,dT + \left(\frac{\partial \mu_2}{\partial x_1}\right)_{T,P,x_2} dx_1 + \left(\frac{\partial \mu_2}{\partial x_2}\right)_{T,P,x_1} dx_2 = 0 \tag{10.240}$$

We can now eliminate dx_2 from these equations, to obtain

$$\left[x_3[\bar{S}_3 - \tilde{S}_3^*(s)] + \left(x_2 + \frac{x_1(\partial\mu_1/\partial x_2)_{T,P,x_1}}{(\partial\mu_2/\partial x_2)_{T,P,x_1}}\right)[\bar{S}_2 - \bar{S}_2^*(s)]\right] dT$$
$$+ x_1\left[\left(\frac{\partial \mu_1}{\partial x_1}\right)_{T,P,x_2} - \frac{(\partial\mu_1/\partial x_2)_{T,P,x_1}(\partial\mu_2/\partial x_1)_{T,P,x_2}}{(\partial\mu_2/\partial x_2)_{T,P,x_1}}\right] dx_1 = 0 \tag{10.241}$$

when $\tilde{S}$ is set equal to $x_1\bar{S}_1 + x_2\bar{S}_2 + x_3\bar{S}_3$. Then

$$\left(\frac{\partial T}{\partial x_1}\right)_P = -\frac{x_1\left(\left(\frac{\partial \mu_1}{\partial x_1}\right)_{T,P,x_2} - \frac{(\partial\mu_1/\partial x_2)_{T,P,x_1}(\partial\mu_2/\partial x_1)_{T,P,x_2}}{(\partial\mu_2/\partial x_2)_{T,P,x_1}}\right)}{x_3[\bar{S}_3 - \tilde{S}_3^*(s)] + \left(x_2 + \frac{x_1(\partial\mu_1/\partial x_2)_{T,P,x_1}}{(\partial\mu_2/\partial x_2)_{T,P,x_1}}\right)[\bar{S}_2 - \bar{S}_2^*(s)]} \tag{10.242}$$

If the liquid phase is an ideal solution, this equation is simplified to

$$\left(\frac{\partial T}{\partial x_1}\right)_P = -\frac{x_1(\partial\mu_1/\partial x_1)_{T,P,x_2}}{x_2[\bar{S}_2 - \tilde{S}_2^{\bullet}(\mathrm{s})] + x_3[\bar{S}_3 - \tilde{S}_3^{\bullet}(\mathrm{s})]} \tag{10.243}$$

or

$$\left(\frac{\partial T}{\partial x_1}\right)_P = -\frac{RT}{x_2[\bar{S}_2 - \tilde{S}_2^{\bullet}(\mathrm{s})] + x_3[\bar{S}_3 - \tilde{S}_3^{\bullet}(\mathrm{s})]} \tag{10.244}$$

because $(\partial\mu_1/\partial x_2)_{T,P,x_2} = (\partial\mu_2/\partial x_1)_{T,P,x_2} = 0$ for an ideal solution.

10.32 Liquid–liquid equilibria

The study of distribution coefficients—the distribution of a solute between two liquid phases—is closely related to the complete study of liquid–liquid equilibria in ternary systems. We limit the discussion to conditions of constant temperature and constant pressure. In the most general case the system is composed of two liquid phases, with all three components existing in each phase. The conditions of equilibria are

$$\mu_1'[T, P, x_1', x_3'] = \mu_1''[T, P, x_1'', x_3''] \tag{10.245}$$

$$\mu_2'[T, P, x_1', x_3'] = \mu_2''[T, P, x_1'', x_3''] \tag{10.246}$$

and

$$\mu_3'[T, P, x_1', x_3'] = \mu_3''[T, P, x_1'', x_3''] \tag{10.247}$$

If we choose to express each chemical potential in terms of mole fractions and choose the standard state of each component to be the pure liquid component, we have the set of equations

$$RT \ln x_1' + \Delta\mu_1^{\mathrm{E}}[x_1', x_3'] = RT \ln x_1'' + \Delta\mu_1^{\mathrm{E}}[x_1'', x_3''] \tag{10.248}$$

$$RT \ln x_2' + \Delta\mu_2^{\mathrm{E}}[x_1', x_3'] = RT \ln x_2'' + \Delta\mu_2^{\mathrm{E}}[x_1'', x_3''] \tag{10.249}$$

and

$$RT \ln x_3' + \Delta\mu_3^{\mathrm{E}}[x_1', x_3'] = RT \ln x_3'' + \Delta\mu_3^{\mathrm{E}}[x_1'', x_3''] \tag{10.250}$$

where we have indicated that each excess chemical potential is dependent on two mole fractions; the conditions of temperature and pressure have not been included. The variables in this set of equations are x_1', x_1'', x_3', and x_3''. The system is univariant and, when a value is chosen for one of the four mole fractions, the values of the other three can be calculated by use of the three equations. The calculations require a knowledge of each of the excess chemical potentials as a function of the two mole fractions

If the two phases are ideal solutions, the mole fraction of a component is the same in each phase. Then the two phases are identical and only one liquid phase actually exists. The solutions must be nonideal when two liquid phases do exist. It is evident that the ratio of the activities of each component in the two phases is always unity. The distribution ratio in terms of mole fractions for each component is given by

$$RT \ln(x_i'/x_i'') = \Delta\mu_i^{\mathrm{E}}[x_1'', x_3''] - \Delta\mu_i^{\mathrm{E}}[x_1', x_3'] \tag{10.251}$$

Here we see that the ratio is dependent upon the difference between the excess chemical potentials of the component in each of the two phases. This ratio is not constant.

The *thermodynamic distribution coefficient* is introduced when one of the components can be considered as a solute in each phase, and when we choose the reference states of that component to be the infinitely dilute solution in each phase. For discussion, we designate the first and second components as those that form the solvents and the third component as the solute. Equations (10.245), (10.246), (10,248), and (10.249) are still applicable when we choose the pure liquid phase as the standard state for each of the two components. When we introduce expressions for the chemical potential of the third component into Equation (10.247), Equation (10.250) becomes

$$RT \ln x_3' + \Delta\mu_3^{\mathrm{E}}[x_1', x_3'] + [\mu_3^{\ominus}]' = RT \ln x_3'' + \Delta\mu_3^{\mathrm{E}}[x_1'', x_3''] + [\mu_3^{\ominus}]'' \tag{10.252}$$

or, in terms of the activity,

$$RT \ln a_3' + [\mu_3^{\ominus}]' = RT \ln a_3'' + [\mu_3^{\ominus}]'' \tag{10.253}$$

Two reference states, one for each phase, must be defined. As we decrease the mole fraction of the third component, we approach the two-liquid-phase binary system composed of the first and second components. We thus define the reference states of the third component as the infinitely dilute solutions of the component in the two liquid phases that are at equilibrium in the 1–2 binary system. Thus, the value of $\Delta\mu_3^{\mathrm{E}}[x_1', x_3']$ approaches zero as x_3' approaches zero, and x_1' approaches its value in the 1–2 system, and the value of $\Delta\mu_3^{\mathrm{E}}[x_1'', x_3'']$ also approaches zero, and x_1'' approaches its value in the 1–2 system. In the limit Equation (10.251) becomes

$$RT \ln[(x_3'/x_3'')^{\infty}] = [\mu_3^{\ominus}]'' - [\mu_3^{\ominus}]' \tag{10.254}$$

where the infinity sign is used to designate the ratio of (x_3'/x_3'') in the infinitely dilute system. Both $[\mu_3^{\ominus}]''$ and $[\mu_3^{\ominus}]'$ are functions of the temperature and pressure, but not of the composition. Therefore, the difference is a constant at constant temperature and pressure. We choose to write this difference as

$$[\mu_3^{\ominus}]'' - [\mu_3^{\ominus}]' = -RT \ln k \tag{10.255}$$

where k is a function of the temperature and pressure, and is a constant at constant temperature and pressure. The quantity k is called the *thermo-*

dynamic distribution coefficient. We see from Equation (10.253) that k can be evaluated by determining the value of the ratio x_3''/x_3' in the limit as x_3' goes to zero, i.e.,

$$k = \lim_{x_3' \to 0} (x_3''/x_3') \tag{10.256}$$

With the definition of the thermodynamic distribution coefficient, Equation (10.252) can be written as

$$RT \ln(x_3'/x_3'') + \Delta\mu_3^{\mathrm{E}}[x_1', x_3'] - \Delta\mu_3^{\mathrm{E}}[x_1'', x_3''] = -RT \ln k \tag{10.257}$$

and Equation (10.253) as simply

$$a_3''/a_3' = k \tag{10.258}$$

We see that the ratio of the activities is always constant at constant temperature and pressure; the ratio of the mole fractions is not constant over a range of concentrations of x_3', and the variation of the value of the ratio depends upon the difference of the excess chemical potentials.

If we had expressed the chemical potential of the third component in terms of its molality and activity coefficient rather than the mole fraction and excess chemical potential, Equation (10.258) would be expressed as

$$\left(\frac{m''\gamma''}{m'\gamma'}\right)_3 = k \tag{10.259}$$

If the solute were a salt, having the formula $M_{\nu_+}A_{\nu_-}$ and was nonionized in one phase and completely ionized in the other phase, the distribution coefficient would be expressed as

$$\left[\frac{(m_\pm \gamma_\pm)^\nu}{m'\gamma'}\right]_3 = k \tag{10.260}$$

The relations that obtain when the solute is partially polymerized or ionized are discussed in Chapter 11.

11

Chemical equilibrium

We turn our attention in this chapter to systems in which chemical reactions occur. We are concerned not only with the equilibrium conditions for the reactions themselves, but also the effect of such reactions on phase equilibria and, conversely, the possible determination of chemical equilibria from known thermodynamic properties of solutions. Various expressions for the equilibrium constants are first developed from the basic condition of equilibrium. We then discuss successively the experimental determination of the values of the equilibrium constants, the dependence of the equilibrium constants on the temperature and on the pressure, and the standard changes of the Gibbs energy of formation. Equilibria involving the ionization of weak electrolytes and the determination of equilibrium constants for association and complex formation in solutions are also discussed.

11.1 Basic considerations

In Section 5.4 we develop the condition of equilibrium for a chemical reaction: for any chemical reaction written as $\sum_k \nu_k B_k = 0$, the condition of equilibrium is $\sum_k \nu_k \mu_k = 0$. In these equations each ν_k represents the stoichiometric coefficient of the substance B_k in the balanced chemical equation; the sign is positive for products and negative for reagents. Expressions for the equilibrium constants are readily obtained from this condition. For a general relation we express the chemical potentials in terms of the activities, so

$$\mu_k[T, P, a] = RT \ln a_k + \mu_k^\ominus[T, P] \tag{11.1}$$

Then, according to the condition of equilibrium,

$$\sum_k \nu_k \mu_k^\ominus[T, P] = -RT \ln \prod_k a_k^{\nu_k} \tag{11.2}$$

when the temperature of every reacting substance is the same. In Equation (11.2), $\prod_k a_k^{\nu_k}$ represents the product of the activities of the reacting substances, each activity being raised to the appropriate power. At constant temperature and pressure, the left-hand side of Equation (11.2) is constant, and hence the right-hand side must be constant. We can therefore define the thermodynamic equilibrium constant for the reaction as

$$\prod_k a_k^{\nu_k} = K \tag{11.3}$$

so

$$\sum_k \nu_k \mu_k^{\ominus}[T, P] = -RT \ln K \tag{11.4}$$

As defined here, the equilibrium constant is a function of the temperature and pressure in general. The form of Equation (11.3) is consistent with the conventional method of expressing the equilibrium constant, in that the activities of the products appear in the numerator and those of the reagents in the denominator of the expression for the equilibrium constant. In many cases $\sum_k \nu_k \mu_k^{\ominus}$ is written as $\Delta G^{\ominus}$ to represent the change of the Gibbs energy for the change of state given by the balanced chemical equation when all reagents and products are in their standard states at the given temperature and pressure.

With this general basis we consider the development of specific expressions for different types of chemical reactions. We first consider a homogenous reaction in the gas phase, and use the virial equation of state accurate to the second virial coefficient. The chemical potential of the kth substance is given by Equation (7.72)

$$\mu_k[T, P, y] = RT \ln Py_k + \sum_i \sum_j (2\beta_{ik} - \beta_{ij}) y_i y_j P + \mu_k^{\ominus}[T] \tag{11.5}$$

For equilibrium,

$$\sum_k \nu_k RT \ln(Py_k) + \sum_k \nu_k \sum_i \sum_j (2\beta_{ik} - \beta_{ij}) y_i y_j P = -\sum_k \nu_k \mu_k^{\ominus}[T] \tag{11.6}$$

and the equilibrium constant in terms of partial pressures is given by

$$K_P = \prod_k P_k^{\nu_k} \exp\left[\sum_k \nu_k \sum_i \sum_j (2\beta_{ik} - \beta_{kj}) y_i y_j P/RT\right] \tag{11.7}$$

where $P_k = Py_k$. The constant is a function of the temperature alone, and is independent of the pressure according to Equation (11.6). If we choose the standard state of each substance as the pure substance in the ideal gas state at the pressure P rather than 1 bar, the chemical potential of the kth substance is given by

$$\mu_k[T, P, y] = RT \ln y_k + \sum_i \sum_j (2\beta_{ik} - \beta_{ij}) y_i y_j P + \mu_k^{\ominus}[T, P] \tag{11.8}$$

We then have for equilibrium

$$\sum_k \nu_k RT \ln y_k + \sum_k \nu_k \sum_i \sum_j (2\beta_{ik} - \beta_{ij}) y_i y_j P = -\sum_k \nu_k \mu_k^{\ominus}[T, P] \tag{11.9}$$

The equilibrium constant in terms of mole fractions for the reaction is thus

$$K_y = \prod_k y_k^{\nu_k} \exp\left[\sum_k \nu_k \sum_i \sum_j (2\beta_{ik} - \beta_{ij}) y_i y_j P/RT\right] \tag{11.10}$$

The right-hand side of Equation (11.9) is now a function of both the temperature and the pressure, and consequently so is K_y. The chemical potential of the kth substance, when expressed in terms of concentrations, becomes

$$\mu_k[T, V, n] = RT\left[\ln \frac{n_k}{V} + \sum_i 2\beta_{ik}\left(\frac{n_i}{V}\right)\right] + \mu_k^{\ominus}[T, \tilde{V} = 1, y_k = 1] \tag{11.11}$$

where the standard state is now the pure substance in the ideal gas state at unit molar volume. The equilibrium constant, K_c, is then

$$K_c = \prod_k (n_k/V)^{\nu_k} \exp\left[\sum_k \nu_k \sum_i 2\beta_{ik}(n_i/V)\right] \tag{11.12}$$

and is a function of the temperature alone. The expression for each of the equilibrium constants are somewhat complex for real gases. The simple relations involving the partial pressures, mole fractions, or concentrations alone are applicable only in the approximation of ideal gas behavior as the pressure of the system becomes small.

Expressions for equilibrium constants for homogenous reactions taking place in solution are readily derived by use of the appropriate expressions of the chemical potentials. When mole fractions are used to express the compositions, no difficulty occurs, and the expression for the equilibrium constant for the generalized change of state $\sum_k \nu_k B_k = 0$ is

$$K_x = \prod_k (x_k \gamma_{kx})^{\nu_k} \tag{11.13}$$

where we have used the activity coefficients rather than the excess chemical potentials in order to obtain symmetry in the expression. Questions concerning the solvent arise when we use molalities or molarities, because the expression for the chemical potential of the solvent differs in form from those for the solutes. If the solvent does not take part in the change of state, there is no problem and the equilibrium constants are

$$K_m = \prod_k (m_k \gamma_{km})^{\nu_k} \tag{11.14}$$

in terms of molalities, and

$$K_c = \prod_k (c_k \gamma_{kc})^{\nu_k} \tag{11.15}$$

in terms of molarity. When the solvent does take part in the change of state, its chemical potential occurs in the expression for the condition of equilibrium. Then, if we use Equation (8.102)

$$\mu_1[T, P, m] = -\frac{RTM_1}{1000} \sum_{i \neq 1} m_k + RT \ln \gamma_1 + \mu_1^{\ominus}[T, P] \tag{11.16}$$

for the chemical potential of the solvent, the expression for the equilibrium constant becomes

$$K_m = \prod_{k \neq 1} (m_k)^{\nu_k} \prod_k \gamma_k^{\nu_k} \exp\left[-\frac{\nu_1 RTM_1}{1000} \sum_{i \neq 1} m_i \right] \tag{11.17}$$

If we choose to express the chemical potential of the solvent in terms of the osmotic coefficient, so that (Eq. (8.119))

$$\mu_1[T, P, m] = -\frac{RTM_1}{1000} \phi_m \sum_{i \neq 1} m_i + \mu_i^{\ominus}[T, P] \tag{11.18}$$

The expression for the equilibrium constant is

$$K_m = \prod_{k \neq 1} (m_k \gamma_k)^{\nu_k} \exp\left[-\frac{RTM_1}{1000} \phi_k \sum_{i \neq 1} m_1 \right] \tag{11.19}$$

Expressions similar to Equations (11.17) and (11.19) are obtained when molarities are used rather than molalities (refer to Eqs. (8.118) and (8.120)). In each case $\sum_k \nu_k \mu_k^{\ominus}$ is a function of the temperature and pressure, and therefore each equilibrium constant is also a function of the same variables.

The condition of equilibrium is also applicable to changes of state that involve heterogenous reactions, and the same methods used for homogenous reactions to obtain expressions of the equilibrium constant are used for heterogenous reactions. One difference is that in many heterogenous reactions one or more of the substances taking part in the change of state is a pure phase at equilibrium. In such cases the standard state of the substance is chosen as the pure phase at the experimental temperature and pressure. The chemical potential of the pure substance in its standard state still appears in $\sum_k \nu_k \mu_k^{\ominus}$, but the activity of the substance is unity and its activity does not appear in the expression for the equilibrium constant.

11.2 The experimental determination of equilibrium constants

Two main problems arise in the experimental determination of thermodynamic equilibrium constants. The first, and probably the simplest, is the determination of the concentrations of all of the reacting species at equilibrium. Several general principles are applicable in such determinations. First, the chemistry of the system must be known, so that we have knowledge of all the species present in the equilibrium system. The condition of equilibrium is applicable to every independent chemical reaction that occurs in the system. Second, the equilibrium system is generally a closed system, so the number of moles of every component used to form the system is a constant. Then the equations expressing mass balances must be satisfied for every component. If ions or charged particles are present in the system, the condition of electroneutrality must be satisfied because no system has a net charge at equilibrium. The experimental determination of the concentrations of the reacting species then requires the determination, directly or indirectly, of the concentrations of a sufficient number of species, so that the concentrations of all of the other species may be calculated by use of the mass balance and electroneutrality equations. In the case of gases, an equation of state must also be satisfied.

The second problem is much more difficult because it requires the determination or knowledge of the excess chemical potentials or activity coefficients of *all* of the species in the reacting mixture. For reactions in the gas phase the problem is not too difficult. If the total pressure is sufficiently small, the system might be assumed to be a mixture of ideal gases, so that the thermodynamic equilibrium constant is obtained as $\prod_k P_k^{\nu_k}$. Otherwise, a series of experiments might be made at different total pressures and values of the nonthermodynamic constant, $\prod_k P_k^{\nu_k}$, obtained at each pressure. The thermodynamic equilibrium constant is then obtained as the limiting value of the nonthermodynamic constant as the total pressure goes to zero. When it is necessary to work at higher pressures, values of the virial coefficients appearing in Equations (11.10) or similar quantities in other equations of state must be known or estimated. Direct experimental studies of the P–V–T properties of appropriate nonreacting mixtures may be made, and the methods for the combination of constants as discussed in Section 7.3 may be used where necessary. For reactions taking place in solution, the nonthermodynamic equilibrium constants are those values of the expressions given in Equations (11.13)–(11.15), (11.17), and (11.19) when the assumption is made that values of the activity coefficients and the osmotic coefficients are all unity. When the reference states of all of the reacting substances are the infinitely dilute solution of all solutes in the solvent, the thermodynamic equilibrium constants can be obtained by determining the limiting value of the nonthermodynamic constant in the infinitely dilute solutions. When the reference states of the reacting substances are the pure substances, extrapolations to a single system in which all activity or osmotic coefficients are unity cannot be made. It is then necessary to determine the values of the coefficients experimentally (when possible), or to estimate them on theoretical grounds. The activity coefficients of many components in binary systems and some in ternary systems have been measured. However, general methods have not been developed for the calculation of the activity coefficients of substances in multicomponent systems from those in binary systems. In some cases it may be necessary to use only the nonthermodynamic equilibrium constants. An appreciation of the effect of activity or osmotic coefficients on the equilibrium constants can come only with experience and with an understanding of the present theories of solutions.

11.3 The effect of temperature on the equilibrium constant

The dependence of the equilibrium constant on the temperature is easily obtained from Equation (11.4) with the use of the relation (Eq. (4.41))

$$\left[\frac{\partial(G/T)}{\partial T}\right]_{P,n} = -\frac{H}{T^2}$$

From Equation (11.4)

$$\left(\frac{\partial(\Delta G^{\ominus}/T)}{\partial T}\right)_P = \left(\frac{\partial(\sum_k \nu_k \mu_k^{\ominus}/T)}{\partial T}\right)_P = -R\left(\frac{\partial \ln K}{\partial T}\right)_P \tag{11.20}$$

However,

$$\left(\frac{\partial(\sum_k \nu_k \mu_k^{\ominus}/T)}{\partial T}\right)_P = \sum_k \nu_k \left(\frac{\partial \mu_k^{\ominus}/T}{\partial T}\right)_P = -\sum_k \nu_k \bar{H}_k^{\ominus}/T^2 \tag{11.21}$$

On combining these equations,

$$\left(\frac{\partial \ln K}{\partial T}\right)_P = \frac{\sum_k \nu_k \bar{H}^{\ominus}}{RT^2} = \frac{\Delta \bar{H}^{\ominus}}{RT^2} \tag{11.22}$$

The quantity $\Delta \bar{H}^{\ominus}$ is the change of enthalpy for the change of state represented by a balanced chemical equation under the condition that all substances are in their standard states at the temperature and pressure in question. The overbar is used here because some of the individual enthalpies, $\bar{H}_k^{\ominus}$, may be partial molar quantities.

Equation (11.22) may be written as

$$\left(\frac{\partial \ln K}{\partial(1/T)}\right)_P = -\frac{\Delta \bar{H}^{\ominus}}{R} \tag{11.23}$$

when the independent variable is changed from T to $1/T$. According to this equation, R times the slope of the curve $\ln K$ versus $1/T$ gives the negative value of $\Delta\bar{H}^\ominus$ at the given pressure and at the temperature at which the slope is determined.

The integration of Equation (11.22) to determine the equilibrium constant as a function of the temperature or to determine its value at one temperature with the knowledge of its value at another temperature is very similar to the integration of the Clausius–Clapeyron equation as discussed in Section 10.2. The quantity $\Delta\bar{H}^\ominus$ must be known as a function of the temperature. This in turn may be determined from the change in the heat capacity for the change of state represented by the balanced chemical equation with the condition that all substances involved are in their standard states.

One complication arises when a change of phase occurs within the range of the temperature variation. In such a case the integration must be carried out in at least two steps, the break occurring when the two states of aggregation are in equilibrium. At this point the change in the heat capacity and the change in the enthalpy are discontinuous, whereas the change of the Gibbs energy is continuous. The problem is very similar to that discussed in Section 9.7 with the addition that $\Delta G^\ominus$, and consequently K, has the same value at the transition point irrespective of the state of aggregation. As a generalized example consider the problem of obtaining $\Delta G^\ominus$ for the change of state

$$aA[\text{g}, T, 1\text{ bar}] + bB[\text{g}, T, 1\text{ bar}] = mM[\text{g}, T, 1\text{ bar}] + nN[\text{g}, T, 1\text{ bar}]$$

knowing it for the change of state

$$\begin{aligned} &aA[\ell, 298\text{ K}, 1\text{ bar}] + bB[\text{g}, 298\text{ K}, 1\text{ bar}] \\ &\quad = mM[\text{g}, 298\text{ K}, 1\text{ bar}] + nN[\text{g}, 298\text{ K}, 1\text{ bar}] \end{aligned}$$

Let T_A represent the boiling point of A at 1 bar pressure. Knowing $\Delta\bar{H}^\ominus$ as a function of temperature when A is in the liquid state, we may integrate from 298 K to T_A. Then, $\Delta G^\ominus$ is known for the change of state

$$\begin{aligned} &aA[\ell, T_A, 1\text{ bar}] + bB[\text{g}, T_A, 1\text{ bar}] \\ &\quad = mM[\text{g}, T_A, 1\text{ bar}] + nN[\text{g}, T_A, 1\text{ bar}] \end{aligned}$$

We may then add to this change of state the change of state

$$aA[\text{g}, T_A, 1\text{ bar}] = aA[\ell, T_A, 1\text{ bar}]$$

for which ΔG is zero and ΔH is equal to the change of enthalpy on condensation. We then have the change of state

$$\begin{aligned} &aA[\text{g}, T_A, 1\text{ bar}] + bB[\text{g}, T_A, 1\text{ bar}] \\ &\quad = mM[\text{g}, T_A, 1\text{ bar}] + nN[\text{g}, T_A, 1\text{ bar}] \end{aligned}$$

for which $\Delta G^\ominus$ is identical to the change of state at the temperature T_A but involving liquid A, but $\Delta\bar{H}^\ominus$ differs by the change of enthalpy on condensation. These two quantities are used to determine the necessary integration constants for the succeeding integrations. The first of these is the obtaining of $\Delta\bar{H}^\ominus$ as a function of temperature when gaseous A is involved using the corresponding $\Delta\bar{C}_P^\ominus$. Having the correct function for $\Delta\bar{H}^\ominus$, the integration of Equation (11.24) is then carried out between the limits of T_A and T. When several changes of phase occur over the given temperature range, it is necessary to carry out the integration in a series of such steps.

11.4 The effect of pressure on the equilibrium constant

The dependence of the equilibrium constant on pressure for a chemical reaction is easily determined from Equation (11.4) with the aid of Equation (4.38), which gives the change of the Gibbs energy with pressure at constant temperature and constant number of moles. Thus, we have

$$\left(\frac{\partial G^\ominus}{\partial P}\right)_T = \left(\frac{\partial \sum_k \nu_k \mu_k^\ominus}{\partial P}\right)_T = -RT\left(\frac{\partial \ln K}{\partial P}\right)_T \tag{11.24}$$

However,

$$\left(\frac{\partial \sum_k \nu_k \mu_k^\ominus}{\partial P}\right)_T = \sum_k \nu_k \left(\frac{\partial \mu_k^\ominus}{\partial P}\right)_T = \sum_k \nu_k \bar{V}_k^\ominus \tag{11.25}$$

On combining these equations, we obtain

$$\left(\frac{\partial \ln K}{\partial P}\right)_T = \frac{\sum_k \nu_k \bar{V}_k^\ominus}{RT} = -\frac{\Delta\bar{V}^\ominus}{RT} \tag{11.26}$$

Here $\Delta\bar{V}^\ominus$ is the change of volume that occurs for the change of state represented by the balanced chemical equation under standard state conditions at a given temperature and pressure.

The integration of Equation (11.26) follows the methods discussed in Section 11.3. The quantity $\Delta\bar{V}^\ominus$ must be known as a function of pressure, which may be obtained from an equation of state or the compressibility of each substance in its standard state. Again, the integration must be taken in steps whenever a change of phase takes place within the range of the pressure change. This is occasioned by a difference in the compressibilities of a substance in different states of aggregation, and by the change in volume that occurs when such a change of phase takes place. The limits of integration in the various steps must be chosen so that the changes of phase take place under equilibrium conditions, because only then does the chemical potential of the substance have the same value in the two different phases. The change of the standard Gibbs energy is continuous at the transition point, whereas both the volume change and the compressibility are discontinuous.

11.5 The standard Gibbs energy of formation

The calculation of the equilibrium constant for a given chemical reaction from thermodynamic data requires the knowledge of $\Delta G^{\ominus}$ for the change of state represented by the balanced chemical equation under standard state conditions. For its definition, the calculation of $\Delta G^{\ominus}$ would appear to require the knowledge of absolute values of the chemical potential of each reacting substance in its standard state. This problem is avoided by making use of the changes of the Gibbs energy for the formation of a compound from the elements, the compound and the elements being in their standard state, as was done in the case for the change of enthalpy for a change of state involving a chemical reaction. Every chemical equation can be considered as a sum of equations representing the formation of each compound from the elements. Then

$$\Delta G^{\ominus}[T, P] = \sum_k \nu_k \mu_k^{\ominus}[T, P] = \sum_k \nu_k (\Delta \tilde{G}_f^{\ominus})_k[T, P] \tag{11.27}$$

where $(\Delta \tilde{G}_f^{\ominus})_k[T, P]$ represents the molar change of the Gibbs energy for the formation of the kth substance from the elements, all substances being in their standard states at the temperature T and pressure P. (In general, the standard state for gases would still be the ideal gas state at a pressure of 1 bar.) The experimental determination of the standard molar Gibbs energies of formation follows from Equations (11.4) and (11.27). The value of $\Delta G^{\ominus}[T, P]$ for a chemical reaction is obtained from the experimentally determined value of the equilibrium constant; then, if values of all but one of the quantities $(\Delta \tilde{G}_f^{\ominus})_k[T, P]$, in the sum in Equation (11.27) are known, the value of the remaining one can be calculated. Some changes of state involve only the formation of a compound from the elements; in such cases the standard molar Gibbs energy of formation of the compound is obtained directly from the equilibrium constant.

Conventions followed in the compilation of standard molar Gibbs energies of formation are similar to and consistent with those used for the standard molar enthalpies of formation. A standard temperature of 298.15 K and a standard pressure of 1 bar is usually used. Values at other temperatures and pressures are calculated by the methods discussed in Sections 11.3 and 11.4. The state of aggregation of a substance is the *naturally* occurring state at the standard temperature and pressure. When a substance exists in two or more allotropic forms in the solid state, one state is chosen as the standard state; for example, graphite is chosen as the standard form of carbon, rather than diamond. An additional definition must be given to the standard molar Gibbs energy of formation of a solute when the reference state for the solute is taken as the infinitely dilute solution. Here we include the difference between the chemical potential of the solute in its standard state in solution and that of the pure compound in its natural state of aggregation, in addition to the standard molar Gibbs energy of formation of the pure compound. Thus, the standard molar Gibbs energy of formation of solute is given by

$$(\Delta\tilde{G}_f^\ominus[T, P, \text{soln}])_k = (\Delta\tilde{G}_f^\ominus[T, P, \text{pure}])_k + \mu_k^\ominus[T, P, \text{soln}] - \mu_k^\ominus[T, P, \text{pure}] \quad (11.28)$$

where, for purposes of compiling data, the temperature would be 298.15 K and the pressure would be 1 bar.

11.6 The standard Gibbs energy of formation of ions in solution

We consider only aqueous solutions here, but the methods used are applicable to any solvent system. The standard Gibbs energy of formation of a strong electrolyte dissolved in water is obtained according to Equation (11.28). In such solutions the ions are considered as the species and we are concerned with the thermodynamic functions of the ions rather than the component itself. We express the chemical potential of the electrolyte, considered to be $M_{\nu_+}A_{\nu_-}$, in its standard state as

$$\mu_2^\ominus[T, P] = \nu_+\mu_+^\ominus[T, P] + \nu_-\mu_-^\ominus[T, P] \quad (11.29)$$

according to Equation (8.185). Here $\mu_+^\ominus$ and $\mu_-^\ominus$ represent the standard chemical potentials of the positive and negative ions, respectively. The equivalent of Equation (11.29) is

$$[\Delta G_f^\ominus]_2 = \nu_+[\Delta G_f^\ominus]_+ + \nu_-[\Delta G_f^\ominus]_- \quad (11.30)$$

in terms of the standard Gibbs energies of formation. Although values for the individual ions cannot be determined from experimental data, the arbitrary definition of the value of $\Delta G_f^\ominus$ for one ionic species determines the values of all other ionic species. *By convention the standard Gibbs energy of formation of the hydrogen ion is defined as zero at all temperatures and pressures.* Thus, $\Delta G_f^\ominus$ is defined as zero for

$$\tfrac{1}{2}H_2 = H^+ + e^- \quad (11.31)$$

where e^- represents 1 mole of electrons. Standard state conditions are implied but not stated in the equation. The standard state of the electron and its chemical potential is not defined. This creates no problem because every solution must be electrically neutral and Equation (11.31) must be used in conjunction with a change of state for the formation of a negative ion requiring 1 mole of electrons as a reagent. The symbol for the electron and its chemical potential thus cancel on the required addition.

As an example, we consider hydrochloric acid in aqueous solution, for which the standard molar Gibbs energy of formation is known. According to Equation (11.30),

$$\Delta\tilde{G}_f^\ominus(HCl) = \Delta\tilde{G}_f^\ominus(H^+) + \Delta\tilde{G}_f^\ominus(Cl^-) \quad (11.32)$$

We define $\Delta\tilde{G}_f^{\ominus}(H^+)$ as zero and, therefore,

$$\Delta\tilde{G}_f^{\ominus}(Cl^-) = \Delta\tilde{G}_f^{\ominus}(HCl) \tag{11.33}$$

that is, the standard molar Gibbs energy of formation of the chloride ion is identical to that of the hydrochloric acid. The change of state to which $\Delta\tilde{G}_f^{\ominus}(HCl)$ refers is

$$\tfrac{1}{2}H_2 + \tfrac{1}{2}Cl_2 = H^+ + Cl^- \tag{11.34}$$

and that to which $\Delta G_f^{\ominus}(Cl^-)$ refers is

$$\tfrac{1}{2}Cl_2 + e^- = Cl^- \tag{11.35}$$

The addition of Equations (11.31) and (11.35) yields Equation (11.34).

The standard molar Gibbs energies of other ions are obtained by use of the same methods. Thus, for sodium chloride solutions we have

$$\Delta\tilde{G}_f^{\ominus}(NaCl) = \Delta\tilde{G}_f^{\ominus}(Na^+) + \Delta\tilde{G}_f^{\ominus}(Cl^-) \tag{11.36}$$

from which $\Delta\tilde{G}_f^{\ominus}(Na^+)$ can be obtained once $\Delta\tilde{G}_f^{\ominus}(Cl^-)$ is determined. Similarly, for sulfuric acid

$$\Delta\tilde{G}_f^{\ominus}(H_2SO_4) = 2\Delta\tilde{G}_f^{\ominus}(H^+) + \Delta\tilde{G}_f^{\ominus}(SO_4^{2-}) \tag{11.37}$$

when we assume that the only species present are hydrogen ions and sulfate ions in addition to water. Then, $\Delta\tilde{G}_f^{\ominus}(SO_4^{2-}) = \Delta\tilde{G}_f^{\ominus}(H_2SO_4)$. Finally, for sodium sulfate

$$\Delta\tilde{G}_f^{\ominus}(Na_2SO_4) = 2\Delta\tilde{G}_f^{\ominus}(Na^+) + \Delta\tilde{G}_f^{\ominus}(SO_4^{2-}) \tag{11.38}$$

The value of $\Delta\tilde{G}_f^{\ominus}(Na_2SO_4)$ calculated from the values of $\Delta\tilde{G}_f^{\ominus}(Na^+)$ and $\Delta\tilde{G}_f^{\ominus}(SO_4^{2-})$ according to Equation (11.38) must be consistent with the value obtained directly from experimental data. Moreover, it must be consistent with the value determined by

$$\Delta\tilde{G}_f^{\ominus}(Na_2SO_4) = 2\Delta\tilde{G}_f^{\ominus}(NaCl) + \Delta\tilde{G}_f^{\ominus}(H_2SO_4) - 2\Delta\tilde{G}_f^{\ominus}(HCl) \tag{11.39}$$

according to Equations (11.32), (11.36), and (11.38).

11.7 Conventions concerning solvated species

When dissolved in a solvent, some solutes combine with the solvent to form solvated species. The two outstanding examples in aqueous solution are carbon dioxide to form CO_2(aq) (carbonic acid) and ammonia to form NH_3(aq) (ammonium hydroxide). In many cases the equilibrium constant for the reaction is unknown or not known with sufficient accuracy for thermodynamic purposes. Conventions have been established for treating such systems thermodynamically. Here we discuss the carbon dioxide–water system. The methods used are applicable to any solute in any solvent. In addition, we discuss the thermodynamic relations between the hydrogen ion and the hydronium ion, because of the similarity of this problem to that of solvated species.

The extent of the reaction of carbon dioxide with water to form carbonic acid is fairly well known—less than 1%. However, for thermodynamic purposes we make no distinction between the two nonionized species, CO_2 and H_2CO_3. We are thus concerned with the sum of the concentration of these species, a quantity that can be determined experimentally. We must therefore develop the methods used to define the standard state of the combined nonionized species and the standard molar Gibbs energies of formation.

Two methods are possible. In the first method we assume the nonionized species to be carbonic acid, H_2CO_3, and define its molality as the sum of the molalities of the actual nonionized species. If we designate the molalities of the actual species by primes, then the chemical potential of the assumed species is written as

$$\mu(H_2CO_3) = RT \ln m(H_2CO_3) + RT \ln \gamma(H_2CO_3) + \mu^{\ominus}(H_2CO_3) \tag{11.40}$$

where

$$m(H_2CO_3) = m'(H_2CO_3) + m'(CO_2) \tag{11.41}$$

We define the reference state in the usual way, so that $\gamma(H_2CO_3)$ goes to unity as $m(H_2CO_3)$ goes to zero. This definition defines the standard state of the *assumed* species. The change of state for the first ionization of carbonic acid may then be written as

$$H_2CO_3 = H^+ + HCO_3^- \tag{11.42}$$

so that at equilibrium

$$\mu(H^+) + \mu(HCO_3^-) - \mu(H_2CO_3) = 0 \tag{11.43}$$

The expression for the equilibrium constant becomes

$$K_1 = \frac{a(H^+)a(HCO_3^-)}{a(H_2CO_3)} \tag{11.44}$$

in terms of activities.

In the second method we assume that the only nonionized species is carbon dioxide, CO_2, dissolved in water, and we equate its molality to the sum of the molalities of the actual nonionized species, so

$$m(CO_2) = m'(H_2CO_3) + m'(CO_2) \tag{11.45}$$

The chemical potential of the *assumed* species is expressed as usual as

$$\mu(CO_2) = RT \ln m(CO_2) + RT \ln \gamma(CO_2) + \mu^{\ominus}(CO_2) \tag{11.46}$$

and we choose the reference state as the infinitely dilute solution, so that $\gamma(CO_2)$ goes to unity as $m(CO_2)$ goes to zero. The standard state is thus fixed. The change of state for the first ionization is now written as

$$CO_2 + H_2O = H^+ + HCO_3^- \tag{11.47}$$

Then, at equilibrium,

$$\mu(H^+) + \mu(HCO_3^-) - \mu(CO_2) - \mu(H_2O) = 0 \tag{11.48}$$

and the equilibrium constant becomes

$$K_1 = \frac{a(H^+)a(HCO_3^-)}{a(CO_2)a(H_2O)} \tag{11.49}$$

The relations between the chemical potential of the carbonic acid given in Equation (11.40) and that of carbon dioxide given in Equation (11.46), between the corresponding chemical potentials, and between the two equilibrium constants still need to be defined in order to make the two methods consistent. By use of the methods developed in Section 8.15, we can show that

$$\mu(H_2CO_3) = \mu'(H_2CO_3) \tag{11.50}$$

$$\mu(H_2CO_3) = \mu'(CO_2) + \mu'(H_2O) \tag{11.51}$$

and

$$\mu(CO_2) = \mu'(CO_2) \tag{11.52}$$

where the primes refer to the actual species. Equations (11.50) and (11.51) are consistent with the condition of equilibrium for the reaction $H_2O + CO_2 = H_2CO_3$. When we write expressions for the chemical potentials given in Equation (11.50), we obtain

$$\begin{aligned} &RT \ln m(H_2CO_3) + RT \ln \gamma(H_2CO_3) + \mu^{\ominus}(H_2CO_3) \\ &\quad = RT \ln m(H_2CO_3)' + RT \ln \gamma(H_2CO_3)' + \mu^{\ominus}(H_2CO_3)' \end{aligned} \tag{11.53}$$

This equation becomes

$$\begin{aligned} &RT \ln m^{\infty}(H_2CO_3) + \mu^{\ominus}(H_2CO_3) = RT \ln m^{\infty}(H_2CO_3)' \\ &\quad + RT \ln \gamma^{\infty}(H_2CO_3)' + \mu^{\ominus}(H_2CO_3)' \end{aligned} \tag{11.54}$$

in the infinitely dilute solution. We now, in principle, define $\gamma^{\infty}(H_2CO_3)'$ in a manner such that

$$m^{\infty}(H_2CO_3)'\gamma^{\infty}(H_2CO_3)' = m^{\infty}(H_2CO_3) \tag{11.55}$$

in the limit as $m^{\infty}(H_2CO_3)$ goes to zero. Then

$$\mu^{\ominus}(H_2CO_3) = \mu^{\ominus}(H_2CO_3)' \tag{11.56}$$

According to Equation (11.51), we have

$$RT \ln m^{\infty}(H_2CO_3) + \mu^{\ominus}(H_2CO_3) = RT \ln m'(CO_2)^{\infty} + RT \ln \gamma'(CO_2)^{\infty} + \mu^{\ominus}(CO_2)' + \mu^{\ominus}(H_2O) \tag{11.57}$$

in the very dilute solution when $\gamma^{\infty}(H_2CO_3)$ is set equal to unity and $\mu(H_2O)'$ becomes equal to $\mu^{\ominus}(H_2O)$, the chemical potential of the water in its standard state. Here we define $\gamma(CO_2)'$ in a manner such that

$$m^{\infty}(CO_2)'\gamma^{\infty}(CO_2)' = m^{\infty}(H_2CO_3) \tag{11.58}$$

in the limit as $m^{\infty}(H_2CO_3)$ goes to zero. Then

$$\mu^{\ominus}(H_2CO_3) = \mu^{\ominus}(CO_2)' + \mu^{\ominus}(H_2O) \tag{11.59}$$

When the same definitions are applied to Equation (11.52) with Equation (11.58) ($m^{\infty}(CO_2) = m^{\infty}(H_2CO_3)$ by definition), we obtain

$$\mu^{\ominus}(CO_2) = \mu^{\ominus}(CO_2)' \tag{11.60}$$

These definitions of the activity coefficients of the actual species yield relations between the chemical potentials of the actual species and the assumed species for the standard states according to Equations (11.56), (11.59), and (11.60). The combination of Equations (11.56) and (11.59) yields

$$\mu^{\ominus}(H_2CO_3)' = \mu^{\ominus}(CO_2)' + \mu_1^{\ominus}(H_2O) \tag{11.61}$$

The result, then, is that the equilibrium constant for the reaction $H_2O + CO_2 = H_2CO_3$ has been defined as unity. Conversely, we could have originally defined the value of the equilibrium constant to be unity and then show that this definition is required in the definitions of the activity coefficients of the actual species as expressed in Equations (11.55) and (11.58). The combination of Equations (11.59) and (11.60) yields

$$\mu^{\ominus}(H_2CO_3) = \mu^{\ominus}(CO_2) + \mu^{\ominus}(H_2O) \tag{11.62}$$

in terms of the assumed species. The relation

$$\Delta\tilde{G}_f^{\ominus}(H_2CO_3) = \Delta\tilde{G}_f^{\ominus}(CO_2) + \Delta\tilde{G}_f^{\ominus}(H_2O) \tag{11.63}$$

is readily obtained from Equation (11.62), so the difference between the standard molar Gibbs energy of formation of carbonic acid and that of carbon dioxide in aqueous solution is simply the standard molar Gibbs energy of formation of water. Finally, the values of the two equilibrium constants given in Equations (11.44) and (11.49) are identical because of the relation given in Equation (11.62).

The relation between the standard molar Gibbs energy of hydrogen ion and that of hydronium ion is obtained by two methods that differ only in the assumption used. The result is the same. In the first method we choose to define the acid species either as hydrogen ion or as hydronium ion, and we do not consider an equilibrium between the two species and the solvent. In this case the chemical potential of the hydrogen ion is related to that of the hydronium ion in aqueous solution by

$$\mu(H^+) = \mu(H_3O^+) - \mu(H_2O) \tag{11.64}$$

with the use of the methods discussed in Section 8.15. We thus can write

$$\begin{aligned} &RT \ln m(H^+) + RT \ln \gamma(H^+) + \mu^\ominus(H^+) \\ &\quad = RT \ln m(H_3O^+) + RT \ln \gamma(H_3O^+) \\ &\qquad + \mu^\ominus(H_3O^+) + \frac{RTM(H_2O)}{1000} \sum_{i \neq 1} m_i - RT \ln \gamma(H_2O) - \mu^\ominus(H_2O) \end{aligned} \tag{11.65}$$

The molality of the hydrogen ion must be equal to the molality of the hydronium ion and both $\gamma(H^+)$ and $\gamma(H_3O^+)$ are defined as going to unity as the molality goes to zero. Therefore,

$$\mu^\ominus(H^+) = \mu^\ominus(H_3O^+) - \mu^\ominus(H_2O) \tag{11.66}$$

and, in terms of the standard Gibbs energies of formation,

$$\Delta\tilde{G}_f^\ominus(H_3O^+) = \Delta\tilde{G}_f^\ominus(H^+) + \Delta\tilde{G}_f^\ominus(H_2O) \tag{11.67}$$

When the standard Gibbs energy of formation of the hydrogen ion is defined as zero, the standard molar Gibbs energy of hydronium ion must equal that of water. The change of state related to $\Delta\tilde{G}_f^\ominus(H_3O^+)$ is

$$\tfrac{3}{2}H_2 + \tfrac{1}{2}O_2 = H_3O^+ + e^- \tag{11.68}$$

Equation (11.66) and, consequently, Equation (11.67) can be obtained on the assumption of an equilibrium between hydrogen ion and hydronium ion with water by the same methods used for carbonic acid. We assume the acid species to be either hydrogen ion or hydronium ion at a molality equal to the sum of the actual molalities of the two species, so

$$m(H^+) = m(H^+)' + m(H_3O^+)' \tag{11.69}$$

or

$$m(H_3O^+) = m(H^+)' + m(H_3O^+) \tag{11.70}$$

where primes are used to denote quantities related to the actual species. The relations between the chemical potentials of the assumed species and those of the actual species are

$$\mu(H^+) = \mu'(H^+) \tag{11.71}$$

or

$$\mu(H_3O^+) = \mu'(H_3O^+) \tag{11.72}$$

together with

$$\mu(H^+) = \mu'(H_3O^+) - \mu'(H_2O) \tag{11.73}$$

which is obtained from Equation (11.71) by use of the condition of equilibrium. Equations (11.71) and (11.72) become

$$\begin{aligned} RT \ln m(H^+) + RT \ln \gamma(H^+) + \mu^\ominus(H^+) \\ = RT \ln m(H^+)' + RT \ln \gamma(H^+)' + \mu^\ominus(H^+)' \end{aligned} \tag{11.74}$$

and

$$\begin{aligned} RT \ln m(H_3O^+) + RT \ln \gamma(H_3O^+) + \mu^\ominus(H_3O^+) \\ = RT \ln m(H_3O^+)' + RT \ln \gamma(H_3O^+)' + \mu^\ominus(H_3O^+)' \end{aligned} \tag{11.75}$$

when written in terms of the molalities. We now let $\gamma(H^+)$ and $\gamma(H_3O^+)$ approach unity as $m(H^+)$ and $m(H_3O^+)$ approach zero, and define $\gamma(H^+)'$ such that $m(H^+)'\gamma(H^+)'$ approaches $m(H^+)$ and $m(H_3O^+)'\gamma(H_3O^+)'$ approaches $m(H_3O^+)$ as $m(H^+)$ and $m(H_3O^+)$ approach zero. Then, we have

$$\mu^\ominus(H^+) = \mu^\ominus(H^+)' \tag{11.76}$$

and

$$\mu^\ominus(H_3O^+) = \mu^\ominus(H_3O^+)' \tag{11.77}$$

We also obtain

$$\mu^\ominus(H^+) = \mu^\ominus(H_3O^+)' - \mu^\ominus(H_2O) \tag{11.78}$$

from Equation (11.73) by the use of the same definitions. Appropriate combination of Equations (11.76)–(11.78) yields Equation (11.66) and

$$\mu^\ominus(H^+)' = \mu^\ominus(H_3O^+)' - \mu^\ominus(H_2O) \tag{11.79}$$

Equation (11.79) indicates that the definitions of $\gamma(H^+)'$ and $\gamma(H_3O^+)'$ given here result in assigning the value of *unity* to the equilibrium constant for the change of state

$$H^+ + H_2O = H_3O^+ \tag{11.80}$$

11.8 Ionization constants of weak acids

The ionization of a weak acid, HA, in aqueous solution can be written as

$$HA = H^+ + A^- \tag{11.81}$$

or

$$HA + H_2O = H_3O^+ + A^- \tag{11.82}$$

depending on the way in which we choose to consider the acid species. The change of the standard Gibbs energy on ionization is

$$\Delta G^\ominus = \Delta G_f^\ominus[A^-] - \Delta G_f^\ominus[HA] \tag{11.83}$$

for Equation (11.81) and

$$\Delta G^\ominus = \Delta G_f^\ominus[H_3O^+] + \Delta G_f^\ominus[A^-] - \Delta G_f^\ominus[HA] - \Delta G_f^\ominus[H_2O] \tag{11.84}$$

for Equation (11.82). Thus, the values of the ionization constants for the two equilibria given in Equations (11.81) and (11.82) are the same according to the conventions established in Section 11.7. The expressions for the ionization constants are

$$K = \frac{m(H^+)m(A^-)\gamma(H^+)\gamma(A^-)}{m(HA)\gamma(HA)} \tag{11.85}$$

and

$$K = \frac{m(H_3O^+)m(A^-)\gamma(H_3O^+)\gamma(A^-)}{m(HA)\gamma(HA)a(H_2O)} \tag{11.86}$$

The molalities and the activity coefficients of the anion and of the nonionized acid must be equal, respectively, in the two expressions. We have chosen also to make $m(H^+)$ and $m(H_3O^+)$ equal. Then the activity coefficients of the hydrogen and hydronium ions must be related by the equation[1]

$$\gamma(H^+) = \frac{\gamma(H_3O^+)}{a(H_2O)} \tag{11.87}$$

If we choose to combine the activity coefficients of the ions to obtain the mean activity coefficients, Equation (11.87) would become

$$(\gamma_\pm)^2_{H^+} = \frac{(\gamma_\pm)^2_{H_3O^+}}{a_{H_2O}} \tag{11.88}$$

[1] We have assumed throughout this discussion that the formula for the hydronium ion is H_3O^+. If the formula were written more generally as $H^+ \cdot \nu H_2O$, then the specific equations discussed here would be changed, but the arguments used to develop the equations would be the same. In fact, there is considerable evidence to indicate that the predominant species in water is $H^+ \cdot 4H_2O$.

where the subscripts are used to designate the formula assumed for the acid species. (We use subscripts in Equation (11.88) for clarity of presentation.) According to either equation, the two activity coefficients approach each other in dilute solution as the activity of water approaches unity.

The difficulties of considering solvated hydrogen ions increase when a weak acid is dissolved in a mixed solvent of two or more components, several of which form solvated species with the hydrogen ion. If, or when, the individual species can be identified and their concentrations determined, the normal thermodynamic methods in terms of all of the independent equilibria involved could be used without any assumptions. Without this information, the assumption of unsolvated acid species is justified and most convenient.

11.9 Aqueous solutions of sulfuric acid

The discussion in the previous sections concerning solvated species indicates that a complete knowledge of the chemical reactions that take place in a system is not necessary in order to apply thermodynamics to that system, provided that the assumptions made are applied consistently. The application of thermodynamics to sulfuric acid in aqueous solution affords another illustration of this fact. We choose the reference state of sulfuric acid to be the infinitely dilute solution. However, because we know that sulfuric acid is dissociated in aqueous solution, we must express the chemical potential in terms of the dissociation products rather than the component (Sect. 8.15). Either we can assume that the only solute species present are hydrogen ion and sulfate ion (we choose to designate the acid species as hydrogen rather than hydronium ion), or we can take into account the weak character of the bisulfate ion and assume that the species are hydrogen ion, bisulfate ion, and sulfate ion. With the first assumption, the effect of the weakness of the bisulfate ion is contained in the mean activity coefficient of the sulfuric acid, whereas with the second assumption, the ionization constant of the bisulfate ion is involved indirectly.

With the first assumption, the chemical potential of the sulfuric acid is written as

$$\mu(H_2SO_4) = 2\mu(H^+) + \mu(SO_4^{2-}) \tag{11.89}$$

where the chemical potentials of the ions are to be expressed in terms of the stoichiometric molality of the sulfuric acid. With the second assumption, the chemical potential of the sulfuric acid is written either as Equation (11.89) or as

$$\mu(H_2SO_4) = \mu(H^+) + \mu(HSO_4^-) \tag{11.90}$$

where in each case the chemical potentials of the ions are to be expressed in terms of the actual molalities of the ionic species. Three equations can then be written for the chemical potential of sulfuric acid. These are

$$\mu(H_2SO_4)) = RT\ln(4m^3) + RT\ln(\gamma_\pm^3) + 2\mu^\ominus(H^+) + \mu^\ominus(SO_4^{2-}) \tag{11.91}$$

$$\mu(H_2SO_4) = RT\ln[m(H^+)']^2[m(SO_4^{2-})'] + RT\ln[\gamma(H^+)']^2[\gamma(SO_4^{2-})'] + 2\mu^\ominus(H^+) + \mu^\ominus(SO_4^{2-}) \tag{11.92}$$

$$\mu(H_2SO_4) = RT\ln[m(H^+)'][m(HSO_4^-)'] + RT\ln[\gamma(H^+)'][\gamma(HSO_4^-)'] + \mu^\ominus(H^+) + \mu^\ominus(HSO_4^-) \tag{11.93}$$

where the primed quantities refer to the actual species. The three equations must be mutually consistent if we wish to use them interchangeably. We have therefore chosen the standard chemical potentials of the hydrogen ion in the three equations to be equal, and those of the sulfate ion in the first two equations. There are then seven undefined quantities—the three standard chemical potentials and four activity coefficients—which are related by the three equations. We can then define the four activity coefficients in terms of the usual reference state by making each one go to unity as the molality of the sulfuric acid goes to zero. With these definitions Equations (11.91) and (11.92) become identical in the reference state.

The equating of Equations (11.92) and (11.93) yields

$$RT\ln\frac{[m(H^+)'][m(SO_4^{2-})']}{[m(HSO_4^-)']} + RT\ln\frac{[\gamma(H^+)'][\gamma(SO_4^{2-})']}{[\gamma(HSO_4^-)']} = -(\mu^\ominus(H^+) + \mu^\ominus(SO_4^{2-}) - \mu^\ominus(HSO_4^-)) = -RT\ln K \tag{11.94}$$

which is the expression for the ionization constant of the bisulfate ion. The thermodynamic value of the ionization constant cannot be obtained experimentally without knowledge of the generalized product of the activity coefficients. However, extrapolation to zero molality of the approximate values obtained by assuming that the value of the product of the activity coefficients is unity yields the thermodynamic value of the constant.

The mean activity coefficient of sulfuric acid is usually calculated in terms of Equation (11.91), where the weakness of the bisulfate ion is ignored. The relationship between the various activity coefficients when the incomplete ionization of the ion is included, and when it is not, is now readily obtained by the combination of the appropriate equations. Thus, when Equations (11.91) and (11.92) are equated,

$$[m(H^+)'][m(HSO_4^-)'][\gamma(H^+)'][\gamma(HSO_+^-)'] = \frac{4m^3\gamma_\pm^3}{K} \tag{11.95}$$

is obtained. Here the contribution of the ionization constant is self-evident.

11.10 Nonstoichiometric solid solutions or compounds

Nonstoichiometric solid solutions are substances whose composition approximates that of stoichiometric compounds, but which have a range of compositions. The problem of applying thermodynamics to such substances is primarily how to express the composition of the solution. The simplest choice would be to use the mole fractions or atom fractions in terms of the components. In such a case the effects of the formation of the compound from the components would be contained in the values of the activity coefficients or excess chemical potentials.

In some systems, particularly metal oxides or nitrides, different states of oxidation of the metal or metals could be assumed or actually determined. Expressions for equilibrium constants related to reactions between the atoms in different oxidation states could be set up in terms of the mole fractions of the reacting species. The expressions for the chemical potentials could also be written in terms of these mole fractions. As an example, consider the substance $U_{1-y}Pu_yO_{2-x}$. The question might be to determine how the pressure of oxygen varies with the value of x at constant temperature and constant y. We assume that the uranium is all in the oxidation state of $+4$ and that the plutonium exists in the $+3$ and $+4$ oxidation states for positive values of x. The equilibrium change of state is

$$Pu^{3+} + \tfrac{1}{4}O_2[g] = Pu^{4+} + \tfrac{1}{2}O^{2-}$$

By the conditions of mass balance and electroneutrality, we determine the mole fractions to be

$$x(Pu^{3+}) = 2x/(3-x) \tag{11.96}$$

$$x(Pu^{4+}) = (y-2x)/(3-x) \tag{11.97}$$

$$x(U^{4+}) = (1-y)/(3-x) \tag{11.98}$$

$$x(O^{2-}) = (2-x)/(3-x) \tag{11.99}$$

With the mole fractions and the expression for the nonthermodynamic equilibrium constant for the postulated reaction, we can obtain the relation between the pressure of oxygen and x.

For the expression for the chemical potentials, we choose to use U^{4+} as an example. We may write

$$\mu[U^{4+}, T, P] = RT\ln[(1-y)/(3-x)] + \Delta\mu^E[U^{4+}, T, P] + c[T, P] \tag{11.100}$$

where c is a constant at constant T and P. We may choose to take the standard state to be the stoichiometric compound, $U_{1-y}Pu_yO_2$. We may then write an equation similar to Equation (11.100) as

$$\mu[\mathrm{U}^{4+} \text{ in } \mathrm{U}_{1-y}\mathrm{Pu}_y\mathrm{O}_2, T, P] = RT\ln[(1-y)/3] + \Delta\mu^{\mathrm{E}}[\mathrm{U}^{4+} \text{ in } \mathrm{U}_{1-y}\mathrm{Pu}_y\mathrm{O}_2, T, P] + c[T, P] \tag{11.101}$$

With the elimination of the constant,

$$\mu[\mathrm{U}^{4+}, T, P] = RT\ln[3/(3-x)] + (\Delta\mu^{\mathrm{E}}[\mathrm{U}^{4+}, T, P] - \Delta\mu^{\mathrm{E}}[\mathrm{U}^{4+} \text{ in } \mathrm{U}_{1-y}\mathrm{Pu}_y\mathrm{O}_2, T, P]) + \mu[\mathrm{U}^{4+} \text{ in } \mathrm{U}_{1-y}\mathrm{Pu}_y\mathrm{O}_2, T, P] \tag{11.102}$$

A suggested third method might be based on the concept of defects in the crystal lattice. Whatever methods are used may be chosen for convenience, but must be consistent with the chemistry or assumed chemistry of the system and with the particular problem of interest.

11.11 Association and complex formation in condensed phases

In many solutions strong interactions may occur between like molecules to form polymeric species, or between unlike molecules to form new compounds or complexes. Such new species are formed in solution or are present in the pure substance and usually cannot be separated from the solution. Basically, thermodynamics is not concerned with detailed knowledge of the species present in a system; indeed, it is sufficient as well as necessary to define the state of a system in terms of the mole numbers of the components and the two other required variables. We can make use of the expressions for the chemical potentials in terms of the components. In so doing all deviations from ideal behavior, whether the deviations are caused by the formation of new species or by the intermolecular forces operating between the molecules, are included in the excess chemical potentials. However, additional information concerning the formation of new species and the equilibrium constants involved may be obtained on the basis of certain assumptions when the experimental data are treated in terms of species. *The fact that the data may be explained thermodynamically in terms of species is no proof of their existence. Extra-thermodynamic studies are required for the proof.*

We consider only binary solutions in this discussion. The standard states of the two components are defined as the pure components, and the chemical potentials of the components are based on the molecular mass of the monomeric species. We designate the components by the subscripts 1 and 2 and the monomeric species by the subscripts A_1 and B_1, respectively. From the discussion given in Section 8.15 we know that the chemical potential of a substance considered in terms of the species present in a solution must be equal to the chemical potential of the same substance in terms of the component. Then, the basic equations relating the chemical potential of the components to the species are

$$\mu_1[T, P, x] = \mu_{A_1}[T, P, x] \tag{11.103}$$

and

$$\mu_2[T, P, x] = \mu_{B_1}[T, P, x] \tag{11.104}$$

We consider only the first of these two equations, because the development of the necessary relations for the two components are identical. We express each of the chemical potentials in terms of the mole fractions, so Equation (11.103) becomes

$$RT \ln x_1 + \Delta\mu_1^{\mathrm{E}} + \mu_1^{\ominus} = RT \ln x_{A_1} + \Delta\mu_{A_1}^{\mathrm{E}} + \mu_{A_1}^{\ominus} \tag{11.105}$$

The chemical potentials in the two standard states are related by the conditions that the chemical potential of the monomeric species in the pure component must equal the chemical potential of the pure component. Then

$$\mu_1^{\ominus} = \mu_{A_1}^* = RT \ln x_{A_1}^* + (\Delta\mu_{A_1}^{\mathrm{E}})^* + \mu_{A_1}^{\ominus} \tag{11.106}$$

where the asterisks refer to the monomeric species in the pure liquid component. Equation (11.105) is now written as

$$\Delta\mu_1^{\mathrm{E}} = RT \ln \frac{x_{A_1}}{x_1 x_{A_1}^*} + \Delta\mu_{A_1}^{\mathrm{E}} - (\Delta\mu_{A_1}^{\mathrm{E}})^* \tag{11.107}$$

Thermodynamic studies yield values of $\Delta\mu_1^{\mathrm{E}}$ as a function of the mole fraction in terms of the components at constant temperature and pressure. We must now assume which species are present, basing the assumptions on our knowledge of the chemical behavior of the component and of the system as a whole, and values of the equilibrium constants relating to the formation of the species from the components. The most convenient independent variable is the mole fraction of the monomeric species, x_{A_1}. A sufficient number of equations to calculate all mole fractions of the species and the components in terms of x_{A_1} are obtained from the expressions for the assumed independent equilibrium constants and the fact that the sum of the mole fractions in terms of species must equal unity. The assumed values of the equilibrium constants are adjusted and the species changed until Equation (11.107) is satisfied. The equivalent equation for the second component must also be satisfied.

For exact calculations, values of the excess chemical potentials of all the species must be known or calculable on some theoretical basis. Unfortunately, this is not generally the case. Under such circumstances it is convenient to assume that the species form an ideal solution and that all deviations based on the components result from the formation of the new species. Such solutions are called *quasi-ideal solutions* or ideal solutions in terms of species.

Several examples of both association and complex formation are discussed in detail in the following paragraphs. Quasi-ideal solutions are assumed in all cases. First we assume that the first component exists in monomeric and dimeric forms only. The equilibrium reaction is

$$2A_1 = A_2 \tag{11.108}$$

for which the expression for the equilibrium constant is

$$K = x_{A_2}/x_{A_1}^2 \tag{11.109}$$

The sum of the mole fractions of the species must equal unity, so

$$x_{A_1} + x_{A_2} + x_{B_1} = 1 \tag{11.110}$$

Then the value of x_{B_1} is given by

$$x_{B_1} = 1 - x_{A_1} - Kx_{A_1}^2 \tag{11.111}$$

with Equation (11.109). We consider the system as one that contains 1 mole of solution in terms of the species. Then the number of moles of the first component contained in the system is

$$n_1 = x_{A_1} + 2x_{A_2} \tag{11.112}$$

$$n_1 = x_{A_1} + 2Kx_{A_1}^2 \tag{11.113}$$

and that of the second component is:

$$n_2 = x_{B_1} = 1 - x_{A_1} - Kx_{A_1}^2 \tag{11.114}$$

The mole fraction of the first component is obtained from

$$x_1 = \frac{x_{A_1} + Kx_{A_1}^2}{1 + Kx_{A_1}^2} \tag{11.115}$$

The activity coefficient of the first component is equal to $x_{A_1}/x_1x_{A_1}^*$ according to Equation (11.107) and, therefore,

$$\gamma_1 = \frac{x_{A_1}}{x_1x_{A_1}^*} = \frac{1 + Kx_{A_1}^2}{(1 + 2Kx_{A_1})x_{A_1}^*} \tag{11.116}$$

By the use of similar methods we find that

$$\gamma_2 = \frac{x_{B_1}}{x_2} = 1 + Kx_{A_1}^2 \tag{11.117}$$

Here we see that, according to Equations (11.115)–(11.117), γ_1 and γ_2 can be calculated as a function of x_1 for various values of K with the use of x_{A_1} as the independent variable. According to Equation (11.117), γ_2 is always greater than unity because both K and $x^2_{A_1}$ are positive. Thus, deviations from ideality in terms of the components are always positive. In the limit as x_1 approaches unity, γ_1 approaches unity and γ_2 approaches $[1 + K(x^*_{A_1})^2]$. In the limit as x_1 approaches zero, γ_1 approaches $1/x^*_{A_1}$ and γ_2 approaches unity.

A second example is one in which we assume that the first component forms of continuous series of polymers, called *chain association*, according to the series of equations

$$\begin{aligned} 2A_1 &= A_2 \\ A_1 + A_2 &= A_3 \\ A_1 + A_3 &= A_4 \\ A_1 + A_4 &= A_5 \\ &\vdots \end{aligned} \tag{11.118}$$

The assumption is usually made that the values of the equilibrium constants for the reactions given in Equation (11.118) are all equal, for reasons of simplicity. With this assumption, we have the set of equations

$$K = \frac{x_{A_2}}{x^2_{A_1}} = \frac{x_{A_3}}{x_{A_1}x_{A_2}} = \frac{x_{A_4}}{x_{A_1}x_{A_3}} = \cdots \tag{11.119}$$

The value of x_{B_1} is given by

$$x_{B_1} = 1 - \sum_{i=1}^{\infty} x_{A_i} = 1 - x_{A_1} \sum_{i=0}^{\infty} K^i x^i_{A_1} \tag{11.120}$$

where the sum is taken to infinity. Now

$$\sum_{i=0}^{\infty} K^i x^i_{A_1} = \frac{1}{1 - Kx_{A_1}} \tag{11.121}$$

so

$$x_{B_1} = \frac{1 - x_{A_1} - Kx_{A_1}}{1 - Kx_{A_1}} \tag{11.122}$$

The number of moles of the first component contained in 1 mole of solution based on the species is

$$n_1 = x_{A_1} + 2x_{A_2} + 3x_{A_3} + \cdots \tag{11.123}$$

$$= x_{A_1}(1 + 2Kx_{A_1} + 3K^2x^2_{A_1} + \cdots) \tag{11.124}$$

$$= \frac{x_{A_1}}{(1 - Kx_{A_1})^2} \tag{11.125}$$

and that of the second component is

$$n_2 = x_{B_1} \tag{11.126}$$

The mole fraction of the first component is then

$$x_1 = \frac{x_{A_1}}{x_{A_1} + (1 - x_{A_1} - Kx_{A_1})(1 - Kx_{A_1})} \tag{11.127}$$

and that of the second component is

$$x_2 = \frac{(1 - x_{A_1} - Kx_{A_1})(1 - Kx_{A_1})}{x_{A_1} + (1 - x_{A_1} - Kx_{A_1})(1 - Kx_{A_1})} \tag{11.128}$$

The activity coefficient of the first component is given by

$$\gamma_1 = [x_{A_1} + (1 - x_{A_1} - Kx_{A_1})(1 - Kx_{A_1})](1 + K) \tag{11.129}$$

according to Equation (11.107), where

$$x^*_{A_1} = 1/(1 + K) \tag{11.130}$$

Similarly,

$$\gamma_2 = \frac{x_{A_1} + (1 - x_{A_1} - Kx_{A_1})(1 - Kx_{A_1})}{(1 - Kx_{A_1})^2} \tag{11.131}$$

These values of γ_1, γ_2, and x_1 can be calculated for various values of K with the use of x_{A_1} as an independent variable. The limiting value of γ_2 as x_1 approaches unity is also $(1 + K)$ in the present case. The deviations from ideal behavior on the basis of components are again positive.

The methods used when complexes are formed between the unlike molecules are the same as those discussed in the previous paragraphs. We consider here two cases, one in which only one complex, A_sB_t, is assumed to be present in solution and the other in which two complexes, A_sB_t and A_uB_v, are assumed to be present. The basic equation is again Equation (11.105), but in the cases discussed here $\mu_1^\ominus$ and $\mu_{A_1}^\ominus$ are equal and, therefore,

$$\Delta\mu_1^{\mathrm{E}} = RT \ln(x_A/x_1) + \Delta\mu_A^{\mathrm{E}} \tag{11.132}$$

where we use the subscript A rather than A_1 because we have assumed that polymeric species are not present. We again assume that the solutions are ideal in terms of the species present. The equilibrium reaction for the first case is

$$sA + tB = A_sB_t \tag{11.133}$$

for which

$$K = x_{A_sB_t}/x_A^s x_B^t \tag{11.134}$$

The equation relating x_A and x_B is

$$x_A + x_B + x_{A_sB_t} = x_A + x_B + Kx_A^s x_B^t = 1 \tag{11.135}$$

which is based on the condition that the sum of the mole fractions of the species must equal unity. This equation permits the calculation of values of x_B for various values of x_A when arbitrary values have been assigned to K. The mole number of the first component per mole of solution in terms of species is given by

$$n_1 = x_A + sKx_A^s x_B^t \tag{11.136}$$

and that of the second component by

$$n_2 = x_B + tKx_A^s x_B^t \tag{11.137}$$

Then the mole fractions of the components are given by

$$x_1 = \frac{x_A + sKx_A^s x_B^t}{1 + (s + t - 1)Kx_A^s x_B^t} \tag{11.138}$$

and

$$x_2 = \frac{x_B + tKx_A^s x_B^t}{1 + (s + t - 1)Kx_A^s x_B^t} \tag{11.139}$$

Finally, the activity coefficients are

$$\gamma_1 = \frac{1 + (s + t - 1)Kx_A^s x_B^t}{1 + sKx_A^{s-1} x_B^t} \tag{11.140}$$

and

$$\gamma_2 = \frac{1 + (s + t - 1)Kx_A^s x_B^t}{1 + tKx_A^s x_B^{t-1}} \tag{11.141}$$

Thus, values of γ_1 and γ_2 can be calculated as functions of x_1 for arbitrarily assumed values of K by the use of Equations (11.135) and (11.138)–(11.141).

Several interesting relations are derivable from Equations (11.140) and (11.141). In the limit as x_1 approaches zero, γ_1 approaches unity when $s > 1$ and $1/(1 + K)$ when $s = 1$. Thus, when $s > 1$, the limiting value of γ_1 is unity both for $x_1 = 1$ and for $x_1 = 0$. Otherwise, the limiting value of γ_1 for $x_1 = 0$ gives the value of K. The limiting values of γ_2 for $x_1 = 1$ are unity when $t > 1$ and $1/(1 + K)$ when $t = 1$. Also, a plot of the values of γ_1 and γ_2 as functions of x_1 consists of two intersecting curves. At the point of intersection $\gamma_1 = \gamma_2$ and

$$x_A/x_B = s/t \tag{11.142}$$

This ratio of x_A/x_B is obtained only when x_1/x_2 also equals s/t; that is, the composition of the solution in terms of the components is the same as the composition of the complex. Thus, when only one complex is assumed to be present in the solution, the graphs of γ_1 and γ_2 as functions of x_1 yield information concerning the composition of the complex and may yield values of the equilibrium constant for the formation of the complex.

The effect of the formation of complex on the activity coefficients is illustrated for three different complexes in Figures 11.1–11.3. The value of the equilibrium constant for each of the three systems has been chosen as 10. For Figure 11.1, the complex AB has been assumed. The deviations from ideality are negative over the entire range of composition. The value of γ_1 at $x_1 = 0$ and that of γ_2 at $x_2 = 0$ is $1/(1 + K)$, and the two curves intersect at a mole fraction of 0.5, in accordance with Equation (11.142). The complex A_3B has been assumed for Figure 11.2. For this system the values of γ_1 are negative over the entire range of composition, whereas γ_2 has both positive and negative values. The value of γ_1 at both $x_1 = 0$ and $x_1 = 1$ is unity. The value of γ_2 at $x_2 = 0$ is $1/(1 + K)$. The two curves intersect at $x_1 = 0.75$. For Figure 11.3 the complex A_3B_3 has been assumed. In this case both γ_1 and γ_2 have positive and negative values and the values at $x_1 = 0$ and $x_1 = 1$ are one for both components. The two curves intersect at the mole fraction of 0.5.

The behavior of systems are much the same when we assume that two

Figure 11.1. Activity coefficients for a binary quasi-ideal solution involving a 1–1 complex.

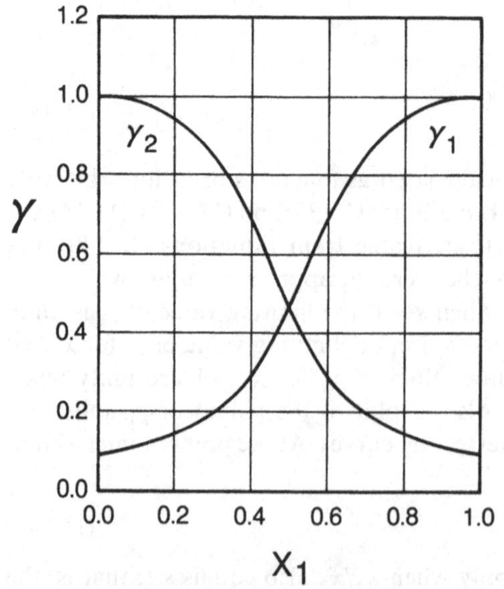

Figure 11.2. Activity coefficients for a binary quasi-ideal solution involving a 3–1 complex.

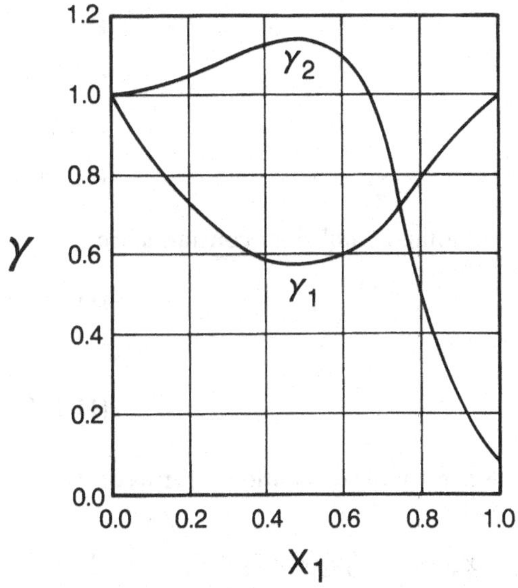

Figure 11.3. Activity coefficients for a binary quasi-ideal solution involving a 3–3 complex.

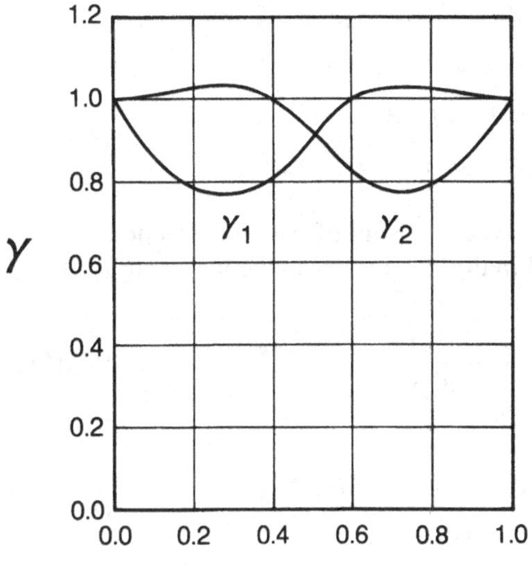

complexes are formed. For discussion we assume that the species present in the solution are A, B, A_sB_t, and A_uB_v. The equilibrium reactions for the formation of the complexes are

$$sA + tB = A_sB_t \tag{11.143}$$

and

$$uA + vB = A_uB_v \tag{11.144}$$

and the expressions for the corresponding equilibrium constants are

$$K_1 = x_{A_sB_t}/x_A^s x_B^t \tag{11.145}$$

and

$$K_2 = x_{A_uB_v}/x_A^u x_B^v \tag{11.146}$$

The mole numbers of the components in 1 mole of solution in terms of species are

$$n_1 = x_A + s x_{A_sB_t} + u x_{A_uB_v} = x_A + sK_1 x_A^s x_B^t + uK_2 x_A^u x_B^v \tag{11.147}$$

and

$$n_2 = x_B + t x_{A_sB_t} + v x_{A_uB_v} = x_B + tK_1 x_A^s x_B^t + vK_2 x_A^u x_B^v \tag{11.148}$$

Then the mole fractions of the components are given by

$$x_1 = \frac{x_A(1 + sK_1 x_A^{s-1} x_B^t + K_2 x_A^{u-1} x_B^v)}{1 + (s + t - 1)K_1 x_A^s x_B^t + (u + v - 1)K_2 x_A^u x_B^v} \tag{11.149}$$

and

$$x_2 = \frac{x_B(1 + tK_1 x_A^s x_B^{t-1} + vK_2 x_A^u x_B^{v-1})}{1 + (s + t - 1)K_1 x_A^s x_B^t + (u + v - 1)K_2 x_A^u x_B^v} \tag{11.150}$$

where we have used the condition that the sum of the mole fractions of the species must be equal to unity. Finally, the expressions for γ_1 and γ_2 are

$$\gamma_1 = \frac{1 + (s + t - 1)K_1 x_A^s x_B^t + (u + v - 1)K_2 x_A^u x_B^v}{1 + sK_1 x_A^{s-1} x_B^t + uK_2 x_A^{u-1} x_B^v} \tag{11.151}$$

and

$$\gamma_2 = \frac{1 + (s + t - 1)K_1 x_A^s x_B^t + (u + v - 1)K_2 x_A^u x_B^v}{1 + tK_1 x_A^s x_B^{t-1} + vK_2 x_A^u x_B^{v-1}} \tag{11.152}$$

The limiting values of γ_1 as x_1 goes to zero are

$$\lim_{x_1 \to 0} \gamma_1 = \begin{cases} \dfrac{1}{1+K_1+K_2} & \text{when } s=1 \text{ and } u=1 \\ \dfrac{1}{1+K_1} & \text{when } s=1 \text{ and } u>1 \\ \dfrac{1}{1+K_2} & \text{when } s>1 \text{ and } u=1 \\ 1 & \text{when } s>1 \text{ and } u>1. \end{cases}$$

Similarly,

$$\lim_{x_1 \to 1} \gamma_2 = \begin{cases} \dfrac{1}{1+K_1+K_2} & \text{when } t=1 \text{ and } v=1 \\ \dfrac{1}{1+K_1} & \text{when } t=1 \text{ and } v>1 \\ \dfrac{1}{1+K_2} & \text{when } t>1 \text{ and } v=1 \\ 1 & \text{when } t>1 \text{ and } v>1 \end{cases}$$

In general the deviations from ideality in terms of components are again negative, as evidenced by the limiting values of the activity coefficients. However, it is not readily apparent from Equations (11.150) and (11.152) whether $\Delta\mu_1^{\mathrm{E}}$ or $\Delta\mu_2^{\mathrm{E}}$ can have positive as well as negative values.

We have discussed the problems concerning association and complex formation separately. There is no reason why species formed by association and by complex formation should not be present in the same solution. The methods used in the thermodynamic treatment of such systems, no matter how complicated, would be the same as those discussed here.

Once the species present in a solution have been chosen and the values of the various equilibrium constants have been determined to give the best fit to the experimental data, other thermodynamic quantities can be evaluated by use of the usual relations. Thus, the excess molar Gibbs energies can be calculated when the values of the excess chemical potentials have been determined. The molar change of enthalpy on mixing and excess molar entropy can be calculated by the appropriate differentiation of the excess Gibbs energy with respect to temperature. These functions depend upon the temperature dependence of the equilibrium constants.

We conclude this section by emphasizing again that, throughout the discussion, the solution has been assumed to be ideal in terms of the species present. This assumption is actually not valid. Although we may be able to estimate the excess chemical potentials of the species in some simple systems, we cannot do so generally. We then need to make the assumption used in this section. The thermodynamic treatment of a solution in terms of species is justified when the species present have been identified by extra-thermodynamic means or when the deviations from ideality are much larger than those related to the intermolecular forces. A knowledge of the behavior of the excess chemical potentials of the components as functions of the mole fraction, as developed here, is helpful, of course, in deciding whether to treat a solution in terms of species or not [31].

11.12 Phase equilibria in terms of species

We may, with appropriate attention to the reference state, develop the relations in terms of components between the intensive variables pertinent to multiphase systems that contain species other than the components. Such relations would be rather complex, because no account would be taken of the effect of the chemical reactions that occur in the system. All deviations from ideality would appear either in the activity coefficients for substances in condensed phases or in the coefficients used in some equations of state for the gas phase. Simpler relations are obtained when the conditions of phase equilibrium are based on species rather than components, once the species have been identified.

The fundamental condition for equilibrium between phases, the equality of the chemical potential of a component in every phase in which the component is present still applies. We then substitute for the chemical potential of a component the equivalent chemical potential of the same substance in terms of species or appropriate sums of chemical potentials of the species, as determined by the methods of Section 8.15 and used in the preceding sections. Several examples are discussed in the following paragraphs.

Many vapors of substances, such as those of certain metals, are mixtures of monomers and dimers and, possibly, of even higher polymers. This must be taken into account when the chemical potential of the substance in a condensed phase, either pure or a solution, is determined by a study of the vapor–liquid equilibrium. We choose to consider only the components in the condensed phase, and use the molecular mass of the monomer to determine the mole numbers. We designate the component whose chemical potential is to be determined by the subscript 1. The condition of equilibrium is then

$$\mu_1[T, P, x] = \mu_A[T, P, y] \tag{11.153}$$

where the subscript A represents the monomeric species in the vapor. The chemical potential of A is expressed in terms of the partial pressure of A. Then

$$\mu_1[T, P, x] = RT \ln P_A + \mu_A^{\ominus}[T, 1 \text{ bar}, y_A = 1] \tag{11.154}$$

where we have assumed, for simplicity, that the vapor follows the ideal gas equation of state in terms of species. When the condensed phase is the pure substance, the condition of equilibrium is

$$\mu_1^{\ominus}[T, P] = \mu_A[T, P, y_A] \tag{11.155}$$

$$\mu_1^{\ominus}[T, P] = RT \ln P_A^{\bullet} + \mu_A^{\ominus}[T, 1 \text{ bar}, y_A = 1] \tag{11.156}$$

where we have again assumed ideal gas behavior. It is to be noted that Raoult's law is approximated, when the effects of total pressure on the condensed phase is ignored, only in terms of the monomeric species. If no monomeric species are present in the vapor but dimeric species are present, the condition of equilibrium becomes

$$\mu_1[T, P, x] = \tfrac{1}{2}\mu_{A_2}[T, P, y_{A_2}] \tag{11.157}$$

where the subscript A_2 represents the dimeric species. In this case the partial pressure of the dimer is used to evaluate the chemical potentials.

The same conditions of equilibrium as given in Equations (11.153) and (11.155) are obtained when the vapor of the substance dissociates incompletely so that some monomeric species are present in the vapor.

When a solute is distributed between two immiscible liquids, different species, formed from the solute, may exist in the two liquids. Thus, when an organic liquid such as benzene or carbon tetrachloride and water are used as the two liquids, a weak acid may dimerize in the organic phase and partially ionize in the aqueous phase. The condition of equilibrium is the equality of the chemical potential of the monomeric, nonionized species in the two phases. If the dimerization is complete, the condition of equilibrium involves half of the chemical potential of the dimer in the organic phase.

An interesting case of solid–liquid equilibrium is one in which a solvent dissociates at least to some extent in the liquid phase and a solute is one of the species formed by the dissociation. We show in Section 10.20 that the experimental temperature–composition curve has a maximum at the composition of the pure solvent. We consider here that the solid phase is the pure, undissociated component, designated by the subscript 1; that this component dissociates in the liquid phase according to the reaction

$$AB = A + B \tag{11.158}$$

and that the second component, the solute in the liquid phase is A. The condition of equilibrium is

$$\mu_{AB}[T, P, x] = \mu_1^{\ominus}[\text{s}, T, P] \tag{11.159}$$

where μ_{AB} represents the chemical potential of the species AB. We assume, again for simplicity, ideal behavior and write

$$RT \ln x_{AB} + \mu^{\ominus}_{AB}[T, P] = \mu^{\ominus}_{1}[\mathrm{s}, T, P] \tag{11.160}$$

We have two alternate developments that depend on whether we choose to use the experimental molar enthalpy of fusion of the solvent or the difference between the molar enthalpy of the pure, liquid species *AB* and that of the solid. In the first case we must introduce the chemical potential of the pure liquid component and eliminate the chemical potential of the *AB* species. In order to do so we consider the pure liquid at the experimental temperature T indicated in Equation (11.160). Then

$$\mu^{\ominus}_{1}[\ell, T, P] = RT \ln x^{\bullet}_{AB} + \mu^{\ominus}_{AB} \tag{11.161}$$

where the solid dot is used to designate the mole fraction of the species *AB* in the pure liquid component. Combination of Equations (11.160) and (11.155) yields

$$RT \ln(x_{AB}/x^{\bullet}_{AB}) = \mu^{\ominus}_{1}[\mathrm{s}, T, P] - \mu^{\ominus}_{1}[\ell, T, P] \tag{11.162}$$

The difference between the two standard chemical potentials is determined in terms of the molar enthalpy of fusion by the methods developed in Section 10.12. For real solutions, the equation corresponding to Equation (11.162) is

$$RT \ln(x_{AB}/x^{\bullet}_{AB}) + (\Delta\mu^{\mathrm{E}}_{AB} - \Delta\mu^{\mathrm{E}\bullet}_{AB}) = \mu^{\ominus}_{1}[\mathrm{s}, T, P] - \mu^{\ominus}_{1}[\ell, T, P] \tag{11.163}$$

This equation requires a knowledge of both x_{AB} and $x^{\bullet}_{AB}$ in order to determine the difference between the excess chemical potentials. The quantity $\Delta\mu^{\mathrm{E}\bullet}_{AB}$ cannot be evaluated generally, and it is the difference $(\Delta\mu^{\mathrm{E}}_{AB} - \Delta\mu^{\mathrm{E}\bullet}_{AB})$ that is determined experimentally.

Equation (11.160) or the corresponding equation

$$RT \ln x_{AB} + \Delta\mu^{\mathrm{E}}_{AB} = \mu^{\ominus}_{1}[\mathrm{s}, T, P] - \mu^{\ominus}_{AB}[T, P] \tag{11.164}$$

for real solutions is used directly in the second case. The difference between the standard chemical potentials is given by

$$\mu^{\ominus}_{1}[\mathrm{s}, T, P] - \mu^{\ominus}_{AB}[T, P] = -T \int_{T}^{T_{\mathrm{mp},AB}} \frac{\tilde{H}^{\bullet}_{AB}(\ell) - \tilde{H}^{\bullet}_{AB}(\mathrm{s})}{T^2}\, \mathrm{d}T \tag{11.165}$$

The difficulty in the use of Equation (11.165) is that both $T_{\mathrm{mp},AB}$, the melting point of the pure species *AB*, and $(\tilde{H}^{\bullet}_{AB}(\ell) - \tilde{H}^{\bullet}_{AB}(\mathrm{s}))$, the difference between the molar enthalpies of the pure species *AB* in the liquid and solid states, can be determined only by extrapolation of the experimental temperature–composition curve. Figure 11.4 illustrates these points. The solid curve represents the experimental curve in the neighborhood of the composition of the species *AB*, and shows a true maximum at this composition. The broken lines represent the extrapolation of the experimental curve consistent with Equation (11.160) or (11.164) and Equation (11.165). The slope of the two branches of the curve must be the same at the composition AB, and this slope yields the value of $(\tilde{H}^{\bullet}_{AB}(\ell) - \tilde{H}^{\bullet}_{AB}(\mathrm{s}))$ at $T_{\mathrm{mp},AB}$. The strict integration of Equation (11.158) requires a knowledge of the molar heat capacity of the solid phase and that of the pure liquid species. This latter quantity cannot be determined experimentally and must be calculated or estimated on some theoretical basis. The choice of which method to use in any given case depends on the complexity of the system and on what information is desired.

A final example concerning phase equilibria in terms of species involves the use of Equation (8.102) or similar equations. We discuss only Equation (8.102) here, which is

$$\mu_1 = -(RTM_1/1000)\sum m + RT \ln \gamma_1 + \mu_1^{\ominus} \tag{11.166}$$

for a solvent that does not dissociate or polymerize. We assume that $(\mu_1 - \mu_1^{\ominus})$ is determined experimentally. This equation can then be used to determine the total molality of solutes, $\sum m$, if the activity coefficient can be estimated or if the solution is sufficiently dilute that it may be set equal to unity. With a knowledge of the total effective molality and the molality of solutes in terms of the components, an estimate of the number of solute species present in the solution is obtained.

11.13 Use of the Gibbs–Duhem equations

The use of the Gibbs–Duhem equations for phase equilibria when no chemical reactions occur in the system is discussed in Section 10.20. Useful expressions for the derivatives of one intensive variable with respect to

Figure 11.4. Alternate curves for the freezing point diagram when a compound dissociates in the liquid phase.

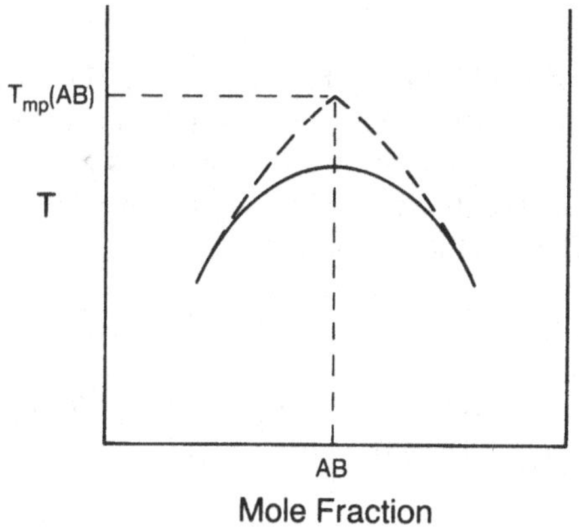

another intensive variable in univariant systems when chemical reactions occur can also be obtained by the solution of the appropriate set of Gibbs–Duhem equations. Two examples are discussed here. In each case we consider the gas–condensed phase equilibria in a binary system at constant temperature. The components are designated by the subscripts 1 and 2.

In the first example we assume that the species in the gas phase are A, B, and A_2B, where A and B represent the same molecular entities as the components 1 and 2, respectively. The Gibbs–Duhem equations are

$$-\tilde{V}(\mathrm{c})\,\mathrm{d}P + x_1\,\mathrm{d}\mu_1 + x_2\,\mathrm{d}\mu_2 = 0 \tag{11.167}$$

and

$$-\tilde{V}(\mathrm{g})\,\mathrm{d}P + y_A\,\mathrm{d}\mu_A + y_B\,\mathrm{d}\mu_B + y_{A_2B}\,\mathrm{d}\mu_{A_2B} = 0 \tag{11.168}$$

These equations are subject to the equilibrium conditions that

$$\mu_{A_2B} - 2\mu_A - \mu_B = 0 \tag{11.169}$$

$$\mu_1 = \mu_A \tag{11.170}$$

and

$$\mu_2 = \mu_B \tag{11.171}$$

There are six variables and five equations, and the system is univariant. We wish to determine the change of the partial pressure of the species A_2B with change of composition of the condensed phase; that is, the derivative $(\partial \ln P_{AB}/\partial x_1)_T$. In the solution of the set of Gibbs–Duhem equations, we therefore must retain μ_{AB} and μ_1 or expressions equivalent to these quantities. The solution can be obtained by first eliminating μ_2 and μ_B from Equations (11.167) and (11.168) by use of Equations (11.169) and (11.171). The expressions

$$\mathrm{d}\mu_1 = -\bar{V}_1(\mathrm{c})\,\mathrm{d}P + (\partial\mu_1/\partial x_1)_{T,P}\,\mathrm{d}x_1 \tag{11.172}$$

and

$$\mathrm{d}\mu_{A_2B} = RT\ln P_{A_2B} \tag{11.173}$$

can then be introduced. (We have assumed ideal gas behavior in Eq. (11.173) for simplicity; for real gases, a term involving dP would be present in Eq. (11.173).) Equations (11.172) and (11.173) introduce the variables x_1 and P_{A_2B}. After the introduction of these equations into the Gibbs–Duhem equations, dP can be eliminated. The resultant equation is

$$\left(\frac{\partial \ln P_{A_2B}}{\partial x_1}\right)_T = -\frac{((y_A - 2y_B)x_2[\bar{V}_2(\mathrm{c}) + 2\bar{V}_1(\mathrm{c})] - (x_1 - 2x_2)\tilde{V}(\mathrm{g}))(\partial\mu_1/\partial x_1)_{T,P}}{x_2RT((y_{A_2B} + y_B)[\bar{V}_2(\mathrm{c}) + 2\bar{V}_1(\mathrm{c})] + (y_A - 2y_B)\bar{V}_1(\mathrm{c}) - \tilde{V}(\mathrm{g}))} \tag{11.174}$$

If the partial molar volumes of the condensed phase are negligibly small with respect to the molar volume of the gas phase, this becomes

$$\left(\frac{\partial \ln P_{A_2B}}{\partial x_1}\right)_T = -\frac{(x_1 - 2x_2)(\partial \mu_1/\partial x_1)_{T,P}}{x_2 RT} \tag{11.175}$$

We see that under these conditions the value of the derivative becomes zero (a maximum) when $x_1 = 2x_2$. The maximum then occurs when the composition of the condensed phase is that of the species A_2B. When the partial molar volumes of the condensed phase are not negligibly small and when the nonideality of the gas is included, the maximum does not occur at this exact composition, but does occur very close to it [32].

In the second example we assume that the species in the gas phase are A, B, and AB, and wish to obtain an expression for $(\partial \ln P_{AB}/\partial \ln P_A)_T$. The Gibbs–Duhem equations are

$$-\tilde{V}(\mathrm{c})\,\mathrm{d}P + x_1\,\mathrm{d}\mu_1 + x_2\,\mathrm{d}\mu_2 = 0 \tag{11.176}$$

and

$$-\tilde{V}(\mathrm{g})\,\mathrm{d}P + y_A\,\mathrm{d}\mu_A + y_B\,\mathrm{d}\mu_B + y_{AB}\,\mathrm{d}\mu_{AB} = 0 \tag{11.177}$$

The condition equations are

$$\mu_{AB} - \mu_A - \mu_B = 0 \tag{11.178}$$

and Equations (11.170) and (11.171). First, $\mathrm{d}\mu_2$ and $\mathrm{d}\mu_B$ are eliminated by use of Equations (11.178) and (11.171). The partial pressures of AB and A are introduced by the equations

$$\mathrm{d}\mu_{AB} = RT\,\mathrm{d}\ln P_{AB} \tag{11.179}$$

and

$$\mathrm{d}\mu_A = \mathrm{d}\mu_1 = RT \ln P_A \tag{11.180}$$

where we have used Equation (11.170) and again have assumed ideality of the gas phase for simplicity. After introduction of Equations (11.179) and (11.180) into the Gibbs–Duhem equations, the quantity $\mathrm{d}P$ can be eliminated. The resultant equation is

$$\left(\frac{\partial \ln P_{AB}}{\partial \ln P_A}\right)_T = \frac{(y_A - y_B)\tilde{V}(\mathrm{c}) - (x_1 - x_2)\tilde{V}(\mathrm{g})}{(y_{AB} + y_B)\tilde{V}(\mathrm{c}) - x_2\tilde{V}(\mathrm{g})} \tag{11.181}$$

The interest in this equation is that it affords the calculation of the composition of the condensed phase in equilibrium with the gas phase when the values of the derivative, y_A, y_B, and y_{AB} can be determined [32].

11.14 Equilibria between pure solids and liquids

Consideration of the condition of equilibrium applied to reactions between pure condensed phases, liquid or solid only, leads to some interesting conclusions. We consider a closed system consisting of P phases, each of which is a pure phase. The temperature and pressure of each phase is the same. Because each phase is pure, there are P species. We assume that one chemical reaction may take place between the species. The condition of equilibrium for the reactions is $\sum_k \nu_k \mu_k = 0$. However, each μ_k is equal to the chemical potential of the kth species in its standard state at the given temperature and pressure, so the condition of equilibrium may be written as

$$\sum_k \nu_k \mu_k^{\ominus}[T, P] = 0 \tag{11.182}$$

This becomes

$$\sum_k \nu_k (\Delta G_f^{\ominus})_k [T, P] = 0 \tag{11.183}$$

with the introduction of the standard molar Gibbs energies of formation. Equations (11.182) and (11.183) are implicit functions of the temperature and pressure; therefore, the temperature at which equilibrium exists is a function of the pressure or, conversely, the pressure is a function of the temperature. Actually, the dependence on pressure is usually small because of the small changes of volume associated with the changes of state. The attainment of equilibrium is solely dependent upon the temperature when the effect of pressure is neglected and, if the expression given in either equation is a single-valued function of the temperature, there is only one possible temperature at which the equilibrium condition is satisfied. There may be systems for which there is no positive root and, hence, no temperature at which the equations are satisfied. In general the value of $\sum_k \nu_k (\Delta G_f^{\ominus})_k$ is either positive or negative, but not zero. Then the reaction proceeds, unless prevented by passive forces, in the direction to minimize the Gibbs energy of the system until one of the species no longer exists in the system.

In a system in which there are P pure condensed phases and one chemical reaction at equilibrium, there are $(P-1)$ components. The system is thus univariant and hence indifferent. The state of the system is defined by assigning a value to at least one extensive variable in addition to the mole numbers of the species. The extent of the reaction taking place within the system is dependent upon the value of the additional extensive variable. A simple example is a phase transition of a pure compound when the change of phase is considered as a reaction. We consider the two phases as two species in the one-component system. In order to define the state of the system, we assign values to the volume of the system in addition to the temperature and mole number of the component. For the given temperature and mole number, the number of moles of the component in each phase is determined by the assigned volume.

11.15 The change of the Gibbs energy for a chemical change of state under arbitrary conditions

We consider a change of state involving a chemical reaction indicated as $\sum_k \nu_k B_k = 0$, but with the condition that the state of each substance is set arbitrarily, and with the exception that the temperature of each substance must be the same. The change of the Gibbs energy for this change of state is given by

$$\Delta G = \sum_k \nu_k \mu_k \tag{11.184}$$

When we express each chemical potential in terms of activities, this becomes

$$\Delta G = \sum_k \nu_i \mu_i^{\ominus} + RT \ln\left(\prod_i a_i^{\nu_i}\right) \tag{11.185}$$

or

$$\Delta G = \Delta G^{\ominus} + RT \ln\left(\prod_i a_i^{\nu_i}\right) \tag{11.186}$$

This equation permits the calculation of ΔG for any arbitrary condition when $\Delta G^{\ominus}$ is known from a study of the equilibrium condition or calculable from standard Gibbs energies of formation.

12

Equilibria in electrochemical systems

In the previous chapters the condition of electroneutrality was applied to all systems that contained charged species. In this chapter we study the results when this condition is relaxed. This leads to studies of electrochemical systems, especially those involving galvanic cells. Cells without transference are emphasized, although simple cells with transference are discussed. At the end of the chapter the conditions of equilibrium across membranes in electrochemical systems are outlined.

12.1 Electrically charged phases and phases of identical composition

The effect of relaxing the condition of electroneutrality in terms of the electrical potential of a charged phase must first be considered. We choose a single-phase system containing 10^{-10} mol of an ionic species with a charge number of $+1$. The phase is assumed to be spherical with a radius of 1 cm and surrounded by empty space. Essentially, all of the excess charge will reside on the surface of the sphere. The electrical potential of the sphere is given by

$$\psi = Q/\varepsilon_0 r \tag{12.1}$$

where Q is the charge on the sphere, ε_0 is the permittivity of free space and r is the radius of the sphere. With the numerical values of these quantities

$Q = 9.6 \times 10^{-6}$ coulombs (C)

$\varepsilon_0 = 1.11 \times 10^{-10}$ coulombs volt^{-1} meter^{-1} (C V^{-1} m^{-1})

$r = 10^{-2}$ meter (m)

ψ becomes 8.6×10^6 V. We conclude that an excess quantity of ions far too small to be detected by chemical methods results in an exceedingly large electrostatic potential. The potentials found under normal conditions are very much smaller. We can thus make use of the concept of two phases having *identical composition* but *different electrical potentials*. The two phases actually have slightly different composition, but the excess quantity of an ion is so small that it can be detected only by electrical methods.

12.2 Conditions of equilibrium

A charge is always associated with matter in real systems. Therefore, the energy of a homogenous system is still expressed as a function of the entropy, volume, and mole numbers of the species, so

$$dE = T\,dS - P\,dV + \sum_i \tilde{\mu}_i\,dn_i \tag{12.2}$$

where the sum is taken over all of the species. The symbol $\tilde{\mu}_i$ is used here for the chemical potential to emphasize that we are now dealing with systems in which the electrical potentials of the phase must be considered.[1]

Equation (12.2) is applicable to every phase in a heterogenous system. Because of the identity of this equation with Equation (4.12) with the exclusion of other work terms, the conditions of equilibrium must be the same as those developed in Chapter 5; in a heterogenous system without restrictions, the temperature of every phase must be the same, the pressure of every phase must be the same, and the chemical potential of a species must be the same in every phase in which the species exists. For phase equilibrium, then,

$$\tilde{\mu}_i' = \tilde{\mu}_i'' = \tilde{\mu}_i''' = \cdots \tag{12.3}$$

where the primes are used to designate the phases. Equation (12.3) is applicable to every species. For a chemical reaction written as $\sum_i \nu_i B_i = 0$,

$$\sum_i \nu_i \tilde{\mu}_i = 0 \tag{12.4}$$

at equilibrium.

12.3 Differences in electrical potentials

The energy of 1 mole of a charged substance in a single, electrically conducting phase which has an electrical potential ϕ is $z_i\mathscr{F}\phi$, relative to zero potential. In this expression z_i is the charge number of the species including the sign and $\mathscr{F}$ represents the value of the faraday. The electrical potential is constant throughout the phase when no current flows. It is customary to express the chemical potential of a charged species as a sum of two terms, one of which is $z_i\mathscr{F}\phi$ and the other a function which is independent of the charge, so

$$\tilde{\mu}_i = \mu_i + z_i\mathscr{F}\phi \tag{12.5}$$

[1] Guggenheim [36] calls this the *electrochemical potential*. However, it is not distinct from the chemical potential, and we prefer to continue the use of the term 'chemical potential.'

For a neutral species

$$\tilde{\mu}_i = \mu_i \tag{12.6}$$

We recognize from our previous experience that μ_i is a function of the entropy, volume, temperature, or pressure in appropriate combinations and the composition variables. The splitting of $\tilde{\mu}_i$ into these two terms is not an operational definition, but its justification is obtained from experiment. The quantity $\tilde{\mu}_i$ is the quantity that is measured experimentally, relative to some standard state, whereas the electrical potential of a phase cannot be determined. Neither can the difference between the electrical potentials of two phases alone at the same temperature and pressure generally be measured. Only if the two phases have identical composition can this be done. If the two phases are designated by primes,

$$\tilde{\mu}_i'' - \tilde{\mu}_i' = \mu_i'' - \mu_i' + z_i\mathscr{F}(\phi'' - \phi') \tag{12.7}$$

We see that $(\phi'' - \phi')$ can be obtained only from known values of $(\tilde{\mu}_i'' - \tilde{\mu}_i')$ when $\mu_i'' = \mu_i'$.

In order to clarify these concepts, we consider a length of copper wire along which there exists a potential gradient, such as a slidewire in a potentiometer. The potentials at two points a and b in the wire are designated as ϕ^a and ϕ^b, so that the potential difference between the two points is given by $(\phi^b - \phi^a)$. If we attach a piece of copper wire at each point, the potential difference between the two pieces of wire is still $(\phi^b - \phi^a)$. The only species that can transfer across the boundary at each junction is electrons, and these electrons are in equilibrium. We have, from Equation (12.3),

$$\tilde{\mu}_e^a(\text{Pot}) = \tilde{\mu}_e^a(\text{Cu}) \tag{12.8}$$

and

$$\tilde{\mu}_e^b(\text{Pot}) = \tilde{\mu}_e^b(\text{Cu}) \tag{12.9}$$

where the terms on the left of these equations refer to the potentiometer wire and those on the right refer to the pieces of copper wire. The difference between the electrical potentials is given by

$$\tilde{\mu}_e^b(\text{Pot}) - \tilde{\mu}_e^a(\text{Pot}) = -\mathscr{F}(\phi^b - \phi^a) \tag{12.10}$$

according to Equation (12.5). Then

$$\tilde{\mu}_e^b(\text{Cu}) - \tilde{\mu}_e^a(\text{Cu}) = -\mathscr{F}(\phi^b - \phi^a) \tag{12.11}$$

because $\mu_e^b(\text{Cu}) = \mu_e^a(\text{Cu})$. The fact that current may be flowing through the slidewire does not alter the condition of equilibrium across the junction between the slidewire and the pieces of copper wire. If we attach a piece of silver wire to each of the copper wires, electrons can cross the boundary and the equilibrium condition for the electrons in the silver and copper wire is achieved. For this equilibrium

$$\tilde{\mu}_e^a(\mathrm{Cu}) = \tilde{\mu}_e^a(\mathrm{Ag}) \tag{12.12}$$

and

$$\tilde{\mu}_e^b(\mathrm{Cu}) = \tilde{\mu}_e^b(\mathrm{Ag}) \tag{12.13}$$

We may write these two conditions as

$$\mu_e^a(\mathrm{Cu}) - \mathscr{F}\phi^a(\mathrm{Cu}) = \mu_e^a(\mathrm{Ag}) - \mathscr{F}\phi^a(\mathrm{Ag}) \tag{12.14}$$

and

$$\mu_e^b(\mathrm{Cu}) - \mathscr{F}\phi^b(\mathrm{Cu}) = \mu_e^b(\mathrm{Ag}) - \mathscr{F}\phi^b(\mathrm{Ag}) \tag{12.15}$$

according to Equation (12.5). On subtraction,

$$\phi^b(\mathrm{Ag}) - \phi^a(\mathrm{Ag}) = \phi^b(\mathrm{Cu}) - \phi^a(\mathrm{Cu}) \tag{12.16}$$

so that the difference between the electrical potentials in the two silver wires is equal to that between the copper wires and between the points a and b of the potentiometer wire. However, the difference of potential between the silver wire and copper wire to which it is attached cannot be determined because we cannot evaluate the difference between $\mu_e(\mathrm{Cu})$ and $\mu_e(\mathrm{Ag})$. Ideally, we assume that the transfer of copper or silver ions or atoms across the boundary is restricted and no transfer can take place. In actuality, the diffusion across the boundary is extremely slow, so the effect of irreversible processes across the boundary is negligible. Finally, we contact two portions of a solution containing silver ions, each portion being in an insulated vessel, with the silver wires. The difference between the electrical potentials of the two solutions is obtained from

$$\tilde{\mu}_{\mathrm{Ag}^+}^b(\mathrm{soln}) - \tilde{\mu}_{\mathrm{Ag}^+}(\mathrm{soln}) = \mu_{\mathrm{Ag}^+}^b - \mu_{\mathrm{Ag}^+}^a + \mathscr{F}[\phi^b(\mathrm{soln}) - \phi^a(\mathrm{soln})] \tag{12.17}$$

because $\mu_{\mathrm{Ag}^+}^b = \mu_{\mathrm{Ag}^+}^a$. That this difference of potentials is equal to that between the silver wires is proved by the fact that the silver ions in solution are in equilibrium with silver ions in the silver metal and these, in turn, are in equilibrium with silver atoms and electrons in the metal. Thus,

$$\tilde{\mu}_{\mathrm{Ag}^+}(\mathrm{soln}) = \tilde{\mu}_{\mathrm{Ag}^+}(\mathrm{Ag}) = \mu_{\mathrm{Ag}}(\mathrm{Ag}) - \tilde{\mu}_e(\mathrm{Ag}) \tag{12.18}$$

and therefore

$$\tilde{\mu}_{\mathrm{Ag}^+}^b(\mathrm{soln}) - \tilde{\mu}_{\mathrm{Ag}^+}^a(\mathrm{soln}) = -[\tilde{\mu}_e^b(\mathrm{Ag}) - \tilde{\mu}_e^a(\mathrm{Ag})] \tag{12.19}$$

Again, we cannot determine the difference in potential between the solution and the silver wire because, from Equation (12.18), the difference $[\mu_{\mathrm{Ag}^+}(\mathrm{soln}) - \mu_{\mathrm{Ag}^+}(\mathrm{Ag})]$ is not known or measurable.

The solution containing silver ions must contain at least one species of negative ion, and the chemical potential of the salt must be considered. If we assume that the negative ions in the solutions containing silver ion are nitrate ions, the chemical potential of the salt is written as

$$\tilde{\mu}_{AgNO_3} = \tilde{\mu}_{Ag^+} + \tilde{\mu}_{NO_3^-} \tag{12.20}$$

according to Equation (11.178). However,

$$\tilde{\mu}_{Ag^+} = \mu_{Ag^+} + \mathscr{F}\phi \tag{12.21}$$

and

$$\tilde{\mu}_{NO_3^-} = \mu_{NO_3^-} - \mathscr{F}\phi \tag{12.22}$$

We therefore see that

$$\tilde{\mu}_{AgNO_3} = \mu_{Ag^+} + \mu_{NO_3^-} = \mu_{AgNO_3} \tag{12.23}$$

i.e., the chemical potential of the salt is independent of the electrical potential of the solution.

12.4 An electrochemical cell

With an understanding of the meaning and measurement of the difference of electrical potential, we can develop the thermodynamics of a galvanic cell. We choose a specific cell, but one in which many of the principles related to the obtaining of thermodynamic data from measurement of the electromotive forces (emf) of the cell are illustrated. The specific cell is depicted as

$$\text{Cu}|\text{Pt, H}_2\left|\begin{matrix}\text{HCl }(m)\\ \text{satd. with H}_2\end{matrix}\right|\begin{matrix}\text{HCl }(m)\\ \text{satd. with AgCl}\end{matrix}\left|\text{AgCl–Ag}|\text{Cu}\right. \tag{12.24}$$

where the vertical lines indicate boundaries between phases. The left-hand electrode consists of a piece of copper wire attached to a piece of platinum which is in contact with hydrogen gas. The combination of platinum and hydrogen constitutes the hydrogen electrode. This electrode in turn is immersed in an aqueous solution of hydrochloric acid at a given molality. The solution is saturated with hydrogen. Another aqueous solution of hydrochloric acid is in contact with the first solution. This second solution has the same molality of hydrochloric acid as the first, but it is saturated with silver chloride rather than hydrogen. A silver wire, on which silver chloride has been deposited, is immersed in the second solution. Finally, a length of copper wire is attached to the silver wire. The state of the thermodynamic system consisting of the cell must be specified by assigning values to the temperature, the molality of the hydrochloric acid, and to the total pressure or the pressure of the hydrogen or both, depending on the restrictions that may be used.

The difference between the electrical potentials in the two copper wires is determined by the difference $[\tilde{\mu}_e''(\text{Cu}) - \tilde{\mu}_e'(\text{Cu})]$ under equilibrium conditions with certain restrictions. (The single prime refers here to all parts of the cell to the left of the boundary between the two solutions, and the double prime to all parts to the right of the boundary.) The restrictions are that the boundaries between the various parts of the cell are permeable only to certain species. Without such restrictions the electrical potential difference of the electrons in the copper wires would be zero at equilibrium. The boundary between the copper and platinum or between the copper and silver is permeable only to electrons; that between the platinum with adsorbed hydrogen and the first solution is permeable to hydrogen ions but not electrons; that between the second solution and the silver chloride is permeable to chloride ions but not electrons; and that between the silver chloride and silver is permeable only to silver ions. We ignore the presence of the boundary between the two solutions, for the present. The conditions of equilibrium in terms of the chemical potentials are then:

$$\begin{aligned}
&\tilde{\mu}_e'(\text{Cu}) = \tilde{\mu}_e(\text{Pt}) \\
&\tfrac{1}{2}\mu_{\text{H}_2}(\text{g}) = \mu_{\text{H}}(\text{Pt}) \\
&\mu_{\text{H}}(\text{Pt}) = \tilde{\mu}_{\text{H}^+}'(m) + \tilde{\mu}_e(\text{Pt}) \\
&\tilde{\mu}_{\text{Cl}^-}''(m) = \tilde{\mu}_{\text{Cl}^-}(\text{AgCl}) \\
&\tilde{\mu}_{\text{Ag}^+}(\text{Ag}) = \tilde{\mu}_{\text{Ag}^+}(\text{AgCl}) \\
&\tilde{\mu}_{\text{Ag}^+}(\text{AgCl}) + \tilde{\mu}_{\text{Cl}^-}(\text{AgCl}) = \mu_{\text{AgCl}}(\text{s}) \\
&\mu_{\text{Ag}}(\text{s}) = \tilde{\mu}_{\text{Ag}^+}(\text{Ag}) + \tilde{\mu}_e(\text{Ag}) \\
&\tilde{\mu}_e(\text{Ag}) = \tilde{\mu}_e''(\text{Cu})
\end{aligned} \tag{12.25}$$

These conditions yield

$$\tilde{\mu}_e''(\text{Cu}) - \tilde{\mu}_e'(\text{Cu}) = \tilde{\mu}_{\text{H}^+}'(m) + \tilde{\mu}_{\text{Cl}^-}''(m) + \mu_{\text{Ag}}(\text{s}) - \mu_{\text{AgCl}}(\text{s}) - \tfrac{1}{2}\mu_{\text{H}_2}(\text{g}) \tag{12.26}$$

on appropriate addition

If the cell is short-circuited, we find that the reaction

$$\begin{aligned}
&\tfrac{1}{2}\text{H}_2[\text{g}, T, P_{\text{H}_2}] + \text{AgCl}[\text{s}, T, P] \\
&\qquad \rightarrow (\text{H}^+)'[m, T, P] + (\text{Cl}^-)''[m, T, P] + \text{Ag}[\text{s}, T, P]
\end{aligned} \tag{12.27}$$

takes place as current passes through the cell without the change of state of the system. The change of the Gibbs energy for the change of state represented by Equation (12.27) is

$$\begin{aligned}
\Delta\tilde{G} = {} & \tilde{\mu}_{\text{H}^+}'[m, T, P] + \tilde{\mu}_{\text{Cl}^-}''[m, T, P] + \mu_{\text{Ag}}[\text{s}, T, P] \\
& - \tfrac{1}{2}\mu_{\text{H}_2}[\text{g}, T, P_{\text{H}_2}] - \mu_{\text{AgCl}}[\text{s}, T, P]
\end{aligned} \tag{12.28}$$

This relation is identical to the right-hand side of Equation (12.26), and therefore

$$\Delta\tilde{G} = \tilde{\mu}_e''(\mathrm{Cu}) - \tilde{\mu}_e'(\mathrm{Cu}) = -\mathscr{F}(\phi'' - \phi') \tag{12.29}$$

This equation gives the relation between the electrical potential difference between the copper wires attached to the electrodes when the cell is at equilibrium and the change of the Gibbs energy for the change of state that would take place in the cell if the cell were short-circuited. We point out here that the chemical potentials of electrons refer to 1 mole of electrons or 1 faraday of electricity. Therefore, $\Delta\tilde{G}$ refers to the change of state per faraday. If the change of state requires n faradays,

$$\Delta G = n[\tilde{\mu}_e''(\mathrm{Cu}) - \tilde{\mu}_e'(\mathrm{Cu})] = -n\mathscr{F}(\phi'' - \phi') \tag{12.30}$$

The electromotive force (emf) of the cell is defined as the difference between the electrical potentials under open-circuit conditions (the requirement of reversibility is discussed in Section 12.6), so $(\phi'' - \phi')$ equals $\mathscr{E}$ and

$$\Delta G = -n\mathscr{F}\mathscr{E} \tag{12.31}$$

in general. The emf of a cell is defined for open-circuit conditions. Its value is therefore the same whether current passes through the cell or not. On the other hand, the value of $(\phi'' - \phi')$ does change when current is passed through the cell (an irreversible process).

The relation expressed in Equation (12.31) is consistent with the relation between a change of the Gibbs energy and the maximum work (excluding pressure–volume work) that a system may do on the surroundings. A galvanic cell may be considered as a closed system operating at constant temperature and pressure. Then, from Equations (7.7) and (12.31),

$$\Delta G = -W_{\mathrm{r}}' = -n\mathscr{F}\mathscr{E}$$

and the quantity $n\mathscr{F}\mathscr{E}$ is the *maximum* electrical work that the system can perform on the surroundings for n faradays of electricity when the values of all of the intensive variables are held constant.

Throughout this discussion we have ignored the effect of the boundary between the two solutions. We have indicated that the solution to the left of the boundary is saturated with hydrogen and to the right of the boundary with silver chloride. The two solutions are not identical and not at equilibrium. There would be an electrical potential difference across the boundary. However, the solubilities of hydrogen and silver chloride in the solutions are very small. The chemical potential of the hydrogen ion is essentially independent of the concentration of hydrogen in the solution, and that of the chloride ion is essentially independent of the concentration of the silver chloride. Therefore, the two solutions are considered to be identical, so that no boundary exists. We may then write

$$\tilde{\mu}'_{H^+}[m] + \tilde{\mu}''_{Cl^-}[m] = \tilde{\mu}_{H^+}[m] + \tilde{\mu}_{Cl^-}[m] \tag{12.32}$$

and, by Equation (12.5),

$$\tilde{\mu}_{H^+}[m] + \tilde{\mu}_{Cl^-}[m] = (\mu_{H^+} + \mathscr{F}\phi) + (\mu_{Cl^-} - \mathscr{F}\phi) = \mu_{H^+} + \mu_{Cl^-} \tag{12.33}$$

Equation (12.28) is then written as

$$\Delta\tilde{G} = \mu_{H^+}[m, T, P] + \mu_{Cl^-}[m, T, P] + \mu_{Ag}[s, T, P] - \tfrac{1}{2}\mu_{H_2}[g, T, P_{H_2}] - \mu_{AgCl}[s, T, P] \tag{12.34}$$

We observe that no term on the right-hand side is dependent upon an electrical potential. In actual practice only one solution is used. There are concentration gradients of hydrogen and silver chloride within the solution, but the effect of the hydrogen and silver chloride on the chemical potentials of the hydrogen and chloride ions is small and negligible.

Cells in which only one electrolytic solution is used are called *cells without liquid junction*. Cells in which it is necessary to use two solutions with a boundary between them are called *cells with liquid junction*. Such cells are discussed in Section 12.12.

12.5 The convention of signs for a galvanic cell[2]

Equation (12.31) relates the emf of a cell to the change of the Gibbs energy for the change of state, based on n faradays, that takes place within the cell when the leads are short-circuited. The change of state is spontaneous, the value of ΔG is negative, and therefore the emf is positive. We conclude that the emf of any actual cell is always positive. However, the *representation* of a cell may be written in two ways. For the particular cell that we have been discussing, we may depict it as either

$$H_2[g, T, P_{H_2}]|HCl[m, T, P]|AgCl[s, T, P] - Ag[s, T, P] \tag{12.35}$$

or

$$Ag[s, T, P] - AgCl[s, T, P]|HCl[m, T, P]|H_2[g, T, P_{H_2}] \tag{12.36}$$

where the presence of platinum and the identical metallic leads is understood. The direction of the flow of current, and hence that of the associated chemical reaction, may also be considered in two directions in each case.

Conventions have been established to obviate these difficulties. For any

The convention discussed here is consistent with those adopted by the Physical Chemistry Section of the Commission on Physico-Chemical Symbols and Terminology of IUPAC in Stockholm, 1953; i.e., the 'Stockholm Convention.' See Christiansen, A. (1960). *J. Am. Chem. Soc.*, **82**, 5517.

cell, depicted as above, *the direction of current within the cell is assumed to be from left to right, so that positively charged species flow from left to right and negatively charged species flow from right to left.* The chemical reaction associated with the flow of current must be consistent with this convention. Thus, the reaction for the cell depicted in Equation (12.35) is

$$\tfrac{1}{2}H_2 + AgCl = H^+ + Cl^- + Ag \tag{12.37}$$

and that depicted in Equation (12.36) is

$$Ag + H^+ + Cl^- = \tfrac{1}{2}H_2 + AgCl \tag{12.38}$$

We have already stated that the change of Gibbs energy for the change of state related to Equation (12.37) is negative and that the emf of the cell depicted in Equation (12.35) is positive. Therefore, the change of the Gibbs energy for the change of state related to Equation (12.38) is positive and the value of the emf of the cell depicted in Equation (12.36) is given a negative sign.

An equivalent statement of the convention is obtained by considering the type of reaction that takes place at each electrode. If we consider negative charges as flowing from right to left through the cell, then electrons enter the cell and reduction occurs at the electrode on the right, whereas electrons leave the cell and oxidation occurs at the electrode on the left. *The cathode is defined as the electrode at which reduction takes place, and the anode is defined as the electrode at which oxidation takes place.* Therefore, the convention is that the electrode on the left of any representation of a cell is taken to be the anode and that on the right to be the cathode. If the emf of the cell so represented is negative, the cell reaction occurs spontaneously in the *opposite* direction.

A third statement is obtained when we wish to assign signs to the electrodes. The electrode on the right is assigned a positive sign, and that on the left is assigned a negative sign.

12.6 Conditions of reversibility

The emf of a cell has been defined as being equal to the difference between the electrical potentials of the two identical leads of the cell under equilibrium and open-circuit conditions. The measurement of the emf must therefore be made under conditions that approach reversible conditions, so that the equilibrium conditions of the cell are not disturbed to any great extent and essentially no current flows through the cell. These conditions are closely approximated by the use of a potentiometer in which the potential difference between the cell leads is balanced by an essentially equal potential drop along the slide wire (or its equivalent) of the potentiometer. Under such conditions no current flows through the cell within the sensitivity of the galvanometer. Modern high-input impedance digital voltmeters essentially satisfy the same conditions for zero current flow, but are not adjustable in the sense of being able to vary the opposing voltage.

In addition, the cell reaction must also be reversible. The concept of equilibrium requires that, for a given system, the same state is obtained when equilibrium is approached from any direction. For a cell, the chemical reaction that takes place when the potential drop along the slide wire of the potentiometer is slightly greater than the emf of the cell must be the same as, but in the opposite direction to, that which takes place when the potential drop is slightly less than the emf of the cell.

Finally, no irreversible process can take place within the cell. All of the reacting species must be present in the cell, but these species must be separated in some way to prevent an irreversible reaction. In the cell that we have been discussing, hydrogen and silver chloride are slightly soluble in the solution but, by suitable construction of the cell, it is possible to avoid a relatively high concentration of hydrogen near the cathode and a relatively high concentration of silver chloride near the anode. Fortunately, the reaction between hydrogen and silver chloride is slow and the effect of the irreversibility is negligible.

12.7 The standard electromotive force of a cell

When we write the cell reaction as $\sum_i \nu_i B_i = 0$, the change of the Gibbs energy for the change of state can be written as

$$\Delta G = \sum_i \nu_i \mu_i = RT \ln \prod_i a_i^{\nu} + \sum_i \nu_i \mu_i^{\ominus} \tag{12.39}$$

according to Equation (11.179) where, for simplicity, we have expressed the chemical potentials in terms of the activities. We now define the standard emf of a cell by

$$\sum_i \nu_i \mu_i^{\ominus} = \Delta G^{\ominus} = -n\mathscr{F}\mathscr{E}^{\ominus} \tag{12.40}$$

corresponding to Equation (12.31). The quantity $\mathscr{E}^{\ominus}$ is the emf of a cell in which *all* reacting species are in their standard states. With the use of Equations (12.31), (12.39), and (12.40), we can express the emf of a cell in terms of its standard emf and the activities of the reacting species as

$$\mathscr{E} = \mathscr{E}^{\ominus} - \frac{RT}{n\mathscr{F}} \ln \prod_i a_i^{\nu_i} \tag{12.41}$$

Thus, for the cells represented in Equations (12.35) and (12.36), Equation (12.41) becomes

$$\mathscr{E} = \mathscr{E}^{\ominus} - \frac{RT}{\mathscr{F}} \ln \frac{a_{H^+} a_{Cl^-} a_{Ag}}{a_{AgCl} P_{H_2}^{1/2}} \tag{12.42}$$

and

$$\mathscr{E} = \mathscr{E}^{\ominus} - \frac{RT}{\mathscr{F}} \ln \frac{a_{AgCl} P_{H_2}^{1/2}}{a_{H^+} a_{Cl^-} a_{Ag}} \tag{12.43}$$

respectively, where we have assumed for simplicity that hydrogen behaves as an ideal gas. We note that the sign of the numerical values $\mathscr{E}$ and $\mathscr{E}^{\ominus}$ are opposite in the two equations. These equations express the dependence of the emf of a cell on the concentration of the species in solution at constant temperature and pressure.

The determination of the value of $\mathscr{E}^{\ominus}$ may be illustrated by the use of Equation (12.42). We choose the standard state of silver and silver chloride to be the solid state at the experimental temperature and pressure, so that the activities are unity. We also express the activities of the hydrogen and chloride ions in terms of their molalities and activity coefficients. Equation (12.41) then becomes

$$\mathscr{E} + \frac{RT}{\mathscr{F}} \ln \frac{m_{H^+} m_{Cl^-}}{P_{H_2}^{1/2}} = \mathscr{E}^{\ominus} - \frac{RT}{\mathscr{F}} \ln \gamma_{H^+} \gamma_{Cl^-} \tag{12.44}$$

where all experimentally measured quantities have been replaced on the left-hand side of the equation. When we choose the reference state of the hydrochloric acid to be the infinitely dilute solution at the experimental temperature and pressure, we find that

$$\lim_{m \to 0} \left[\mathscr{E} + \frac{RT}{\mathscr{F}} \ln \frac{m_{H^+} m_{Cl^-}}{P_{H_2}^{1/2}} \right] = \mathscr{E}^{\ominus} \tag{12.45}$$

Thus, $\mathscr{E}^{\ominus}$ is the value of the left-hand side of Equation (12.44) in the limit of zero molality.

12.8 The temperature dependence of the emf of a cell

Equation (12.31) gives the relation between the emf of a cell and the change of the Gibbs energy for the cell reaction. Temperature differentiation gives

$$n\mathscr{F} \left(\frac{\partial \mathscr{E}}{\partial T} \right)_{P,n} = - \left(\frac{\partial \, \Delta G}{\partial T} \right)_{P,n} \tag{12.46}$$

where the subscripts indicate that the system is a closed system in which the pressure on each phase is constant. For all reversible cells $(\partial \, \Delta G/\partial T)_{P,n}$ may be equated to the negative of a change of entropy, so

$$n\mathscr{F} \left(\frac{\partial \mathscr{E}}{\partial T} \right)_{P,n} = \Delta S \tag{12.47}$$

From Equations (12.31) and (12.47), we also have

$$\mathscr{E} - T\left(\frac{\partial \mathscr{E}}{\partial T}\right)_{P,n} = -\frac{\Delta H}{n\mathscr{F}} \tag{12.48}$$

or

$$\left(\frac{\partial(\mathscr{E}/T)}{\partial T}\right)_{P,n} = \frac{\Delta H}{n\mathscr{F}T^2} \tag{12.49}$$

This equation is known as the *Gibbs–Helmholtz equation.*

Care must be taken to use or determine the correct change of state to which the change of entropy and enthalpy refers. The difficulty arises when one or more equilibrium reactions are present in addition to those used in determining the cell reaction. Such conditions occur when two or more phases are in equilibrium; for example, the electrolytic solution may be saturated with a solid phase, or one of the electrodes may consist actually of a liquid solution that is saturated with a solid solution or with another liquid solution. Such equilibria do not alter the change of the Gibbs energy or the emf. Let the cell reaction be represented as $\sum_i \nu_i B_i$ and an equilibrium reaction as $\sum_j \nu_j B_j$. The usual algebraic sum of these two reactions yields a third reaction, represented as $\sum_k \nu_k B_k$. The change of the Gibbs energy for the first reaction is $\Delta G = \sum_i \nu_i \mu_i$, but that for the equilibrium reaction is zero. Therefore, the change of the Gibbs energy for the third reaction is the same as that for the first. However, the changes of the entropy or the enthalpy for the equilibrium reaction are not zero. Therefore, the changes of the entropy for the first and third reactions are not equal, neither are the changes of enthalpy.

With this understanding, we can derive the correct expression for Equation (12.47) from Equation (12.46). We emphasize that the cell is a closed system at equilibrium (open-circuit) at constant temperature and with a constant pressure on each phase. We let the cell reaction as determined from inspection of the representation of the cell in accordance with the established conventions be represented as $\sum_i \nu_i B_i$. Then

$$\Delta G = \sum_i \nu_i \mu_i \tag{12.50}$$

and

$$\left(\frac{\partial\, \Delta G}{\partial T}\right)_{P,n} = \sum_i \nu_i \left(\frac{\partial \mu_i}{\partial T}\right)_{P,n} \tag{12.51}$$

The chemical potentials are functions of the temperature and the composition variables in general, so each $(\partial \mu_k/\partial T)_{P,n}$ is given by

$$\left(\frac{\partial \mu_k}{\partial T}\right)_{P,n} = \left(\frac{\partial \mu_k}{\partial T}\right)_{P,x} + \sum_j^{C-1} \left(\frac{\partial \mu_k}{\partial x_j}\right)_{T,P} \left(\frac{\partial x_j}{\partial T}\right)_{P,n} \tag{12.52}$$

where the sum is taken over all mole fractions except one. When the composition of each phase in the cell is independent of the temperature, each $(\partial x_i/\partial T)_{P,n}$ is zero and

$$\left(\frac{\partial \mu_k}{\partial T}\right)_{P,n} = \left(\frac{\partial \mu_k}{\partial T}\right)_{P,x} = -\bar{S}_k \tag{12.53}$$

In this case

$$\left(\frac{\partial \Delta G}{\partial T}\right)_{P,n} = -\sum_i \nu_i \bar{S}_i = -\Delta S \tag{12.54}$$

Here ΔS refers to the change of state associated with the simple cell reaction.

When two or more phases are in equilibrium in the cell so that other equilibrium reactions must be considered, some of the derivatives $(\partial x_i/\partial T)_{P,n}$ will not be zero. Thus, if we have a solution saturated with respect to a pure solid phase or one of constant composition, a change of temperature will change the solubility of the solid in the liquid. If a liquid solution and a solid solution are in equilibrium, then a change of temperature will cause a redistribution of the components between the two phases, resulting in a change of composition of each phase. The cell considered as a system is univariant under the given conditions. The phases between which matter is transferred with a change of temperature may be considered as a subsystem that also must be univariant. The derivatives $(\partial x_i/\partial T)_{P,n}$ are evaluated by solution of the appropriate set of Gibbs–Duhem equations applicable to the subsystem. The result is a difference in entropies which, when substituted into Equations (12.52) and (12.51), yields a change of entropy for a specific change of state.

We discuss two different examples. Consider the cell

$$\mathrm{M}\left|\begin{matrix}\mathrm{MA}(m)\\ \mathrm{MA}\cdot p\mathrm{H_2O(s)}\end{matrix}\right|\mathrm{A} \tag{12.55}$$

where the electrolyte is assumed to be an aqueous solution of the salt MA at molality m, which is saturated with respect to the solid hydrate $\mathrm{MA}\cdot p\mathrm{H_2O}$. We assume that the simple cell reaction per faraday for ΔG and $\mathscr{E}$ is

$$\mathrm{M} + \mathrm{A} = \mathrm{MA}(m) \tag{12.56}$$

so that $\Delta G = \mu_{\mathrm{MA}} - \mu_{\mathrm{M}} - \mu_{\mathrm{A}}$ and

$$\left(\frac{\partial \Delta G}{\partial T}\right)_{P,n} = \left(\frac{\partial \mu_2}{\partial T}\right)_{P,n} - \left(\frac{\partial \mu^{\bullet}_{\mathrm{M}}}{\partial T}\right)_{P,n} - \left(\frac{\partial \mu^{\bullet}_{\mathrm{A}}}{\partial T}\right)_{P,n} \tag{12.57}$$

where the subscript 2 refers to the solute MA. M and A are pure phases, so

$$\left(\frac{\partial \mu^{\bullet}_{\mathrm{M}}}{\partial T}\right)_{P,n} = \left(\frac{\partial \mu^{\bullet}_{\mathrm{M}}}{\partial T}\right)_{P} = -\bar{S}^{\bullet}_{\mathrm{M}} \tag{12.58}$$

and

$$\left(\frac{\partial \mu_A^\bullet}{\partial T}\right)_{P,n} = \left(\frac{\partial \mu_A^\bullet}{\partial T}\right)_P = -\tilde{S}_A^\bullet \tag{12.59}$$

However,

$$\left(\frac{\partial \mu_2}{\partial T}\right)_{P,n} = \left(\frac{\partial \mu_2}{\partial T}\right)_{P,m} + \left(\frac{\partial \mu_2}{\partial m}\right)_{T,P}\left(\frac{\partial m}{\partial T}\right)_P = -\bar{S}_2 + \left(\frac{\partial \mu_2}{\partial m}\right)_{T,P}\left(\frac{\partial m}{\partial T}\right)_P \tag{12.60}$$

We are concerned with intensive variables and the requirements of the closed system for the cell has no meaning in Equation (12.60). The Gibbs–Duhem equations for the two-phase subsystem are

$$S(\ell)\,dT + n_1\,d\mu_1 + n_2\,d\mu_2 = 0 \tag{12.61}$$

and

$$\tilde{S}^\bullet(s)\,dT + d\mu(s) = 0 \tag{12.62}$$

where the solid is $MA \cdot pH_2O$ and the subscript 1 refers to the solvent. The equilibrium condition is

$$\mu(s) - \mu_2 - p\mu_1 = 0 \tag{12.63}$$

The solution of Equations (12.61)–(12.63) is

$$\left(\frac{\partial \mu_2}{\partial m}\right)_{T,P}\left(\frac{\partial m}{\partial T}\right)_P = -\frac{p}{p - n_1/n_2}\tilde{S}(\ell) + \frac{n_1/n_2}{p - n_1/n_2}\tilde{S}^\bullet(s) + \bar{S}_2 \tag{12.64}$$

where $\tilde{S}(\ell)$ is the entropy of the solution *per mole of solute*. The result is in terms of the mole ratio (n_1/n_2) in the solution. Equation (12.57) then becomes

$$\left(\frac{\partial\,\Delta G}{\partial T}\right)_{P,n} = -\left(\frac{p}{p - n_1/n_2}\tilde{S}(\ell) - \frac{n_1/n_2}{p - n_1/n_2}\tilde{S}^\bullet(s) - \tilde{S}_M^\bullet - \tilde{S}_A^\bullet\right) \tag{12.65}$$

The expression in the parentheses is the change of entropy for the change of state

$$M + A + \frac{n_1/n_2}{p - n_1/n_2}(MA \cdot pH_2O)(s) = \frac{p}{p - n_1/n_2}(MA \cdot (n_1/n_2)H_2O)(\ell) \tag{12.66}$$

or, if $(n_1/n_2) > p$,

$$M + A + \frac{p}{(n_1/n_2) - p}(MA \cdot (n_1/n_2)H_2O)(\ell) = \frac{n_1/n_2}{(n_1/n_2) - p}(MA \cdot pH_2O)(s) \tag{12.67}$$

Either Equation (12.66) or Equation (12.67) is the correct change of state to which ΔS in Equation (12.47) refers.

For the second example, consider the cell

$$\mathrm{M}\left|\begin{matrix}\mathrm{MA}(m)\\ \mathrm{NB(s)}\end{matrix}\right|\mathrm{A}$$

in which the aqueous electrolytic solution is saturated with respect to a salt NB, which is different from the salt formed by the cell reaction. The cell reaction is given by Equation (12.56), and $(\partial\,\Delta G/\partial T)_{P,n}$, by Equation (12.57). The problem is to evaluate $(\partial\mu_2/\partial T)_{P,n}$ for the two-phase subsystem of the liquid solution and the pure solid. The two Gibbs–Duhem equations are

$$S(\ell)\,\mathrm{d}T + n_1\,\mathrm{d}\mu_1 + n_2\,\mathrm{d}\mu_2 + n_3\,\mathrm{d}\mu_3 = 0 \tag{12.68}$$

and

$$\tilde{S}_3^{\bullet}(\mathrm{s})\,\mathrm{d}T + \mathrm{d}\mu_3 = 0 \tag{12.69}$$

where the subscripts 1, 2, and 3 refer to H_2O, MA, and NB, respectively. It would appear that the subsystem is divariant rather than univariant. However, the molality of MA is constant and the two variables T and m_3 are related by

$$[S(\ell) - n_3\tilde{S}^{\bullet}(\mathrm{s})]\,\mathrm{d}T + n_1\,\mathrm{d}\mu_1 + n_2\,\mathrm{d}\mu_2 = 0 \tag{12.70}$$

obtained after elimination of $\mathrm{d}\mu_3$ from Equations (12.68) and (12.69). Now

$$\mathrm{d}\mu_1 = -\bar{S}_1\,\mathrm{d}T + \left(\frac{\partial\mu_1}{\partial m_3}\right)_{T,P,m_2}\mathrm{d}m_3 \tag{12.71}$$

and

$$\mathrm{d}\mu_2 = -\bar{S}_2\,\mathrm{d}T + \left(\frac{\partial\mu_2}{\partial m_3}\right)_{T,P,m_2}\mathrm{d}m_3 \tag{12.72}$$

Moreover,

$$S(\ell) = n_1\bar{S}_1 + n_2\bar{S}_2 + n_3\bar{S}_3 \tag{12.73}$$

Equation (12.70) becomes

$$n_3[\bar{S}_3 - \tilde{S}_3^{\bullet}(\mathrm{s})]\,\mathrm{d}T + \left[n_1\left(\frac{\partial\mu_1}{\partial m_3}\right)_{T,P,m_2} + n_2\left(\frac{\partial\mu_2}{\partial m_3}\right)_{T,P,m_2}\right]\mathrm{d}m_3 = 0 \tag{12.74}$$

when Equations (12.71)–(12.73) are substituted into it, so

$$\left(\frac{\partial m_3}{\partial T}\right)_P = -\frac{n_3[\bar{S}_3 - \tilde{S}_3^{\bullet}(\mathrm{s})]}{n_1(\partial\mu_1/\partial m_3)_{T,P,m_2} + n_2(\partial\mu_2/\partial m_3)_{T,P,m_2}} \tag{12.75}$$

At constant temperature and pressure,

$$n_1\left(\frac{\partial \mu_1}{\partial m_3}\right)_{T,P,m_2} + n_2\left(\frac{\partial \mu_2}{\partial m_3}\right)_{T,P,m_2} = -n_3\left(\frac{\partial \mu_3}{\partial m_3}\right)_{T,P,m_2} \tag{12.76}$$

Then Equation (12.60) may be written as

$$\left(\frac{\partial \mu_2}{\partial T}\right)_P = -\bar{S}_2 + \frac{[\bar{S}_3 - \tilde{S}_3^\bullet(\mathrm{s})](\partial \mu_2/\partial m_3)_{T,P,m_2}}{(\partial \mu_3/\partial m_3)_{T,P,m_2}} \tag{12.77}$$

This equation can be simplified to some extent. We know that

$$\left(\frac{\partial \mu_2}{\partial n_3}\right)_{T,P,n_1,n_2} = \left(\frac{\partial \mu_3}{\partial n_2}\right)_{T,P,n_1,n_3} \tag{12.78}$$

from the calculation of exactness based on

$$\mathrm{d}G = -S\,\mathrm{d}T + V\,\mathrm{d}P + \mu_1\,\mathrm{d}n_1 + \mu_2\,\mathrm{d}n_2 + \mu_3\,\mathrm{d}n_3 \tag{12.79}$$

Equation (12.78) may be written as

$$\left(\frac{\partial \mu_2}{\partial m_3}\right)_{T,P,m_2} = \left(\frac{\partial \mu_3}{\partial m_2}\right)_{T,P,m_3} \tag{12.80}$$

Therefore, the ratio $(\partial \mu_2/\partial m_3)_{T,P,m_2}/(\partial \mu_3/\partial m_3)_{T,P,m_2}$ is equal to $(\partial \mu_3/\partial m_2)_{T,P,m_3}/(\partial \mu_3/\partial m_3)_{T,P,m_2}$. However,

$$\frac{(\partial \mu_3/\partial m_2)_{T,P,m_3}}{(\partial \mu_3/\partial m_3)_{T,P,m_2}} = -\left(\frac{\partial m_3}{\partial m_2}\right)_{T,P,\mu_3} \tag{12.81}$$

The derivative $(\partial m_3/\partial m_2)_{T,P,\mu_3}$ is the change of the solubility of NB as the molality of MA is changed at constant temperature and pressure. Equation (12.77) is then written as

$$\left(\frac{\partial \mu_2}{\partial T}\right)_P = -\bar{S}_2 - [\bar{S}_3 - \tilde{S}_3^\bullet(\mathrm{s})]\left(\frac{\partial m_3}{\partial m_2}\right)_{T,P,\mu_3} \tag{12.82}$$

Finally, Equation (12.51) becomes

$$\left(\frac{\partial\,\Delta G}{\partial T}\right)_P = -\left[\bar{S}_2 + [\bar{S}_3 - \tilde{S}_3^\bullet(\mathrm{s})]\left(\frac{\partial m_3}{\partial m_2}\right)_{T,P,\mu_3} - \tilde{S}_{\mathrm{M}}^\bullet - \tilde{S}_{\mathrm{A}}^\bullet\right] \tag{12.83}$$

The expression in the brackets is the change of entropy for the change of state

$$\mathrm{M} + \mathrm{A} + \left(\frac{\partial m_3}{\partial m_2}\right)_{T,P,\mu_3}\mathrm{NB(s)} = \mathrm{MA}(m_2) + \left(\frac{\partial m_3}{\partial m_2}\right)_{T,P,\mu_3}\mathrm{NB}(m_3) \tag{12.84}$$

These two cases illustrate the problem associated with the use of Equation (12.47). Although this equation is valid for all reversible cells, we must ascertain the correct change of state to which ΔS refers. This change of state can always be determined by the methods used here.

12.9 The pressure dependence of the emf of a cell

The problems related to the change of the emf of a cell with a change of pressure are idenitcal to those discussed in Section 12.8. The general equations are easily derived from Equation (12.31), so

$$n\mathscr{F}\left(\frac{\partial \mathscr{E}}{\partial P}\right)_{T,n} = -\left(\frac{\partial\, \Delta G}{\partial P}\right)_{T,n} \tag{12.85}$$

When we write ΔG as $\sum_i \nu_i \mu_i$,

$$n\mathscr{F}\left(\frac{\partial \mathscr{E}}{\partial P}\right)_{T,n} = -\sum_i \nu_i \left(\frac{\partial \mu_i}{\partial P}\right)_{T,n} \tag{12.86}$$

Each partial derivative of the chemical potentials can be equated to a partial molar volume or a change of volume, so we may write in general

$$\left(\frac{\partial \mathscr{E}}{\partial P}\right)_{T,n} = \frac{\Delta V}{n\mathscr{F}} \tag{12.87}$$

However, when we do so we must be careful to make use of the correct ΔV and to associate this ΔV with the correct change of state.

A cell may be considered as a heterogenous system at equilibrium with restrictions. In most cells the pressure on each phase is the same and a change of pressure of the system would cause the same change of pressure on all phases. However, it is possible to construct a cell so that the various phases may have different pressures. Then the pressures of some phases may be held constant while the pressures of other phases are changed. In such cases some of the derivatives of the chemical potentials in Equation (12.86) would be zero unless matter would have to be transported across the boundary between phases in order to maintain the equilibrium conditions with a change of pressure.

When matter must be transported from one phase to another with a change of pressure in order to maintain the equilibrium conditions, calculations similar to those discussed in Section 12.8 must be carried out in order to obtain the appropriate ΔV and the corresponding correct change of state. In such cases the systems considered must be univariant, and the chemical potentials are considered as functions of the pressure and those concentration variables whose values can be changed.

12.10 Half-cell potentials[3]

We have defined the emf of a cell in terms of the difference of the electrical potential in two identical pieces of metal connected to the two electrodes when the cell is at equilibrium. In the same manner we can define the standard emf of the cell as the difference of electrical potential in two identical pieces of metal connected to the electrodes when the cell is at equilibrium and all reacting species are in their standard states, so

$$\mathscr{E}^{\ominus} = (\phi^{\ominus})'' - (\phi^{\ominus})' \tag{12.88}$$

where $\phi^{\ominus}$ represents the potential under the standard state conditions. The division of the standard emf into two potentials is nonoperational and it is only the standard emf that can be determined, directly or indirectly, from the experimental studies of cells.

The cell reaction for cells without liquid junction can be written as the sum of an oxidation reaction and a reduction reaction, the so-called *half-cell reactions*. If there are C oxidation reactions, and therefore C reduction reactions, there are $\frac{1}{2}C(C-1)$ possible cells. Not all such cells could be studied because of irreversible phenomena that would take place within the cell. Still, a large number of cells are possible. It is therefore convenient to consider half-cell reactions and to associate a potential with each such reaction or electrode. Because of Equation (12.88), there would be $(C-1)$ independent potentials. We can thus assign an arbitrary value to the potential associated with one half-cell reaction or electrode. By convention, and for *aqueous* solutions, *the value of zero has been assigned to the hydrogen half-cell* when the hydrogen gas and the hydrogen ion are in their standard states, independent both of the temperature and of the pressure on the solution.

However, in addition to this assignment we must adopt the sign of the half-cell potentials and the terminology to be used. Two conventions based on the representations of cells have been used. In each case we consider a cell consisting of a hydrogen electrode in which the reacting substances are in their standard states (a standard hydrogen electrode) and any second electrode in which the reacting species are at arbitrarily chosen states. In the first case the hydrogen electrode is always placed on the left-hand side of the representation. The emf of the cell is defined as the difference $(\phi_{\text{right}} - \phi_{\text{left}})$ or $(\phi_c - \phi_a)$, where the subscripts c and a refer to the cathode and the anode, respectively. Thus, $\mathscr{E} = \phi_c - \phi_a$. However, ϕ_a for the standard hydrogen electrode has been *defined* as zero, and therefore ϕ_c equals the emf of the cell. The symbol ϕ represents the *electrode potential* of the given electrode. This is the only meaning associated with the term *electrode*

[3] The conventions discussed here are consistent with those adopted by the Physical Chemical Section of the Physico-Chemical Symbols and Terminology of IUPAC in Stockholm, 1953 (Compt. rend. 17th conf. union intern. chim. pure et appl., 1953, pp. 82–85). Also, see [22].

potential. If all of the reacting substances of the chosen electrode were in their standard states, the value of $\phi^\ominus$ would be called the *standard electrode potential*. For this convention, which is the current international convention, the electrode on the right of the representation is always taken as the cathode, the electrode at which reduction takes place. Therefore, the term *reduction potential* is an alternative to the term *electrode potential* and emphasizes the nature of the reaction taking place at the electrode.

In the second case the standard hydrogen electrode is placed on the right-hand side of the representation of the cell, and the other electrode would be placed on the left-hand side. The emf of the cell would then be written as $\mathscr{E} = \psi_a - \psi_c$. The value of $\psi_c^\ominus$ is defined to be zero and the potential of the electrode on the left, ψ_a, is the emf of the cell. The symbol ψ is called the *oxidation potential*. When all of the reacting substances of the electrode are in their standard states, then ψ would become $\psi^\ominus$ and would be called the *standard oxidation potential*. This terminology is that of Latimer and emphasizes the nature of the reaction taking place at the electrode. We present it here for completeness, knowing that reduction potentials are now the standard convention, but that some of the older literature used oxidation potentials.

As examples of these conventions, we again consider the cell

$$H_2(1\ \text{bar})|HCl(a = 1)|AgCl(s)\text{–}Ag(s)$$

for which $\mathscr{E}^\ominus$ is $+0.222$ V at 298 K. The standard *electrode potential* of the silver chloride–silver electrode is therefore $+0.222$ V at 298 K. When we represent the cell as

$$Ag(s)\text{–}AgCl(s)|HCl(a = 1)|H_2(1\ \text{bar})$$

the standard emf is -0.222 V at 298 K. The standard *oxidation potential* of the silver–silver chloride is therefore -0.222 V at 298 K. We may represent the electrodes and the values of the potentials at 298 K as

$$\left.\begin{array}{c} Cl^-|AgCl\text{–}Ag \\ \text{or} \\ AgCl(s) + e^- = Ag(s) + Cl^-(a = 1) \end{array}\right\} \phi^\ominus = +0.222\ \text{V}$$

and (12.89)

$$\left.\begin{array}{c} Ag\text{–}AgCl|Cl^- \\ \text{or} \\ Ag(s) + Cl^-(a = 1) = AgCl(s) + e^- \end{array}\right\} \psi^\ominus = -0.222\ \text{V}$$

With these conventions, the standard emf of any cell without liquid junction may be determined by the relations

$$\mathscr{E}^{\ominus}_{\text{cell}} = \phi^{\ominus}_{\text{c}} - \phi^{\ominus}_{\text{a}} = \psi^{\ominus}_{\text{a}} - \psi^{\ominus}_{\text{c}} \tag{12.90}$$

A third possible convention is due to Gibbs. Here the emphasis is on the cell itself, rather than the depiction of the cell. The emf of the cell is always positive and the difference between the standard half-cell potentials is taken to yield positive values. Then

$$\mathscr{E}^{\ominus}_{\text{cell}} = \phi^{\ominus}_{\text{cathode}} - \phi^{\ominus}_{\text{anode}} \tag{12.91}$$

where the cathode and anode potentials refer to the actual cathode and anode of the cell. However, Gibbs did not separate the potentials of the electrode and did not discuss half-cell potentials.

Throughout this discussion we have considered cells in which the electrolytic solution is an aqueous solution. The same methods can be used to define standard half-cell potentials in any solvent system. However, it is important to remember that when the reference state is defined as the infinitely dilute solution of a solute in a particular solvent, the standard state depends upon that solvent. The values so obtained are *not* interchangeable between the different solvent systems. Only if the standard states could all be defined independently of the solvent would the values be applicable to all solvent systems.

12.11 Some galvanic cells without liquid junction

A large amount of thermodynamic data has been and can be obtained from the experimental study of galvanic cells. We have already related the emf of a cell to the change of the Gibbs energy for the change of state corresponding to the cell reaction. The determination of the emf of the cell as a function of the temperature yields values of the changes of enthalpy and entropy for the change of state associated with the cell reaction. The same relations hold between the standard emf of the cell and $\Delta G^{\ominus}$, $\Delta H^{\ominus}$, and $\Delta S^{\ominus}$ for the standard change of state. By Equations (12.40) and (11.4), we also have

$$\mathscr{E}^{\ominus} = \frac{RT}{n\mathscr{F}} \ln K \tag{12.92}$$

a relation between the standard emf of the cell and the equilibrium constant for the cell reaction. From a knowledge of $\Delta G^{\ominus}$ and values of $\Delta \tilde{G}^{\ominus}_{\text{f}}$ for all but one of the reacting species, a value of $\Delta \tilde{G}^{\ominus}_{\text{f}}$ can be calculated for that species. Values of $\Delta \tilde{G}^{\ominus}_{\text{f}}$ of some compounds can be obtained directly from the emf of some cells; for example, the cell reaction of the cell

$$\text{Ag–AgCl}|\text{NaCl}(m)|\text{Cl}_2(\text{g, 1 bar})$$

is the formation of AgCl from the elements.

Values of activity coefficients can also be obtained from the emf of cells. For the purposes of discussion we continue with the cell

$$H_2(g, 1\ bar)|HCl(m)|AgCl\text{–}Ag$$

for which we have already obtained the expression (Eq. (12.42))

$$\mathscr{E} = \mathscr{E}^{\ominus} - \frac{RT}{\mathscr{F}} \ln \frac{a_{H^+} a_{Cl^-} a_{Ag}}{a_{AgCl} P_{H_2}^{1/2}}$$

and have shown how to determine the value of $\mathscr{E}^{\ominus}$ (Eq. (12.44)). Equation (12.44) can be written as

$$\mathscr{E} = \mathscr{E}^{\ominus} - \frac{RT}{\mathscr{F}} \ln \frac{m_{H^+} m_{Cl^-}}{P_{H_2}^{1/2}} - RT \ln \gamma_{H^+} \gamma_{Cl^-} \tag{12.93}$$

We cannot determine values of the activity coefficients of the individual ions, but by definition of the mean activity coefficients (Eq. (11.182)), we have

$$\mathscr{E} = \mathscr{E}^{\ominus} - \frac{RT}{\mathscr{F}} \ln \frac{m_{H^+} m_{Cl^-}}{P_{H_2}^{1/2}} - 2RT \ln \gamma_{\pm} \tag{12.94}$$

Then, when values of $\mathscr{E}$ for various molalities and $\mathscr{E}^{\ominus}$ are known, values of the mean activity coefficient can be calculated.

The same type of cell has been used to determine values of the ionization constants of weak acids and weak bases. We use as an example a weak acid, HA. The representation of the cell is

$$H_2(g, 1\ bar)\left|\begin{matrix} HA(m_1) \\ NaA(m_2) \\ NaCl(m_3) \end{matrix}\right| AgCl\text{–}Ag$$

The electrolyte solution is a buffered solution of the acid and one of its salts, and contains sodium chloride to make the cell reversible. The emf of the cell is given by Equation (12.93). The activity of the hydrogen ion can be expressed as

$$a_{H^+} = m_{H^+} \gamma_{H^+} = \frac{K m_{HA} \gamma_{HA}}{m_{A^-} \gamma_{A^-}}$$

so that Equation (12.93) becomes

$$\mathscr{E} = \mathscr{E}^{\ominus} - \frac{RT}{\mathscr{F}} \ln \frac{m_{HA} m_{Cl^-}}{m_{A^-} P_{H_2}^{1/2}} - RT \ln \frac{\gamma_{HA} \gamma_{Cl^-}}{\gamma_{A^-}} - RT \ln K \tag{12.95}$$

The value of K can be calculated provided that we know the molalities and activity coefficients of the species which are indicated. The molality of HA is $(m_1 - m_{H^+})$ and that of A^- is $(m_2 + m_{H^+})$. When K is sufficiently small,

m_{H^+} may be negligible with respect to m_1 and m_2. When this condition is not satisfied, an iterative calculation may be carried out. We may initially neglect m_{H^+} and obtain a first approximation to K. This value of K may be used to obtain a first approximation of m_{H^+}, which in turn may be used to obtain a second approximation of K. The calculation is repeated until a constant value of K is obtained to the desired precision. The activity coefficient of HA is very close to unity in dilute solutions, and the approximation that it is unity is usually made. The activity coefficients of Cl^- and A^- may be calculated according to some theory. On the other hand, values of K may be calculated at various molalities of the solutes with the assumption that all of the activity coefficients are unity. The values so obtained are not thermodynamic values but are functions of the molalities. The thermodynamic value is then obtained by extrapolation of the nonthermodynamic values to the infinitely dilute solution, where all of the activity coefficients are unity by the choice of the infinitely dilute solution as the reference state for all of the solutes.

A different type of cell is one represented by

$$\mathrm{M}|\mathrm{MA}|\mathrm{M(B)}$$

in which M is a metal that may be liquid or solid, MA is a compound of M that may be pure or in solution and may be liquid or solid, and M(B) is a liquid or solid alloy. The change of state associated with the cell reaction is

$$\mathrm{M}(\ell \text{ or s}) = \mathrm{M}(\text{alloy}, \ell \text{ or s})$$

so

$$\mathscr{E} = -\frac{RT}{n\mathscr{F}} \ln a_{\mathrm{M}} \tag{12.96}$$

where a_M is the activity of M in the alloy referred to pure M as the standard state and n is the number of faradays required for the change of state.

12.12 Galvanic cells with liquid junction

Cells in which at least two electrolytic solutions are in contact are known as cells with liquid junction or with transference. Such cells are inherently *irreversible* and a complete thermodynamic development of them is beyond the scope of this book. However, cells with liquid junction are of sufficient importance that we discuss here the type that *approximates* a reversible cell most closely.

We consider a cell having the same electrodes and in which the electrolytic solutions differ only in the concentration of the single solute; such a cell may be represented as

$$\mathrm{M}|\mathrm{M}_{\nu_+}\mathrm{A}_{\nu_-}(m_1)|\mathrm{M}_{\nu_+}\mathrm{A}_{\nu_-}(m_2)|\mathrm{M} \tag{12.97}$$

where M is a metal and $M_{\nu_+}A_{\nu_-}$ the solute. The charge of the positive ion is z_+ and that of the negative ion is z_-. The irreversible process of diffusion of both the solute and solvent actually takes place across the boundary between the two solutions. This effect can be made negligible within the time of an experiment by proper design of the cell.

The conditions of equilibrium applied to the electrodes and their adjacent solutions yield

$$\tilde{\mu}_e(\mathrm{M}, l) - \tilde{\mu}_e(\mathrm{M}, r) = (1/z_+)[\tilde{\mu}(\mathrm{M}_+, r) - \tilde{\mu}(\mathrm{M}_+, l)] \tag{12.98}$$

where the letters r and l refer to right and left, and the symbol M_+ represents the positive ion without indication of the charge. From Equation (12.29) we also have

$$\tilde{\mu}_e(\mathrm{M}, l) - \tilde{\mu}_e(\mathrm{M}, r) = \mathscr{F}[\phi(r) - \phi(l)] = \mathscr{F}\mathscr{E} \tag{12.99}$$

so

$$\mathscr{F}\mathscr{E} = (1/z_+)[\tilde{\mu}(\mathrm{M}_+, r) - \tilde{\mu}(\mathrm{M}_+, l)] \tag{12.100}$$

The calculation of the emf in terms of the two concentrations of the solute therefore requires the determination of $[\tilde{\mu}(\mathrm{M}_+, r) - \tilde{\mu}(\mathrm{M}_+, l)]$. We consider the cell

$$\mathrm{M}|\mathrm{M}_{\nu_+}\mathrm{A}_{\nu_-}(m)|\mathrm{M}_{\nu_+}\mathrm{A}_{\nu_-}(m + \mathrm{d}m)|\mathrm{M} \tag{12.101}$$

in which the molality of the solute is only infinitesimally different in the two solutions. The differential of the emf of the cell is given by

$$\mathscr{F}\,\mathrm{d}\mathscr{E} = (1/z_+)\,\mathrm{d}\tilde{\mu}(\mathrm{M}_+) \tag{12.102}$$

but we must be able to express $\mathrm{d}\tilde{\mu}(\mathrm{M}_+)$ in terms of the chemical potential of the salt. We associate a system of rectangular coordinates with the cell, let the concentration gradient be in the direction of the x-axis, and assert that there is no concentration gradient in the y- and z-directions. We then assume that the flow of an ionic species across a unit cross section per unit of time is proportional to the gradient of the chemical potential, so

$$J_i = -L_i\left(\frac{\partial \mu_i}{\partial x}\right)_{T,P} \tag{12.103}$$

Under open-circuit conditions, no current can flow through the cell, and therefore

$$z_+J_+ + z_-J_- = -\left[z_+L_+\left(\frac{\partial \tilde{\mu}(\mathrm{M}_+)}{\partial x}\right)_{T,P} + z_-L_-\left(\frac{\partial \tilde{\mu}(\mathrm{A}_-)}{\partial x}\right)\right] = 0 \tag{12.104}$$

When we multiply Equation (12.102) written as

$$\mathscr{F}\left(\frac{\partial \mathscr{E}}{\partial x}\right)_{T,P} = \frac{1}{z_+}\left(\frac{\partial \tilde{\mu}(\mathrm{M}_+)}{\partial x}\right)_{T,P}$$

by $(z_+^2 L_+ + z_-^2 L_-)$ and add Equation (12.104) to the result, we obtain

$$(z_+^2 L_+ + z_-^2 L_-)\mathscr{F}\left(\frac{\partial \mathscr{E}}{\partial x}\right)_{T,P} = z_-^2 L_-\left[\frac{1}{z_+}\left(\frac{\partial \tilde{\mu}(\mathrm{M}_+)}{\partial x}\right)_{T,P} - \frac{1}{z_-}\left(\frac{\partial \tilde{\mu}(\mathrm{A}_-)}{\partial x}\right)_{T,P}\right] \tag{12.105}$$

The quantity $[z_k^2 L_k/\sum_i z_i^2 L_i]$ is the *transference number* of the kth ion as determined by the Hittorf method and, therefore, Equation (12.105) may be written as

$$\mathscr{F}\left(\frac{\partial \mathscr{E}}{\partial x}\right)_{T,P} = t_-\left[\frac{1}{z_+}\left(\frac{\partial \tilde{\mu}(\mathrm{M}_+)}{\partial x}\right)_{T,P} - \frac{1}{z_-}\left(\frac{\partial \tilde{\mu}(\mathrm{A}_-)}{\partial x}\right)_{T,P}\right] \tag{12.106}$$

We assume that the electrolyte is a strong electrolyte, so

$$\mathrm{d}\mu(\mathrm{M}_{\nu_+}\mathrm{A}_{\nu_-}) = \nu_+\,\mathrm{d}\tilde{\mu}(\mathrm{M}_+) + \nu_-\,\mathrm{d}\tilde{\mu}(\mathrm{A}_-) \tag{12.107}$$

On substitution into Equation (12.106), we obtain

$$\mathscr{F}\left(\frac{\partial \mathscr{E}}{\partial x}\right)_{T,P} = -\frac{t_-}{z_-\nu_-}\left(\frac{\partial \mu(\mathrm{M}_{\nu_+}\mathrm{A}_{\nu_-})}{\partial x}\right)_{T,P} = \frac{t_-}{z_+\nu_+}\left(\frac{\partial \mu(\mathrm{M}_{\nu_+}\mathrm{A}_{\nu_-})}{\partial x}\right)_{T,P} \tag{12.108}$$

where we have used the relation $z_+\nu_+ + z_-\nu_- = 0$. The chemical potential is in terms of 1 mole of salt and division by $z_+\nu_+$ or $z_-\nu_-$ converts the unit of mass into equivalents.

We can now return to the cell depicted in Equation (12.97). We must assume that the concentration of the solute is continuous across the boundary between the two solutions. Then the cell may be regarded as a series of cells given in Equation (12.101). In such a series the potential of the intermediate electrodes cancel, leaving only the potentials of the two end-electrodes. The emf of the cell is given by the integral of Equation (12.108) between the limits of x_1 and x_2, where x_1 locates some plane within the solution on the left where the concentration is definitely m_2, and x_2 locates a plane on the right where the concentration is definitely m_2. Thus,

$$\mathscr{F}\mathscr{E} = \mathscr{F}\int_{x_1}^{x_2}\left(\frac{\partial \mathscr{E}}{\partial x}\right)_{T,P}\mathrm{d}x = \frac{1}{z_+\nu_+}\int_{x_1}^{x_2} t_-\left(\frac{\partial \mu(\mathrm{M}_{\nu_+}\mathrm{A}_{\nu_-})}{\partial x}\right)_{T,P}\mathrm{d}x \tag{12.109}$$

For a single solute the transference number is a function of the molality but not of x_1, and the integral is a proper integral. Then

$$\mathscr{F}\mathscr{E} = \frac{1}{z_+ \nu_+} \int t_- \, \mathrm{d}\mu(\mathrm{M}_{\nu_+}\mathrm{A}_{\nu_-}) \tag{12.110}$$

When t_- is independent of the molality or when the difference between m_1 and m_2 is sufficiently small that t_- may be considered to be independent of the molality, the integration is easily performed and

$$\mathscr{F}\mathscr{E} = \frac{t_-}{z_+ \nu_+} [\mu(\mathrm{M}_{\nu_+}\mathrm{A}_{\nu_-}, m_2) - \mu(\mathrm{M}_{\nu_+}\mathrm{A}_{\nu_-}, m_1)] \tag{12.111}$$

This equation provides a means of determining the transference number of the negative ions from measurement of the emf of the cell with the conditions that all of the assumptions made in obtaining the equation are valid and that the values of the mean activity coefficients in the solutions are known. An equation can be derived by use of the same methods for the case in which the solutions contain several solutes. When the electrodes are reversible with respect to the M_+ ion, the equation is

$$\mathscr{F}\mathscr{E} = \int_{x_1}^{x_2} \sum_i t_i \left[\frac{1}{z_{\mathrm{M}_+}} \left(\frac{\partial \mu(\mathrm{M}_+)}{\partial x} \right)_{T,P} - \frac{1}{z_i} \left(\frac{\partial \mu_i}{\partial x} \right)_{T,P} \right] \mathrm{d}x \tag{12.112}$$

When the electrodes are reversible to the A_- ion, the equation is

$$\mathscr{F}\mathscr{E} = \int_{x_1}^{x_2} \sum_i t_i \left[\frac{1}{z_{\mathrm{A}_-}} \left(\frac{\partial \mu(\mathrm{A}_-)}{\partial x} \right)_{T,P} - \frac{1}{z_i} \left(\frac{\partial \mu_i}{\partial x} \right)_{T,P} \right] \mathrm{d}x \tag{12.113}$$

These equations can be expressed in terms of the chemical potentials of the salts when the usual definition of the chemical potentials of strong electrolytes is used. The transference numbers may be a function of x as well as the molality. Arguments which are not thermodynamic must be used to evaluate the integrals in such cases (see Kirkwood and Oppenheim [33]). One special type of cell to which either Equation (12.112) or Equation (12.113) applies is one in which a strong electrolyte is present in both solutions at concentrations that are large with respect to the concentrations of the other solutes. Such a cell, based on that represented in Equation (12.97), is

$$\mathrm{M} \left| \begin{matrix} \mathrm{M}_{\nu_+}\mathrm{A}_{\nu_-}(m_1) \\ \mathrm{KCl}(m_3) \end{matrix} \right| \begin{matrix} \mathrm{M}_{\nu_+}\mathrm{A}_{\nu_-}(m_2) \\ \mathrm{KCl}(m_3) \end{matrix} \left| \mathrm{M} \right. \tag{12.114}$$

in which m_3 is large with respect to both m_1 and m_2. The chemical potential of the potassium chloride is approximately the same in both solutions. Moreover, the transference numbers of M_+ and A_- ions are small. The emf of the cell is given by

$$\mathscr{F}\,\mathrm{d}\mathscr{E} = \frac{t_{\mathrm{A}^-}}{z_{\mathrm{M}_+}\nu_{\mathrm{M}_+}}\,\mathrm{d}\mu(\mathrm{M}_{\nu_+}\mathrm{A}_{\nu_-}) + \frac{t_{\mathrm{K}^+} + t_{\mathrm{Cl}^-}}{z_{\mathrm{M}_+}}\,\mathrm{d}\mu(\mathrm{MCl}_{z\mathrm{M}_+}) \tag{12.115}$$

where $\mathrm{MCl}_{z_{\mathrm{M}_+}}$ is the formula for the chloride of M. With the approximation that $t_{\mathrm{A}_-} = 0$ and that $(t_{\mathrm{K}^+} + t_{\mathrm{Cl}^-}) = 1$, Equation (12.115) becomes

$$\mathscr{F}\,\mathrm{d}\mathscr{E} = \frac{1}{z_{\mathrm{M}_+}}\,\mathrm{d}\mu(\mathrm{MCl}_{z\mathrm{M}_+}) \tag{12.116}$$

and

$$\mathscr{F}\mathscr{E} = \frac{1}{z_{\mathrm{M}_+}}\int_{m_1}^{m_2} \mathrm{d}\mu(\mathrm{MCl}_{z\mathrm{M}_+}) \tag{12.117}$$

The integrand is not dependent upon x and the integration is easily performed.

The emf of a cell with transference of the type discussed but with different electrodes is readily obtained. Consider the cell

$$\mathrm{H_2(g,\ 1\ bar)|HCl}(m_1)|\mathrm{HCl}(m_2)|\mathrm{AgCl–Ag} \tag{12.118}$$

This cell is identical to the two cells

$$\mathrm{H_2(g,\ 1\ bar)|HCl}(m_1)|\mathrm{AgCl–Ag} \tag{12.119}$$

and

$$\mathrm{Ag–AgCl|HCl}(m_1)\mathrm{HCl}(m_2)|\mathrm{AgCl–Ag} \tag{12.120}$$

connected in series. The emf of the cell shown in Equation (12.118) is the sum of the emfs of the two cells shown in Equations (12.119) and (12.120).

12.13 Membrane equilibrium

An electrochemical system, important particularly in biological systems, is one in which the species are ions and the system is separated into two parts by a rigid membrane that is permeable to some but not all of the species. We are interested in the conditions attained at equilibrium, the *Donnan equilibrium*. Two cases, one in which the membrane is not permeable to the solvent (nonosmotic equilibrium) and the other in which the membrane is permeable to the solvent (osmotic equilibrium), are considered. The system is at constant temperature and, for the purposes of discussion, we take sodium chloride, some salt NaR, and water as the components. The membrane is assumed to be permeable to the sodium and chloride ions, but not to the R-ions. We designate the quantities pertinent to the solution on one side of the membrane by primes and those pertinent to the solution on the other side by double primes.

When the membrane is not permeable to the solvent, the pressures on the two parts of the system are independent and need not be the same. The conditions of equilibrium for this case are

$$\tilde{\mu}'[\mathrm{Na}^+, m_1, P'] = \tilde{\mu}''[\mathrm{Na}^+, m_4, P''] \tag{12.121}$$

and

$$\tilde{\mu}'[\mathrm{Cl}^-, m_2, P'] = \tilde{\mu}''[\mathrm{Cl}^-, m_5, P''] \tag{12.122}$$

These equations imply that a difference of electrical potential exists across the membrane. However, we can add the two equations to give

$$\tilde{\mu}[\mathrm{Na}^+, m_1, P'] + \tilde{\mu}[\mathrm{Cl}^-, m_2, P'] = \tilde{\mu}[\mathrm{Na}^+, m_4, P''] + \tilde{\mu}[\mathrm{Cl}^-, m_5, P''] \tag{12.123}$$

from which we obtain

$$\begin{aligned} RT \ln m_1 m_2 + 2RT \ln \gamma'_{\pm}[P'] + \mu^{\ominus}[\mathrm{NaCl}, P'] \\ = RT \ln m_4 m_5 + 2RT \ln \gamma'_{\pm}[P''] + \mu^{\ominus}[\mathrm{NaCl}, P''] \end{aligned} \tag{12.124}$$

The standard states are eliminated by

$$\mu^{\ominus}[\mathrm{NaCl}, P''] = \mu^{\ominus}[\mathrm{NaCl}, P'] + \int_{P'}^{P''} \bar{V}^{\ominus}(\mathrm{NaCl})\,\mathrm{d}P \tag{12.125}$$

Moreover, if we choose to make use of the mean activity coefficients at the same pressure P', we have

$$2RT \ln \gamma''_{\pm}[P''] = 2RT \ln \gamma''_{\pm}[P'] + \int_{P'}^{P''} [\bar{V}^{\ominus''}(\mathrm{NaCl}) - \bar{V}^{\ominus}(\mathrm{NaCl})]\,\mathrm{d}P \tag{12.126}$$

by Equation (8.129). Substitution of Equations (12.125) and (12.126) into Equation (12.124) yields

$$RT \ln \frac{m_1 m_2}{m_4 m_5} + 2RT \ln \frac{\gamma'_{\pm}[P']}{\gamma''_{\pm}[P']} = \int_{P'}^{P''} \bar{V}''[\mathrm{NaCl}]\,\mathrm{d}P \tag{12.127}$$

When the two pressures are identical,

$$m_1 m_2 (\gamma'_{\pm})^2 = m_4 m_5 (\gamma''_{\pm})^2 \tag{12.128}$$

Equations (12.127) and (12.128) give the relation between the molalities of the sodium and chloride ions at equilibrium under the appropriate conditions. The molalities, m_1 and m_2, are not equal but are related by the condition of electroneutrality, so

$$m_1 = m_2 + m'_3(\mathrm{R}^-) \tag{12.129}$$

Similarly,

$$m_4 = m_5 + m_3''(\mathrm{R}^-) \tag{12.130}$$

Also, we denote that the mean activity coefficients are determined at the ionic strengths of the solutions.

When the membrane is permeable to the solvent, the two pressures are no longer independent. The relation between the two pressures is obtained by use of the added condition of equilibrium

$$\mu'(\mathrm{H_2O}, m', P') = \mu''(\mathrm{H_2O}, m'', P'') \tag{12.131}$$

where m' and m'' refer to the sets of molalities in the two parts at equilibrium. If we use Equation (8.102) to express the chemical potential of the solvent, we have

$$-\frac{RTM_1}{1000}\sum_i m_i' + RT\ln\gamma'[\mathrm{H_2O}, P'] + \mu^\ominus[\mathrm{H_2O}, P']$$

$$= -\frac{RTM_1}{1000}\sum_i m_i'' + RT\ln\gamma''[\mathrm{H_2O}, P''] + \mu^\ominus[\mathrm{H_2O}, P''] \tag{12.132}$$

The elimination of the standard states yields

$$-\frac{RTM_1}{1000}\sum_i m_i' + RT\ln\gamma'[\mathrm{H_2O}, P'] + \frac{RTM_1}{1000}\sum_i m_i''$$

$$- RT\ln\gamma''[\mathrm{H_2O}, P''] = \int_{P'}^{P''} \bar{V}^\ominus[\mathrm{H_2O}]\,\mathrm{d}P \tag{12.133}$$

This equation gives the relation between P' and P'' when the equilibrium molalities of the ionic species and the activity coefficient of the water at the two pressures are known. If we choose to make use of the activity coefficient of water at the same pressure P', the equation becomes

$$-\frac{RTM_1}{1000}\sum_i (m_i' - m_i'') + RT\ln\frac{\gamma'[\mathrm{H_2O}, P']}{\gamma''[\mathrm{H_2O}, P']} = \int_{P'}^{P''} \bar{V}[\mathrm{H_2O}]\,\mathrm{d}P \tag{12.134}$$

Equation (12.134) gives the required relation between the equilibrium molalities, the activity coefficients, and the two pressures for osmotic equilibrium. It is evident that the two pressures are not independent. We could write P'' as $P' + \Pi$, where Π is the difference of the osmotic pressures of the two solutions referred to the pure solvent. The solution of the two equations would then give a value of Π.

The difference of the electrical potential across the membrane can be determined from either Equation (12.121) or Equation (12.122). We choose to use Equation (12.122). This equation may be written as

$$\mu''[\mathrm{Cl}^-, m_5, P''] - \mu'[\mathrm{Cl}^-, m_2, P'] = -\mathscr{F}(\phi'' - \phi') \tag{12.135}$$

where $(\phi'' - \phi')$ gives the difference of the electrical potential. When the chemical potentials are written in terms of the molalities and activity coefficients, we have

$$RT \ln \frac{m_5}{m_2} + RT \ln \frac{(\gamma_-)_5}{(\gamma_-)_2} + \mu^\ominus[\mathrm{Cl}^-, P''] - \mu^\ominus[\mathrm{Cl}^-, P'] = -\mathscr{F}(\phi'' - \phi') \tag{12.136}$$

We note, however, that this equation contains quantities related to the individual ions rather than salts. Nonthermodynamic methods have been developed to estimate these properties, but a general discussion is not included in this book [34]. Equation (12.136) can be simplified to

$$RT \ln \frac{m_5}{m_2} + RT \ln \frac{(\gamma_{\mathrm{Cl}^-})_5}{(\gamma_{\mathrm{Cl}^-})_2} = -\mathscr{F}(\phi'' - \phi') \tag{12.137}$$

when the two pressures are the same. Also, the activity coefficient of the chloride ion may be approximated by the mean activity coefficient of the salt, NaCl, consistent with the ionic strength of the solutions. Then

$$RT \ln \frac{m_5}{m_2} + RT \ln \frac{(\gamma_\pm)_5}{(\gamma_\pm)_2} = -\mathscr{F}(\phi'' - \phi') \tag{12.138}$$

In this discussion we have emphasized the difference of electrical potential across a membrane without reference to a galvanic cell. Cells can be devised by which such differences can be experimentally determined.

12.14 Some final comments

In all of the discussion of this chapter we have used an aqueous solution as the electrolyte, and electrodes suitable to those aqueous solutions. However, cells are not limited to aqueous solutions. Indeed, other solvents have been used for which liquid ammonia would be an example. Molten salts, such as mixtures of lithium chloride and potassium chloride, have been used for the study of cells at high temperatures. Some studies have been made at higher temperatures, in which solid electrolytes were used. Electrodes compatible with such solvents have also been devised. For example, a zirconium–zirconium oxide electrode stabilized with calcium oxide was used to measure the oxygen potential in nonstoichiometric metal oxides. However, no matter what the electrolytes or the electrodes are, the principles discussed in this chapter such as reversibility and proper measurement must be followed.

13

Surface effects

In all of the preceding discussion we have neglected the effect of surfaces on the thermodynamic properties of a system. The boundary between two phases has been considered to be a mathematically defined surface at which certain properties, such as the density, are discontinuous. Actually, the surface is a thin region of matter (approximately 10^{-7} cm) across which the properties vary continuously.

Surface effects are negligible in many cases. However, when the surface-to-volume ratio of the system is large, surface effects may become appreciable. Moreover, there are phenomena associated with surfaces that are important in themselves. Only an introduction to the thermodynamics of surfaces can be given here, and the discussion is limited to fluid phases and the surfaces between such phases. Thus, consideration of solid–fluid interfaces are omitted, although the basic equations that are developed are applicable to such interfaces provided that the specific face of the crystal is designated. Also, the thermodynamic properties of films are omitted.

We first consider two-phase systems in which the surface is planar, and treat such systems in terms of the properties of the system as a whole. Then we consider the thermodynamic properties of the surface by use of a defined surface as used by Gibbs. We finally consider certain properties of systems in which the surface is curved [35].

13.1 Surface phenomena related to the system

We consider a two-phase, multicomponent system in which there is one planar surface. The state of the system is defined by assigning values to the entropy and volume of the system, the area of the surface, and the mole numbers of the components. The differential of the energy of the system is then written as

$$dE = \left(\frac{\partial E}{\partial S}\right)_{V,a,n} dS + \left(\frac{\partial E}{\partial V}\right)_{S,a,n} dV + \left(\frac{\partial E}{\partial a}\right)_{S,V,n} da$$
$$+ \sum_i \left(\frac{\partial E}{\partial n_i}\right)_{S,V,a,n} dn_i \qquad (13.1)$$

where a represents the area of the surface. Under equilibrium conditions the value of the temperature is the same in all parts of the system. This is also true for the values of the pressure and of the chemical potentials. (These conditions are proved in the next section.) Equation (13.1) can then be written as

$$dE = T\,dS - P\,dV + \gamma\,da + \sum_i \mu_i\,dn_i \qquad (13.2)$$

where

$$\gamma = \left(\frac{\partial E}{\partial a}\right)_{S,V,n} \qquad (13.3)$$

The quantity γ is usually called the surface tension for liquid–gas interfaces and the interfacial tension for liquid–liquid interfaces. We see from Equation (13.2) that $\gamma\,da$ is the differential quantity of work that must be done reversibly on the system to increase the area of the system by the differential amount da at constant entropy, volume, and mole numbers.

The additional equations

$$dH = T\,dS + V\,dP + \gamma\,da + \sum_i \mu_i\,dn_i \qquad (13.4)$$

$$dA = -S\,dT - P\,dV + \gamma\,da + \sum_i \mu_i\,dn_i \qquad (13.5)$$

$$dG = -S\,dT + V\,dP + \gamma\,da + \sum_i \mu_i\,dn_i \qquad (13.6)$$

are obtained from the definition of the additional thermodynamic functions. We observe from these equations that

$$\gamma = \left(\frac{\partial H}{\partial a}\right)_{S,P,n} = \left(\frac{\partial A}{\partial a}\right)_{T,V,n} = \left(\frac{\partial G}{\partial a}\right)_{T,P,n} \qquad (13.7)$$

in addition to Equation (13.3). Moreover, $\gamma\,da$ is always the differential quantity of work that must be done on the system to increase the area of the surface by the differential amount da.

Many relations between the surface tension or the area and other variables can be obtained by the application of the mathematical properties of

Equations (13.3)–(13.6). Six of the more important relations are obtained by application of the conditions of exactness to Equations (13.5) and (13.6). These six are

$$\left(\frac{\partial \gamma}{\partial T}\right)_{V,a,n} = -\left(\frac{\partial S}{\partial a}\right)_{T,V,n} \tag{13.8}$$

$$\left(\frac{\partial \gamma}{\partial V}\right)_{T,a,n} = -\left(\frac{\partial P}{\partial a}\right)_{T,V,n} \tag{13.9}$$

and

$$\left(\frac{\partial \gamma}{\partial n_j}\right)_{T,V,a,n} = \left(\frac{\partial \mu_j}{\partial a}\right)_{T,V,n} \tag{13.10}$$

from Equation (13.5), and

$$\left(\frac{\partial \gamma}{\partial T}\right)_{P,a,n} = -\left(\frac{\partial S}{\partial a}\right)_{T,P,n} \tag{13.11}$$

$$\left(\frac{\partial \gamma}{\partial P}\right)_{T,a,n} = \left(\frac{\partial V}{\partial a}\right)_{T,P,n} \tag{13.12}$$

and

$$\left(\frac{\partial \gamma}{\partial n_j}\right)_{T,P,a,n} = \left(\frac{\partial \mu_j}{\partial a}\right)_{T,P,n} \tag{13.13}$$

from Equation (13.6). Each of the quantities on the left-hand side of these equations can presumably be measured experimentally. We therefore have a means of determining the change of the entropy, pressure, or volume of the system, and the chemical potential of a component of the system with a change of area of the surface under the appropriate constant conditions. We must emphasize that these quantities are properties of the entire system under the condition of constant mole numbers. A change of area under the appropriate conditions presumably results in a redistribution of the components between the phases and the surface. We cannot interpret, for example, the change of volume given in Equation (13.12) to a change of the volume of the surface.

If we multiply Equations (13.8) and (13.11) by the temperature, we find that $T(\partial\gamma/\partial T)_{V,a,n}$ and $T(\partial\gamma/\partial T)_{P,a,n}$ give the differential quantity of heat absorbed by the system for a reversible differential increase in the area for the specified conditions of constant T, V, and n or constant T, P, and n, respectively.

Equations (13.10) and (13.13) yield relations that are of considerable interest. We have

$$\left(\frac{\partial \mu_j}{\partial a}\right)_{T,V,n} = -\left(\frac{\partial \mu_j}{\partial n_i}\right)_{T,V,a,n}\left(\frac{\partial n_i}{\partial a}\right)_{T,V,n,\mu_j} \tag{13.14}$$

and

$$\left(\frac{\partial \mu_j}{\partial a}\right)_{T,P,n} = -\left(\frac{\partial \mu_j}{\partial n_i}\right)_{T,P,A,n}\left(\frac{\partial n_i}{\partial a}\right)_{T,P,n,\mu_j} \tag{13.15}$$

After substitution and rearrangement, we obtain

$$\left(\frac{\partial n_i}{\partial a}\right)_{T,V,n,\mu_j} = -\frac{(\partial \gamma/\partial n_j)_{T,V,a,n}}{(\partial \mu_j/\partial n_i)_{T,V,a,n}} \tag{13.16}$$

and

$$\left(\frac{\partial n_i}{\partial a}\right)_{T,P,n,\mu_j} = -\frac{(\partial \gamma/\partial n_j)_{T,P,a,n}}{(\partial \mu_j/\partial n_i)_{T,P,a,n}} \tag{13.17}$$

The derivatives on the right-hand side of Equations (13.16) and (13.17) can presumably be determined experimentally; thus, the left-hand side can be evaluated. These quantities give the differential quantity of the ith component that must be added to the system for a differential increase in the area under the appropriate conditions. The ith component may be any of the components, including the jth.

The dependence of the energy and enthalpy of the system is easily derived from Equations (13.2) and (13.4). From these equations we have

$$\left(\frac{\partial E}{\partial a}\right)_{T,V,n} = T\left(\frac{\partial S}{\partial a}\right)_{T,V,n} + \gamma \tag{13.18}$$

and

$$\left(\frac{\partial H}{\partial a}\right)_{T,P,n} = T\left(\frac{\partial S}{\partial a}\right)_{T,P,n} + \gamma \tag{13.19}$$

With the use of Equations (13.8) and (13.11), we obtain

$$\left(\frac{\partial E}{\partial a}\right)_{T,V,n} = \gamma - T\left(\frac{\partial \gamma}{\partial T}\right)_{V,a,n} \tag{13.20}$$

and

$$\left(\frac{\partial H}{\partial a}\right)_{T,P,n} = \gamma - T\left(\frac{\partial \gamma}{\partial T}\right)_{P,a,n} \tag{13.21}$$

Here, again, the right-hand side of Equations (13.20) and (13.21) can be determined experimentally. Thus, the left-hand side can be evaluated.

13.2 Surface phenomena related to a defined surface

We may be concerned with the thermodynamic properties of the surface itself, rather than those of the system. In order to develop the necessary relations we follow the method used by Gibbs, who used the device of defining a two-dimensional surface that lies within the actual surface, attributing the changes of the thermodynamic properties of the system resulting from the surface to the thermodynamic properties of the defined surface.

We consider a multicomponent system consisting of two phases separated by a planar surface in a container of fixed volume. The surface has some thickness, as shown by the slant lines in Figure 13.1. We have already stated that some properties, such as the density or the concentration of the components, change rapidly but continuously across the surface. Such behavior is illustrated by the curve in Figure 13.2, where l is measured along a line normal to the surface. Imaginary boundaries (a and b in Figs. 13.1 and 13.2) are placed in the system so that each boundary lies close to the real surface but at a position within the bulk phases where the properties are those of the bulk phases. The system is thus made to consist of three

Figure 13.1. Illustration of a real and a defined surface.

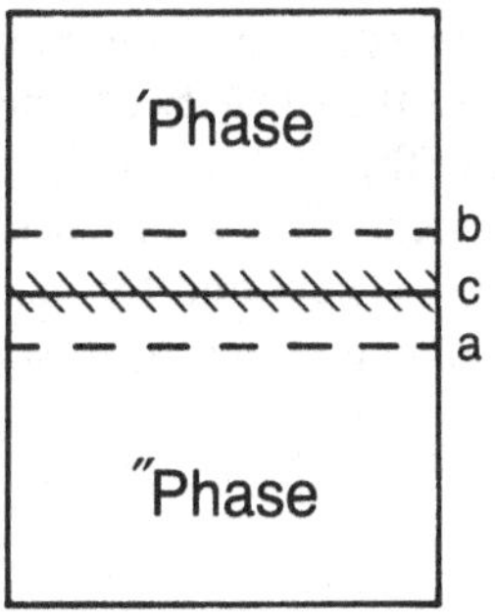

Figure 13.2. Definition of a defined surface in a one-component system.

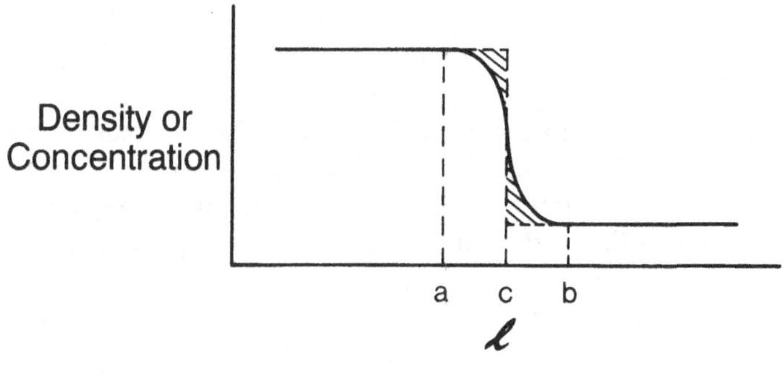

parts: the two bulk phases and a part that contains the surface. The cross section of the container is kept constant in the region that contains the surface. We know that the thermodynamic properties of the system are functions of the entropy, volume, and mole numbers of the components, and therefore the properties of each part are functions of the entropy and volume of the part, and the mole numbers in each part. The differential of the energy of each part is given by the usual expression

$$dE = T\,dS - P\,dV + \sum_i \mu_i\,dn_i \tag{13.22}$$

and the differential of the energy of the entire system is given by the sum of the differentials of the three parts. For equilibrium, the variation of the energy of the system must be equal to or greater than zero. Here, then, we have the same condition that we had when we ignored the surface, and the conditions of equilibrium are that the temperature of each part, the pressure of each part, and the chemical potential of a component in each part must be the same.

We now move the two surfaces a and b toward each other so that they coincide at some position c to give one two-dimensional surface lying wholly within the real surface. The system is thus divided into two parts, and we assume that the properties of each of the two parts are continuous and identical to the properties of the bulk parts up to the single two-dimensional surface. Certain properties of the system are then discontinuous at the surface. The extensive properties of the two-dimensional surface are defined as the difference between the values of the total system and the sum of the values of the two parts. Thus, we have for the energy, entropy, and mole number of the c components:

$$E^\sigma = E - E' - E'' \tag{13.23}$$

$$S^\sigma = S - S' - S'' \tag{13.24}$$

$$n_1^\sigma = n_1 - n_1' - n_1'' \tag{13.25}$$

$$n_2^\sigma = n_2 - n_2' - n_2'' \tag{13.26}$$

$$\vdots$$

$$n_c^\sigma = n_c - n_c' - n_c'' \tag{13.27}$$

where the superscript σ refers to the defined surface and the primes to the bulk parts. The volume of the surface is zero. When we combine Equations (13.2) and (13.20) with Equations (13.23)–(13.27) subject to the conditions of equilibrium, we obtain the fundamental thermodynamic equation for the surface

$$dE^\sigma = T\,dS^\sigma + \gamma\,da + \sum_i \mu_i\,dn_i^\sigma \tag{13.28}$$

The surface energy, E^σ, is a homogenous function of the first degree in S^σ, a, and n_i^σ, so we can write

$$E^\sigma = TS^\sigma + \gamma a + \sum_i n_i^\sigma \mu_i \tag{13.29}$$

Differentiation of Equation (13.29) and comparison with Equation (13.28) gives

$$S^\sigma \,\mathrm{d}T + a\,\mathrm{d}\gamma + \sum_i n_i^\sigma \,\mathrm{d}\mu_i = 0 \tag{13.30}$$

which is the equivalent of the Gibbs–Duhem equation applied to the surface. This equation, when divided by the area, becomes

$$S_\sigma \,\mathrm{d}T + \mathrm{d}\gamma + \sum_i \Gamma_i \,\mathrm{d}\mu_i = 0 \tag{13.31}$$

where S_σ is the entropy density of the surface and Γ_i is the surface concentration defined by Gibbs. Equations (13.30) and (13.31) are used further in Section 13.3..

The enthalpy function for the surface is obtained from the defining equation

$$H^\sigma = H - H' - H'' \tag{13.32}$$

similar to Equation (13.23). From this equation we find that

$$\mathrm{d}H^\sigma = T\,\mathrm{d}S^\sigma + \gamma\,\mathrm{d}a + \sum_i \mu_i \,\mathrm{d}n_i^\sigma \tag{13.33}$$

This equation is identical to Equation (13.28). Consequently, we conclude that the energy function and the enthalpy function are identical for the surface. By similar arguments we find that

$$\mathrm{d}A^\sigma = \mathrm{d}G^\sigma = -S^\sigma \,\mathrm{d}T + \gamma\,\mathrm{d}a + \sum_i \mu_i \,\mathrm{d}n_i^\sigma \tag{13.34}$$

where A^σ is the Helmholtz energy of the surface and G^σ is the Gibbs energy of the surface. We note that these two functions are identical for the surface. Comparison of Equations (13.34) and (13.28) shows that

$$A^\sigma = G^\sigma = E^\sigma - TS^\sigma \tag{13.35}$$

so

$$A^\sigma = G^\sigma = \gamma a + \sum_i n_i^\sigma \mu_i \tag{13.36}$$

The Gibbs energy and Helmholtz energy densities are obtained by dividing this equation by a, so

$$A_\sigma = G_\sigma = \gamma + \sum_i \Gamma_i \mu_i \tag{13.37}$$

Differentiation gives

$$\mathrm{d}A_\sigma = \mathrm{d}G_\sigma = \mathrm{d}\gamma + \sum_i \Gamma_i \,\mathrm{d}\mu_i + \sum_i \mu_i \,\mathrm{d}\Gamma_i \tag{13.38}$$

and when we eliminate $\mathrm{d}\mu_i$ by the use of Equation (13.31) we obtain

$$\mathrm{d}A_\sigma = \mathrm{d}G_\sigma = -S_\sigma \,\mathrm{d}T + \sum_i \mu_i \,\mathrm{d}\Gamma_i \tag{13.39}$$

Equations (13.33) and (13.34) are both exact differential expressions and the conditions of exactness can be applied. One expression, obtained from Equation (13.34), is

$$\left(\frac{\partial \gamma}{\partial T}\right)_{a,n^\sigma} = -\left(\frac{\partial S^\sigma}{\partial a}\right)_{T,n^\sigma} \tag{13.40}$$

We find from Equation (13.33) that

$$\left(\frac{\partial H^\sigma}{\partial a}\right)_{T,n^\sigma} = T\left(\frac{\partial S^\sigma}{\partial a}\right)_{T,n^\sigma} + \gamma \tag{13.41}$$

Combination of these two equations gives

$$\left(\frac{\partial H^\sigma}{\partial a}\right)_{T,n^\sigma} = \gamma - T\left(\frac{\partial \gamma}{\partial T}\right)_{a,n^\sigma} \tag{13.42}$$

which may be compared with Equations (13.20) and (13.21). In making this comparison we note that the temperature, pressure, and the volume of each part must be constant to make each n_i^σ constant when the area is variable.

We have developed the basic equations for the thermodynamic functions of the defined surface in the preceding paragraphs, but have not discussed the determination of the position of the boundary. Actually, the position is somewhat arbitrary, and as a result we must also discuss the dependence of the properties of the surface on the position. The position can be fixed by assigning the value of zero to one of Equations (13.25)–(13.27); that is, by making one of the n_i^σ equal to zero. For a one-component system there is only one such equation. For multicomponent systems we have to choose one of the components for which n_i^σ is made zero. The value of n_i^σ for the other components then would not be zero in general. The most appropriate choice for dilute solutions would be the solvent. The position of the surface for a one-component system is illustrated in Figure 13.2, where the line c is determined by making the areas of the two shaded portions equal.

In order to determine the dependence of the thermodynamic quantities of the surface on the position of the surface, we consider two positions, indicated by subscripts 1 and 2, and determine the difference between values of the quantities for these two positions. Position 2 is obtained by moving the surface toward the prime phase (Fig. 13.1). The state of the real system is maintained constant and the area of the surface is also constant. Then

$$\Delta E^\sigma = E_2^\sigma - E_1^\sigma = -(E_2' - E_1') - (E_2'' - E_1'') \tag{13.43}$$

from Equation (13.23). Now

$$E_2' - E_1' = -(E'/V')\,\Delta V \tag{13.44}$$

and

$$E_2'' - E_1'' = +(E''/V'')\,\Delta V \tag{13.45}$$

where V' and V'' are the initial volumes of the two phases and ΔV (equal to $a\,\Delta l$) is the change of volume of the two parts. We then have

$$\Delta E^\sigma = \left(\frac{E'}{V'} - \frac{E''}{V''}\right)\Delta V \tag{13.46}$$

By similar arguments we obtain

$$\Delta S^\sigma = \left(\frac{S'}{V'} - \frac{S''}{V''}\right)\Delta V \tag{13.47}$$

$$\Delta n_i^\sigma = \left(\frac{n_i'}{V'} - \frac{n_i''}{V''}\right)\Delta \mathrm{V} \tag{13.48}$$

We conclude that the values of the surface energy, surface entropy, surface mole numbers, and surface concentrations are in general dependent upon the position of the surface and independent of it only when the value of the corresponding quantity within the parentheses is zero.

We use Equation (13.29) for the surface tension, so we write

$$\Delta\gamma = \frac{1}{a}\left(\Delta E^\sigma - T\,\Delta S^\sigma - \sum_i \mu_i\,\Delta n_i^\sigma\right) \tag{13.49}$$

When Equations (13.45)–(13.47) are substituted into this equation we obtain

$$\Delta\gamma = \left[\frac{(E' - TS' - \sum_i \mu_i n_i')}{V'} - \frac{(E'' - TS'' - \sum_i \mu_i n_i'')}{V''}\right]\Delta l \tag{13.50}$$

Each term within the brackets is equal to the negative of the pressure, so this equation may be written as

$$\Delta\gamma = -(P' - P'')\,\Delta l \tag{13.51}$$

For plane surfaces the two pressures are equal, and therefore the surface tension is independent of the position of the surface.

The dependence of the derivatives of the surface tension with respect to the properties of the bulk phases on the position of the surface can be determined with the use of Equation (13.31). We assume that $\Gamma_1 = 0$ at the position 1. Then

$$S_{\sigma(1)}\,\mathrm{d}T + \mathrm{d}\gamma + \sum_{i=2}^{C} \Gamma_i\,\mathrm{d}\mu_i = 0 \tag{13.52}$$

from which we obtain

$$\left(\frac{\partial \gamma}{\partial \mu_2}\right)_{T,\mu_3,\ldots,\mu_C} = -\Gamma_{2(1)} \tag{13.53}$$

At the second position Equation (13.31) may be written as

$$\mathrm{d}\gamma + \Gamma_{1(2)}\,\mathrm{d}\mu_1 + \Gamma_{2(2)}\,\mathrm{d}\mu_2 = 0 \tag{13.54}$$

at constant temperature and constant values of $\mu_3, \ldots, \mu_C$. Application of the Gibbs–Duhem equations for each of the two phases yields

$$\mathrm{d}\mu_1 = \frac{(n_2''/V'') - (n_2'/V')}{(n_1'/V') - (n_1''/V'')}\,\mathrm{d}\mu_2 \tag{13.55}$$

so Equation (13.54) becomes

$$\left(\frac{\partial \gamma}{\partial \mu_2}\right)_{T,\mu_3,\ldots,\mu_C} = -\Gamma_{2(2)} - \left(\frac{(n_2''/V'') - (n_2'/V')}{(n_1'/V') - (n_1''/V'')}\right)\Gamma_{1(2)} \tag{13.56}$$

The distance between the two surfaces is $\Gamma_{1(2)}/[(n_1'/V') - (n_1''/V'')]$ according to Equation (13.48), so $\Gamma_{2(2)}$ is given by

$$\Gamma_{2(2)} = \Gamma_{2(1)} - \left(\frac{(n_2''/V'') - (n_2'/V')}{(n_1'/V') - (n_1''/V'')}\right)\Gamma_{1(2)} \tag{13.57}$$

Substitution of Equation (13.57) into Equation (13.54) gives

$$\left(\frac{\partial \gamma}{\partial \mu_2}\right)_{T,\mu_3,\ldots,\mu_C} = -\Gamma_{2(1)} \tag{13.58}$$

The value of the derivative is thus the same at the two positions, and we conclude that the value of all such derivatives is independent of the position of the surface.

13.3 The surface tension and surface concentrations

We are interested in this section in the dependence of the surface tension on the temperature, the pressure, and the concentration of one of the phases as related to the thermodynamic properties of the surface. We are also concerned with the determination of the surface concentrations. The basic equation that we use is Equation (13.31) subject to the condition that the two phases are in equilibrium. We choose the position of the surface such that $\Gamma_1 = 0$. Equation (13.31) then becomes

$$d\gamma = -S_{\sigma(1)}\, dT - \sum_{i=2}^{C} \Gamma_{i(1)}\, d\mu_i \tag{13.59}$$

We choose to use the chemical potentials for the double-primed phase, so that, on substituting the usual expression for the chemical potentials, we obtain

$$d\gamma = -\left(S_{\sigma(1)} - \sum_{i=2}^{C} \Gamma_{i(1)}\bar{S}_i''\right) dT - \sum_{i=2}^{C} \Gamma_{i(1)}\bar{V}_i''\, dP - \sum_{i=2}^{C} \Gamma_{i(1)} \sum_{j=2}^{C} \left(\frac{\partial \mu_i''}{\partial x_j''}\right)_{T,P,x} dx_j'' \tag{13.60}$$

On eliminating the chemical potential of the first component between the two Gibbs–Duhem equations applicable to the two phases, we obtain

$$\left(S' - \frac{n_1' S''}{n_1''}\right) dT - \left(V' - \frac{n_1' V''}{n_1''}\right) dP + \sum_{i=2}^{C}\left(n_i' - \frac{n_i'' n_1'}{n_1''}\right) d\mu_i = 0 \tag{13.61}$$

When we introduce the usual expressions for the differentials of the chemical potentials for the double-primed phase, Equation (13.61) becomes

$$\sum_{i=1}^{C} n_i'(\bar{S}_i' - \bar{S}_i'')\, dT - \sum_{i=1}^{C} n_i'(\bar{V}_i' - \bar{V}_i'')\, dP + \sum_{i=2}^{C}\left(n_i' - \frac{n_i'' n_1'}{n_1''}\right) \sum_{j=2}^{C} \left(\frac{\partial \mu_i''}{\partial x_j''}\right)_{T,P,x} dx_j'' = 0 \tag{13.62}$$

after some simplification. This equation is the condition equation that must be used in conjunction with Equation (13.60). In so doing we must apply sufficient additional conditions to make the system univariant.

We first consider the determination of the surface concentrations. We eliminate the pressure between the two equations under the condition of constant temperature. When this is done, the single equation

$$d\gamma = -\frac{\sum_{i=2}^{C} \Gamma_{i(1)}\bar{V}_i''}{\sum_{i=1}^{C} x_i'(\bar{V}_i' - \bar{V}_i'')}\left[\sum_{i=2}^{C}\left(x_i' - \frac{x_i'' x_1'}{x_1''}\right) \sum_{j=2}^{C} \left(\frac{\partial \mu_i''}{\partial x_j''}\right)_{T,P,x''} dx_j''\right] - \sum_{i=2}^{C} \Gamma_{i(1)} \sum_{j=2}^{C} \left(\frac{\partial \mu_i''}{\partial x_j''}\right)_{T,P,x''} dx_j'' \tag{13.63}$$

is obtained. For a two-component system this equation becomes

$$d\gamma = -\Gamma_{2(1)}\left(\frac{(x_2' - x_2'')\bar{V}_2''}{x_i''[x_1'(\bar{V}_1' - \bar{V}_1'') + x_2'(\bar{V}_2' - \bar{V}_2'')]} + 1\right)\left(\frac{\partial \mu_2''}{\partial x_2''}\right)_{T,P} dx_2'' \tag{13.64}$$

The quantity within the large parentheses and the derivative, $(\partial \mu_2''/\partial x_2'')_{T,P}$, depend upon the properties of the bulk phases and can therefore be evaluated. If we then determine the change of the surface tension with the composition of the double-primed phase, the value of $\Gamma_{2(1)}$ can be determined. When the primed phase is a gas and the molar volume of the gas is very large with respect to that of the other phase, Equation (13.64) simplifies to

$$d\gamma = -\Gamma_{2(1)}\left(\frac{\partial \mu_2''}{\partial x_2''}\right)_{T,P} dx_2'' \tag{13.65}$$

In general, for a multicomponent system, there are $(C-1)$ different $\Gamma_{i(1)}$ to be evaluated. We can measure the surface tension of a series of solutions in each of which the mole fraction of one of the components is slightly different from that in all of the other solutions while the values of the other $(C-2)$ independent mole fractions are the same. There would be C solutions. By taking the differences between the surface tensions and between the mole fractions and applying these differences to Equation (13.63), we obtain $(C-1)$ independent equations, from which the value of each $\Gamma_{i(1)}$ can be calculated.

The derivative of the surface tension with respect to the temperature at constant composition of the double-primed phase is

$$\left(\frac{\partial \gamma}{\partial T}\right)_{x''} = -\left(S_{\sigma(1)} - \sum_{i=2}^{C} \Gamma_{2(1)}\bar{S}_i''\right) - \sum_{i=2}^{C} \Gamma_{i(1)}\bar{V}_i''\left(\frac{\partial P}{\partial T}\right)_{x'',\text{sat}} \tag{13.66}$$

From Equation (13.61) we have

$$\left(\frac{\partial P}{\partial T}\right)_{x'',\text{sat}} = \frac{\sum_{i=1}^{C} x_i'(\bar{S}_i' - \bar{S}_i'')}{\sum_{i=1}^{C} x_i'(\bar{V}_i' - \bar{V}_i'')} \tag{13.67}$$

so

$$\left(\frac{\partial \gamma}{\partial T}\right)_{x''} = -\left(S_{\sigma(1)} - \sum_{i=2}^{C} \Gamma_{i(1)}\bar{S}_i''\right) - \sum_{i=2}^{C} \Gamma_{i(1)}\bar{V}_i'' \frac{\sum_{i=1}^{C} x_i'(\bar{S}_i' - \bar{S}_i'')}{\sum_{i=1}^{C} x_i'(\bar{V}_i' - \bar{V}_i'')} \tag{13.68}$$

With a knowledge of the value of each $\Gamma_{i(1)}$ and the properties of the bulk phases, the quantity $(S_{\sigma(1)} - \sum_{i=2}^{C} \Gamma_{i(1)}\bar{S}_i'')$ can be evaluated. For a one-component system this equation reduces to

$$\left(\frac{\partial \gamma}{\partial T}\right)_{x''} = -S_\sigma \tag{13.69}$$

so that the entropy density of the surface is given by the value of the derivative. For a two-component system the equation becomes

$$\left(\frac{\partial \gamma}{\partial T}\right)_{x''} = -(S_{\sigma(1)} - \Gamma_{2(1)}\bar{S}''_2) - \Gamma_{2(1)}\bar{V}''_2 \frac{\bar{S}'_2 - \bar{S}''_2}{\bar{V}'_2 - \bar{V}''_2} \tag{13.70}$$

If the primed phase is a gas and the temperature is sufficiently below the critical temperature that the molar volume of the double-primed phase is small with respect to that of the primed phase, Equation (13.70) becomes

$$\left(\frac{\partial \gamma}{\partial T}\right)_{x''} = -(S_{\sigma(1)} - \Gamma_{2(1)}\bar{S}''_2) \tag{13.71}$$

We could evaluate the derivative $(\partial\gamma/\partial T)_{P,x''(C-2)}$ where the pressure is kept constant. In order to make the system univariant, $(C-2)$ mole fractions would have to be kept constant, and the remaining mole fraction would be dependent rather than independent. Such experiments would be somewhat difficult and no additional knowledge would be obtained. This derivative therefore has little importance.

We obtain the equation, for a change of pressure,

$$\left(\frac{\partial \gamma}{\partial P}\right)_{x''} = -\sum_{i=2}^{C} \Gamma_{i(1)}\bar{V}''_i - \left(S_{\sigma(1)} - \sum_{i=2}^{C} \Gamma_{i(1)}\bar{S}''_i\right) \frac{\sum_{i=1}^{C} x'_i(\bar{V}'_i - \bar{V}''_i)}{\sum_{i=1}^{C} x'_i(\bar{S}'_i - \bar{S}''_i)} \tag{13.72}$$

by use of the arguments similar to those used to obtain Equation (13.68). Here the temperature is a dependent variable and would have to be adjusted at each experimental pressure to make the composition of the double-primed phase constant. Although the experimental evaluation of the derivative may be difficult, the derivative can be evaluated because we observe that

$$\left(\frac{\partial \gamma}{\partial P}\right)_{x''} = \left(\frac{\partial \gamma}{\partial T}\right)_{x''} \frac{\sum_{i=1}^{C} x'_i(\bar{V}'_i - \bar{V}''_i)}{\sum_{i=1}^{C} x'_i(\bar{S}'_i - \bar{S}''_i)} \tag{13.73}$$

from comparison of Equations (13.68) and (13.72).

The comments that were made concerning the derivative $(\partial\gamma/\partial T)_{P,x''(C-2)}$ are also applicable to the derivative $(\partial\gamma/\partial P)_{T,x''(C-2)}$.

We finally consider the change of the surface tension with respect to the change of the mole fraction of one component when the mole fractions of all the other $(C-2)$ components are held constant at constant pressure or at constant temperature. Equations (13.60) and (13.62) become

$$\mathrm{d}\gamma = -\left(S_{\sigma(1)} - \sum_{i=2}^{C} \Gamma_{i(1)}\bar{S}_i''\right)\mathrm{d}T - \sum_{i=2}^{C} \Gamma_{i(1)}\bar{V}_i''\,\mathrm{d}P$$
$$- \sum_{i=2}^{C} \Gamma_{i(1)}\left(\frac{\partial \mu_i''}{\partial x_j''}\right)_{T,P,x''} \mathrm{d}x_j'' \qquad (13.74)$$

and

$$\sum_{i=1}^{C} x_i'(\bar{S}_i' - \bar{S}_i'')\,\mathrm{d}T - \sum_{i=1}^{C} x_i'(\bar{V}_i' - \bar{V}_i'')\,\mathrm{d}P$$
$$+ \sum_{i=2}^{C}\left(x_i' - \frac{x_i''x_1'}{x_1''}\right)\left(\frac{\partial \mu_i''}{\partial x_j''}\right)_{T,P,x''} \mathrm{d}x_j'' = 0 \qquad (13.75)$$

We obtain from these two equations the two relations

$$\left(\frac{\partial \gamma}{\partial x_j''}\right)_{T,x''(C-2)} = -\frac{\left(\sum_{i=2}^{C} \Gamma_{i(1)}\bar{V}_i''\right)\left[\sum_{i=2}^{C}\left(x_i' - \frac{x_i''x_1'}{x_1''}\right)\right]\left(\frac{\partial \mu_i''}{\partial x_j''}\right)}{\sum_{i=2}^{C} x_i'(\bar{V}_i' - \bar{V}_i'')}$$
$$- \sum_{i=2}^{C} \Gamma_{i(1)}\left(\frac{\partial \mu_i''}{\partial x_j''}\right)_{T,P,x''} \qquad (13.76)$$

and

$$\left(\frac{\partial \gamma}{\partial x_j''}\right)_{P,x''(C-2)} = \frac{\left(S_{\sigma(1)} - \sum_{i=2}^{C} \Gamma_{i(1)}\bar{S}_i''\right)\sum_{i=2}^{C}\left(x_i' - \frac{x_i''x_1'}{x_1''}\right)\left(\frac{\partial \mu_i''}{\partial x_j''}\right)}{\sum_{i=2}^{C} x_i'(\bar{S}_i' - \bar{S}_i'')}$$
$$- \sum_{i=2}^{C} \Gamma_{i(1)}\left(\frac{\partial \mu_i''}{\partial x_j''}\right)_{T,P,x''} \qquad (13.77)$$

Equation (13.76) may be written as

$$\left(\frac{\partial \gamma}{\partial x_j''}\right)_{T,x''(C-2)} = -\sum_{i=2}^{C} \Gamma_{i(1)}\left(\frac{\partial \mu_i''}{\partial x_j''}\right)_{T,P,x''} \qquad (13.78)$$

for a liquid–gas system when the molar volume of the gas is very large with respect to the molar volume of the liquid. For a binary system Equation (13.78) is

$$\left(\frac{\partial \gamma}{\partial x_2''}\right)_{T,P} = -\Gamma_{2(1)}\left(\frac{\partial \mu_2''}{\partial x_2''}\right)_{T,P} \qquad (13.79)$$

which is identical to Equation (13.65). Equation (13.77) cannot be simplified easily, but we have seen from our previous discussion that all quantities on the right-hand side can be evaluated, and thus the value of the derivatives can be calculated.

13.4 Curved surfaces

We have considered only planar surfaces in the previous discussion. Here we consider curved surfaces and discuss two effects related to the curved surface. The defined surface is constructed in a fashion similar to that used for planar surfaces. It lies wholly within the real surface and parallel to it. We assume that the principal curvatures of the surface are uniform. We further assume that the surface tension is independent of the curvature. Experiments have shown that this assumption is valid when the two radii of curvature are very large with respect to the thickness of the real surface. We have already stated that this thickness is approximately 10^{-7} cm. Great care must be used when this second assumption is not valid.

With these assumptions, the thermodynamic quantities for the defined surface are defined in the same way as for planar surfaces. The conditions of equilibrium are determined by the use of the three equations

$$\delta E' = T'\,\mathrm{d}S' - P'\,\delta V' + \sum_i \mu_i'\,\mathrm{d}n_i' \tag{13.80}$$

$$\delta E'' = T''\,\mathrm{d}S'' - P''\,\delta V'' + \sum_i \mu_i''\,\mathrm{d}n_i'' \tag{13.81}$$

$$\delta E^\sigma = T^\sigma\,\delta S^\sigma + \gamma\,\delta a + \sum_i \mu_i^\sigma\,\delta n_i^\sigma \tag{13.82}$$

As usual, the condition of equilibrium is that

$$\delta E = \delta E' + \delta E'' + \delta E^\sigma \geqslant 0 \tag{13.83}$$

at constant entropy, volume, and mole numbers. By the usual arguments we find that the temperatures of the three parts must be equal and that the chemical potentials of the separate components must be the same in the three parts. We are then left with the condition

$$\gamma\,\delta a - P'\,\delta V' - P''\,\delta V'' = 0 \tag{13.84}$$

where we have used the equality sign because all variations may be either positive or negative. We now vary the system by moving all parts of the defined surface by an equal distance δl normal to the surface. The variation in the area is given by

$$\delta a = \left(\frac{1}{r_1} + \frac{1}{r_2}\right) a\,\delta l \tag{13.85}$$

where r_1 and r_2 are the two radii of curvature and the variations of the two volumes are

$$\delta V'' = -\delta V' = a\,\delta l \tag{13.86}$$

When these quantities are substituted into Equation (13.84) we obtain

$$\left[\gamma\left(\frac{1}{r_1}+\frac{1}{r_2}\right)+P'-P''\right]a\,\delta l = 0 \tag{13.87}$$

The value of δl may be positive or negative, and therefore

$$\gamma\left(\frac{1}{r_1}+\frac{1}{r_2}\right) = P''-P' \tag{13.88}$$

This equation is the new condition of equilibrium, and we observe that the pressures on the two sides of the surfaces are not equal, the difference being given by Equation (13.88). The difference $(P''-P')$ is positive when the centers of the radii are in the double-primed phase. Equations similar to those obtained in Section 13.3 can be obtained subject to this difference of pressure, but we observe that only one pressure is independent. Both r_1 and r_2 are equal to infinity for a planar surface, and thus for such a surface the pressures are equal.

The second subject is the effect of the surface on the chemical potential of a component contained in a small drop. We consider a multicomponent system in which one phase is a bulk phase and the second phase is kept constant with the conditions that the interface between the two phases is contained wholly within the bulk phase and does not affect the external pressure. The differential of the Gibbs energy of a two-phase system may be written as

$$\mathrm{d}G = -S\,\mathrm{d}T + V\,\mathrm{d}P + \gamma\,\mathrm{d}a + \sum_i \mu_i\,\mathrm{d}n_i \tag{13.89}$$

For a planar surface between two phases, material can be added to one of the phases at constant area as well as constant temperature and constant pressure. However, if we add material to the small drop at constant temperature and constant pressure on the large phase, the area of the drop must change. Now

$$\mathrm{d}a = \left(\frac{1}{r_1}+\frac{1}{r_2}\right)\mathrm{d}V \tag{13.90}$$

and

$$\mathrm{d}V = \sum_i \bar{V}_i\,\mathrm{d}n_i \tag{13.91}$$

Substitution of Equations (13.90) and (13.91) into Equation (13.89) gives

$$dG = -S\,dT + V\,dP + \sum_i \left[\mu_i + \left(\frac{1}{r_1} + \frac{1}{r_2}\right)\gamma \bar{V}_i \right] dn_i \tag{13.92}$$

We therefore find that the chemical potential of a component within the drop is given by

$$\mu_i = \mu_i^{P} + \left(\frac{1}{r_1} + \frac{1}{r_2}\right)\gamma \bar{V}_i \tag{13.93}$$

where μ_i^{P} is the chemical potential of the component when the surface is planar. Thus, the chemical potential of a component in a small drop is greater than that in a phase of the same composition with a planar surface.

14

Equilibrium conditions in the presence of an external field

In all of the studies of thermodynamic equilibrium that have been presented in the previous chapters, we have neglected the effects of an external field on the equilibrium properties of a system. This has been justified because the field may be present only in specific cases, the effect of the field may be negligible, or the position of the system in the field may be unchanged. The conditions of equilibrium in the presence of a gravitational or centrifugal field, an electrostatic field, and a magnetic field are developed in this chapter.

A distinction must be made between gravitational effects for which the presence of material in the field does not change the intensity of the field, and the electrostatic and magnetic effects for which the presence of material within the field does alter the intensity. A complete treatment of electrostatic and magnetic effects would require a discussion of electromagnetic theory and the use of Maxwell's equations. However, we wish only to illustrate the thermodynamic effects of electric and magnetic fields. We therefore accept the results of a complete treatment and apply the results to simple systems.

THE GRAVITATIONAL AND CENTRIFUGAL FIELD

14.1 The gravitational and centrifugal potential

The gravitational field of the Earth is characterized by a potential, Φ, that has a definite value at each point in the field. For all practical purposes this field is independent of the presence of matter in the quantities used in normal thermodynamic systems. Within this approximation the field is independent of the state of a thermodynamic system within it. The potential can be written as

$$\Phi(r) = -gr \tag{14.1}$$

where g is the acceleration due to gravity and r is the distance from the center of the Earth to the position of the thermodynamic system. A centrifugal field has the same properties as a gravitational field. The potential for a centrifugal field is given as

$$\Phi(r) = -\tfrac{1}{2}\omega^2 r^2 \tag{14.2}$$

where ω is the angular velocity of rotation and r is the distance from the axis of rotation to the system.

The thermodynamic properties of a system in a gravitational field are emphasized in the following sections. However, because of the similarity of the two fields, the concepts and equations for the centrifugal field are the same.

14.2 The fundamental equations

The work done on a system of mass m in raising the system from r_1 to r_2 in the gravitational field is

$$W = m[\Phi(r_2) - \Phi(r_1)] \tag{14.3}$$

and the difference in the energy of the system at r_2 and that at r_1 is

$$E(r_2) - E(r_1) = m[\Phi(r_2) - \Phi(r_1)] \tag{14.4}$$

when all other variables that define the state of the system are kept constant.

Equations (14.3) and (14.4) are the basic new equations with which we will be concerned. However, we must distinguish between two cases. We know that certain properties of a system depend upon the position in the gravitational field. When the system extends over a sufficient distance in the field, this dependence is quite evident. However, when the extension of the system in the direction of the field is small, the difference in the properties in the direction of the field is quite small and usually negligible. In the latter case we may consider the entire system to be at a single position in the field. Any change in the position of the entire system simply changes the energy by the amount of work done in moving the system from one position to the other according to Equation (14.4). This change of energy is the change of the potential energy of the entire system, and is trivial from the point of view of the conditions of equilibrium in a thermodynamic system. In fact, it is this latter case that has been studied in all of the material presented in previous chapters.

For the first case we consider any isolated system having extension in the gravitational field. The system may be homogenous or heterogenous; that is, it may have one or more phases in the sense that a phase is a region of matter whose properties are either constant or continuously varying. Certain properties of any single phase are continuously varying in the gravitational field. We then consider the phase to be divided into a continuous sequence of regions, each region differing infinitesimally from its neighbors. The properties of each region are assumed to be constant in the direction of the field. This concept is illustrated in Figure 14.1, where r_0 designates any base from which a distance may be measured and dr is the distance between adjacent boundaries separating the phases. The distance dr is conceived to be infinitesimal when referred to macroscopic dimensions but large compared with molecular dimensions or mean distance between molecules. With this concept the entire system is divided into a large number of homogenous regions whose volumes are fixed.

The differential quantity of work done in moving dn_i moles of a component of the system from one homogenous region to another is given by

$$\mathrm{d}W = M_i(\Phi'' - \Phi')\,\mathrm{d}n_i \tag{14.5}$$

where M_i is the molecular mass of the component and Φ'' and Φ' are the potentials in the double-primed and primed regions, respectively. The increase in the value of the energy function of the double-primed region due *solely* to the work done in transferring dn_i moles of the component to the region is thus $M_i\Phi''\,\mathrm{d}n_i$. If now we consider the energy of the region to be a function of its entropy, volume, and mole numbers, the differential of the energy is

$$\mathrm{d}E'' = T''\,\mathrm{d}S'' - P''\,\mathrm{d}V'' + \sum_{i=1}^{C} (\mu_i'' + M_i\Phi'')\,\mathrm{d}n_i'' \tag{14.6}$$

This equation is the fundamental equation for the energy function, and one such equation is applicable to each homogenous region.

The quantity $(T''\,\mathrm{d}S'' - P''\,\mathrm{d}V'' + \sum_i \mu_i''\,\mathrm{d}n_i'')$ represents the differential of the energy function in the absence of the field. Yet it is this quantity which has been used throughout this book to express the differential of the energy

Figure 14.1. Division of a phase in a gravitational field into homogenous regions.

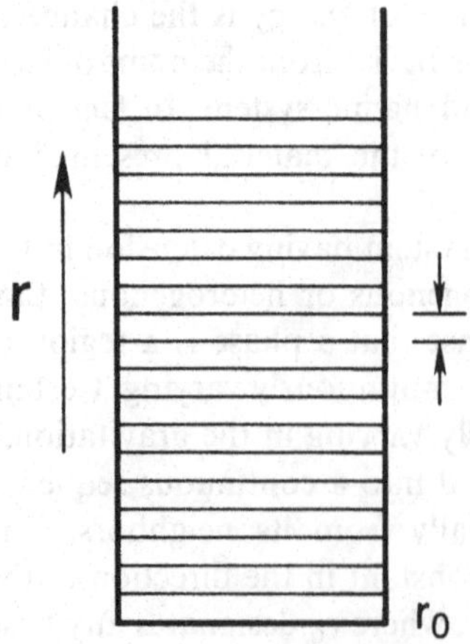

of a thermodynamic system at a fixed position in the gravitational field of the Earth rather than at a position where the field is zero. Actually, there is no problem. First, we do not know the absolute value of the energy for any system and can only determine the difference between the energies of a system in different states. Then, for any substance the quantity $M\Phi$ at a fixed position in the field is constant, and this quantity would cancel in obtaining the differences. Second, when we consider a gravitational field we are concerned only with the differences between the potentials at different positions in the field as indicated by Equations (14.3) and (14.5). Therefore, we are not concerned with the absolute value of the field, but can assign the zero value of the field at any arbitrary position in the field. Such a position might be depicted generally as r_0 in Figure 14.1. Such positions could be sea level or any arbitrary position in the laboratory. In the case of centrifugal fields, the zero point can easily be taken as the axis of rotation.

The fundamental equations for the enthalpy and the Gibbs and Helmholtz energies are obtained from Equation (14.6) by the same methods used previously. Then we have, for each homogenous region,

$$dH'' = T''\,dS'' + V''\,dP + \sum_{i=1}^{C} (\mu_i'' + M_i\Phi'')\,dn_i'' \tag{14.7}$$

$$dA'' = -S''\,dT'' - P''\,dV'' + \sum_{i=1}^{C} (\mu_i'' + M_i\Phi'')\,dn_i'' \tag{14.8}$$

and

$$dG'' = -S''\,dT'' + V''\,dP'' + \sum_{i=1}^{C} (\mu_i'' + M_i\Phi'')\,dn_i'' \tag{14.9}$$

Other equations can now be obtained. The energy of a region is still a homogenous function of first degree in the entropy, volume, and mole numbers. Then

$$E'' = T''S'' - P''V'' + \sum_{i=1}^{C} (\mu_i'' + M_i\Phi'')n_i'' \tag{14.10}$$

From this equation we obtain

$$H'' = T''S'' + \sum_{i=1}^{C} (\mu_i'' + M_i\Phi'')n_i'' \tag{14.11}$$

$$A'' = -P''V'' + \sum_{i=1}^{C} (\mu_i'' + M_i\Phi'')n_i'' \tag{14.12}$$

and

$$G'' = \sum_{i=1}^{C} (\mu_i'' + M_i\Phi'')n_i'' \tag{14.13}$$

The equation corresponding to the Gibbs–Duhem equation is

$$S''\,dT'' - V''\,dP'' + \sum_{i=1}^{C} n_i''\,d\mu_i'' + \sum_{i=1}^{C} n_i'' M_i\,d\Phi'' = 0 \tag{14.14}$$

Note that each of Equations (14.6)–(14.14) reduces to the same previous equation when we choose a fixed position in the field and define Φ'' to be zero at that position.

The partial molar entropies and volumes in a region are independent of the position in the gravitational field, because the field is an external field and independent of the state of a system in the field. Thus, by use of the conditions of exactness applied to Equation (14.9), we obtain for the partial molar entropy

$$\left(\frac{\partial S''}{\partial n_k''}\right)_{T,P,n} = -\left(\frac{\partial(\mu_k'' + M_k\Phi'')}{\partial T}\right)_{P,n} = -\left(\frac{\partial \mu_k''}{\partial T}\right)_{P,n} = \bar{S}_k'' \tag{14.15}$$

Similarly,

$$\left(\frac{\partial V''}{\partial n_k''}\right)_{T,P,n} = \left(\frac{\partial(\mu_k'' + M_k\Phi'')}{\partial P}\right)_{T,n} = \left(\frac{\partial \mu_k''}{\partial P}\right)_{T,n} = \bar{V}_k'' \tag{14.16}$$

14.3 Conditions of equilibrium

The same methods used in Chapter 5 are used here. The basic condition of equilibrium is that the variation of the energy of an isolated, closed system at equilibrium must be zero or greater than zero for any possible variation of the system. The total energy of the system is the sum of the energies of all of the homogenous regions. Similarly, the entropy and volume of the system is the sum of the entropies and volumes of all of the regions. Then, for equilibrium,

$$(\delta E)_{S,V,n} = \left(\sum' \delta E'\right)_{S,V,n} \geqslant 0 \tag{14.17}$$

where the sum is over all of the regions. The variation of the energy of any region is given by

$$\delta E' = T'\,\delta S' + \sum_{i=1}^{C} (\mu_i' + M_i\Phi')\,\delta n_i' \tag{14.18}$$

A term $P'V'$ does not appear in Equation (14.18), because the volumes of the regions are fixed by definition and $\delta V' = 0$. The condition that the entropy of the system is constant for any possible variation leads to the determination that the temperature is uniform throughout the system. The condition of constant mass of each component in the system leads to the new condition of equilibrium that

$$\mu_i' + M_i\Phi' = \mu_i'' + M_i\Phi'' = \mu x_i''' + M_i\Phi''' = \cdots \tag{14.19}$$

where the primes refer to the various regions in the system. This equality is valid for each region in which the component exists. It is the quantity $(\mu_i + M_i\Phi)$ rather than μ_i alone that determines the equilibrium condition related to the components. When a component does not exist in a given region, then the quantity $(\mu_i + M_i\Phi)$ in this region must have a value greater than that in a region in which the component does exist.

If the mole numbers in Equation (14.18) referred to species rather than components, then the condition of constant mass of components would lead to the condition for chemical equilibrium

$$\sum_i \nu_i(\mu_i' + M_i\Phi') = 0 \tag{14.20}$$

for each independent chemical reaction written as $\sum_i \nu_i B_i = 0$, where B_i refers to the individual species entering the reaction. The expansion of Equation (14.20) yields

$$\sum_i \nu_i\mu_i' + \Phi_i' \sum_i \nu_i M_i = 0 \tag{14.21}$$

However, $\sum_i \nu_i M_i = 0$ because of the conservation of mass and, therefore, we have for chemical equilibrium the condition

$$\sum_i \nu_i\mu_i' = 0 \tag{14.22}$$

We thus determine that the condition for chemical equilibrium is independent of the field and is the same throughout the entire system.

14.4 Dependence of pressure on the potential

We were not able to obtain a condition of equilibrium with regard to the pressure in the previous section, because a change of volume of a region was not considered to be a possible variation. We consider the problem of the dependence of pressure on the potential in this section.

Let us consider a closed, single-phase system containing C components in a gravitational field. The state of the system is fixed and the system is in equilibrium. The condition of equilibrium is that for each component the quantity $(\mu_i + M_i\Phi)$ must have the same value in each homogenous region of the phase. In general, $(\mu_i + M_i\Phi)$ is a function of the temperature, pressure, $(C-1)$ mole fractions, and the potential, so the differential of $(\mu_i + M_i\Phi)$ may be written as

$$d(\mu_i + M_i\Phi) = -\bar{S}_i\, dT + \bar{V}_i\, dP + \sum_{j=1}^{C-1} \left(\frac{\partial \mu_i}{\partial x_j}\right)_{T,P,x,\Phi} dx_j + M_i\, d\Phi \tag{14.23}$$

With the knowledge that the pressure and each mole fraction are continuous functions of r and the temperature is independent of r, we can obtain the derivative

$$\left(\frac{\partial(\mu_i + M_i\Phi)}{\partial r}\right)_T = \bar{V}_i\left(\frac{\partial P}{\partial r}\right)_T + \sum_{j=1}^{C-1}\left(\frac{\partial \mu_i}{\partial x_j}\right)_{T,P,x,\Phi}\left(\frac{\partial x_j}{\partial r}\right)_T + M_i\frac{\mathrm{d}\Phi}{\mathrm{d}r} \tag{14.24}$$

However, $[\partial(\mu_i + M_i\Phi)/\partial r]_T = 0$. Therefore,

$$\bar{V}_i\left(\frac{\partial P}{\partial r}\right)_T + \sum_{j=1}^{C-1}\left(\frac{\partial \mu_i}{\partial x_j}\right)_{T,P,x,\Phi}\left(\frac{\partial x_j}{\partial r}\right)_T + M_i\frac{\mathrm{d}\Phi}{\mathrm{d}r} = 0 \tag{14.25}$$

When we multiply Equation (14.25) by x_i and sum over all of the components we obtain

$$\tilde{V}\left(\frac{\partial P}{\partial r}\right)_T + \sum_{i=1}^{C} x_i M_i\frac{\mathrm{d}\Phi}{\mathrm{d}r} = 0 \tag{14.26}$$

with the knowledge that $\tilde{V} = \sum_{i=1}^{C} x_i\bar{V}_i$ and $\sum_{i=1}^{C} x_i \sum_{j=1}^{C-1} (\partial\mu_i/\partial x_j)_{T,P,x,\Phi}\,\mathrm{d}x_j = 0$ from the Gibbs–Duhem equation.

Equation (14.26) gives the dependence of the pressure in the phase as a function of r, the variable that determines the position in the field, in terms of the molar volume and the average molecular mass. Both of these latter quantities are themselves dependent on the position of the field.

In a system that contains one or more phases, Equation (14.26) is applicable to each phase. The pressure of the system must be a continuous function of r even in a heterogenous system, but the derivative $(\partial P/\partial r)_T$ is discontinuous at a phase boundary. Then the pressure must be the same on either side of a phase boundary (neglecting surface effects) when the boundary is wholly within a single region, no matter how small the thickness of the region. The derivative $(\partial P/\partial r)_T$ however, will have a different value on either side of the boundary.

14.5 Dependence of composition on the potential

We first consider a closed, single-phase system containing C components in a gravitational field. The state of the system is fixed and the system is in equilibrium. Equation (14.25) is applicable to each component in the system. We have then a set of equations, C in number, relating $(\partial P/\partial r)_T$ and the $(C-1)$ derivatives $(\partial x_j/\partial r)_T$. The solution of this set of equations then gives equations for each of the derivatives. Equation (14.26) might be used to replace $(\partial P/\partial r)_T$, but this equation is not independent of the set of equations given by Equation (14.25) because it is derived from this set. If Equation (14.26) is used, then one of the C equations from Equation (14.25) would be omitted.

The complete solution of the problem would require the integration of the derivatives to obtain P and each independent x_i as a function of r, with the knowledge of the required integration constants. These would be obtained by fixing the value of the pressure at some position in the field and also fixing the value of each mole fraction at some position in the field.

As an illustration let us consider a two-component system, make use of Equation (14.26), and use x_1 as the independent mole fraction. The set of equations are

$$\bar{V}_1\left(\frac{\partial P}{\partial r}\right)_T + \left(\frac{\partial \mu_1}{\partial x_1}\right)_{T,P,\phi}\left(\frac{\partial x_1}{\partial r}\right)_T + M_1\frac{\mathrm{d}\Phi}{\mathrm{d}r} = 0 \qquad (14.27)$$

and

$$\bar{V}\left(\frac{\partial P}{\partial r}\right)_T + [M_2 + (M_1 - M_2)x_1]\frac{\mathrm{d}\Phi}{\mathrm{d}r} = 0 \qquad (14.28)$$

This set may be solved in principle to yield P and x_1, each as a function of r, with the knowledge of a value of P and x_1 for fixed values of r. The solution is easily obtained on the assumption that the ideal solution laws are applicable and that either the ideal gas equation is followed in the case of a gas phase or the volumes are independent of the pressure. Then, for an ideal liquid solution, Equation (14.27) becomes

$$\tilde{V}_1^{\bullet}\left(\frac{\partial P}{\partial r}\right)_T + \frac{RT}{x_1}\left(\frac{\partial x_1}{\partial r}\right)_T + M_1\frac{\mathrm{d}\Phi}{\mathrm{d}r} = 0 \qquad (14.29)$$

The elimination of $(\partial P/\partial r)_T$ between Equations (14.28) and (14.29) yields

$$\left(\frac{\tilde{V}_1^{\bullet}}{x_2} + \frac{\tilde{V}_2^{\bullet}}{x_1}\right)\left(\frac{\partial x_1}{\partial r}\right)_T = -\frac{(M_1\tilde{V}_2^{\bullet} - M_2\tilde{V}_1^{\bullet})}{RT}\frac{\mathrm{d}\Phi}{\mathrm{d}r} \qquad (14.30)$$

which on integration gives

$$\tilde{V}_2^{\bullet}\ln\frac{x_1}{x_1^0} - \tilde{V}_1^{\bullet}\ln\frac{1-x_1}{1-x_1^0} = -\left(\frac{M_1\tilde{V}_2^{\bullet} - M_2\tilde{V}_1^{\bullet}}{RT}\right)[\Phi(r) - \Phi(r_0)] \qquad (14.31)$$

where x_1^0 is the value of x_1 at r_0. The substitution of this equation for x_1 in Equation (14.28) and subsequent integration gives P as a function of r, given P_0 at r_0.

Similar calculations would be carried out for heterogenous systems provided the phase boundary was wholly confined in a single region. The pressure at the phase boundary would be the same on either side of the boundary, as described in Section 14.4. The composition on one side of the boundary would be determined from knowledge of the composition on the other side of the boundary by the condition that $\mu_i^\alpha = \mu_i^\beta$ for each component or species, where α and β designate the two phases.

14.6 The Gibbs–Duhem equation and the phase rule

The Gibbs–Duhem equation for a homogenous region in any phase is given by Equation (14.14). This equation, expressed in molar quantities, is

$$\tilde{S}' \, \mathrm{d}T - \tilde{V}' \, \mathrm{d}P' + \sum_i x_i' \, \mathrm{d}(\mu_i + M_i\Phi) = 0 \tag{14.32}$$

with the use of the conditions of equilibrium. Such an equation can be written for one arbitrarily chosen homogenous region in each phase, so we would have P such equations for a system containing P phases. The pressures of the homogenous regions are not independent, but are determined by Equation (14.26) and one pressure, P', in one region. This one pressure, then, is the pressure variable. The set of P such equations are now equivalent to the set of equations used in Section 5.10 with the intensive variables being T, P', and the C variables $(\mu_i + M_i\Phi)$ rather than T, P, and μ_i. The phase rule is therefore unchanged when $(\mu_i + M_i\Phi)$ is used in place of μ_i.

Each $(\mu_i + M_i\Phi)$is a function of $(C-1)$ mole fractions and Φ. Therefore, in terms of independent variables, the mole fraction of a component at a fixed position in the potential field, x_i, may be substituted for the corresponding $(\mu_i + M_i\Phi)$.

14.7 The definition of the state of the system

We have discussed only the dependence of the intensive variables, temperature, pressure, and mole fractions, on the position in the field. With this knowledge we can now discuss the definition of the state of a system in a potential field. We consider a heterogenous system containing C components and divide the entire system into homogenous regions of infinitesimal depth. The energy of a given region is given by Equation (14.10). The energy of a phase, indicated by superscript Greek letters, is the sum of the energies of all of the regions in the phase. Then

$$E^\alpha = \sum^{\prime} E' = T \sum^{\prime} S' - \sum^{\prime} P'V' + \sum_{i=1}^{C} (\mu_i + M_i\Phi) \sum^{\prime} n_i' \tag{14.33}$$

where we have used the knowledge that $T$and $(\mu_i + M_i\Phi)$ are both constant throughout the system. The entropy of the phase is the sum of the entropies of the regions, and the moles of a component in each phase is the sum of the moles of the component in each region. Also, if the volumes of each region are infinitesimal we can write

$$P'V' = \int_{V^\alpha} P \, \mathrm{d}V \tag{14.34}$$

where the integral is over the volume of the phase. Then Equation (14.34) can be written as

$$E^{\alpha} = TS^{\alpha} - \int_{V^{\alpha}} P\,\mathrm{d}V + \sum_{i=1}^{C} (\mu_i + M_i\Phi)n_i \tag{14.35}$$

The energy of the total system is the sum of the energies of the phases, so

$$E_{\mathrm{sys}} = TS_{\mathrm{sys}} - \int_{V} P\,\mathrm{d}V + \sum_{i=1}^{C} (\mu_i + M_i\Phi)n_{i,\mathrm{sys}} \tag{14.36}$$

where

$$S_{\mathrm{sys}} = \sum^{\alpha} S^{\alpha}$$

$$n_{i,\mathrm{sys}} = \sum^{\alpha} n_i^{\alpha}$$

and the integral is over the entire volume of the system.

Equation (14.35) shows that the energy of the system is a function of the entropy, volume, and mole numbers, as before, but with one addition. Either from Equation (14.4) or from the fact that

$$\left(\frac{\partial E_{\mathrm{sys}}}{\partial n_i}\right)_{S,V,n} = \mu_i + M_i\Phi$$

it is evident that the energy of the system depends also upon the potential. The potential is a function of r by Equations (14.1) and (14.2) and, therefore, to define the potential we need only to fix the position of one homogenous region in the field. It is accepted that all of the homogenous regions are contiguous.

The concept of the volume as an independent variable needs further discussion. We are interested in the extension of the system in the field. Here we are actually concerned with the variable r rather than the volume. In so doing we need to have information concerning the cross section of the system perpendicular to the field. The cross section may or may not be a function of the extension in the field. This requirement is consistent with Equation (14.10) and the equations that have been developed in this section from Equation (14.10), because we have used the concept of the volume of a homogenous region but have defined only the thickness, dr, of the region in the direction of the field. A knowledge of the cross section is thus required.

We then see that the state of a system is defined by assigning values to the entropy, volume, mole numbers, and the position of a single homogenous region in the field. However, in so doing we need also to have a knowledge of the cross section of the system.

The possibility of substituting an intensive variable for an extensive variable is determined by the number of degrees of freedom that are permitted for the system, as discussed in Chapter 5. Thus, when such substitutions are permitted, the temperature may be used in place of the entropy, and the pressure at some fixed position in the field may be used instead of the volume.

As an illustration consider a single-component, two-phase system in a potential field. Let the phases be a gas and a liquid phase. Such a system has one degree of freedom, and we then choose the temperature, volume, and moles of the component to be the variables that define the state of the system. The cross section of the container, A, is uniform. Let the bottom of the container be at the position r_1 in the field. Figure 14.2 illustrates this system where r_0 is the position of the phase boundary. With knowledge of the volume of the system and the cross section of the container, $(r_2 - r_1)$ is known. The unknown quantities are r_0 and the moles of the component in each phase. The pressure at the phase boundary is the vapor pressure of the liquid at the chosen temperature. Then the number of moles of the component in the gas phase is

$$n(\mathrm{g}) = \int_{r_0}^{r_2} \frac{A\,dr}{\tilde{V}(\mathrm{g})} \tag{14.37}$$

and the moles of the component in the liquid phase is

$$n(\ell) = \int_{r_1}^{r_0} \frac{A\,dr}{\tilde{V}(\ell)} \tag{14.38}$$

Finally, the sum of the moles of the component in the two phases must equal the original number of moles of component, n. Therefore

$$n = \int_{r_1}^{r_0} \frac{A\,dr}{\tilde{V}(\ell)} + \int_{r_0}^{r_2} \frac{A\,dr}{\tilde{V}(\mathrm{g})} \tag{14.39}$$

Figure 14.2. A two-phase, one-component system in a gravitational field.

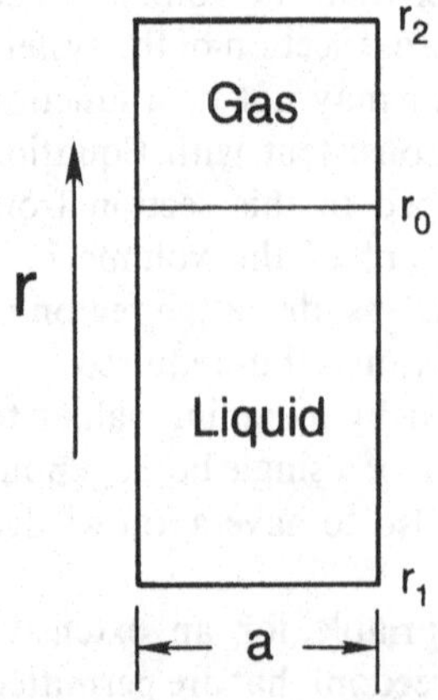

This equation is sufficient to determine r_0 and hence $n(\mathrm{g})$ and $n(\ell)$ when the molar volumes of each phase are known as a function of T, P, and r. The pressure can be determined as a function of r by means of Equation (14.26) and the knowledge of the pressure at r_0.

SYSTEMS IN AN ELECTROSTATIC FIELD

In studying systems in an electrostatic field, we must consider two systems because of the dependence of the field on matter within the field. One system is a parallel-plate condenser in empty space. The area of the plates is designated by A, and the distance between the plates by l. The other is an identical condenser immersed in an isotropic, homogenous, dielectric medium. The conductivity of the medium is zero, so no free charges are present in the medium. Edge effects are neglected and rational units are used throughout.

14.8 The condenser in empty space

We consider a parallel-plate condenser that has charges $+Q$ and $-Q$ on the plates. A *potential difference*, $\Delta\Phi$, is defined so that the work required to move a differential quantity of positive charge from the negative to the positive plate is given by $\Delta\Phi\, dQ$. The electric field strength, $\mathbf{E}$, is given by $\Delta\Phi/l$. The *permittivity of empty space*, ε_0, is given by

$$\varepsilon_0 \mathbf{E} = Q/A \tag{14.40}$$

where Q/A is the charge density on either one of the plates.

14.9 The condenser in a dielectric medium

We now consider the same parallel-plate condenser immersed in the dielectric medium. The charges on the two plates are represented again by the symbols $+Q$ and $-Q$. The *permittivity of the fluid*, ε, is defined by

$$\varepsilon \mathbf{E} = Q/A \tag{14.41}$$

similar to Equation (14.40). If we write this as

$$\varepsilon \mathbf{E} = Ql/V_c \tag{14.42}$$

where V_c is the volume of the condenser, we see that $\varepsilon\mathbf{E}$ is the electric moment of the charged condenser per unit volume of the condenser. When edge effects are neglected, V_c is also the volume of the liquid contained between the plates of the condenser.

The dielectric medium is polarized when it is contained between the charged plates of the condenser. The electric moment of the condenser then may be considered as the sum of the electric moment of the condenser itself and the polarization of the medium. Thus, Equation (14.42) may be written as

$$Ql/V_c = \varepsilon \mathbf{E} = \varepsilon_0 \mathbf{E} + \mathbf{p} \tag{14.43}$$

where $\mathbf{p}$ is the *polarization of the medium per unit volume.* The *electric displacement*, $\mathbf{D}$, is defined as

$$\mathbf{D} = \varepsilon \mathbf{E} = \varepsilon_0 \mathbf{E} + \mathbf{p} \tag{14.44}$$

We see that $\mathbf{p}$ may be written as

$$\mathbf{p} = \left(\frac{\varepsilon}{\varepsilon_0} - 1\right)\varepsilon_0 \mathbf{E} \tag{14.45}$$

The quantity $\varepsilon/\varepsilon_0$ is called the *dielectric constant* when $\mathbf{E}$ is independent of the field, and $[(\varepsilon/\varepsilon_0) - 1]$ is called the *electric susceptibility*, whose usual symbol is χ_e.

Of the three quantities $\mathbf{D}$, $\mathbf{E}$, and $\mathbf{p}$, only one is independent according to Equation (14.44). In the case of the parallel-plate condenser and an isotropic medium, all of the vectors are parallel and normal to the plates of the condenser. We are primarily concerned with their scalar values; however, we continue to use the vector symbols for clarity.

14.10 Work associated with electrostatic fields

The differential quantity of work done on the system when a differential quantity of charge is transferred from the positive to the negative plate is given by

$$\mathrm{d}W = \Delta\Phi \, \mathrm{d}Q \tag{14.46}$$

where

$$\Delta\Phi = \mathbf{E}l = Ql/A\varepsilon \tag{14.47}$$

Then

$$\mathrm{d}Q = A \, \mathrm{d}(\varepsilon \mathbf{E}) \tag{14.48}$$

and

$$\mathrm{d}W = V_c \mathbf{E} \, \mathrm{d}(\varepsilon \mathbf{E}) = V_c \mathbf{E} \, \mathrm{d}\mathbf{D} \tag{14.49}$$

In this derivation it is assumed that the field and displacement are uniform throughout the volume of the condenser, and that the field is zero outside the condenser. This is true for the isotropic dielectric medium with the edge effects neglected. If this were not true, then the work done on differential volumes would have to be considered and the total work would be obtained by integration over the volume of the condenser.

14.11 Thermodynamics of the total system

We choose the total system to be the condenser and the entire dielectric medium. The condenser is immersed in the medium which, for purposes of this discussion, is taken to be a single-phase, multicomponent system. The pressure on the system is the pressure exerted by the surroundings on a surface of the dielectric. In setting up the thermodynamic equations we omit the properties of the metal plates, because these remain constant except for a change of temperature. The differential change of energy of the system is expressed as a function of the entropy, volume, and mole numbers, but with the addition of the new work term. Thus,

$$dE = T\,dS - P\,dV + \sum_i \mu_i\,dn_i + V_c\mathbf{E}\,d\mathbf{D} \tag{14.50}$$

where V_c is the volume of the condenser. The differentials of the Helmholtz and Gibbs energies are expressed by

$$dA = -S\,dT - P\,dV + \sum_i \mu_i\,dn_i + V_c\mathbf{E}\,d\mathbf{D} \tag{14.51}$$

and

$$dG = -S\,dT + V\,dP + \sum_i \mu_i\,dn_i + V_c\mathbf{E}\,d\mathbf{D} \tag{14.52}$$

It is evident that μ_i represents the chemical potential of the ith component when $\mathbf{E} = 0$. However, we must determine the expression for the chemical potential of the components that are contained in the volume of the condenser. We use the symbol μ_i^ε for this chemical potential. With the use of Equation (14.51) we extend the definition of the chemical potential, so

$$\mu_i^\varepsilon = \left(\frac{\partial A}{\partial n_i}\right)_{T,V,n,\mathbf{D}} \tag{14.53}$$

We see from Equations (14.41) and (14.44) that $\mathbf{D}$ is actually the charge density, and it is the quantity that is held constant. However, $\mathbf{D} = \varepsilon\mathbf{E}$ and it is this product that is held constant while both ε and $\mathbf{E}$ depend upon the mole numbers. Thus

$$d\mathbf{D} = d(\varepsilon\mathbf{E}) = \varepsilon\,d\mathbf{E} + \mathbf{E}\,d\varepsilon \tag{14.54}$$

and, when $\mathbf{D}$ is constant,

$$\mathbf{E}\,d\varepsilon = -\varepsilon\,d\mathbf{E} \tag{14.55}$$

Now, assuming ε is constant,

$$V_c\mathbf{E}\,d\mathbf{D} = V_c\mathbf{E}\,d(\varepsilon\mathbf{E}) = \tfrac{1}{2}V_c\,d(\varepsilon\mathbf{E}^2) \tag{14.56}$$

so

$$\mu_i^{\varepsilon} = \mu_i^{c} + \tfrac{1}{2}V_c\left(\frac{\partial(\varepsilon\mathbf{E}^2)}{\partial n_i}\right)_{T,V,n,\mathbf{D}} \tag{14.57}$$

or, by Equation (14.55),

$$\mu_i^{\varepsilon} = \mu_i^{c} - \tfrac{1}{2}V_c\mathbf{E}^2\left(\frac{\partial\varepsilon}{\partial n_i}\right)_{T,V,n,\mathbf{D}} \tag{14.58}$$

where μ_i^c refers to the chemical potential of the ith component in the fluid within the condenser excluding the contribution of the term involving the field. The condition of equilibrium between the material within the condenser and that outside of the condenser must be that the chemical potentials, μ_i^{ε} and μ_i, must be equal.

There are several relations, similar to the Maxwell relations, that can be obtained from Equations (14.51) and (14.52). Upon application of the conditions of exactness to each of the Equations (14.51) and (14.52) we obtain

$$\left(\frac{\partial S}{\partial \mathbf{D}}\right)_{T,V,n} = -V_c\left(\frac{\partial \mathbf{E}}{\partial T}\right)_{V,n,\mathbf{D}} = -(\varepsilon/\varepsilon_0)V_c\mathbf{E}\left[\frac{\partial(\varepsilon_0/\varepsilon)}{\partial T}\right]_{V,n,\mathbf{D}} \tag{14.59}$$

$$\left(\frac{\partial P}{\partial \mathbf{D}}\right)_{T,V,n} = -V_c\left(\frac{\partial \mathbf{E}}{\partial V}\right)_{T,n,\mathbf{D}} = -(\varepsilon/\varepsilon_0)V_c\mathbf{E}\left[\frac{\partial(\varepsilon_0/\varepsilon)}{\partial V}\right]_{T,n,\mathbf{D}} \tag{14.60}$$

$$\left(\frac{\partial \mu_k}{\partial \mathbf{D}}\right)_{T,V,n} = V_c\left(\frac{\partial \mathbf{E}}{\partial n_k}\right)_{T,V,n,\mathbf{D}} = (\varepsilon/\varepsilon_0)V_c\mathbf{E}\left[\frac{\partial(\varepsilon_0/\varepsilon)}{\partial n_k}\right]_{T,V,n,\mathbf{D}} \tag{14.61}$$

from Equation (14.51) and

$$\left(\frac{\partial S}{\partial \mathbf{D}}\right)_{T,P,n} = -V_c\left(\frac{\partial \mathbf{E}}{\partial T}\right)_{P,n,\mathbf{D}} = -(\varepsilon/\varepsilon_0)V_c\mathbf{E}\left[\frac{\partial(\varepsilon_0/\varepsilon)}{\partial T}\right]_{P,n,\mathbf{D}} \tag{14.62}$$

$$\left(\frac{\partial V}{\partial \mathbf{D}}\right)_{T,P,n} = V_c\left(\frac{\partial \mathbf{E}}{\partial P}\right)_{T,n,\mathbf{D}} = (\varepsilon/\varepsilon_0)V_c\mathbf{E}\left[\frac{\partial(\varepsilon_0/\varepsilon)}{\partial P}\right]_{T,n,\mathbf{D}} \tag{14.63}$$

$$\left(\frac{\partial \mu_k}{\partial \mathbf{D}}\right)_{T,P,n} = V_c\left(\frac{\partial \mathbf{E}}{\partial n_k}\right)_{T,P,n,\mathbf{D}} = (\varepsilon/\varepsilon_0)V_c\mathbf{E}\left[\frac{\partial(\varepsilon_0/\varepsilon)}{\partial n_k}\right]_{T,P,n,\mathbf{D}} \tag{14.64}$$

from Equation (14.52). The far right-hand side of each of these equations is expressed in terms of the volume of the condenser, the electric field, and the properties of the dielectric constant (relative permittivity) of the fluid. For sufficiently low field strengths the left-hand side of the equations could be expressed in terms of the variable **E** or **p** by use of Equation (14.44).

The dielectric constant of most substances decreases with increasing temperature, so the entropy of the system generally decreases with an increase of the displacement according to Equations (14.59) and (14.62). The volume of the system usually decreases with increasing displacement according to Equation (14.63), because the dielectric constant usually increases with increasing pressure. The effect is called *electrostriction*.

14.12 Thermodynamics of the dielectric medium

It is of interest to obtain thermodynamic relations that pertain to the dielectric medium alone. The system is identical to that described in Section 14.11. However, in developing the equations we exclude the electric moment of the condenser in empty space. We are concerned, then, with the work done on the system in polarizing the medium. Instead of $\mathbf{D}$ we use $(\mathbf{D}-\varepsilon_0\mathbf{E})$, which is equal to the polarization per unit volume of the medium, $\mathbf{p}$. Finally, we define $\mathbf{P}$, the *total polarization*, to be equal to $V_c\mathbf{p}$. Now the equation for the differential of the energy is

$$dE = T\,dS - P\,dV + \sum_i \mu_i\,dn_i + \mathbf{E}\,d\mathbf{P} \tag{14.65}$$

The differentials of the Helmholtz and Gibbs energies could be written in the usual manner. However, it is convenient to use the functions $(A-\mathbf{EP})$ and $(G-\mathbf{EP})$ in order to change the independent variable from $\mathbf{P}$ to $\mathbf{E}$. When this is done we obtain

$$d(A-\mathbf{EP}) = -S\,dT - P\,dV + \sum_i \mu_i\,dn_i - \mathbf{P}\,d\mathbf{E} \tag{14.66}$$

and

$$d(G-\mathbf{EP}) = -S\,dT + V\,dP + \sum_i \mu_i\,dn_i - \mathbf{P}\,d\mathbf{E} \tag{14.67}$$

The dependence of the total polarization on the field can be expressed in either of two ways. From Equation (14.44) we have

$$\mathbf{P} = V_c(\varepsilon-\varepsilon_0)\mathbf{E} \tag{14.68}$$

and from Equation (14.45), with the definition of χ_e

$$\mathbf{P} = V_c\varepsilon_0\chi_e\mathbf{E} \tag{14.69}$$

From the conditions of exactness we obtain the six equations

$$\left(\frac{\partial S}{\partial \mathbf{E}}\right)_{T,V,n} = \left(\frac{\partial \mathbf{P}}{\partial T}\right)_{V,n,\mathbf{E}} = V_c\varepsilon_0\mathbf{E}\left[\frac{\partial(\varepsilon/\varepsilon_0)}{\partial T}\right]_{V,n,\mathbf{E}} = V_c\varepsilon_0\mathbf{E}\left(\frac{\partial \chi_e}{\partial T}\right)_{V,n,\mathbf{E}} \tag{14.70}$$

$$\left(\frac{\partial P}{\partial \mathbf{E}}\right)_{T,V,n} = -\left(\frac{\partial \mathbf{P}}{\partial V}\right)_{T,n,\mathbf{E}} = -V_c\varepsilon_0\mathbf{E}\left[\frac{\partial(\varepsilon/\varepsilon_0)}{\partial V}\right]_{T,n,\mathbf{E}} = -V_c\varepsilon_0\mathbf{E}\left(\frac{\partial \chi_e}{\partial V}\right)_{T,n,\mathbf{E}} \tag{14.71}$$

$$\left(\frac{\partial \mu_k}{\partial \mathbf{E}}\right)_{T,V,n} = -\left(\frac{\partial \mathbf{P}}{\partial n_k}\right)_{T,V,n,\mathbf{E}} = -V_c\varepsilon_0\mathbf{E}\left[\frac{\partial(\varepsilon/\varepsilon_0)}{\partial n_k}\right]_{T,V,n,\mathbf{E}}$$

$$= -V_c\varepsilon_0\mathbf{E}\left(\frac{\partial \chi_e}{\partial n_k}\right)_{T,V,n,\mathbf{E}} \tag{14.72}$$

from Equation (14.66) and

$$\left(\frac{\partial S}{\partial \mathbf{E}}\right)_{T,P,n} = \left(\frac{\partial \mathbf{P}}{\partial T}\right)_{P,n,\mathbf{E}} = V_c\varepsilon_0\mathbf{E}\left[\frac{\partial(\varepsilon/\varepsilon_0)}{\partial T}\right]_{P,n,\mathbf{E}} = V_c\varepsilon_0\mathbf{E}\left(\frac{\partial \chi_e}{\partial T}\right)_{P,n,\mathbf{E}} \tag{14.73}$$

$$\left(\frac{\partial V}{\partial \mathbf{E}}\right)_{T,P,n} = -\left(\frac{\partial \mathbf{P}}{\partial P}\right)_{T,n,\mathbf{E}} = -V_c\varepsilon_0\mathbf{E}\left[\frac{\partial(\varepsilon/\varepsilon_0)}{\partial P}\right]_{T,n,\mathbf{E}}$$

$$= -V_c\varepsilon_0\mathbf{E}\left(\frac{\partial \chi_e}{\partial P}\right)_{T,n,\mathbf{E}} \tag{14.74}$$

$$\left(\frac{\partial \mu_k}{\partial \mathbf{E}}\right)_{T,P,n} = -\left(\frac{\partial \mathbf{P}}{\partial n_k}\right)_{T,P,n,\mathbf{E}} = -V_c\varepsilon_0\mathbf{E}\left[\frac{\partial(\varepsilon/\varepsilon_0)}{\partial n_k}\right]_{T,P,n,\mathbf{E}}$$

$$= -V_c\varepsilon_0\mathbf{E}\left(\frac{\partial \chi_e}{\partial n_k}\right)_{T,P,n,\mathbf{E}} \tag{14.75}$$

from Equation (14.67). Again, we observe that both the entropy and the volume decrease with an increase of the field.

The chemical potential for the material within the condenser is obtained from

$$\mathrm{d}A = -S\,\mathrm{d}T - P\,\mathrm{d}V + \sum_i \mu_i\,\mathrm{d}n_i + \mathbf{E}\,\mathrm{d}\mathbf{P} \tag{14.76}$$

and the definition of $(\partial A/\partial n_i)_{T,V,n,\mathbf{P}} = \mu_i^{\varepsilon}$. The same methods used in Section 14.11 are used in this section, with the result that

$$\mu_i^{\varepsilon} = \mu_i^{c} - \tfrac{1}{2}V_c\mathbf{E}^2\left(\frac{\partial(\varepsilon - \varepsilon_0)}{\partial n_i}\right)_{T,V,n,\mathbf{P}} = \mu_i^{c} - \tfrac{1}{2}V_c\mathbf{E}^2\left(\frac{\partial \varepsilon}{\partial n_i}\right)_{T,V,n,\mathbf{P}} \tag{14.77}$$

Partial molar quantities can be defined as the change of an extensive variable with respect to the mole number of one component at constant temperature, pressure, *electric field*, and mole numbers of all other components. Then, with Equations (14.73) and (14.74), the change of the partial molar entropy and partial molar volume with the electric field is given by

$$\left(\frac{\partial \bar{S}_k}{\partial \mathbf{E}}\right)_{T,P,n} = V_c\varepsilon_0\mathbf{E}\left[\frac{\partial}{\partial T}\left(\frac{\partial \chi_e}{\partial n_k}\right)_{T,P,n,\mathbf{E}}\right]_{P,n,\mathbf{E}} \tag{14.78}$$

and

$$\left(\frac{\partial \bar{V}_k}{\partial \mathbf{E}}\right)_{T,P,n} = V_c\varepsilon_0\mathbf{E}\left[\frac{\partial}{\partial P}\left(\frac{\partial \chi_e}{\partial n_k}\right)_{T,P,n,\mathbf{E}}\right]_{T,n,\mathbf{E}} \tag{14.79}$$

The change of the chemical potential of a component with the field at constant temperature, pressure, and mole numbers is given by Equation (14.75). We note that the electric susceptibility is a function of the temperature, pressure, and mole numbers. It is also a function of the field, but may be taken as independent of the field except for high fields when saturation effects may occur.

It is of interest to discuss the changes in the dielectric fluid with a change of the field. We continue to consider a single-phase, closed system composed of C components with the condenser immersed in the fluid. The temperature and pressure on the fluid external to the condenser are constant. We treat the fluid as consisting of two parts, the material contained within the volume of the condenser and the material external to the condenser. Single primes refer to quantities applicable to the material within the condenser. When the field is increased, the number of moles of material within the condenser will change because of the electrostriction effect, and the mole fractions of the components in both parts may change. The problem is to determine each of these effects. The variables are C mole numbers for the material within the condenser and C mole numbers for the external material. There is no special designation for quantities that refer to the material outside the condenser. The volume of the condenser may be considered as a variable even though it is constant for a given condenser. There are thus $(2C+1)$ variables. The condition equations are C equations expressing the equality of the chemical potential of a component in each part, C mass balance equations, and the equation expressing the volume of the material within the condenser, which is constant, in terms of the mole numbers. There are $(2C+1)$ such condition equations, and the problem can be solved. We consider only one- and two-component systems.

For a one-component system the chemical potential of the material outside the condenser is constant, because the temperature and pressure are constant. We assume for simplicity that the pressure within the condenser is constant and equal to the pressure on the material outside the condenser. (For a condensed phase, $(\partial\tilde{V}/\partial P)_T$ is very small and, even if the pressure is different within reason, the effect may be neglected.) We are then concerned only with the electrostriction. When Equation (14.74) is integrated between the limits 0 and $\mathbf{E}$, we obtain

$$\Delta V = \tilde{V}\,\Delta n = -\tfrac{1}{2}V_c\varepsilon_0\mathbf{E}^2\left(\frac{\partial \chi_e}{\partial P}\right)_{T,P,\mathbf{E}} \tag{14.80}$$

We emphasize that ΔV is the change of volume of the entire system and therefore Δn is the change of the number of moles, and $\tilde{V}$ is the molar volume of the material outside the condenser. Thus, Δn can be determined. Then, from the condition that V_c must be constant, the change of the molar volume of the material inside the condenser may be calculated.

For a two-component system we have

$$\mu_1 = \mu_1' = \mu_1^c - \tfrac{1}{2}V_c\mathbf{E}^2\left(\frac{\partial \varepsilon}{\partial n_1}\right)_{T,V,n,\mathbf{D}} \tag{14.81}$$

$$\mu_2 = \mu_2' = \mu_2^c - \tfrac{1}{2}V_c\mathbf{E}^2\left(\frac{\partial \varepsilon}{\partial n_2}\right)_{T,V,n,\mathbf{D}} \tag{14.82}$$

$$n_1^0 = n_1' + n_1 \tag{14.83}$$

$$n_2^0 = n_2' + n_2 \tag{14.84}$$

and

$$V_c = \text{constant} \tag{14.85}$$

together with Equation (14.74). These equations are sufficient to determine the distribution of the components between the condenser and the volume outside the condenser and the partial molar volumes of the two components within the charged condenser.

SYSTEMS IN A MAGNETIC FIELD

The treatment of systems in a magnetic field follows very closely the development of the thermodynamics of systems in an electrostatic field. We again consider two systems. One is a long solenoid with uniform windings, in empty space. The second is the same solenoid in which the total volume within the solenoid is filled with isotropic matter. Edge effects are neglected. Ferromagnetic effects and hysteresis are omitted.

14.13 The solenoid in empty space

If i is the current in the windings of the solenoid and l the length of the solenoid per turn, the vector $\mathbf{B}$, called the *magnetic induction*, is defined by

$$\mathbf{B} = \mu_0 i/l \tag{14.86}$$

where μ_0 is the *permeability of vacuum.*[1] This relation is written in terms of

[1] Distinction must be made in this section in the use of the symbol μ for both the chemical potential and the permeability.

the volume within the solenoid as

$$\frac{\mathbf{B}}{\mu_0} = \frac{niA}{V_S} \tag{14.87}$$

where n is the total number of turns, A is the cross section of the solenoid, and $V_S = nlA$. The product niA may be called the *magnetic moment* and $\mathbf{B}/\mu_0$ is thus the *magnetic moment per unit volume* of the solenoid. The *magnetic field strength* is given by

$$\mathbf{H} = \mathbf{B}/\mu_0 \tag{14.88}$$

The vectors are parallel to the axis of the solenoid.

14.14 The solenoid filled with isotropic matter

We now consider the same solenoid, the volume of which is filled with some isotropic matter. We discuss only the case in which this material is homogenous and multicomponent. A *magnetic moment per unit volume*, **m**, is induced in the substance within the solenoid by the current. The net magnetic moment per unit volume is then given by

$$\frac{\mathbf{B}}{\mu_0} - \mathbf{m} = \frac{i}{l} = \frac{niA}{V_S} \tag{14.89}$$

The *magnetic strength* is then expressed as

$$\mathbf{H} = \frac{\mathbf{B}}{\mu_0} - \mathbf{m} \tag{14.90}$$

With the omission of hysteresis, **B** may be taken to be proportional to **H**, so

$$\mathbf{H} = \mathbf{B}/\mu \tag{14.91}$$

where μ is called the *permeability* and is a function of the temperature, pressure, and composition of the matter. It may be a function of the magnetic field strength, but for our purpose we assume that it is independent of the field strength. When **B** is eliminated between Equations (14.90) and (14.91), we obtain

$$\mathbf{m} = \left(\frac{\mu}{\mu_0} - 1\right)\mathbf{H} = \chi_M \mathbf{H} \tag{14.92}$$

where χ_M is the *magnetic susceptibility*. The ratio μ/μ_0 is called the *relative permeability*. Diamagnetic substances have values of μ/μ_0 less than unity and, for such substances, the induced moment opposes the vector **H**. Paramagnetic substances have values of the ratio greater than unity and the induced moment adds to the vector.

For the system that we consider here, the vectors **H**, **B**, and **m** are all parallel, and scalar values could be used for these quantities. However, we continue to use the vector symbolism for clarity.

14.15 Work associated with magnetic effects

The work done on the total system when **B** is increased from zero to some value **B** is given by

$$W = \int_{V_S} \mathrm{d}V \int_0^{\mathbf{B}} \mathbf{H}\,\mathrm{d}\mathbf{B} \tag{14.93}$$

where the integration is taken over the volume of the system. For the simple case which we have chosen, **B** and **H** are uniform throughout the volume. The integration over the volume then can be made so that Equation (14.92) becomes

$$W = V_S \int_0^{\mathbf{B}} \mathbf{H}\,\mathrm{d}\mathbf{B} \tag{14.94}$$

The differential quantity of work done on the system for a differential change in **B** is then

$$\mathrm{d}W = V_S \mathbf{H}\,\mathrm{d}\mathbf{B} \tag{14.95}$$

We could make use of Equation (14.91) and change the variable from **B** to **H** for our simple system. However, we continue to use **B** because μ may be dependent on the field in the general case.

14.16 Thermodynamics of the total system

When we assume that the mole numbers of the materials that compose the solenoid are constant, the energy of the total system is a function of the entropy, volume, mole numbers of the material within the solenoid, and the magnetic induction. Thus, we have for the differential of the energy of the system

$$\mathrm{d}E = T\,\mathrm{d}S - P\,\mathrm{d}V + \sum_i \mu_i\,\mathrm{d}n_i + V_S \mathbf{H}\,\mathrm{d}\mathbf{B} \tag{14.96}$$

where we have accepted the fact that both the temperature and the pressure of the solenoid have the same values as those of the substance within the solenoid. Differential expressions for the total enthalpy and the Gibbs and Helmholtz energies can be obtained from the usual definitions of these functions. Then the application of the condition of exactness to such expressions and Equation (14.96) yields a number of expressions involving the magnetic induction as the independent variable. However, the interest is in the thermodynamic properties of the material within the solenoid.

14.17 Thermodynamics of a dia- or paramagnetic substance

The same method used in Section 14.12 can be applied here to obtain a differential expression for the energy of the substance. In order to do so we subtract from the magnetic induction of the total system, **B**, the magnetic induction of the empty solenoid when the current is the same as that of the filled solenoid. Thus,

$$\mathrm{d}E = T\,\mathrm{d}S - P\,\mathrm{d}V + \sum_i \mu_i\,\mathrm{d}n_i + V_S\mathbf{H}\,\mathrm{d}(\mathbf{B} - \mu_0\mathbf{H}) \tag{14.97}$$

by use of Equations (14.90) and (14.92). However,

$$\mathbf{B} - \mu_0\mathbf{H} = \mu_0\mathbf{m} \tag{14.98}$$

We define $\mathbf{M} = V_S\mathbf{m}$, where **M** is the *total magnetic moment*. Then Equation (14.97) becomes

$$\mathrm{d}E = T\,\mathrm{d}S - P\,\mathrm{d}V + \sum_i^C \mu_i\,\mathrm{d}n_i + \mu_0\mathbf{H}\,\mathrm{d}\mathbf{M} \tag{14.99}$$

We can obtain expressions for the differentials of the enthalpy and the Gibbs and Helmholtz energies from the usual definitions. In such expressions **M** is an independent variable. Because it is more convenient to use **H** as an independent variable, the new functions $(H - \mu_0\mathbf{HM})$, $(A - \mu_0\mathbf{HM})$, and $(G - \mu_0\mathbf{HM})$ are used. The differential expressions for these functions are

$$\mathrm{d}(H - \mu_0\mathbf{HM}) = T\,\mathrm{d}S + V\,\mathrm{d}P + \sum_{i=1}^C \mu_i\,\mathrm{d}n_i - \mu_0\mathbf{M}\,\mathrm{d}\mathbf{H} \tag{14.100}$$

$$\mathrm{d}(A - \mu_0\mathbf{HM}) = -S\,\mathrm{d}T - P\,\mathrm{d}V + \sum_{i=1}^C \mu_i\,\mathrm{d}n_i - \mu_0\mathbf{M}\,\mathrm{d}\mathbf{H} \tag{14.101}$$

and

$$\mathrm{d}(G - \mu_0\mathbf{HM}) = -S\,\mathrm{d}T + V\,\mathrm{d}P + \sum_{i=1}^C \mu_i\,\mathrm{d}n_i - \mu_0\mathbf{M}\,\mathrm{d}\mathbf{H} \tag{14.102}$$

Again, the condition of exactness may be applied to such equations. Particularly, from Equation (14.102) we have

$$\left(\frac{\partial V}{\partial \mathbf{H}}\right)_{T,P,n} = -\mu_0\left(\frac{\partial \mathbf{M}}{\partial P}\right)_{T,n,\mathbf{H}} = -\mu_0 V\mathbf{H}\left(\frac{\partial \chi_{\mathbf{M}}}{\partial P}\right)_{T,n,\mathbf{H}} \tag{14.103}$$

and

$$\left(\frac{\partial \mu_k}{\partial \mathbf{H}}\right)_{T,P,n} = -\mu_0\left(\frac{\partial \mathbf{M}}{\partial n_k}\right)_{T,P,n,\mathbf{H}} = -\mu_0 V\mathbf{H}\left(\frac{\partial \chi_{\mathbf{M}}}{\partial n_k}\right)_{T,P,n,\mathbf{H}} \tag{14.104}$$

with the use of Equation (14.92). These two equations give the change of the volume and the chemical potential, respectively, with a change of the magnetic field strength. Again, partial molar quantities may be defined as the derivative of an extensive variable with respect to the mole number of a component at constant temperature, pressure, mole numbers of the other components, and magnetic field strength.

An expression of particular interest is obtained from Equation (14.100). We have

$$\left(\frac{\partial T}{\partial \mathbf{H}}\right)_{S,P,n} = -\mu_0 \left(\frac{\partial \mathbf{M}}{\partial S}\right)_{T,P,n,\mathbf{H}} \tag{14.105}$$

With the use of Equation (14.92) and the introduce of the temperature,

$$\left(\frac{\partial T}{\partial \mathbf{H}}\right)_{S,P,n} = -\mu_0 V\mathbf{H} \frac{(\partial \chi_{\mathbf{M}}/\partial T)_{P,n,\mathbf{H}}}{(\partial S/\partial T)_{P,n,\mathbf{H}}} \tag{14.106}$$

At low temperatures $(\partial \chi_{\mathbf{M}}/\partial T)_{P,n,\mathbf{H}}$ is large and negative, so the value of the derivative $(\partial T/\partial \mathbf{H})_{S,P,n}$ is positive, i.e., the temperature decreases with an isentropic decrease of the field strength. This equation provides the basis for the attainment of low temperatures by adiabatic demagnetization.

15
The third law of thermodynamics

The energy and entropy functions have been defined in terms of differential quantities, with the result that the absolute values could not be known. We have used the difference in the values of the thermodynamic functions between two states and, in determining these differences, the process of integration between limits has been used. In so doing we have avoided the use or requirement of integration constants. The many studies concerning the possible determination of these constants have culminated in the *third law of thermodynamics.*

We can obtain a concept of the problem by referring to the discussion in Section 7.12. There we derive expressions for the thermodynamic functions for an ideal gas by expressing the energy and entropy in terms of the heat capacities and the required integration constants. The result is that the expression for the Gibbs energy contains the term $(\tilde{e} - T\tilde{s})$, where $\tilde{e}$ and $\tilde{s}$ are the required integration constants for the energy and entropy, respectively. When we consider the difference between the values of the Gibbs energy of the ideal gas at two temperatures, $\tilde{e}$ cancels but the quantity $(T_1 - T_2)\tilde{s}$ remains. Therefore, in order to make the calculation, we need to know the value of $\tilde{s}$. The result is the same when we consider substances other than an ideal gas. We are, therefore, concerned with the possible determination of the absolute value of the entropy.

The same problem arises when we consider the question of being able to determine equilibrium conditions of thermodynamic systems from thermal measurements alone; that is, by measurements of heat capacities, changes of enthalpy for changes of phase, and changes of enthalpy for chemical reactions. In all cases of equilibrium we need knowledge of the Gibbs energy, the Helmholtz energy, or the chemical potentials. This in turn requires knowledge of $\tilde{e}$ and $\tilde{s}$ or their equivalents. For a chemical reaction given by $\sum_i \nu_i B_i$, we shall see that we can obtain the equivalent of $\sum_i \nu_i \tilde{e}_i$ from thermal measurements but we need information concerning $\sum_i \nu_i \tilde{s}_i$. We might choose a standard state for each element and assign a value of zero to the entropy of the element in this state. We would then need to determine the change of entropy for the formation of a compound from the elements under standard state conditions. However, the determination of such values requires a study of equilibrium conditions and cannot be done from thermal measurements alone.

We see from this discussion that a third generalization is necessary. Up to this point in the text we have been concerned solely with the macroscopic properties. However, in order to obtain an understanding of the third law we must use some concepts concerning the entropy function, based on statistical mechanics. We do so in this chapter with the assumption that the basic concepts that are used are familiar to the reader.

15.1 The preliminary concepts of the third law

Richards, in 1902, determined from a study of the emf of cells that the change of the Gibbs energy for an isothermal change of state involving a chemical reaction between pure substances and the corresponding change of enthalpy approached each other asymptotically as the temperature was decreased [16]. Although the studies were not made at low temperatures, the results gave some indication that the slopes of the curves obtained by plotting ΔG and ΔH as a function of the Kelvin temperature approached zero as the temperature approached zero. We would expect ΔG and ΔH to become equal as the temperature approached zero, because of the relation $\Delta G = \Delta H - T\,\Delta S$. That they do so asymptotically indicates that the values of $(\partial\,\Delta G/\partial T)_{P,n}$ and $(\partial\,\Delta H/\partial T)_{T,n}$ become equal as the temperature decreases. The second observation indicates that the values of these derivatives approach zero as the temperature is decreased.

In 1906 Nernst proposed the principle that, for any change of state in a condensed system, the values of $(\partial\,\Delta G/\partial T)_{P,n}$ and $(\partial\,\Delta H/\partial T)_{P,n}$ become zero at absolute zero [16]. This principle became known as the *Nernst heat theorem*. The consequences of the theorem are that

$$(\Delta C_P)_{T=0} = 0 \tag{15.1}$$

and

$$(\Delta S)_{T=0} = 0 \tag{15.2}$$

Equation (15.2) is the essential statement of Nernst, because Equation (15.1) can be derived from Equation (15.2) provided that $(\Delta S)_{T=0}$ is not infinitely large. We have

$$\left(\frac{\partial\,\Delta G}{\partial T}\right)_{P,n,T=0} = -\Delta S_{T=0} = \left(\frac{\Delta G - \Delta H}{T}\right)_{T=0} \tag{15.3}$$

The last term is this equation is indeterminant at $T = 0$, but by l'Hôpital's rule

$$\left(\frac{\Delta G - \Delta H}{T}\right)_{T=0} = \left[\frac{(\partial\,\Delta G/\partial T)_{P,n} - (\partial\,\Delta H/\partial T)_{P,n}}{1}\right]_{T=0}$$

$$= -(\Delta S)_{T=0} - (\Delta C)_{T=0} \tag{15.4}$$

Combination of Equations (15.3) and (15.4) gives

$$(\Delta S)_{T=0} = (\Delta S)_{T=0} + (\Delta C_P)_{T=0} \tag{15.5}$$

from which we conclude that $(\Delta C_P)_{T=0}$ must be zero if $(\Delta S)_{T=0}$ is not infinitely large. Of course, if $(\Delta S)_{T=0} = 0$, then $(\Delta C_P)_{T=0}$ must also be zero.

The postulates of Nernst are those that are required when we wish to determine equilibrium conditions for chemical reactions from thermal data alone. In order to calculate the equilibrium conditions, we need to know the value of $\Delta G^{\ominus}$ for the change of state involved. We take the standard states of the individual substances to be the pure substances at the chosen temperature and pressure. The value of $\Delta H^{\ominus}$ can be determined from measurements of the heat of reaction. We now have

$$\Delta S^{\ominus} = \int_0^T \frac{\Delta C_P^{\ominus}}{T}\,\mathrm{d}T \tag{15.6}$$

where we have set $\Delta S^{\ominus}$ at 0 K as zero according to Equation (15.2). The integral can be evaluated from knowledge of the heat capacities of the reactants, because $\Delta C_P^{\ominus}/T$ remains finite as the temperature approaches zero. Because all substances (except liquid helium) are solids at 0 K, Equation (15.6) is strictly applicable only to reactions taking place between the solid reactants. Additional terms including the changes of entropy for a change of phase would be included if we needed values of $\Delta S^{\ominus}$ when some or all of the reactants are in the liquid or gaseous state.

Planck, in 1912, postulated that the value of the entropy function for all pure substances in condensed states was zero at 0 K. This statement may be taken as a preliminary statement of the third law. The postulate of Planck is more extensive than, but certainly is consistent with, the postulate of Nernst.

15.2 Experimental determination of absolute entropies

The stable phase of all substances, except helium, at sufficiently low temperatures is the solid phase. We therefore consider the solid phase as the condensed state whose entropy is zero at 0 K, and exclude helium from the discussion for the present. The absolute entropy of a pure substance in some state at a given temperature and pressure is the value of the entropy function for the given state taking the value of the entropy of the solid phase at 0 K as zero independent of the pressure. In so doing, we have assigned the value of the integration constant in integrating the relation $(\partial S/\partial T)_P = C_P/T$.

Let us consider the change of state

$$B[\text{s}, 0\ \text{K}, P] \rightarrow B[\text{g}, T, P] \qquad (15.7)$$

The change of entropy for this change of state is $S[\text{g}, T, P] - S^{\ominus}[\text{s}, 0\ \text{K}, P]$ but $S^{\ominus}[\text{s}, 0\ \text{K}, P]$ is zero according to the third law. Thus, the absolute value of the entropy in the state $B[\text{g}, T, P]$ is equal to the change of entropy for the given change of state. The change of state is equivalent to the sum of the changes of state

$$\begin{aligned} &B[\text{s}, 0\ \text{K}, P] \rightarrow B[\text{s}, T_{\text{mp}}, P] \\ &B[\text{s}, T_{\text{mp}}, P] \rightarrow B[\ell, T_{\text{mp}}, P] \\ &B[\ell, T_{\text{mp}}, P] \rightarrow B[\ell, T_{\text{b}}, P] \qquad (15.8) \\ &B[\ell, T_{\text{b}}, P] \rightarrow B[\text{g}, T_{\text{b}}, P] \end{aligned}$$

and

$$B[\text{g}, T_{\text{b}}, P] \rightarrow B[\text{g}, T, P]$$

where T_{mp} and T_{b} represents the melting and boiling points of the substance, respectively, at the pressure P. We have assumed that only one solid phase exists between 0 K and the melting point. The absolute entropy in the state $B[\text{g}, T, P]$ is then obtained from

$$S[\text{g}, T, P] = \int_0^{T_{\text{mp}}} \frac{C_P[\text{s}]}{T}\,\text{d}T + \frac{\Delta H_{\text{mp}}}{T_{\text{mp}}} + \int_{T_{\text{mp}}}^{T_{\text{b}}} \frac{C_P[\ell]}{T}\,\text{d}T + \frac{\Delta H_{\text{vap}}}{T_{\text{b}}} + \int_{T_{\text{b}}}^{T} \frac{C_P[\text{g}]}{T}\,\text{d}T \qquad (15.9)$$

where ΔH_{mp} and ΔH_{vap} are the changes of enthalpy on melting and evaporation at the melting and boiling points, respectively at the pressure P. We observe that the thermal data required are the heat capacities at the chosen constant pressure and changes of enthalpy for the changes of phase at the same pressure.

Equation (15.9) illustrates the type of calculations that must be made to obtain the value of the entropy function of a substance in a given state. We must include all increases in entropy for the increase of temperature of a single phase and for changes of phase that are required to reach the desired state from the state at 0 K.

Two points concerning the evaluation of the first integral in Equation (15.9) require further discussion. In most experimental determinations of absolute entropies, the lowest temperature attained ranges from 1 to 15 K;

we then write the integral as

$$\int_0^T \frac{C_P[\mathrm{s}]}{T}\,\mathrm{d}T = \int_0^{T'} \frac{C_P[\mathrm{s}]}{T}\,\mathrm{d}T + \int_{T'}^T \frac{C_P[\mathrm{s}]}{T}\,\mathrm{d}T \tag{15.10}$$

where T' represents the lowest experimental temperature. The evaluation of the first integral on the right-hand side of this equation requires the extrapolation of the integrand from T' to zero. Moreover, the integrand must remain finite at 0 K. The extrapolation is generally made according to the Debye theory of the heat capacity of solids, in which the heat capacity of a solid is shown to be proportional to the cube of the temperature; the proportionality constant is determined from the experimental data in the neighborhood of T'. We observe that with this dependence of temperature, the value of $C_P(\mathrm{s})/T$ becomes zero at 0 K and thus is finite. As experimental methods have improved and lower temperatures attained, the Debye theory extrapolation accounts for only a very small part of absolute entropy measurements.

15.3 Confirmation of the third law

Over the years, many experiments have been carried out which confirm the third law. The experiments have generally been of two types. In one type the change of entropy for a change of phase of a pure substance or for a standard change of state for a chemical reaction has been determined from equilibrium measurements and compared with the value determined from the absolute entropies of the substances based on the third law. In the other type the absolute entropy of a substance in the state of an ideal gas at a given temperature and pressure has been calculated on the basis of statistical mechanics and compared with those based on the third law. Except for well-known, specific cases the agreement has been within the experimental error. The specific cases have been explained on the basis of statistical mechanics or further experiments. Such studies have led to a further understanding of the third law as it is applied to chemical systems.

15.4 Understanding the third law

According to statistical mechanics, the value of zero for the entropy function of a system is obtained when all molecules comprising the system are in the same quantum state. This statement applies to any state of aggregation: gas, liquid, or solid. Now all substances[1] have an infinite number of possible quantum states, and consequently the state in which all the

---

[1] We do not consider here the case in which a nuclear magnetic subsystem or any other subsystem that has a limited number of quantum states may be considered as a thermodynamic system separate from the other parts of the total molecular system.

molecules of a system have the same quantum state (the lowest possible quantum state) is approached only as the temperature is decreased. At sufficiently low temperatures all substances become solid except helium (at pressures of <25 bar) and, therefore, the third law applies primarily to the solid phase.

The condition discussed in the previous paragraph demands certain care in the experimental determination of absolute entropies, particularly in the cooling of the sample to the lowest experimental temperature. In order to approach the condition that all molecules are in the same quantum state at 0 K, we must cool the sample under the condition that thermodynamic equilibrium is maintained within the sample at all times. Otherwise some state may be obtained at the lowest experimental temperature that is metastable with respect to another state and in which all the molecules may not be in the same quantum state at 0 K.

We must also consider the conditions that are implied in the extrapolation from the lowest experimental temperature to 0 K. The Debye theory of the heat capacity of solids is concerned only with the linear vibrations of molecules about the crystal lattice sites. The integration from the lowest experimental temperature to 0 K then determines the decrease in the value of the entropy function resulting from the decrease in the distribution of the molecules among the quantum states associated solely with these vibrations. Therefore, if all of the molecules are not in the same quantum state at the lowest experimental temperature, excluding the lattice vibrations, the state of the system, figuratively obtained on extrapolating to 0 K, will not be one for which the value of the entropy function is zero.

We can now discuss certain systems for which we expect the third law to be valid, other systems for which the law is not valid at all, and some specific systems that appear to deviate from the law.

In the first discussion of equilibrium (Ch. 5) we recognized that there may be states of a system that are actually metastable with respect to other states of the system but which appear to be stable and in equilibrium over a time period. Let us consider, then, a pure substance that can exist in two crystalline states, α and β, and let the α phase be metastable with respect to the β phase at normal temperatures and pressures. We assume that, on cooling the α phase to the lowest experimental temperature, equilibrium can be maintained within the sample, so that on extrapolation the value of the entropy function becomes zero. If, now, it is possible to cool the β phase under the conditions of maintaining equilibrium with no conversion to the α phase, such that all molecules of the phase attain the same quantum state excluding the lattice vibrations, then the value of the entropy function of the β phase also becomes zero on the extrapolation. The molar absolute entropy of the α phase and of the β phase at the equilibrium transition temperature, T_{tr}, for the chosen pressure are given by

$$\tilde{S}[\alpha, T_{tr}, P] = \int_0^{T_{tr}} \frac{\tilde{C}_P(\alpha)}{T} \, dT \tag{15.11}$$

and

$$\tilde{S}[\beta, T_{tr}, P] = \int_0^{T_{tr}} \frac{\tilde{C}_P(\beta)}{T} \, dT \tag{15.12}$$

respectively. Then the change of entropy for the transition of 1 mole of the β phase to 1 mole of the α phase at the transition temperature is

$$\tilde{S}[\alpha, T_{tr}, P] - \tilde{S}[\beta, T_{tr}, P] = \int_0^{T_{tr}} \frac{\tilde{C}_P(\alpha) - \tilde{C}_P(\beta)}{T} \, dT \tag{15.13}$$

according to the third law. This quantity must agree with the value determined experimentally by measuring the change of enthalpy for the transition by calorimetric means or from the Clapeyron equation when the third law is valid for both phases. Examples of this behavior are monoclinic and rhombic sulfur, different crystalline forms of phosphine, and diamond and graphite.

Liquid helium presents an interesting case leading to further understanding of the third law. When liquid ^{4}He, the abundant isotope of helium, is cooled at pressures of <25 bar, a second-order transition takes place at approximately 2 K to form liquid HeII. On further cooling HeII remains liquid to the lowest observed temperature at 10^{-5} K. HeII does become solid at pressures greater than about 25 bar. The slope of the equilibrium line between liquid and solid helium apparently becomes zero at temperatures below approximately 1 K. Thus, dP/dT becomes zero for these temperatures and therefore $\Delta\tilde{S}$, the difference between the molar entropies of liquid HeII and solid helium, is zero because $\Delta\tilde{V}$ remains finite. We may assume that liquid HeII remains liquid as 0 K is approached at pressures below 25 bar. Then, if the value of the entropy function for solid helium becomes zero at 0 K, so must the value for liquid HeII. Liquid ^{3}He apparently does not have the second-order transition, but like ^{4}He it appears to remain liquid as the temperature is lowered at pressures of less than approximately 30 bar. The slope of the equilibrium line between solid and liquid ^{3}He appears to become zero as the temperature approaches 0 K. If, then, the slope is zero at 0 K, the value of the entropy function of liquid ^{3}He is zero at 0 K if we assume that the entropy of solid ^{3}He is zero at 0 K. Helium is the only known substance that apparently remains liquid as absolute zero is approached under appropriate pressures. Here we have evidence that the third law is applicable to liquid helium and is not restricted to crystalline phases.

Solutions and glasses do not follow the third law. If a solid solution continues to persist on cooling a sample to the lowest possible temperature, then at this temperature the molecules of the several components are distributed in some fashion in the same crystal lattice. Under such conditions all of the molecules of the substance could not attain the same quantum state on further cooling to 0 K in the sense of the required extrapolation. Only if the solid solution was separated into the pure components would the value of zero be obtained for the entropy function at 0 K. If the molecules of the components were randomly distributed in the crystal lattice, as in an ideal solution, then the entropy of the substance at absolute zero would be equal to the ideal entropy of mixing, so

$$S^{\ominus}[\text{solid soln., } 0\text{ K}, P] = -\sum_i n_i R \ln x_i \tag{15.14}$$

If the distribution were not random, the entropy at 0 K would be less than the value calculated from Equation (15.14) but it would not be zero. Thus, *the third law can be applied only to pure substances.*

Some pure substances from a glass-like solid, on cooling, rather than a crystalline solid. In such solids the molecules have a certain randomness in their spatial distribution, more like a liquid than a crystal. This is an example in which a metastable state rather than an equilibrium state is obtained on cooling. If such a solid is obtained on cooling to the lowest experimental temperature, then the value of the entropy function at 0 K, obtained on extrapolation, will be greater than zero.

Another type of nonequilibrium state that may be obtained on cooling a sample is a certain amount of randomness of orientation of the molecules in the crystal. Some of the substances that exhibit this effect are CO, NO, N_2O, N_2O_2, $H_2O(D_2O)$, 1-olefins with more than 10 carbon atoms, and certain hydrates such as $Na_2SO_4 \cdot 10H_2O$. In each case the molecules may have different states of orientations with respect to their neighbors, each state having almost the same energy. Consequently, when a temperature is reached at which rotational motion no longer exists, the distribution of the molecules among the possible states of orientation becomes fixed and may be somewhat random. Further lowering of the temperature will not change the distribution to one in which all of the molecules are in the same state of orientation and the value of the entropy function will not be zero at 0 K.

The presence of isotopes presents a different problem, the solution of which depends on our interests and our choice of the components of the system. If the system contains a pure substance without regard to its isotopic composition, then we may choose that pure substance with a fixed isotopic composition as the component or we may choose the individual molecular species that have the same isotopes as the components. In the latter case the actual system is considered as a solution of the isotopic molecular species, and there is an entropy of mixing in forming the solution from its components. Now, in the determination of the entropy of the substance, the distribution of the isotopes among the individual isotopic molecules is essentially random and remains constant on cooling the substance to the lowest experimental temperature in ordinary cases. In the extrapolation from this temperature to 0 K, we do not consider any change of the distribution. Thus, the entropy of mixing of the isotopic molecular species to obtain the solution is not decreased on approaching 0 K and remains positive with respect to the value of the entropy function of the system that consists of the unmixed isotopic species. The value of the entropy function of each of the isotopic molecular species presumably does become zero at 0 K. (Here we ignore any possible spatial orientation of the molecules on the lattice sites identified only by the different isotopes.) In the general case we may take the value of the entropy function of an isotopic solution of constant isotopic composition to be zero at 0 K provided that, in any use of the absolute value of the entropy function based on this zero, the isotopic composition remains unchanged. We are permitted to do this because the contribution of the entropy of mixing is the same at 0 K as at any higher temperature. However, when the changes of state in which we are interested involve changes of the isotopic composition or the distribution of the isotopes among the isotopic molecular species, we must take the isotopic molecular species as the components. The state of the system whose value of the entropy function at 0 K is zero is then that of the unmixed isotopic molecular species at 0 K.

The effect of nuclear spin on the values of the entropy function is the final effect to be considered. When the atoms that are contained in a molecule have nuclear spin, there are various possible orientations of the spins. The differences between the energies associated with the different orientations are generally small, and consequently the orientations are randomly distributed among the possible states. The distribution remains random on cooling a substance to the lowest experimental temperature and, consequently, the value of the entropy function at 0 K obtained by the usual extrapolation is greater than that of a system in which the orientations are not randomly distributed. This situation is similar to what we encountered in considering the presence of isotopes. The value of the entropy function associated with the distribution of the nuclear spins for a given substance remains constant independent of the temperature within the temperature range of interest. Also, for a given set of atoms with nuclear spin, it is independent of the chemical compounds in which the atoms occur. Consequently, we can assign the value of zero to the entropy function for chemical substances in the state obtained on cooling the substances to the lowest experimental temperature and performing the usual extrapolation.

Hydrogen substance (including the three isotopes) is the one substance in which the equilibrium distribution of the orientation of the nuclear spins is not random, and hence is an exception to the previous statements. We briefly discuss here only the light isotope $^1H^1H$. The spin quantum number is $\frac{1}{2}$ and the spins of the two nuclei can orient in two ways, either parallel or antiparallel. The net result is that two species of hydrogen exist, one of which has only even values of the rotational quantum number J (antiparallel orientation) and the other only odd values of J (parallel orientation). The first species is called *parahydrogen* and the second is called *orthohydrogen*. The equilibrium mixture at sufficiently high temperatures is one-fourth *para* and three-fourths *ortho*. In contrast to other substances, the differences of energy between the rotational states is large, with the result that the transition of hydrogen molecules between the even and odd states is extremely slow. The lowest possible rotational quantum state is the state for which $J = 0$. Yet, when solid hydrogen is obtained at low temperatures merely by cooling, the lowest rotational quantum state for *parahydrogen* has a J-value of zero and that for *orthohydrogen* has a J-value of unity with the consequence that not all of the molecules are in the same quantum state. However, if the transition from the odd to even quantum states is catalyzed by an appropriate catalyst, all of the molecules become *para* at sufficiently low temperatures and the lowest possible quantum state for all of the molecules is the state for which $J = 0$. Consequently, the value of the entropy function of the solid hydrogen which is a mixture of *orthohydrogen* and *parahydrogen* is greater than that which is composed of only *parahydrogen*.

It is extremely difficult to make a single, unequivocal statement of the third law. Instead, we have attempted to discuss the problem of understanding the third law from the viewpoint that the value of the entropy function for a substance is zero at 0 K. From a statistical mechanical point of view we consider the contribution of the translational, electronic, vibrational, and rotational degrees of freedom, and of the intermolecular forces to the value of the entropy function. However, we exclude all contributions of the nuclei of the atoms. From a macroscopic and experimental point of view we are required to reduce the temperature of the experimental sample of a substance to the lowest possible experimental temperature in such a way that equilibrium conditions are maintained within the sample in the process of cooling. We are finally forced to extrapolate the entropy function from the lowest experimental temperature to 0 K according to some theoretical equation. In so doing we ignore any contribution of the nuclear degrees of freedom to the value of the entropy function. Thus, any concept of a zero value of the entropy function as used at present excludes the contributions of the nuclei. It is interesting to recall at this point the arbitrariness of the definition of temperature scales. Instead of the Kelvin scale we could have used an exponential scale for which the lowest possible temperature is $-\infty$. On the exponential scale to the base e, 10^{-5} K is only -11.5 degrees. What new phenomena might be observed as the temperature is decreased from

−11.5 to minus infinity can only be surmised. We can assume on the basis of our present knowledge of physical systems that such phenomena would be associated with the nuclei. *With the exclusion of nuclear phenomena, we can state that the value of the entropy function for most pure crystalline substances and for pure liquid* ^{4}*He and* ^{3}*He is zero at* 0 K. Glasses and solutions are excluded. Substances in the crystalline state, which may have some randomness in the orientation of the molecules on the lattice sites, would not have a zero value at 0 K. Finally, in the strict sense, substances composed of isotopes would not be pure but actually solutions, and would not have a zero value of the entropy function at 0 K.

15.5 Practical basis for absolute values of the entropy

We can now return to the original problem of calculating the equilibrium properties of a system from thermal data alone. In Section 15.2 we show that absolute values of the entropy can be obtained at given temperatures and pressures, provided that we can take the value of the entropy function as zero at 0 K. We have seen in Section 15.4 that this is not always true; however, the answer to the dilemma is approached in the same section. Let us assume that some effect results in a nonzero value of the entropy of one or more substances at 0 K. If the contribution of this effect to the value of the entropy at some temperature and pressure of interest is the same as that at 0 K and, further, if the contribution remains the same for any change of state at the temperature of interest, such contributions cancel, with the result that we can assign the value of zero to the entropy function of such substances at 0 K. In order to illustrate this argument, we consider the standard change of entropy for some isothermal change of state, $\Delta S^\ominus[T]$, as the sum of two terms, one of which includes all contributions to the entropy of the reacting substances excluding the effects that result in nonzero values at 0 K, $\Delta S^{\ominus\prime}[T]$, and the other of which includes only the contributions resulting from these effects, $\Delta S^{\ominus\prime\prime}[T]$. Then

$$\Delta S^\ominus[T] = \Delta S^{\ominus\prime}[T] + \Delta S^{\ominus\prime\prime}[T] \tag{15.15}$$

We write $\Delta S^{\ominus\prime}(T)$ as $\sum_i \nu_i \Delta S_i^{\ominus\prime}[T]$ for the change of state. The quantity that is actually observed from the thermal measurements on each substance as discussed in Section 15.2 is $(S_i^{\ominus\prime}[T] - S_i^{\ominus\prime\prime}[0])$, where $S_i^{\ominus\prime\prime}[0]$ is the nonzero value of the entropy of the ith substance at 0 K. Equation (15.15) is then written as

$$\Delta S^\ominus[T] = \sum_i \nu_i (S_i^{\ominus\prime}[T] - S_i^{\ominus\prime\prime}[0]) + \Delta S^{\ominus\prime\prime}[T] - \sum_i \nu_i S_i^{\ominus\prime\prime}[0] \tag{15.16}$$

If we can express $\Delta S^{\ominus\prime\prime}[T]$ as (condition 1)

$$\Delta S^{\ominus\prime\prime}[T] = \sum_i \nu_i S_i^{\ominus\prime\prime}[T] \tag{15.17}$$

then

$$\Delta S^{\ominus}[T] = \sum_i \nu_i (S_i^{\ominus\prime}[T] - S_i^{\ominus\prime\prime}[0]) + \sum_i \nu_i (S_i^{\ominus\prime\prime}[T] - S_i^{\ominus\prime\prime}[0]) \tag{15.18}$$

When $S_i^{\ominus\prime\prime}[T] = S_i^{\ominus\prime\prime}[0]$ (condition 2), $\Delta S^{\ominus}[T]$ can be expressed as $\sum_i \nu_i (S_i^{\ominus\prime}[T] - S_i^{\ominus\prime\prime}[0])$; that is, in terms of the observed quantities. We use the difference $(S_i^{\ominus\prime}[T] - S_i^{\ominus\prime\prime}[0])$ as the absolute value of the entropy, which is equivalent to assigning the value of zero to $S_i^{\ominus\prime\prime}[0]$. The two effects for which this assignment is valid are: (1) the nuclear effects including those of nuclear spin, provided that the isothermal change of state does not involve a nuclear reaction; and (2) the isotopic effects, provided there is no change in the isotopic composition of the substances.

We thus can obtain a consistent set of absolute values of the entropy function for pure substances from thermal measurements alone on the practical basis of assigning the value of zero to the entropy function at 0 K with the exclusion of nuclear and isotopic effects, within the understanding of the third law as discussed in Section 15.4. The calculation of the entropy function of pure substances in the ideal gas state by the methods of statistical mechanics must be consistent with the practical basis. In addition to obtaining absolute values by the methods that have been discussed, values can also be obtained from equilibrium measurements from which $\Delta S^{\ominus}$ can be determined for some change of state. If all but one of the absolute values in the equivalent sum $\sum_i \nu_i S_i^{\ominus}$ are known, then the value of that one can be calculated.

15.6 Thermodynamic functions based on the third law

We have seen that absolute values of the entropy function based on the third law can be obtained from measurements of the heat capacity and heats of transitions. A more general equation than Equation (15.9) may be written as

$$S^{\ominus}[T, P] = \int_0^{T'} \frac{C_P}{T'}\, \mathrm{d}T + \sum_i \int_{T_i'}^{T_i''} \frac{C_{P_i}}{T}\, \mathrm{d}T + \sum_j \frac{\Delta H_j}{T_j} \tag{15.19}$$

The first term on the right-hand side of this equation represents the required extrapolation from 0 K to the lowest experimental temperature. The second term represents the sum of the contributions to the entropy for the change of temperature of a single phase, and the subscript i refers to the different phases. The third term represents the sum of the contributions for all of the phase transitions, where the subscript j refers to the different transitions. Values of the enthalpy function can be calculated by the similar equation

$$H^{\ominus}[T, P] - H^{\ominus}(0)[\mathrm{s}, 0, P] = \int_0^{T'} C_P \, \mathrm{d}T + \sum_i \int_{T_i'}^{T_i''} C_{P_i} \, \mathrm{d}T + \sum_j \Delta H_j \tag{15.20}$$

so the symbol $H^{\ominus}(0)[\mathrm{s}, 0, P]$ refers to the unknown value of the enthalpy of the pure substance at 0 K and the experimental pressure. The left-hand side of this equation is usually written as $[H^{\ominus} - H^{\ominus}(0)]_T$ or as $[H^{\ominus}(T) - H^{\ominus}(0)]$ (which is preferred) where the 0 refers to 0 K. The molar values of $\tilde{S}^{\ominus}(T)$ and $[\tilde{H}^{\ominus}(T) - \tilde{H}^{\ominus}(0)]$ are combined to give the values of the function $[\tilde{G}^{\ominus}(T) - \tilde{H}^{\ominus}(0)]$ by

$$[\tilde{G}^{\ominus}(T) - \tilde{H}^{\ominus}(0)] = [\tilde{H}^{\ominus}(T) - \tilde{H}^{\ominus}(0)] - T\tilde{S}^{\ominus}(T) \tag{15.21}$$

Alternately, values for the same function can be obtained by the appropriate integration of either Equation (15.19) or Equation (15.20). An equivalent function is $[\tilde{G}^{\ominus}(T) - \tilde{H}^{\ominus}(0)]/T$. We thus have the functions $\tilde{S}^{\ominus}(T)$, $[\tilde{H}^{\ominus}(T) - \tilde{H}^{\ominus}(0)]$, and $[\tilde{G}^{\ominus}(T) - \tilde{H}^{\ominus}(0)]$ or the equivalent $[\tilde{G}^{\ominus}(T) - \tilde{H}^{\ominus}(0)]/T$.

For any compilation of values of these functions the standard states must be completely defined. In essentially all cases the pressure is 1 bar and for gases the state is usually the ideal gas at 1 bar. For compilations at a single temperature, the temperature is usually 298.15 K. When values at various temperatures are required, it is easier to compile values at convenient temperature intervals rather than to obtain algebraic functions of the value. In such cases the function $[\tilde{G}^{\ominus}(T) - \tilde{H}^{\ominus}(0)]/T$ is used because the interpolation between the listed values is more nearly linear than between the values of $[\tilde{G}^{\ominus}(T) - \tilde{H}^{\ominus}(0)]$.

Another set of thermodynamic functions is based on the values of the enthalpy at 298.15 K. With a knowledge of the heat capacities and changes of enthalpy for changes of phase, values of the function $[\tilde{H}^{\ominus}(T) - \tilde{H}^{\ominus}(298)]$ are obtained. The corresponding function for the Gibbs energy is $[\tilde{G}^{\ominus}(T) - \tilde{H}^{\ominus}(298)]$ or $[\tilde{G}^{\ominus}(T) - \tilde{H}^{\ominus}(298)]/T$. Values for these functions are obtained by

$$[\tilde{G}^{\ominus}(T) - \tilde{H}^{\ominus}(298)] = [\tilde{H}^{\ominus}(T) - \tilde{H}^{\ominus}(298)] - T\tilde{S}^{\ominus}(T) \tag{15.22}$$

These functions are particularly useful when the low-temperature thermal data are not available. Under such circumstances the values of $S^{\ominus}(T)$ must be obtained from equilibrium measurements as indicated in Section 15.5.

The two sets of functions must be related to each other. Thus,

$$[\tilde{H}^{\ominus}(T) - \tilde{H}^{\ominus}(298)] = [\tilde{H}^{\ominus}(T) - \tilde{H}^{\ominus}(0)] - [\tilde{H}^{\ominus}(298) - \tilde{H}^{\ominus}(0)] \tag{15.23}$$

and

$$[\tilde{G}^{\ominus}(T) - \tilde{H}^{\ominus}(298)] = [\tilde{G}^{\ominus}(T) - \tilde{H}^{\ominus}(0)] - [\tilde{H}^{\ominus}(298) - \tilde{H}^{\ominus}(0)] \tag{15.24}$$

15.7 Uses of the thermodynamic functions

We consider a change of state involving a chemical reaction, and wish to calculate the equilibrium conditions. We assume that the change of state is isothermal and that the standard states of all of the reacting substances are the pure substances at the temperature and 1 bar pressure. When the pressure is not 1 bar, the appropriate corrections are easily made. When the change of state is written as $\sum_i \nu_i B_i = 0$, the condition of equilibrium becomes

$$\Delta G^{\ominus}(T) = \textstyle\sum_i \nu_i \tilde{G}^{\ominus}(T)_i = -RT \ln K$$

The values of $\Delta G^{\ominus}(T)$ are obtained by

$$\Delta G^{\ominus}(T) = \sum_i \nu_i[\tilde{G}^{\ominus}(T) - \tilde{H}^{\ominus}(0)]_i + \sum_i \nu_i \tilde{H}^{\ominus}(0)_i \tag{15.25}$$

or

$$\Delta G^{\ominus}(T) = \sum_i \nu_i[\tilde{G}^{\ominus}(T) - \tilde{H}^{\ominus}(298)]_i + \sum_i \nu_i \tilde{H}^{\ominus}(298)_i \tag{15.26}$$

The calculation then reverts to the determination of $\sum_i \nu_i \tilde{H}^{\ominus}(0)_i$ or $\sum_i \nu_i \tilde{H}^{\ominus}(298)_i$. These can be obtained from the knowledge of the change of enthalpy for the same change of state because

$$\Delta H^{\ominus}(T) = \sum_i \nu_i \tilde{H}^{\ominus}(T)_i = \sum_i \nu_i[\tilde{H}^{\ominus}(T) - \tilde{H}^{\ominus}(0)]_i + \sum_i \nu_i \tilde{H}^{\ominus}(0)_i \tag{15.27}$$

or

$$\Delta H^{\ominus}(T) = \sum_i \nu_i[\tilde{H}^{\ominus}(T) - \tilde{H}^{\ominus}(298)]_i + \sum_i \nu_i \tilde{H}^{\ominus}(298)]_i \tag{15.28}$$

Thus, in order to make the calculations we need values of $\Delta H^{\ominus}(T)$ for the change of state, of $[\tilde{H}^{\ominus}(T) - \tilde{H}^{\ominus}(0)]$ or $[\tilde{H}^{\ominus}(T) - \tilde{H}^{\ominus}(298)]$ and of $[\tilde{G}^{\ominus}(T) - \tilde{H}^{\ominus}(0)]$ or $[\tilde{G}^{\ominus}(T) - \tilde{H}^{\ominus}(298)]$ for each of the substances taking part in the reaction. Similar calculations of the temperature–pressure relations for a two-phase, one-component system can be made.

The concepts of the standard changes of enthalpy of formation and standard changes of the Gibbs energy of formation are developed in Chapters 9 and 11. These functions can be combined with the set of functions based on the third law. From the definitions of the standard changes on formation, we have for any pure substance j

$$\Delta\tilde{G}_{\rm f}(T)_j = \tilde{G}(T)_j - \sum_i \nu_i \tilde{G}^{\ominus}(T)_i \tag{15.29}$$

where $\tilde{G}^{\ominus}(T)_j$ refers to the Gibbs energy of the compound, $\tilde{G}^{\ominus}(T)_i$ refers to the elements, and the sum is taken over the elements. We can then write Equation (15.29) as

$$\Delta\tilde{G}_{\rm f}^{\ominus}(T)_j = [\tilde{G}^{\ominus}(T) - \tilde{H}^{\ominus}(0)]_j - \sum_i \nu_i[\tilde{G}^{\ominus}(T) - \tilde{H}^{\ominus}(0)]_i + \left(\tilde{H}^{\ominus}(0)_j - \sum_i \nu_i \tilde{H}^{\ominus}(0)_i\right) \tag{15.30}$$

The quantity $[\tilde{H}^{\ominus}(0)_j - \sum_i \nu_i \tilde{H}^{\ominus}(0)_i]$ can be written as $[\Delta\tilde{H}^{\ominus}(0)_j]$, which would be the standard change of the enthalpy on formation of the compound at 0 K. For the standard enthalpies of formation we have

$$\Delta H_{\rm f}^{\ominus}(T)_j = [\tilde{H}^{\ominus}(T) - \tilde{H}^{\ominus}(0)]_j - \sum_i \nu_i[\tilde{H}^{\ominus}(T) - \tilde{H}^{\ominus}(0)]_i + \Delta H_{\rm f}^{\ominus}(0)_j \tag{15.31}$$

The quantity $\Delta\tilde{H}_{\rm f}^{\ominus}(0)_j$ required in Equation (15.30) is thus obtained from Equation (15.31). If the basis of the enthalpy functions is at 298 K, the equivalent of Equation (15.30) is

$$\Delta\tilde{G}_{\rm f}^{\ominus}(T)_j = [\tilde{G}^{\ominus}(T) - \tilde{H}^{\ominus}(298)]_j - \sum_i \nu_i[\tilde{G}^{\ominus}(T) - \tilde{H}^{\ominus}(298)]_i + \Delta H_{\rm f}^{\ominus}(298)_j \tag{15.32}$$

where $\Delta\tilde{H}_{\rm f}^{\ominus}(298)_j$ is the standard change of the enthalpy of formation at 298 K. Similarly, the equation corresponding to Equation (15.31) is

$$\Delta\tilde{H}_{\rm f}^{\ominus}(T)_j = [\tilde{H}^{\ominus}(T) - \tilde{H}^{\ominus}(298)]_j - \sum_i \nu_i[\tilde{G}^{\ominus}(T) - \tilde{H}^{\ominus}(298)]_i + \Delta H_{\rm f}^{\ominus}(298)_j \tag{15.33}$$

Equations (15.32) and (15.33) illustrate another advantage of the use of functions based on the enthalpy at 298 K, in that many more values of $\Delta\tilde{H}_{\rm f}^{\ominus}(298)$ are known than of $\Delta\tilde{H}_{\rm f}^{\ominus}(0)$. The standard change of entropy for formation of a compound can be obtained from the absolute values as

$$\Delta\tilde{S}_{\rm f}^{\ominus}(T)_j = \tilde{S}(T)_j - \sum_i \nu_i \tilde{S}^{\ominus}(T)_i \tag{15.34}$$

We finally have

$$\Delta\tilde{G}_{\rm f}^{\ominus}(T)_j = \Delta\tilde{H}_{\rm f}^{\ominus}(T)_j - T\,\Delta\tilde{S}_{\rm f}^{\ominus}(T)_j \tag{15.35}$$

so one quantity in this equation can be determined when the other two are known.

Equations (15.25) and (15.26) afford a means of determining the validity of experimental data when values of $[G^{\ominus}(T) - H^{\ominus}(0)]$ or $[G^{\ominus}(T) - H^{\ominus}(298)]$ are known for each of the reacting substances. Assume that values of $\Delta G^{\ominus}(T)$ have been determined at various temperatures for some standard change of state from the experimental measurement of the equilibrium conditions of the system. Then the value of $\sum_i \nu_i \tilde{H}^{\ominus}(0)_i$ or $\sum_i \nu_i \tilde{H}^{\ominus}(298)_i$ is calculated from the value of $\Delta\tilde{G}^{\ominus}(T)$ for each experimental point. All of the values of $\sum_i \nu_i \tilde{H}^{\ominus}(0)_i$ or $\sum_i \nu_i \tilde{H}^{\ominus}(298)_i$ must be equal within the experimental error, because such values must be independent of the temperature for a fixed pressure.

Appendix A

SI units and fundamental constants

Table A.1. Base SI units.

Quantity	Name	Symbol
Length	meter	m
Mass	kilogram	kg
Time	second	s
Electric current	ampère	A
Thermodynamic temperature[1]	kelvin	K
Amount of substance[2]	mole	mol
Luminous intensity	candela	cd

[1] The 13th General Conference on Weights and Measures in 1967 recommended that the kelvin, symbol K, be used both for thermodynamic temperature and for thermodynamic temperature interval, and that the unit symbols °K and deg be abandoned. The kelvin is defined as 1/273.16 of the thermodynamic temperature of the triple point of water.
[2] The amount of substance should be expressed in units of moles, 1 mole being Avogadro's constant number of designated particles or group of particles, whether these are electrons, atoms, molecules, or the reactants and products specified by a chemical equation.

Table A.2. SI-derived units.

Quantity	Name	Symbol	Expression in terms of other units
Force	newton	N	$m\,kg\,s^{-2}$
Pressure	pascal	Pa	$N\,m^{-2}$
	bar	bar	$N\,m^{-2}$
Energy, work, quantity of heat	joule	J	N m
Power	watt	W	$J\,s^{-1}$
Electric charge	coulomb	C	A s
Electric potential, electromotive force	volt	V	$W\,A^{-1}$, $J\,C^{-1}$
Celsius temperature	degree celsius	°C	K
Heat capacity, entropy	joule per kelvin		$J\,K^{-1}$

Table A.3. 1986 recommended values of the fundamental physical constants.

Quantity	Symbol	Value	Unit	Relative uncertainty (ppm)
Speed of light in vacuum	c	299 792 458	$\mathrm{m\,s^{-1}}$	(exact)
Permeability of vacuum	μ_0	$4\pi \times 10^{-7}$	$\mathrm{N\,A^{-2}}$	
		$= 12.566\,370\,614\cdots$	$10^{-7}\ \mathrm{N\,A^{-2}}$	(exact)
Permittivity of vacuum	ε_0	$1/\mu_0 c^2$		
		$= 8.854\,187\,817$	$10^{-12}\ \mathrm{F\,m^{-1}}$	(exact)
Newtonian constant of gravitation	G	6.672 59(85)	$10^{-11}\ \mathrm{m^3\,kg^{-1}\,s^{-2}}$	128
Planck constant	h	6.626 075 5(40)	$10^{-34}\ \mathrm{J\,s}$	0.60
$h/2\pi$	$\hbar$	1.054 572 66(63)	$10^{-34}\ \mathrm{J\,s}$	0.60
Elementary charge	e	1.602 177 33(49)	$10^{-19}\ \mathrm{C}$	0.30
Electron mass	m_e	9.109 389 7(54)	$10^{-31}\ \mathrm{kg}$	0.59
Proton mass	m_p	1.672 623 1(10)	$10^{-27}\ \mathrm{kg}$	0.59
Proton–electron mass ratio	m_p/m_e	1836.152 701(37)		0.020
Fine-structure constant, $\frac{1}{2}\mu_0 ce^2/h$	α	7.297 353 08(33)	10^{-3}	0.045
inverse fine-structure constant	α^{-1}	137.035 989 5(61)		0.045
Rydberg constant, $\frac{1}{2}m_e c\alpha^2/h$	R_∞	10 973 731.534(13)	$\mathrm{m^{-1}}$	0.001 2
Avogadro constant	N_A, L	6.022 136 7(36)	$10^{23}\ \mathrm{mol^{-1}}$	0.59
Faraday constant, $N_A e$	$\mathscr{F}$	96 485.309(29)	$\mathrm{C\,mol^{-1}}$	0.30
Molar gas constant	R	8.314 510(70)	$\mathrm{J\,mol^{-1}\,K^{-1}}$	8.4
Boltzmann constant, R/N_A	k	1.380 658(12)	$10^{-23}\ \mathrm{J\,K^{-1}}$	8.5
Stefan–Boltzmann constant, $(\pi^2/60)k^4/\hbar^3 c^2$	σ	5.670 51(19)	$10^{-8}\ \mathrm{W\,m^{-2}\,K^{-4}}$	34
		Non-SI units used with SI		
Electron-volt, $(e/C)\ \mathrm{J} = \{e\}\ \mathrm{J}$	eV	1.602 177 33(49)	$10^{-19}\ \mathrm{J}$	0.30
(Unified) atomic mass unit, $1\ \mathrm{u} = m_u = \frac{1}{12}m(^{12}\mathrm{C})$	u	1.660 540 2(10)	$10^{-27}\ \mathrm{kg}$	0.59

Appendix B

Molar heat capacities at constant pressure for selected substances[1]

$\tilde{C}_P/\text{cal mol}^{-1} = A + B(T/\text{K}) + C(T/\text{K})^2 + D(T/\text{K})^3$

Substance	A	$10^3\,B$	$10^6\,C$	$10^9\,D$
Argon	4.969	−0.007 67	0.012 34	
Hydrogen	6.483	2.215	−3.298	1.826
Nitrogen	7.440	−3.42	6.400	−2.790
Oxygen	6.713	−0.000 879	4.170	−2.544
Carbon monoxide	7.373	−3.07	6.662	−3.037
Carbon dioxide	4.728	17.54	−13.38	4.097
Chlorine	6.432	8.082	−9.241	3.695
Ammonia	6.524	5.692	4.078	−2.830
Water	7.701	0.459 5	2.521	−0.859
Methane	4.598	12.45	2.860	−2.703
Ethane	1.292	42.54	−16.57	2.081
Propane	−1.009	73.15	−37.89	7.678
n-Butane	2.266	79.13	−26.47	−0.674
n-Pentane	−0.866	116.4	−61.63	12.67
Ethyne	6.406	18.10	−11.96	3.373
Benzene	−8.101	113.3	−72.06	17.03
Cyclohexane	−13.027	146.0	−60.27	3.156
Toluene	−5.817	122.4	−66.05	11.73

Reid, R. C., Prausnitz, J. M., and Sherwood, T. K. (1977). *The Properties of Gases and Liquids*, 3rd edn. New York: McGraw-Hill. These values are for substances in the vapor or gas state. $\text{cal}_{\text{th}} = 4.184$ J.

Appendix C

Thermodynamic data

Table C.1. Enthalpies and Gibbs energies of formation at 298.15 K.[1]

Substance	$\Delta\tilde{H}_f^\ominus$/kJ mol^{-1}	$\Delta\tilde{G}_f^\ominus$/kJ mol^{-1}
H (g)	217.965	203.263
O (g)	249.17	231.731
N (g)	472.679	454.805
C (g)	716.677	671.285
Cl (g)	121.679	105.680
Br (g)	111.884	82.396
CO (g)	−110.53	−137.168
CO_2 (g)	−393.51	−394.359
NH_3 (g)	−45.94	−16.45
H_2O (g)	−241.818	−228.512
H_2O (ℓ)	−285.830	−237.129
CH_4 (g)	−74.52	−50.49
C_2H_6 (g)	−83.82	−31.92
C_3H_8 (g)	−104.68	−24.39
n-C_4H_{10} (g)	−125.79	−16.70
n-C_5H_{12} (g)	−146.76	−8.81
C_2H_2 (g)	227.48	209.97
Benzene (ℓ)	49.08	124.42
Benzene (g)	82.93	129.65
Cyclohexane (g)	−29.43	7.60
Toluene (g)	50.17	122.42

[1] Various sources: The NBS Tables of Chemical Thermodynamic Properties (1982), *J. Phys. Chem. Ref. Data*, **11** (Supplement No. 2); *TRC Thermodynamics Tables—Hydrocarbons*. College Station, TX: Thermodynamics Research Center.

Table C.2. Absolute entropies at 298.15 K.[1]

Substance	$\tilde{S}^{\ominus}$/J mol^{-1} K^{-1}	Substance	$\tilde{S}^{\ominus}$/J mol^{-1} K^{-1}
He	126.150	H_2O(g)	188.711
Ne	146.328	H_2O(ℓ)	69.915
Ar	154.843	CH_4	186.27
Kr	164.082	C_2H_6	229.12
Xe	169.683	C_3H_8	270.20
		n-C_4H_{10}	309.91
C (gr)	5.734	n-C_5H_{12}	349.45
H_2	130.566	C_2H_2	200.79
N_2	191.464	Benzene (g)	269.45
O_2	205.15	Benzene (ℓ)	173.45
CO	197.56	Cyclohexane (ℓ)	298.24
CO_2	213.68	Toluene	320.13
Cl_2 (g)	222.066		
Br_2 (ℓ)	152.231		
NH_3	192.77		

Various sources including the *TRC Thermodynamic Tables—Hydrocarbons*. College Station, TX: Thermodynamics Research Center.

Table C.3. Enthalpy increment function $T^{-1}[H^{\ominus}(T) - H^{\ominus}(0)]$.[1,2]

Substance	$\Delta\tilde{H}^{\ominus}_{f,0}$/kJ mol^{-1}	$T^{-1}[H^{\ominus}(T) - H^{\ominus}(0)]$/J mol^{-1} K^{-1}					
		298.15	400	600	800	1000	1500
H_2 (g)		28.401	28.504	28.794	28.958	29.146	29.836
O_2 (g)		29.124	29.272	29.879	30.649	31.386	32.854
N_2 (g)		29.08	29.10	29.28	29.65	30.13	31.38
C (gr)		21.921	21.642	21.363	21.221	21.135	21.023
CO (g)	−113.82	29.109	29.137	29.366	29.817	30.365	31.679
CO_2 (g)	−393.14	31.415	33.423	37.119	40.211	42.758	47.371
H_2O (g)	−238.915	33.213	33.380	33.999	34.865	35.882	38.664
Methane (g)	−66.505	33.60	34.72	38.65	43.51	48.49	59.54
Ethane (g)	−68.20	39.83	44.69	55.73	66.53	76.36	96.11
Propane (g)	−81.67	49.45	58.20	76.19	92.72	107.19	135.48
Ethyne (g)	227.313	33.560	37.041	42.727	47.066	50.585	57.296
Benzene (g)	100.416	47.74	60.29	85.69	107.78	126.19	160.00
Toluene (g)	73.220	60.42	76.02	106.94	134.01	156.69	198.74

[1] *TRC Thermodynamic Tables—Hydrocarbons*. College Station, TX: Thermodynamics Research Center.
[2] These values are calculated for a standard state of 1 atm. The difference with 1 bar is not significant.

Table C.4. Gibbs energy increment function $T^{-1}[G^{\ominus}(T) - H^{\ominus}(0)]$.[1,2]

Substance	$T^{-1}[G^{\ominus}(T) - H^{\ominus}(0)]$/J mol^{-1} K^{-1}					
	298.15	400	600	800	1000	1500
H_2 (g)	102.165	−110.550	−122.177	−130.486	−136.959	−148.904
O_2 (g)	176.024	−184.500	−196.572	−205.273	−212.192	−225.213
N_2 (g)	−162.53	−171.08	−182.90	−191.37	−198.04	−210.50
C (g)	−136.07	−142.47	−151.18	−157.31	−162.03	−170.58
CO (g)	−168.44	−177.00	−188.85	−197.37	−204.09	−216.69
CO_2 (g)	−182.26	−191.77	−206.04	−217.15	−226.41	−244.69
H_2O (g)	−155.503	−165.285	−178.925	−188.816	−196.707	−211.777
Methane (g)	−152.67	−162.67	−177.44	−189.20	−199.45	−221.29
Ethane (g)	−189.28	−201.67	−221.84	−239.37	−255.31	−290.20
Propane (g)	−220.75	−236.48	−263.47	−287.73	−309.99	−359.15
Ethyne (g)	−167.260	−177.615	−193.774	−206.690	−217.589	−239.455
Benzene (g)	−221.45	−237.24	−266.79	−294.87	−321.24	−379.82
Toluene (g)	−260.06	−280.16	−317.33	−352.29	−385.02	−457.65

[1] *TRC Thermodynamic Tables—Hydrocarbons*. College Station, TX: Thermodynamics Research Center.
[2] These values are calculated for a standard state of 1 atm. The difference with 1 bar is negligible.

Appendix D

Standard reduction potentials in aqueous solutions[1,2]

Reaction	$\mathscr{E}^{\ominus}$/V	$(d\mathscr{E}^{\ominus}/dT)$/mV K^{-1}
$Li^+(aq) + e^- = Li(s)$	−3.045	−0.534
$Ca^{2+}(aq) + 2e^- = Ca(s)$	−2.84	−0.175
$Na^+(aq) + e^- = Na(s)$	−2.714	−0.772
$Mg^{2+}(aq) + 2e^- = Mg(s)$	−2.356	+0.103
$Zn^{2+}(aq) + 2e^- = Zn(s)$	−0.762 6	+0.099
$Cr^{3+}(aq) + 3e^- = Cr(s)$	−0.744	+0.468
$Fe^{2+}(aq) + 2e^- = Fe(s)$	−0.447	+0.052
$Cr^{3+}(aq) + e^- = Cr^{2+}(aq)$	−0.424	
$Cd^{2+}(aq) + 2e^- = Cd(s)$	−0.402 5	−0.093
$PbSO_4(s) + 2e^- = Pb(s) + SO_4^{2-}(aq)$	−0.350 5	−1.015
$Co^{2+}(aq) + 2e^- = Co(s)$	−0.277	−0.060
$Pb^{2+}(aq) + 2e^- = Pb(s)$	−0.125 1	−0.451
$2H^+(aq) + 2e^- = H_2(g)$	0	0
$Cu^{2+}(aq) + e^- = Cu^+(aq)$	0.159	+0.073
$AgCl(s) + e^- = Ag(s) + Cl^-(aq)$	0.222 3	−0.543
$Hg_2Cl_2(s) + 2e^- = 2Hg(\ell) + 2Cl^-(aq)$	0.268 0	−0.317
$Cu^{2+}(aq) + 2e^- = Cu(s)$	0.340	+0.008
$Fe^{3+}(aq) + e^- = Fe^{2+}(aq)$	0.771	+1.188
$Hg_2^{2+}(aq) + 2e^- = 2Hg(\ell)$	0.796 0	
$Ag^+(aq) + e^- = Ag(s)$	0.799 1	−1.000
$2Hg^{2+}(aq) + 2e^- = Hg_2^{2+}(aq)$	0.911 0	
$O_2(g) + 4H^+(aq) + 4e^- = 2H_2O(\ell)$	1.229	−0.846
$Cl_2(g) + 2e^- = 2Cl^-(aq)$	1.358 3	−1.260
$PbO_2(s) + 4H^+(aq) + 2e^- = Pb^{2+}(aq) + 2H_2O(\ell)$	1.468	−0.238
$Co^{3+}(aq) + e^- = Co^{2+}(aq)$	1.92	
$F_2(g) + 2e^- = 2F^-(aq)$	2.87	−1.830

[1] Acidic solutions. See (2) below for more-complete compilations.

[2] From: Bard, A. J., Parsons, R., and Jordan, J. (1985). *Standard Potentials in Aqueous Solutions.* New York: Marcel Dekker; Milazzo, G., Caroli, S., and Sharma, V. K. (1978). *Tables of Standard Electrode Potentials.* New York: John Wiley.

Cited refernces and selected bibliography

I. Cited references

1. Bridgman, P. W. (1936). *Condensed Collection of Thermodynamic Formulas*. Cambridge, MA: Harvard University Press.
2. Shaw, A. M. (1935). *Phil. Trans. Roy. Soc.* (*London*), **A234**, 299. See also: Pinkerton, R. C. (1952). *J. Phys. Chem.*, **56**, 799.
3. Gibbs, J. W. (1961). *Collected Works, Thermodynamics*, Vol. 1, p. 56. New York: Dover Publications.
4. A complete discussion of the sufficiency and necessity of these conditions is given in: Gibbs, J. W., *op. cit.*, pp. 56–62.
5. Gibbs, J. W., *op. cit.*, p. 67.
6. Prigogine, I., and DeFay, R. (1954). *Chemical Thermodynamics* (transl. D. H. Everett), pp. 175, 468–96. New York: Longmans, Green and Co.
7. A very complete discussion of indifferent states of thermodynamic systems is given in: Prigogine, I., and DeFay, R., *op. cit.*, pp. 450–509.
8. Prigogine, I., and DeFay, R., *op. cit.*, pp. 450–5, 479–86.
9. Rowlinson, J. S. (1955). *Trans. Faraday Soc.*, **51**, 1317. Data for Figure 7.1 are based on the *P–V–T* relations of the heavier inert gases.
10. Schneider, G. (1950). *J. Chem. Phys.*, **18**, 1269. Figure 7.2 gives the second virial coefficient of carbon dioxide as a function of the temperature.
11. Dymond, J. H., and Smith, E. B. (1980). *The Virial Coefficients of Pure Gases and Mixtures*. Oxford: Oxford University Press.
12. Cholinski, J., Szafranski, A., and Wyrzykowska-Stankiewicz, D. (1986). *Computer-aided Second Virial Coefficient Data for Organic Individual Compounds and Binary Systems*. Warszawa: Polish Scientific Publishers.
13. Mason, E. A., and Spurling, T. H. (1969). *The Virial Equation of State*. Oxford: Pergamon Press.
14. Stryjek, R., and Vera, J. H. (1986). An improved equation of state. In: *Equations of State. Theories and Applications*, eds. K. C. Chao, and R. L. Robinson, Jr., pp. 561–70. Washington: American Chemical Society.
15. Gibbs, J. W., *op. cit.*, pp. 155.
16. Lewis, G. N., and Randall, M. (1923). *Thermodynamics* (revised by Pitzer, K. S., and Brewer, L. (1961) as 2nd edn.). New York: McGraw-Hill. p. 191 (1st edn.); p. 153 (2nd edn.).

17. Pitzer, K. S., Lippmann, D. Z., Curl, R. F., Huggins, C. M., and Peterson, D. E. (1955). *J. Am. Chem. Soc.*, **77**, 3433.
18. Emister, W. C., and Lee, B. I. (1984). *Applied Hydrocarbon Thermodynamics*, Vol. 1 (2nd edn.). Houston: Gulf Publishing Company.
19. Prigogine, I., and DeFay, R., *op. cit.*, p. 88.
20. Freeman, R. D. (1984). *J. Chem. Eng. Data.*, **29**, 105–11.
21. Freeman, R. D. (1985). *J. Chem. Ed.*, **62**, 681–6.
22. IUPAC (1979). Manual of Symbols and Terminology for Physicochemical Quantities and Units, *Pure Appl. Chem.* Appendix IV, **51**, 1–41; (1982) *Pure Appl. Chem.*, **54**, 1239–50; (1982) *J. Chem. Thermodyn.*, **14**, 805–15.
23. IUPAC (1979). *op. cit.* See also: Mills, I., Critas, T., Homann, K., Kallay, N., and Kuchitsu, K. (1988). *Units and Symbols in Physical Chemistry.* Oxford: Blackwell Scientific Publications.
24. Lewis, G. N., and Randall, M., *op. cit.*, pp. 254–9, 328–9 (1st edn.); pp. 242–3, 249–52 (2nd edn.). The term 'reference state' as used by Lewis and Randall/Pitzer and Brewer has an entirely different meaning from that used in this book.
25. Callen, H. B. (1985). *Thermodynamics and an Introduction to Thermostatics* (2nd edn.). New York: John Wiley. Prigogine, I., and DeFay, R., *op. cit.*, p. 305.
26. Examples are found in: (a) Scatchard, G. (1940). *J. Am. Chem. Soc.*, **62**, 2426; (b) Wood, S. E. (1946). *J. Am. Chem. Soc.*, **68**, 1962; and (c) Wood, S. E. (1950). *Ind. Eng. Chem.*, **42**, 660.
27. Darken, L. S. (1950). *J. Am. Chem. Soc.*, **72**, 2909.
28. Gokcen, N. A. (1960). *J. Phys. Chem.*, **64**, 401.
29. Wagner, C. (1952). *Thermodynamics of Alloys*, Reading, MA: Addison-Wesley Publishing Co.
30. Schuhmann, R. Jr. (1955). *Acta Met.*, **3**, 219.
31. References for illustrative purposes are: Hogfeldt, E. (1954). *Arkiv for Kemi*, **7**, 315; (1956) *Rec. Trav. Chim. Pays-Bas*, **45**, 790; Fischer, A. K., Johnson, S. A., and Wood, S. E. (1967). *J. Phys. Chem.*, **71**, 1465; Wood, S. E., Fine, B. D., and Isaacson, L. M. (1957). *J. Phys. Chem.*, **61**, 1605.
32. Edwards, R. K. (1969). *High Temp. Sci.*, **1**, 232; see also: Edwards, R. K., Chandrasekharich, M. S., and Danielson, P. M. (1969). *High Temp. Sci.*, **1**, 98.
33. Kirkwood, J. G., and Oppenheim, I. (1961). *Chemical Thermodynamics.* New York: McGraw-Hill.
34. For example, see: Scatchard, G. (1953). *J. Am. Chem. Soc.*, **75**, 2883; Keenan, A. G., Notz, K., and Wilcox, F. L. (1968). *J. Phys. Chem.*, **72**, 1085.
35. An excellent review and discussion of the main features of the thermodynamic and mechanical treatment of surfaces is given by: Melrose, J. C. (1968). *Ind. Eng. Chem.*, **60**, 53. This article contains numerous references to more-complete studies. See also: Adamson, A. W. (1982). *The Physical Chemistry of Surfaces* (4th edn.). New York: John Wiley.
36. Guggenheim, E. A. (1929). *J. Phys. Chem.*, **33**, 842.

II. Bibliography

—— (1941–). *Temperature. Its Measurement and Control in Science and Industry*, New York: Reinhold Publishing.

Acree, W. E. Jr. (1984). *Thermodynamic Properties of Nonelectrolyte Solutions*, New York: Academic Press.

Adamson, A. W. (1982). *The Physical Chemistry of Surfaces* (4th edn.). New York: John Wiley.

Beattie, J. A., and Oppenheim, J. (1980). *Principles of Thermodynamics.* Amsterdam: Elsevier.

Ben-Naim, A. (1987). *Solvation Thermodynamics*. New York: Plenum Publishing Co.

Bett, K. E., Rowlinson, J. S., and Saville, G. (1975). *Thermodynamics for Chemical Engineers*. Cambridge, MA: MIT Press.

Blander, M. (ed.) (1964). *Molten Salt Chemistry*, New York: John Wiley.

Bockris, J. O'M., *et al.* (eds.) (1980 *et seq.*). *Comprehensive Treatise of Electrochemistry*. New York: Plenum Publishing Co.

Callen, H. B. (1985). *Thermodynamics and an Introduction to Thermostatics* (2nd edn.). New York: John Wiley.

Chao, K. C., and Robinson, R. L. (eds.) (1979). *Equations of State in Engineering and Research*. Washington: American Chemical Society.

Chao, K. C., and Robinson, R. L. (eds.) (1986). *Equations of State. Theories and Applications*. Washington: American Chemical Society.

Chattoraj, D. K., and Birdi, K. S. (1984). *Adsorption and the Gibbs Surface Energy*. New York: Plenum Publishing Co.

DeFay, R., and Prigogine, I. 1966. *Surface Tension and Adsorption*, Ann Arbor: Books Demand UMI.

Denbigh, K. G. (1981). *The Principles of Chemical Equilibrium* (4th edn.). Cambridge: Cambridge University Press.

Edmister, W. C., and Lee, B. I. (1985). *Applied Hydrocarbon Thermodynamics* (2nd edn.), Vol. 1. Houston: Gulf Publishing Co.

Everett, D. H. (1960). *Introduction to the Study of Chemical Thermodynamics*. London: Longmans.

Gibbs, J. W. (1961). *Collected Works*, Vol. 1, *Thermodynamics*. New York: Dover Publications.

Gokcen, N. A. (1975). *Thermodynamics*. Hawthorne, CA: Techscience Inc.

Guggenheim, E. A. (1968). *Thermodynamics* (5th edn.). New York: Wiley Interscience.

Guggenheim, E. A. (1985). *Thermodynamics: An Advanced Treatise for Chemists and Physicists*. Amsterdam: Elsevier.

Guggenheim, E. A., and Stokes, R. H. (1969). *Equilibrium Properties of Aqueous Solutions of Single Strong Electrolytes*. Oxford: Pergamon Press.

Harned, H. S., and Owen, B. B. (1959). *The Physical Chemistry of Electrolytic Solutions* (3rd edn.). New York: Reinhold Publishing Corp.

Harned, H. S., and Robinson, R. A. (1969). *Multicomponent Electrolyte Solutions*. Oxford: Pergamon Press.

Hatsopoulos, G. M., and Keenan, J. H. (1981). (reprint of 1961 edn.) *Principles of General Thermodynamics*. Melbourne, FL: Krieger.

Hildebrand, J. H., Scott, R. L., and Prausnitz, J. M. (1970). *Regular and Related Solutions*. Melbourne, FL: Krieger.

Honig, J. M. (1982). *Thermodynamics*. Amsterdam: Elsevier.

Hougen, O. A., Watson, K. M., and Ragatz, R. A. (1959). *Chemical Process Principles. Part 2. Thermodynamics* (2nd edn.). New York: John Wiley.

Janz, G. J., *et al.* (1968–). *Molten Salts*. Washington: US National Bureau of Standards (available Supt. of Docs., US Gov't. Print. Off.).

Jones, M. N. (1988). *Biochemical Thermodynamics*. Amsterdam: Elsevier.

Kehiaian, H. V., and Renon, H. (1986). *Measurement, Evaluation and Prediction of Phase Equilibria*. Amsterdam: Elsevier.

Kestin, J. J. (ed.) (1976). *The Second Law of Thermodynamics*. New York: Academic Press.

Kestin, J. (1979). *A Course in Thermodynamics*, 2 Vols. Waltham, MA: Blaisdell Publishing Co.

Kirkwood, K. G., and Oppenheim, I. (1961). *Chemical Thermodynamics*. New York: McGraw-Hill.

Klotz, I. M., and Rosenberg, R. M. (1986). *Chemical Thermodynamics: Basic Theory and Methods* (4th edn.). Menlo Park: Benjamin/Cummings.

Kubaschewski, O., Evans, L., and Alcock, C. B. (1979). *Metallurgical Thermochemistry* (5th edn.). Oxford: Pergamon Press.

Lehninger, A. L. (1971). *Bioenergetics. The Molecular Basis of Biological Energy Transformation* (2nd edn.). Menlo Park: Benjamin/Cummings.

Lewis, G. N., and Randall, M. (1923). *Thermodynamics*. New York: McGraw-Hill. Revised edn.: Pitzer, K. S., and Brewer, L. (1961). New York: McGraw-Hill.

Lupis, C. H. P. (1983). *Chemical Thermodynamics of Materials*. Amsterdam: Elsevier.

McGlashan, M. L. (1980). *Chemical Thermodynamics*. New York: Academic Press.

Mamantov, G. (ed.) (1969). *Molten Salts: Characterization and Analysis*. New York: Marcel Dekker.

Mason, E. A., and Spurling, T. H. (1969). *The Viral Equation of State*. Oxford: Pergamon Press.

Oonk, H. A. 1981. *Phase Theory: The Thermodynamics of Heterogeneous Equilibria*. Amsterdam: Elsevier.

Planck, M. (1945). *Treatise on Thermodynamics* (3rd ed.) (transl. A. Ogg). New York: Dover Publications.

Prausnitz, J. M., Lichtenthaler, R. N., and Gomes de Azevedo, E. (1986). *Molecular Thermodynamics of Fluid-Phase Equilibria* (2nd edn.). Englewood Cliffs, NJ: Prentice–Hall.

Prigogine, I. (1957). *The Molecular Theory of Solutions*. New York: Interscience Publishers.

Prigogine, I., and DeFay, R. (1954). *Chemical Thermodynamics* (transl. D. H. Everett). New York: Longmans, Green and Co.

Redlich, O. (1976). *Thermodynamics Fundamentals, Applications*. Amsterdam: Elsevier.

Reid, R. C., Prausnitz, J. M., and Sherwood, T. K. (1977). *The Properties of Gases and Liquids* (3rd edn.). New York: McGraw-Hill.

Robinson, R. A., and Stokes, R. H. (1959). *Electrolyte Solutions* (2nd edn.). London: Butterworths.

Rock, P. A. (1983). *Chemical Thermodynamics*. Mill Valley, CA: University Science Books.

Rowlinson, J. S., and Swinton, F. L. (1982). *Liquids and Liquid Mixtures* (3rd edn.). London: Butterworth:

Scatchard, G. (1976). *Equilibrium in Solutions, Surface and Colloid Chemistry*. Cambridge, MA: Harvard University Press.

Van Ness, H. C. (1983). *Understanding Thermodynamics*. New York: Dover Publications.

Van Ness, H. C., and Abbott, M. M. (1982). *Classical Thermodynamics of Non-electrolyte Solutions*. New York: McGraw-Hill.

Van Wylen, G. J., and Sonntag, R. E. (1986). *Fundamentals of Classical Thermodynamics* (3rd edn.). New York: John Wiley.

Wagner, C. (1952). *Thermodynamics of Alloys*. Reading, MA: Addison-Wesley.

Wall, F. T. (1974). *Chemical Thermodynamics* (3rd edn.). San Francisco: W. H. Freeman and Co.

Zemansky, M. W., and Dittman, R. (1981). *Heat and Thermodynamics* (6th edn.). New York: McGraw-Hill.

III. Compilations of thermodynamic data[1]

——(1937). *Contributions to the Data on Theoretical Metallurgy.* Bureau of Mines Bulletins. Washington, DC: Supt. of Documents.

—— (1940). *Selected Values of Properties of Hydrocarbons and Related Compounds: Thermodynamics Research Center Hydrocarbon Project*, API 44 Tables. College Station, TX: Thermodynamics Research Center.

—— (1947). *Selected Values of the Properties of Hydrocarbons.* National Bureau of Standards Circular C461.

—— (1952). *Selected Values of Chemical Thermodynamic Properties.* National Bureau of Standards Circular 500 plus supplements.

—— (1956, 1959). *Phase Diagrams for Ceramicists.* Columbus, OH: American Ceramic Society.

—— (1979). *Solubility Data Series.* International Union of Pure and Applied Chemistry, Commission on Solubility Data. Oxford: Pergamon Press.

—— (1981). *Engineering Data Book* (9th edn.). Tulsa: Pennwell Books.

—— (n.d.). *Selected Values for the Thermodynamic Properties of Metals and Alloys.* Minerals Research Laboratory, Berkeley, CA: Institute of Engineering Research.

—— (n.d.) *Selected Values of the Properties of Chemical Compounds.* College Station, TX: Thermodynamics Research Center.

Angus, S. *International Thermodynamics Tables of the Fluid State*, series. Oxford: Pergamon Press.

Bard, A. J., Parsons, R., and Jordan, J. (1985). *Oxidation–Reduction Potentials in Aqueous Solutions: A Selective and Critical Source Book.* New York: Marcel Dekker.

Bard, A. J., Parsons, R., and Jordan, J. (eds.) (1985). *Standard Potentials in Aqueous Solutions.* New York: Marcel Dekker.

Barin, I., and Knacke, O. (1973). *Thermochemical Properties of Inorganic Substances.* Berlin: Springer-Verlag.

Benedek, P., and Olti, F. (1985). *Computer Aided Chemical Thermodynamics of Gases and Liquids.* New York: John Wiley.

deBethune, A. D., and Loud, N. A. S. (1964). *Standard Aqueous Electrode Potentials and Temperature Coefficients at 25°C*, Skokie, IL: C.A. Hampel Publishing Co.

Boublik, T., Fried, V., and Hala, E. (eds.) (1984). *Vapor Pressure of Pure Substances* (2nd revised edn.). Amsterdam: Elsevier.

Cholinski, J., Szafranski, A., and Wyrzykowski-Stankiewicz, D. (1986). *Computer-aided Second Virial Coefficient Data for Organic Individual Compounds and Binary Systems.* Warsaw: PWN–Polish Scientific Publishers.

Clark, W. M. (1960). *Oxidation–Reduction Potentials of Organic Systems.* Baltimore: Williams and Wilkins.

CODATA: see their publications for data compilations and data handling. CODATA Secretariat, 51 Blvd. de Montmorency, 75016 Paris, France.

Cox, J. C., and Pilcher, G. (1969). *Thermochemistry of Organic and Organometallic Compounds.* New York: Academic Press.

Dreisbach, R. R. (Vol. I, 1955), (Vol. II, 1959), (Vol. III, 1961). *Physical Properties of Chemical Compounds.* Washington: American Chemical Society.

[1] Many data are also currently available in databases that are accessible via phone lines or purchasable floppy disks. None are cited here.

Dymond, J. H., and Smith, E. B. (1980). *The Virial Coefficients of Pure Gases and Mixtures*. Oxford: Clarendon Press.

Gmehling, J., and Onken, U. *Vapor–Liquid Equilibrium Data Collection*, series. Great Neck, NY: Scholium International.

Horvath, A. L. 1982. *Halogenated Hydrocarbons. Solubility–Miscibility with Water*. New York: Marcel Dekker.

Hultgren, R. H., Orr, R., Anderson, P., and Kelley, K. K. (1963). *Selected Values for the Thermodynamic Properties of Metals and Alloys*. New York: John Wiley–Books Demand UMI.

Ives, D. G., and Janz, G. J. (eds.) (1969). *Reference Electrodes: Theory and Practice*. New York: Academic Press.

Janz, G. J. (1967). *Thermodynamic Properties of Organic Compounds: Estimation Methods, Principles and Practice*. New York: Academic Press.

Janz, G. J. (1967). *Molten Salts Handbook*. New York: Academic Press.

Janz, G. J., and Tomkins, R. P. T. (1972, 1974). *Non-aqueous Electrolytes Handbook*. New York: Academic Press.

Jordan, T. E. (1954). *Vapor Pressure of Organic Compounds*. New York: Interscience Publishers.

Kubachewski, O., and Catterall, J. A. (1956). *Thermodynamic Data of Alloys*. Oxford: Pergamon Press.

Landolt, H. H., *et al.* (1950). *Landolt-Bornstein: Zahlenwerte und Funktionen aus Physik, Chemie, Astronomie, Geophysik und Technik*. Berlin: Springer-Verlag.

Latimer, W. M. (1952). *The Oxidation States of the Elements and Their Potentials in Aqueous Solutions*. New York: Prentice–Hall.

Levin, E. M., Robbins, C. R., and McMurdie, H. F. (1964). *Phase Diagrams for Ceramists*. Columbus, OH: The American Ceramic Society.

Lyman, W. J., Reehl, W. F., and Rosenblatt, D. H. (1982). *Handbook of Chemical Property Estimation Methods*. New York: McGraw-Hill.

Majer, V., and Svoboda, V. (1985). *Enthalpies of Vaporization of Organic Compounds*. Oxford: Blackwells Scientific Publications.

Milazzo, G., Caroli, S., and Sharma, V. K. (1978). *Tables of Standard Electrode Potentials*. New York: John Wiley.

Nesmeeilanov, A. M. (1963). *Vapor Pressure of the Elements*. New York: Academic Press.

Pedley, J. B., *et al.* (1985). *Thermochemical Data of Organic Compounds*. London: Chapman and Hall.

Perry, R. H., and Chilton, C. H. (1984). *Chemical Engineers Handbook* (6th edn.). New York: McGraw-Hill.

Selover, T. B., Jr. (ed.) (1987). *National Standard Reference Data Services of the USSR: A Series of Property Tables*. New York: Hemisphere Publishing Corp.

Sohnel, O., and Novotny, P. (1985). *Densities of Aqueous Solutions in Inorganic Substances*. Amsterdam: Elsevier.

Stull, D. R., Sinke, G. C., and Gerard, C. (eds.) (1956). *Thermodynamic Properties of the Elements*. Washington: American Chemical Society.

Stull, D. R., and Prophet, H. (1960–). *JANAF Thermochemical Tables*. Midland, MI: Dow Chemical Co. (Available Supt. Docs., US Govt. Printing Office.)

Stull, D. R., Westrum, E. F., Jr., and Sinke, G. C. (1986). *The Chemical Thermodynamics of Organic Compounds*. Melbourne, FL: Krieger.

Tamir, A., Tamir, E., and Stephan, K. (1983). *Heats of Phase Change of Pure Compounds and*

Mixtures. A Literature Guide. Amsterdam: Elsevier.

Timmermans, J. (1950–65). *Physico-chemical Constants of Pure Organic Compounds*. Amsterdam: Elsevier.

Timmermans, J. (1959–). *The Physico-chemical Constants of Binary Systems in Concentrated Solutions*. New York: Interscience Publishers.

Touloukian, Y. S., and Ho, C. Y. (1971–). *Thermophysical Properties of Matter*, 13 Vols. New York: Plenum Publishing.

Wichterle, I., Linek, J., and Hala, E. (1973–). *Vapor–Liquid Equilibrium Data Bibliography*. Amsterdam: Elsevier.

Wicks, C. E., and Block, F. E. (1963). *Thermodynamic Properties of 65 Elements—Their Oxides, Halides, Carbides, and Nitrides*. Bureau of Mines Bulletin 605. Washington, DC: US Govt. Printing Office.

Wisniak, J. (1981). *Phase Diagrams*. Amsterdam: Elsevier; (1986). Supplement 1.

Wisniak, J., and Herskowitz, M. (1984). *Solubility of Gases and Solids: A Literature Source Book*. Amsterdam: Elsevier.

Wisniak, J., and Tamir, A. (1980, 1981). *Liquid–Liquid Equilibrium and Extraction: A Literature Source Book*. Amsterdam: Elsevier.

Wisniak, J., and Tamir, A. (1982, 1986). *Mixing and Excess Thermodynamic Properties: A Literature Source Book*. Amsterdam: Elsevier.

Index